# *Elementary Differential Equations with Linear Algebra*

# Elementary Differential Equations with Linear Algebra

**ROSS L. FINNEY**
University of Illinois-Urbana

**DONALD R. OSTBERG**
Northern Illinois University

with the assistance of
**ROBERT G. KULLER**
Northern Illinois University

**ADDISON-WESLEY PUBLISHING COMPANY**
Reading, Massachusetts · Menlo Park, California
London · Amsterdam · Don Mills, Ontario · Sydney

This book is in the
ADDISON-WESLEY SERIES IN MATHEMATICS
Lynn H. Loomis
*Consulting Editor*

*Second printing, January 1981*

The first edition of this text was published under the title *Elementary Differential Equations*, by Donald L. Kreider, Robert G. Kuller, and Donald R. Ostberg.

ISBN 0-201-05515-5
GHIJKLMNOP-MA-8987654321

# *Preface*

Times change, and with time come changes in the perspectives of authors and the expectations of readers. The world of mathematics is very different from what it was eight years ago, when *Elementary Differential Equations*, by Donald L. Kreider, Robert G. Kuller, and Donald R. Ostberg, was published. The present edition of that text is a different book from the first edition. This is not to say that the underlying point of view toward the subject adopted in the first edition has changed, for in a large measure it has not. But it has been tempered by time and enlarged by experience, and it is less austere than before.

Thus we begin this edition with a careful but traditional chapter on first-order equations in which applications of our subject to other areas of science are emphasized wherever possible. Only then do we turn to the theme of linearity, with which we began the first edition, and develop that portion of linear algebra which we feel is indispensable for a real understanding of the techniques and results that involve linear differential equations. Thereafter, the subject is developed much as it was in the previous edition, although the order of topics has been changed and a chapter on numerical methods has been added immediately before the chapter on existence and uniqueness theorems.

One of the most significant differences between this edition and its predecessor is its treatment of applications. They now appear in almost every chapter and are varied enough to convince even the most confirmed sceptic that differential equations are one of the basic tools used in man's continuing effort to describe the world in which he lives. We hope that these applications will deepen the reader's appreciation of the part differential equations play in this description, and will stimulate him to explore the subject more deeply. Should this occur, we will have been well rewarded.

In preparing this edition we have eliminated the errors that were discovered in the first edition, without (we hope) producing new ones to take their place. At the same time we have continued to number definitions, theorems, and equations consecutively within each chapter, and to use the mark ∎ to signify the end of a formal proof. Sections marked with an asterisk may be omitted without loss of continuity, and exercises so marked constitute optional challenges to the more ambitious student.

As authors we are aware that in writing this book we have incurred many debts: to the scientific community that brought our subject into being, to the students and teachers who taught us much that we know about that subject, and to our professional colleagues whose criticism has sharpened our insights into what we were writing. To all we extend our sincere, though inadequate, thanks. We give thanks also to the staff of the Addison-Wesley Publishing Company, who encouraged the preparation of this edition, and who worked faithfully to see it to completion. And we give thanks of a very special kind to Rebecca Finney, who did a splendid job of typing the manuscript for this edition, and whose understanding and patience contributed so much to its production.

*January 1976*                                                      R. L. F.
                                                                    D. R. O.

# Logical Interdependence of Chapters

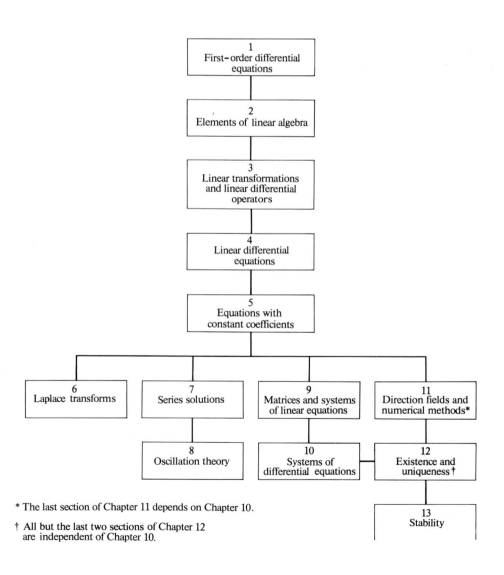

1
First-order differential equations

2
Elements of linear algebra

3
Linear transformations and linear differential operators

4
Linear differential equations

5
Equations with constant coefficients

6
Laplace transforms

7
Series solutions

9
Matrices and systems of linear equations

11
Direction fields and numerical methods*

8
Oscillation theory

10
Systems of differential equations

12
Existence and uniqueness †

13
Stability

* The last section of Chapter 11 depends on Chapter 10.

† All but the last two sections of Chapter 12 are independent of Chapter 10.

# *Contents*

# *First-Order Differential Equations* **1**

## 1–1 INTRODUCTION

A **differential equation** is an equation involving an unknown function and one or more of its derivatives. The order of the highest derivative appearing in such an equation is known as the **order** of the equation, and the equation is said to be an **ordinary differential equation** if all of its derivatives are ordinary derivatives. Otherwise, it is called a **partial differential equation**. Thus,

$$\frac{dy}{dx} + xy = 1,$$

$$\frac{d^2\theta}{dt^2} + k^2 \sin\theta = 0, \qquad k \text{ a constant,}$$

$$x\frac{d^3y}{dx^3} - \left(\frac{dy}{dx}\right)^2 + e^x y = 0$$

are ordinary differential equations of orders one, two, and three, respectively, while

$$\frac{\partial^2 u}{\partial x^2} + \frac{\partial^2 u}{\partial y^2} + \frac{\partial^2 u}{\partial z^2} = 0$$

is a second-order partial differential equation. In this text we shall be concerned almost exclusively with ordinary differential equations.

Differential equations have long fascinated mathematicians, both because of their intrinsic interest and because of their applications in such disciplines as physics, chemistry, astronomy, economics, and psychology. The fact that differential equations arise in such a variety of fields is hardly surprising when we recall that the derivative $y' = dy/dx$ of a function $y = f(x)$ can be interpreted as the rate of change of $y$ with respect to $x$. Thus, descriptions of natural phenomena that involve change are often given in terms of differential equations.

One of the most pervasive phenomena in the physical universe is motion, and the most famous differential equation of all time is Newton's *second law of motion*:

$$\mathbf{F} = \frac{d}{dt}(m\mathbf{v}) \qquad (1-1)$$

which states that the force $\mathbf{F}$ acting on an object of mass $m$ that is moving with velocity $\mathbf{v}$ is equal to the time derivative of the **momentum** $m\mathbf{v}$ of the object. In this equation $\mathbf{F}$ and $\mathbf{v}$ are vector quantities,* that is, quantities which have both magnitude and direction, while $m$ is a **scalar** or number. In the case of motion in a straight line, where the directions of $\mathbf{F}$ and $\mathbf{v}$ are determined by whether they are positive or negative, Eq. $(1-1)$ can be rewritten in scalar form as

$$F = \frac{d}{dt}(mv); \qquad (1-2)$$

and when $m$ is a constant, it can be further rewritten as

$$F = m\frac{dv}{dt} \qquad \text{or} \qquad F = ma,$$

where $a$ is the acceleration of the object.

**Free Fall**    An object that falls toward the earth under the influence of gravity alone is said to be in **free fall**.

Example 1    Describe the motion of an object falling freely to the surface of the earth.

*Solution.*    Objects near the surface of the earth experience a downward force of magnitude $mg$, where $m$ is the mass of the object and $g$ is the acceleration due to gravity ($g \approx 32$ ft/sec$^2$ or 980 cm/sec$^2$). In the case of free fall this gravitational pull is assumed to be the only force acting on the object. Thus, if we measure distance from the surface of the earth as shown in Fig. 1–1, then Newton's second law becomes

$$-mg = m\frac{d^2y}{dt^2}$$

or

$$\frac{d^2y}{dt^2} = -g, \qquad (1-3)$$

where the minus sign was introduced because the force acts in the negative $y$-direction. The problem requires us to find $y$ as a function of $t$, that is, to *solve* the differential equation $(1-3)$.

---

* Throughout this book vectors will be denoted by boldface type.

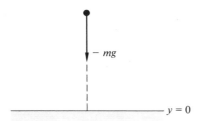

**Figure 1–1**

To do this, we first integrate the equation to obtain

$$v = \frac{dy}{dt} = -gt + c_1, \tag{1–4}$$

where $c_1$ is an arbitrary constant. Another integration then yields

$$y = -\tfrac{1}{2}gt^2 + c_1 t + c_2, \tag{1–5}$$

where $c_2$ is a second arbitrary constant. Equation (1–3) has now been solved: we have found a function $y$ whose second derivative with respect to $t$ is $-g$. Note, however, that this "solution" actually comprises infinitely many different solutions, one for each pair of values for $c_1$ and $c_2$.

To complete the description of the motion, suppose that the progress of the falling body is measured in such a way that when $t = 0$ the body has position $y_0$ and velocity $v_0$. A substitution of 0 for $t$ in (1–4) and (1–5) then shows that $c_1 = v_0$ and $c_2 = y_0$. In short,

$$v = -gt + v_0,$$
$$y = -\tfrac{1}{2}gt^2 + v_0 t + y_0,$$

and the motion has been completely described. Conditions like

$$v = v_0 \quad \text{and} \quad y = y_0 \qquad \text{when } t = 0,$$

that select a particular solution of a differential equation from all its solutions, are called **initial conditions**.

**Population Growth**   As a second example of how differential equations arise in the study of natural phenomena, we consider the problem of determining the future size of a population when its rate of growth is known.

**Example 2**   Describe the growth of a population under the assumption that its rate of change at time $t$ is proportional to the number of individuals present at that time.

*Solution.*   Let $y(t)$ denote the population at time $t$. Then

$$\frac{dy}{dt} = ky,$$

where $k$ is a constant that is positive if the population is increasing and negative if it is decreasing. To solve this equation, we rewrite it to get

$$\frac{dy}{y} = k \, dt,$$

and then integrate to obtain

$$\int \frac{dy}{y} = k \int dt,$$

which yields (remember that $y$ is positive)

$$\ln y = kt + c_1,$$

where $c_1$ is a positive constant. It follows that

$$y = e^{kt + c_1}$$

or

$$y = ce^{kt},$$

where $c = e^{c_1}$. Finally, if $y_0$ denotes the population at time $t = 0$, then $c = y_0$ and

$$y = y_0 e^{kt}. \tag{1-6}$$

Thus $y$ obeys the well-known *law of exponential growth*. In a later chapter we shall consider situations that result in something more realistic than prolonged exponential growth of a population.

The technique used to solve the differential equation in the preceding example is known as the method of **separation of variables**. It can be described as follows. If a first-order differential equation can be written in the form

$$M(x) \, dx + N(y) \, dy = 0, \tag{1-7}$$

where $M$ is a function of $x$ alone and $N$ is a function of $y$ alone, then its "general" solution is

$$\int M(x) \, dx + \int N(y) \, dy = c,$$

where $c$ is an arbitrary constant.

**Example 3**   Solve the equation

$$\frac{dy}{dx} = \frac{x}{y}$$

by the method of separation of variables.

*Solution.*   This equation can be rewritten as

$$x \, dx - y \, dy = 0.$$

It follows that

$$\int x \, dx - \int y \, dy = c.$$

Thus,

$$\frac{x^2}{2} - \frac{y^2}{2} = c$$

or equivalently

$$x^2 - y^2 = c,$$

where $c$ in each case is an arbitrary constant.

This technique will be discussed in more detail in Section 1–3.

### EXERCISES

Solve the differential equations in Exercises 1 through 8 by separating the variables. In each case, check your solution by substituting $y$ and its derivative in the original equation.

1. $y' = xy$
2. $y' = \dfrac{y^2 - 1}{x}$
3. $e^x y' = -1$
4. $y' = 2(y^2 + 1)$
5. $y' = \dfrac{1 - x}{y^2 + 1}$
6. $y' = e^{x-y}$
7. $y' = (xy)^2$
8. $e^{x^2} y' + y^2 x = 0$

9. a) Find the equation of motion of a ball that is thrown upward at a speed of 40 ft/sec from a height of 144 ft above the ground. Neglect air resistance and let $g = 32$ ft/sec$^2$.
   b) How long after the ball is thrown will it reach its maximum height?
   c) How high will the ball rise?
   d) When will the ball strike the ground?

10. The number of bacteria in a colony increases at a rate proportional to the number present. If the colony starts with 2 bacteria, and triples every hour thereafter, how many bacteria will the colony contain at the end of 24 hours?

11. Bacteria in a colony are born and die at rates proportional to the number present, so so that the equation governing the growth of the colony is

$$\frac{dy}{dt} = (k_1 - k_2)y.$$

   a) Solve this equation in terms of $k_1$ and $k_2$.
   b) Find $k_1$ and $k_2$ if it is known that the colony doubles in size every 24 hours and that its size would be halved in 8 hours if there were no births.

12. The decay of a radioactive element can be described by Eq. (1–6), because the rate of decay is proportional to the number of radioactive nuclei present. The **half-life** of a radioactive element is the time required for half of the radioactive nuclei originally present in any sample to decay. Show that the half-life of a radioactive element is a constant that does not depend on the number of radioactive nuclei initially present in the sample.

13. The half-lives of radioactive elements (see Exercise 12) can sometimes be used to date events from the earth's past. The ages of rocks more than 2 billion years old have been

measured by the extent of the radioactive decay of uranium (half-life 4.5 billion years!). In a living organism, the ratio of radioactive carbon, Carbon 14, to ordinary carbon stays fairly constant during the lifetime of the organism, being approximately equal to the ratio in the organisms's surroundings at the time. After the organism's death, however, no new carbon is ingested, and the proportion of Carbon 14 in the organism's remains changes as the Carbon 14 decays. Since the half-life of Carbon 14 is known to be about 5700 years, it is possible to estimate the age of organic remains by comparing the proportion of Carbon 14 they contain with the proportion assumed to have been in the organism's environment at the time it lived. Archeologists have dated shells (which contain $CaCO_3$), seeds, and wooden artifacts this way. The estimate of 15,500 years for the age of the cave paintings at Lascaux, France, is based on Carbon 14 dating.
a) Find $k$ in Eq. (1–6) for Carbon 14.
b) What is the age of a sample of charcoal in which 90% of the Carbon 14 has decayed?
c) The charcoal from a tree killed in the volcanic eruption that formed Crater Lake in Oregon contained $44\frac{1}{2}\%$ of the Carbon 14 found in living matter. About how old is Crater Lake?

14. With time measured in years, the value of $k$ in Eq. (1–6) for Cobalt 60 is about $-0.13$. Estimate the half-life of Cobalt 60.

15. A gram of radium takes 10 years to diminish to 0.997 gm. How long does it take a gram of radium to diminish to 0.5 gm?

16. Find a differential equation that describes the motion of an object of mass $m$ that encounters air resistance directly proportional to its speed as it falls to the surface of the earth. (See Example 1 on page 38.)

17. A curve in the $xy$-plane has the property that its slope at each point is equal to the sum of the squares of the coordinates of the point. Find a differential equation that describes the curve.

18. The sum of the $x$- and $y$-intercepts of the tangent to a curve in the $xy$-plane is a constant regardless of the point of tangency. Find a differential equation for the curve.

19. The graph of a nonnegative function has the property that the length of the arc between any two points on the graph is equal to the area of the region under the arc. Find a differential equation for the curve.

## 1–2 SOLUTIONS OF FIRST-ORDER EQUATIONS

With only a few exceptions, we shall restrict our attention in this chapter to the general first-order differential equation

$$F(x, y, y') = 0 \qquad\qquad (1-8)$$

involving an unknown function $y = y(x)$ and its first derivative $y'$. We begin by defining what we mean by a solution of such an equation.

**Definition 1–1**  A real-valued function $y = y(x)$ is said to be a **solution** of Eq. (1–8) on an interval $I$ if $y$ is differentiable everywhere on $I$ and

$$F(x, y(x), y'(x)) = 0$$

for all $x$ in $I$. The graph of $y$ is then called a **solution curve** of the equation.

**Example 1**   Show that every function of the form

$$y = \frac{1}{x} e^{cx},$$ (1–9)

where $c$ is a constant, is a solution of the equation

$$xy' + y - y \ln(xy) = 0$$ (1–10)

for all $x \neq 0$.

*Solution.*   We first differentiate (1–9) to obtain

$$y' = \frac{1}{x}\left(c - \frac{1}{x}\right) e^{cx}$$

and then substitute for $y$ and $y'$ in (1–10). This gives

$$xy' + y - y \ln(xy) = \left(c - \frac{1}{x}\right)e^{cx} + \frac{1}{x}e^{cx} - \frac{1}{x}e^{cx} \ln e^{cx}$$

$$= \left(c - \frac{1}{x}\right)e^{cx} + \left(\frac{1}{x} - c\right)e^{cx}$$

$$= 0,$$

showing that (1–10) is satisfied. The graphs of several of these solutions are shown in Fig. 1–2.

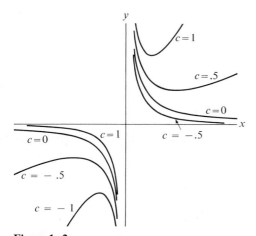

**Figure 1–2**

Of course, there is no reason to believe that a first-order differential equation must have solutions, and indeed many do not. And even though an equation is known to have solutions, it may be impossible to express them in terms of elementary functions. These issues will be considered in later chapters, but first we shall consider a number of equations that do not raise such difficulties.

The systematic study of first-order differential equations begins with equations that can be solved for $y'$ and written in so-called **normal form** as

$$y' = f(x, y), \tag{1–11}$$

where $f$ is continuous in one or more regions of the $xy$-plane.* The continuity of $f$ implies that each solution $y = y(x)$ of $(1–11)$ has a continuous first derivative on some interval $I$ of the $x$-axis. (In view of this fact, we say that $y$ is **continuously differentiable** on $I$.) Thus the graph of each solution $y$ is a smooth curve lying in the domain of $f$, as suggested by Fig. 1–3. Indeed, "smooth" is often used as a synonym for "continuously differentiable."

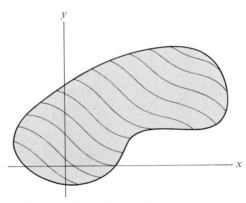

Solution curves in the domain of $f$

**Figure 1–3**

The most familiar normal first-order differential equations are those of the form

$$y' = f(x), \tag{1–12}$$

where $f$ is continuous on one or more open intervals of the $x$-axis. The continuity of $f$ guarantees that $(1–12)$ has solutions, and they can be expressed in integral form as

$$y = \int f(x)\, dx + c, \tag{1–13}$$

---

* In order to make the theorems of this and later chapters be true, the word **region** is given the technical meaning, "connected open set." A subset of the plane is **open** if each of its points is the center of a circle whose interior lies entirely in the set. An *open* subset of the plane is **connected** if every two of its points can be joined by a path that lies entirely in the set. You need not be concerned with these technicalities now. The domains on which most of the functions of this book are defined really are regions or unions of regions, and it will not be necessary to test every function that comes along. But without these conditions, or others like them, many of the most useful theorems of differential equations would become false.

where $c$ is an arbitrary (real) constant. In fact, we can say more: if $x_0$ is any point in the domain of $f$ and $y_0$ is an arbitrary real number, then there exists a *unique* solution curve of (1–12) through the point $(x_0, y_0)$. This is really a restatement of the *fundamental theorem of calculus*, which asserts that the solution in question is

$$y = y_0 + \int_{x_0}^{x} f(t) \, dt. \tag{1–14}$$

Thus the most important theorem in elementary calculus can now be interpreted as a *uniqueness* theorem for the differential equation

$$y' = f(x),$$

subject to the initial condition

$$y = y_0 \qquad \text{when} \qquad x = x_0.$$

Although the solution curve defined by (1–14) is uniquely determined by $y_0$ when $x_0$ is fixed, different values of $y_0$ give solution curves whose equations differ by constants. Thus the solutions of $y' = f(x)$ form a one-parameter family of plane curves, the parameter being the constant $c$ in (1–13). We call this family the **general solution** of $y' = f(x)$, even though it comprises infinitely many distinct solutions, while (1–14) is called the **particular solution** of the equation that satisfies the initial condition

$$y(x_0) = y_0.$$

**Example 2**   The equation

$$y' = 2x$$

is defined for all $x$ and has

$$y = x^2 + c$$

as its general solution. This "solution" is a one-parameter family of parabolas. (See Fig. 1–4.) To find the particular solution that satisfies the initial condition

$$y(2) = 3,$$

we set $x = 2$ and $y = 3$ in the general solution and solve for $c$. The result is $c = -1$, and we conclude that

$$y = x^2 - 1.$$

**Example 3**   The equation

$$y' = -\frac{x}{y} \tag{1–15}$$

is defined everywhere in the $xy$-plane except at points on the $x$-axis. To solve this equation, we rewrite it as

$$x \, dx + y \, dy = 0$$

and integrate to obtain

$$\frac{x^2}{2} + \frac{y^2}{2} = c,$$

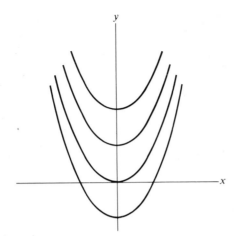

**Figure 1–4**

where $c$ is an arbitrary *positive* constant. (We still require $y \neq 0$.) Thus we are led to the family of circles

$$x^2 + y^2 = c^2, \qquad c \neq 0, \tag{1–16}$$

as possible solutions of (1–15). Here we are confronted with a difficulty: (1–16) is not in the form $y = y(x)$ required of a solution of a first-order differential equation. Hence to obtain the solutions of (1–15) from (1–16), we must solve the latter equation for $y$. When this is done, we obtain *two* families of solutions,

$$y = \sqrt{c^2 - x^2} \qquad \text{and} \qquad y = -\sqrt{c^2 - x^2},$$

as shown in Fig. 1–5. These semicircles, rather than the circles themselves, are the solution curves of (1–15).

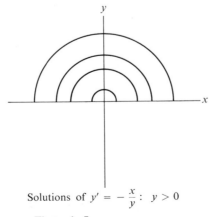

Solutions of $y' = -\dfrac{x}{y}$ : $y > 0$

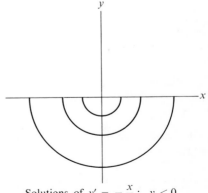

Solutions of $y' = -\dfrac{x}{y}$ : $y < 0$

**Figure 1–5**

The difficulty encountered in the preceding example is typical in that most of the techniques used to solve first-order differential equations lead to expressions of the form $F(x, y) = c$ rather than the type required by Definition 1–1. Thus we are confronted with the problem of determining conditions under which such an expression will serve to define $y$ as a function of $x$. The answer to this problem is given by a result known as the *implicit function theorem*.

> ***Theorem 1–1\**** Let $F(x, y)$ and its derivatives $\partial F/\partial x$, $\partial F/\partial y$ be continuous in a region $R$ of the xy-plane. Let $(x_0, y_0)$ be any point in $R$ where $\partial F/\partial y \neq 0$, and let $F(x_0, y_0) = c$. Then there exists a unique continuously differentiable function $y = y(x)$ defined in an open interval $I$ about $x_0$ such that
>
> i)  $y(x_0) = y_0$,
>
> ii)  $F(x, y(x)) \equiv c$ on $I$, and
>
> iii)  $y'(x) = -\dfrac{\partial F/\partial x}{\partial F/\partial y}$.

This result can be stated less precisely but more vividly as follows. Let

$$F(x, y) = c \qquad (1-17)$$

be given, and suppose that $F$ and its first partial derivatives are continuous in $R$. Then the graph of (1–17) will be a smooth curve in $R$. Moreover, if $(x_0, y_0)$ is any point on the curve where the tangent line is not vertical (i.e., where $\partial F/\partial y \neq 0$), then (1–17) defines a unique function $y = y(x)$ that is continuously differentiable in an open interval about $x_0$. Intuitively this means that (1–17) can be solved for $y$ in a neighborhood of $x_0$, although in practice it may be impossible to find an explicit expression for the solution. In any event, we say that $F(x, y) = c$ defines $y$ *implicitly* as a function of $x$ at those points where the hypotheses of the theorem are satisfied.

## EXERCISES

In Exercises 1 through 5, show that each given function is a solution of the accompanying differential equation, and specify the interval or intervals in which it is a solution.

1.  $y = (c^2 - x^2)^{1/2}$;     $yy' + x = 0$

2.  $y = c(x + c)$;     $(y')^2 + xy' - y = 0$

3.  $y = (c + \sin x)^2$;     $(y')^2 - 4y \cos^2 x = 0$

4.  $y = \tan^{-1}(x + c)$;     $y' - \cos^2 y = 0$

\* 5.  $y = \dfrac{\sin x}{c + \cos x}$;     $y' \sin x - y(\cos x + y \sin x) = 0$

---

\* For a proof of Theorem 1–1 see *Advanced Calculus* by R. C. Buck, McGraw-Hill, New York, 1965.

In Exercises 6 through 13 find (a) the general solution of each equation, and (b) the particular solution that satisfies the given initial conditions.

6. $y' = xe^x$;    $y(0) = -1$

7. $y' = \dfrac{1}{x^2 - 1}$;    $y(0) = \dfrac{1}{2}$

8. $y' = \sin^{-1} x$;    $y(1) = 0$

9. $y' = \sqrt{4 - x^2}$;    $y(-2) = 0$

10. $y^2 y' + x = 0$;    $y(2) = 3$

11. $y' = xy^3$;    $y(2) = 1$

12. $x^2 y' = 1 - y$;    $y(3) = 0$

13. $y' \tan y + x \cos^2 y = 0$;    $y(0) = \pi/4$

14. a) Find the general solution of

$$y' = 2\sqrt{y}$$

and discuss the behavior of the solution curves.

b) Repeat (a) for the equation

$$(y')^2 = 4y.$$

15. Show that in general it is not possible to write every solution of $y' = f(x)$ in the form

$$y = \int_a^x f(t)\, dt,$$

and compare this result with the fundamental theorem of calculus.

## 1–3  SEPARATION OF VARIABLES; EQUATIONS WITH HOMOGENEOUS COEFFICIENTS

In the next several sections we shall develop techniques for solving special types of normal first-order differential equations. In most of the discussion it will be convenient to use differential notation and write equations in the form

$$M(x, y)\, dx + N(x, y)\, dy = 0. \tag{1–18}$$

To avoid misunderstandings, we agree that this equation is just an alternative version of

$$\frac{dy}{dx} = -\frac{M(x, y)}{N(x, y)} \tag{1–19}$$

and hence is not defined where $N(x, y) = 0$. But notice that (1–18) can also be rewritten as $dx/dy = -N/M$, in which case $y$ is assumed to be the independent variable. Although we shall avoid this form, it is worth observing that differential notation does not commit one in advance to a choice of independent variable. This is one of its advantages.

**The Method of Separation of Variables**   As we have seen, any first-order differential equation of the form

$$M_1(x)M_2(y)\, dx + N_1(x)N_2(y)\, dy = 0 \tag{1–20}$$

can be converted into an equation that can be integrated by multiplying its coefficients by $1/(M_2 N_1)$. This yields

$$\frac{M_1(x)}{N_1(x)} dx + \frac{N_2(y)}{M_2(y)} dy = 0.$$

The general solution of (1–20) is then determined by the expression

$$\int \frac{M_1(x)}{N_1(x)} dx + \int \frac{N_2(y)}{M_2(y)} dy = c. \tag{1–21}$$

This method of solving a first-order differential equation is known as the method of **separation of variables**. Note, however, that the solutions of (1–20) which satisfy the equation $M_2(y) = 0$ do not appear in (1–21). Solutions like these that are inadvertently lost as an equation is solved, are known as **suppressed solutions**. They can be recovered by examining the steps that lead to the solution of the equation.

    **Example 1**   Solve the equation

$$x(y + 1)^2 \, dx + (x^2 + 1) y e^y \, dy = 0. \tag{1–22}$$

    *Solution.*   We divide the equation by $(x^2 + 1)(y + 1)^2$ to obtain

$$\frac{x}{x^2 + 1} dx + \frac{y e^y}{(y + 1)^2} dy = 0, \tag{1–23}$$

and integrate. This gives

$$\frac{1}{2} \ln (x^2 + 1) + \frac{e^y}{y + 1} = c,$$

and a straightforward application of the implicit function theorem shows that this expression determines the general solution of (1–22) except where $y = 0$ and $y = -1$. Finally, the function $y = -1$, suppressed in passing from (1–22) to (1–23), is also a solution of (1–22), and we are done.

**Equations with Homogeneous Coefficients**   Occasionally an equation whose variables cannot be separated can be converted into one whose variables are separable by making a substitution. One instance where this method will always work is when the coefficients of the equation are homogeneous *of the same degree* according to the following definition.

    ***Definition 1–2***   *A continuous function $f(x, y)$ is said to be **homogeneous of degree $\lambda$** if*

$$f(tx, ty) = t^\lambda f(x, y).$$

    Thus the functions $x^2 + y^2$ and $xy$ are homogeneous of degree 2, while $e^{x/y}$ and $(2x + y)/y$ are homogeneous of degree 0, and $1/\sqrt{x + y}$ is homogeneous of degree $-\frac{1}{2}$.

Now suppose that $M(x, y)$ and $N(x, y)$ are both homogeneous of degree $\lambda$, and consider the equation

$$M(x, y) \, dx + N(x, y) \, dy = 0. \tag{1–24}$$

We assert that the substitution $y = vx$ will convert this equation into one in which the variables are separable. Indeed, we then have

$$M(x, vx) \, dx + N(x, vx)(v \, dx + x \, dv) = 0,$$

which, because of the assumed homogeneity of $M$ and $N$, can be written as

$$x^\lambda M(1, v) \, dx + x^\lambda N(1, v)(v \, dx + x \, dv) = 0$$

or

$$M_1(v) \, dx + N_1(v)(v \, dx + x \, dv) = 0,$$

where $M_1$ and $N_1$ are functions of $v$ alone. We can now separate the variables to obtain

$$\frac{dx}{x} + \frac{N_1(v)}{M_1(v) + vN_1(v)} \, dv = 0,$$

and our assertion is proved.

*Remark.* The normal form of Eq. (1–24) is

$$y' = -\frac{M(x, y)}{N(x, y)},$$

where the function $-M/N$ is homogeneous of degree 0. Thus the substitution $y = vx$ will convert $y' = f(x, y)$ into an equation whose variables are separable whenever $f$ is homogeneous of degree 0.

**Example 2**   Solve the equation

$$y' = \frac{x + y}{x - y}. \tag{1–25}$$

*Solution.*   Since $(x + y)/(x - y)$ is homogeneous of degree 0, we make the substitution $y = vx$, obtaining

$$v + x \frac{dv}{dx} = \frac{1 + v}{1 - v}.$$

We can now separate the variables to obtain

$$\frac{dx}{x} + \frac{v - 1}{v^2 + 1} \, dv = 0,$$

and it follows that

$$\ln |x| + \tfrac{1}{2} \ln (v^2 + 1) - \tan^{-1} v = \ln c, \qquad c > 0.$$

Hence

$$x\sqrt{v^2 + 1} = ce^{\tan^{-1} v}, \qquad c \neq 0$$

and since $v = y/x$, we have

$$\sqrt{x^2 + y^2} = ce^{\tan^{-1}(y/x)}, \qquad c \neq 0,$$

or, in polar coordinates,

$$r = ce^{\theta}.$$

Since (1–25) is not defined along the line $y = x$, the solution curves of the equation form the two one-parameter families suggested by Fig. 1–6. Note that the line $y = x$ consists of precisely those points where the spirals $r = ce^{\theta}$ have vertical tangents.

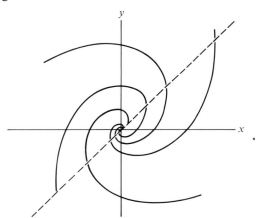

**Figure 1–6**

**Example 3**   Solve the equation

$$(x + y + 1)\, dx + (y - x - 3)\, dy = 0. \tag{1–26}$$

*Solution.*   Although the technique just introduced cannot be applied to the equation as it stands, we observe that if the constants 1 and $-3$ were eliminated from the coefficients, the equation could be solved. This suggests a change of variables of the form

$$x = u + \alpha, \qquad y = v + \beta,$$

where $\alpha$ and $\beta$ are chosen so as to eliminate the constants 1 and $-3$. Substituting into (1–26), we obtain

$$(u + v + \alpha + \beta + 1)\, du + (v - u - \alpha + \beta - 3)\, dv = 0, \tag{1–27}$$

which in turn yields

$$\alpha + \beta = -1 \qquad \text{and} \qquad -\alpha + \beta = 3.$$

Thus $\alpha = -2$, $\beta = 1$, and (1–27) becomes

$$(u + v)\, du + (v - u)\, dv = 0.$$

Hence

$$\frac{dv}{du} = \frac{u + v}{u - v}.$$

The results of Example 2 now imply that

$$\sqrt{u^2 + v^2} = ce^{\tan^{-1}(v/u)}, \qquad c \neq 0.$$

It follows that

$$\sqrt{(x + 2)^2 + (y - 1)^2} = ce^{\tan^{-1}(y - 1)/(x + 2)}, \qquad c \neq 0.$$

### EXERCISES

Solve the following equations.

1. $(y + 1) dx - (x + 1) dy = 0$      2. $\cos y \, dx + x \sin y \, dy = 0$

3. $4y \, dx - x(y - 3) \, dy = 0$      4. $x \, dx + e^{(x + y)} \, dy = 0$

5. $(y + 1) dx + (x^2 - 4) dy = 0$      6. $xy \, dx + (x^2 + 1) dy = 0$

7. $(y + 1)(1 - x) dx - xy \, dy = 0$

8. $(x + 1)(y - 1) dx + (x - 1)(y + 1) dy = 0$

9. $(x^3 + y^3) dx - 3xy^2 \, dy = 0$      10. $(y + \sqrt{x^2 + y^2}) dx - x \, dy = 0$

11. $(x^2 + y^2) dx + 2xy \, dy = 0$      12. $(1 + 2e^{x/y}) dx + 2e^{x/y}\left(1 - \dfrac{x}{y}\right) dy = 0$

13. $\left(x \sin \dfrac{y}{x} - y \cos \dfrac{y}{x}\right) dx + x \cos \dfrac{y}{x} \, dy = 0$

14. $(x - y) dx + (2y - x) dy = 0$      15. $(y^2 - x^2) dx + xy \, dy = 0$

16. $(4x + 2y + 1) dx - (2x - y - 1) dy = 0$

17. $(2x - y - 4) dx - (x - 2y + 1) dy = 0$

18. $y' = -\dfrac{x + 2y}{2x + 3y + 1}$      19. $y' = \dfrac{1 - x - y}{x}$

Find the indicated particular solution of each of the following equations.

20. $dx + (1 - x^2) \cot y \, dy = 0; \qquad y(0) = \pi/4$

21. $y \, dx + \cot x \, dy = 0; \qquad y(0) = 2$      22. $(y^2 + xy) dx - x^2 \, dy = 0; \qquad y(1) = 1$

23. $y' = \dfrac{y^2}{x}; \qquad y(1) = 0$      24. $y' = -\dfrac{4x + 11y - 42}{11x - 9y - 37}; \qquad y(0) = -1$

*25. Solve $y' = y^n/x^m$, where $m$ and $n$ are positive integers.

26. a) Show that the substitution

$$u = ax + by + c$$

reduces

$$y' = f(ax + by + c)$$

to an equation in which the variables are separable.

b) Use the substitution in (a) to solve

$$y' = (x + y)^2.$$

27. a) Show that the substitution

$$u = a_1 x + b_1 y$$

     reduces

$$y' = \frac{a_1 x + b_1 y + c_1}{a_2 x + b_2 y + c_2}$$

to an equation in which the variables are separable whenever $a_1 b_2 - a_2 b_1 = 0$ and $a_1 \neq 0$ or $b_1 \neq 0$. [*Hint*: Show that there exists a constant $k$ such that $a_2 x + b_2 y = k(a_1 x + b_1 y)$.]

   b) Find a substitution that will reduce the equation of part (a) to one in which the variables are separable when $a_1 = b_1 = 0$.

Use the results of Exercise 27 to solve the following equations.

28. $y' = -\dfrac{x + y + 1}{x + y}$

29. $y' = y - x + 1$

30. $y' = \dfrac{6x + 3y - 5}{2x + y}$

31. $y' = \dfrac{1}{2x + y}$

## 1–4   EXACT EQUATIONS

Our next technique for solving normal first-order differential equations is based on the notion of the **total differential** $dF$ of a function of two variables. We recall that when $F$ has continuous first partial derivatives, $dF$ is defined by the rule

$$dF = \frac{\partial F}{\partial x}\, dx + \frac{\partial F}{\partial y}\, dy.$$

Since $\partial F/\partial x$ and $\partial F/\partial y$ are themselves functions of $x$ and $y$, we can rewrite $dF$ as

$$dF = M(x, y)\, dx + N(x, y)\, dy. \tag{1–28}$$

Hence, if

$$M(x, y)\, dx + N(x, y)\, dy = 0, \tag{1–29}$$

then

$$dF = 0,$$

and

$$F(x, y) = c, \tag{1–30}$$

where $c$ is a constant, defines the general solution of (1–29). (See Exercise 18.) These facts lead to the following definition.

**Definition 1–3**   *The expression*

$$M(x, y)\, dx + N(x, y)\, dy$$

*is an **exact differential** in a region R of the xy-plane if there exists a function F such that*

$$dF = M(x, y)\, dx + N(x, y)\, dy$$

*everywhere in R. In this case the differential equation*

$$M(x, y)\, dx + N(x, y)\, dy = 0$$

*is said to be **exact**, and the equation*

$$F(x, y) = c,$$

*where c is a constant, is called the **general integral** of the equation.*

**Example 1**   The expression

$$x\, dx + y\, dy$$

is an exact differential in the entire $xy$-plane, since

$$x\, dx + y\, dy = d\left(\frac{x^2 + y^2}{2}\right).$$

Hence

$$x\, dx + y\, dy = 0$$

is exact, and the general integral of this equation is

$$x^2 + y^2 = c^2, \qquad c \neq 0.$$

(See Example 3 on p. 9.)

**Example 2**   The equation

$$(2x + e^y)\, dx + xe^y\, dy = 0$$

is exact, because

$$\frac{\partial}{\partial x}[x(x + e^y)] = 2x + e^y \quad \text{and} \quad \frac{\partial}{\partial y}[x(x + e^y)] = xe^y.$$

Thus its general integral is

$$x(x + e^y) = c, \qquad c \text{ arbitrary.}$$

These examples illustrate how easy it is to solve the equation

$$M\, dx + N\, dy = 0$$

when we know that the left-hand side is the differential of a function $F$. Thus we are led to look for a test that will enable us to tell whether $M\, dx + N\, dy$ is an exact differential *without knowing F in advance*. With this in mind, let us suppose that $M\, dx + N\, dy$ is the differential of an (unknown) function $F$ in a region $R$, and suppose also that $M$ and $N$ have continuous first partial derivatives in $R$. Then since

$$M = \frac{\partial F}{\partial x} \quad \text{and} \quad N = \frac{\partial F}{\partial y},$$

we have

$$\frac{\partial M}{\partial y} = \frac{\partial^2 F}{\partial y\, \partial x}, \qquad \frac{\partial N}{\partial x} = \frac{\partial^2 F}{\partial x\, \partial y},$$

and the continuity assumptions in effect imply that these derivatives are equal.

Thus, **if** $M\,dx + N\,dy$ *is an exact differential whose coefficients have continuous first partial derivatives,* **then**

$$\frac{\partial M}{\partial y} = \frac{\partial N}{\partial x}. \tag{1-31}$$

It turns out that with a mild restriction on the geometric nature of the region $R$, the converse of this statement is also true, and (1–31) can be used to test $M\,dx + N\,dy$ for exactness. Specifically, we require that the region be simply connected, which is just another way of saying that there should be no "holes" in $R$. More formally, a region $R$ of the plane is said to be **simply connected** if it is impossible to draw a simple closed curve in $R$ that surrounds points not in $R$. (See Fig. 1–7.) Having said this, we now state a theorem.

Not connected                Simply connected                Multiply connected

**Figure 1–7**

***Theorem 1–2\**    *Let $M(x, y)$ and $N(x, y)$ have continuous first partial derivatives in a simply connected region $R$ of the xy-plane. Then*

$$M\,dx + N\,dy$$

*is an exact differential whenever*

$$\frac{\partial M}{\partial y} = \frac{\partial N}{\partial x}.$$

**Example 3**    In the previous example we showed that

$$(2x + e^y)\,dx + xe^y\,dy$$

is exact by producing a function $F$ such that

$$dF = (2x + e^y)\,dx + xe^y\,dy.$$

We now show how to obtain this result without knowing $F$ in advance.

In the first place, $(2x + e^y)\,dx + xe^y\,dy$ is an exact differential in the entire plane, because

$$\frac{\partial}{\partial y}(2x + e^y) = \frac{\partial}{\partial x}(xe^y) = e^y.$$

---

\*  A proof of this result, which requires the use of line integrals, may be found in R. C. Buck's *Advanced Calculus*, McGraw-Hill, New York, 1965.

Thus Theorem 1–2 guarantees the existence of a function $F$ with

$$\frac{\partial F}{\partial x} dx + \frac{\partial F}{\partial y} dy = (2x + e^y) dx + xe^y dy.$$

Hence

$$\frac{\partial F}{\partial x} = 2x + e^y \qquad \text{and} \qquad \frac{\partial F}{\partial y} = xe^y.$$

We now integrate the first of these equations "with respect to $x$," treating $y$ as a constant, to obtain

$$F = x^2 + xe^y + \phi(y), \tag{1–32}$$

where $\phi$ is a function of $y$ alone and appears here as a "constant" of integration. Equation (1–32) implies that

$$\frac{\partial F}{\partial y} = xe^y + \phi'(y);$$

and since

$$\frac{\partial F}{\partial y} = xe^y,$$

it follows that $\phi'(y) = 0$. Hence $\phi(y)$ is a constant, and

$$F(x, y) = x^2 + xe^y + c.$$

**Example 4**   Find the general integral of

$$(3x^2 + 2xy + y^2) dx + (x^2 + 2xy + 3y^2) dy = 0. \tag{1–33}$$

*Solution.*   Since

$$\frac{\partial}{\partial y}(3x^2 + 2xy + y^2) = 2x + 2y,$$

$$\frac{\partial}{\partial x}(x^2 + 2xy + 3y^2) = 2x + 2y,$$

the equation is exact. Hence there exists a function $F$ such that

$$\frac{\partial F}{\partial x} = 3x^2 + 2xy + y^2,$$

$$\frac{\partial F}{\partial y} = x^2 + 2xy + 3y^2.$$

By integrating the first equation with respect to $x$, we obtain

$$F = x^3 + x^2y + xy^2 + \phi(y),$$

so that

$$\frac{\partial F}{\partial y} = x^2 + 2xy + \phi'(y).$$

We now use the known value of $\partial F/\partial y$ to deduce that

$$\phi'(y) = 3y^2.$$

Thus

$$\phi(y) = y^3$$

(we will pick up an arbitrary constant when we set $F$ equal to $c$),

$$F = x^3 + x^2 y + xy^2 + y^3$$

and the general integral of (1–33) is

$$x^3 + x^2 y + xy^2 + y^3 = c.$$

## EXERCISES

Show that the equations in Exercises 1 through 15 are exact and find their general integrals.

1. $2xy\, dx + (x^2 + 4y)\, dy = 0$

2. $y(y^2 - 3x^2)\, dx + x(3y^2 - x^2)\, dy = 0$

3. $\left(\dfrac{1}{y} + \dfrac{y}{x^2}\right) dx - \left(\dfrac{1}{x} + \dfrac{x}{y^2}\right) dy = 0$

4. $(\sin xy + xy \cos xy)\, dx + x^2 \cos xy\, dy = 0$

5. $\dfrac{y\, dx - x\, dy}{(x + y)^2} + \dfrac{1}{y}\, dy = 0$

6. $x^2\, dx + y^2\, dy + \dfrac{y\, dx - x\, dy}{x^2 + y^2} = 0$

\* 7. $\dfrac{y\, dx - x\, dy}{xy} + \dfrac{x\, dy + y\, dx}{\sqrt{1 + (xy)^2}} = 0$

8. $(1 + \ln xy)\, dx + \left(1 + \dfrac{x}{y}\right) dy = 0$

9. $\left(\dfrac{y}{x} + \ln y\right) dx + \left(\dfrac{x}{y} + \ln x\right) dy = 0$

10. $(ye^x + e^y)\, dx + (e^x + xe^y)\, dy = 0$

11. $(\cos xy - \sin xy)(y\, dx + x\, dy) = 0$

12. $2 \sec^2 (x^2 + y^2)(x\, dx + y\, dy) = 0$

13. $(1 + \tan xy)\, dx + (\sec xy \tan xy + x \sec^2 xy)(y\, dx + x\, dy) = 0$

14. $y(e^{xy} + y)\, dx + x(e^{xy} + 2y)\, dy = 0$

15. $\dfrac{y\, dx + x\, dy}{1 + (xy)^2} = 0$

16. Show that a normal first-order differential equation in which the variables are separated is exact, and find its general integral by the method of this section.

\* 17. a) Prove that

$$y' = \dfrac{ax + by}{cx + dy}, \qquad ad - bc \neq 0$$

is exact if and only if $b + c = 0$. Find the general integral of this equation when it is exact, and discuss the behavior of its solution curves.

b) Sketch the solution curves of

$$y' = \frac{3x - y}{x + y} \quad \text{and} \quad y' = \frac{3x + y}{x - y}$$

\* 18. Let

$$M(x, y)\, dx + N(x, y)\, dy = 0$$

be defined in a region $R$ of the $xy$-plane, and suppose that $F$ is differentiable in $R$ with

$$\frac{\partial F}{\partial x} = M \quad \text{and} \quad \frac{\partial F}{\partial y} = N.$$

Suppose in addition that $F(x_0, y_0) = c$, where $(x_0, y_0)$ is any point in $R$. Prove that $F(x, y) = c$ defines a solution of $M\, dx + N\, dy = 0$ in an open interval $I$ about $x_0$. [*Hint*: Apply the implicit function theorem.]

## 1–5 INTEGRATING FACTORS

The technique introduced in the preceding section can sometimes be used to solve

$$M(x, y)\, dx + N(x, y)\, dy = 0 \tag{1–34}$$

even when the equation is not exact. This is done by multiplying the coefficients of (1–34) by a nonzero function $\mu = \mu(x, y)$, chosen so that the new equation,

$$\mu M\, dx + \mu N\, dy = 0, \tag{1–35}$$

is exact. This new equation is then solved, and the solution of (1–34) is completed by adjoining any suppressed solutions arising from the equation $\mu(x, y) = 0$. The function $\mu$ is called an **integrating factor** for Equation (1–34).

**Example 1**  Find the general solution of

$$y(xy + 1)\, dx - x\, dy = 0. \tag{1–36}$$

*Solution.*  Since none of our earlier techniques apply, we look for an integrating factor. To this end, we rewrite the equation as

$$xy^2\, dx + y\, dx - x\, dy = 0, \tag{1–37}$$

and recall that the combination $y\, dx - x\, dy$ occurs in a large number of exact differentials:

$$d\left(\frac{x}{y}\right) = \frac{y\, dx - x\, dy}{y^2},$$

$$d\left(-\frac{y}{x}\right) = \frac{y\, dx - x\, dy}{x^2},$$

$$d\left(\ln\frac{x}{y}\right) = \frac{y\, dx - x\, dy}{xy},$$

and so forth. Thus we seek an integrating factor that will make $xy^2\,dx$ an exact differential and at the same time produce one of the known differentials involving $y\,dx - x\,dy$. Clearly, multiplication by $1/y^2$ will do just that. Equation (1–37) then becomes

$$x\,dx + \frac{y\,dx - x\,dy}{y^2} = 0 \tag{1–38}$$

or

$$d\left(\frac{x^2}{2}\right) + d\left(\frac{x}{y}\right) = 0,$$

and it follows that the general integral of (1–37) is

$$\frac{x^2}{2} + \frac{x}{y} = c.$$

Thus the solutions of (1–36) are

$$y = 0 \quad \text{[suppressed in passing from (1–37) to (1–38)],}$$

$$y = \frac{2x}{c - x^2}, \quad c \text{ arbitrary.}$$

The graphs of several of these solution curves in the region $x > 0$ are shown in Fig. 1–8. Those in the region $x < 0$ are obtained by reflecting the figure in the origin.

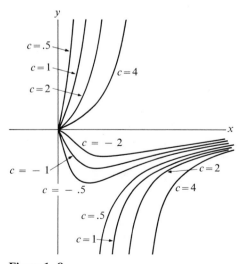

**Figure 1–8**

**Example 2**   Solve the equation

$$y(x^2 + 1)\, dx + x\, dy = 0. \tag{1--39}$$

*Solution.*   We first rewrite the equation as

$$x^2 y\, dx + y\, dx + x\, dy = 0,$$

and then try to exploit the combination $y\, dx + x\, dy$ which arises in differentials involving the product $xy$. The form of the first term in the equation suggests that we first try $1/xy$ as an integrating factor. When we do so, we obtain

$$x\, dx + \frac{y\, dx + x\, dy}{xy} = 0,$$

an equation which is exact. Since the last equation can be written as

$$d\left(\frac{x^2}{2}\right) + d\, (\ln|xy|) = 0,$$

the general integral of (1--39) is

$$\frac{x^2}{2} + \ln|xy| = c.$$

**Example 3**   Solve the equation

$$(x + 2y)\, dx - x\, dy = 0. \tag{1--40}$$

*Solution.*   As it stands, this equation is not exact, but it will become exact if we multiply it by $1/x^3$. Then we have

$$\frac{x + 2y}{x^3}\, dx - \frac{1}{x^2}\, dy = 0$$

and

$$\frac{\partial M}{\partial y} = \frac{\partial N}{\partial x} = \frac{2}{x^3}.$$

Thus $1/x^3$ is an integrating factor for (1--40), and the general integral of (1--40) can be found by the technique introduced in the preceding section. But here again we can proceed by inspection, since (1--40) can be rewritten as

$$\frac{dx}{x^2} - \frac{x\, dy - 2y\, dx}{x^3} = 0$$

or

$$d\left(-\frac{1}{x}\right) - d\left(\frac{y}{x^2}\right) = 0,$$

from which we obtain

$$\frac{1}{x} + \frac{y}{x^2} = c$$

or

$$y = x(cx - 1).$$

Theoretically, the method of solving a normal first-order differential equation by finding an integrating factor is extremely general, since it can be proved that whenever $M$ and $N$ are continuously differentiable in a simply connected region of the plane, the equation $M\,dx + N\,dy = 0$ has an integrating factor. In fact, it has infinitely many, and once one of them is known, others can be found without difficulty. (See Exercise 26.) Unfortunately, there is no general technique by which even a single integrating factor can be produced for an arbitary differential equation, and the search for one can be a frustrating experience. Table 1–1 lists a number of integrating factors and exact differentials that are frequently encountered.

**Table 1–1**

| With the combination | Use the integrating factor | To obtain the exact differential |
|---|---|---|
| $y\,dx + x\,dy$ | $1/xy$ | $\dfrac{y\,dx + x\,dy}{xy} = d\,(\ln xy)$ |
|  | $1/(xy)^n, \qquad n \neq 1$ | $\dfrac{y\,dx + x\,dy}{(xy)^n} = -d\left(\dfrac{1}{(n-1)(xy)^{n-1}}\right)$ |
| $x\,dx + y\,dy$ | $\dfrac{1}{x^2 + y^2}$ | $\dfrac{x\,dx + y\,dy}{x^2 + y^2} = \dfrac{1}{2}d\,(\ln(x^2 + y^2))$ |
|  | $\dfrac{1}{(x^2 + y^2)^n}, \qquad n \neq 1$ | $\dfrac{x\,dx + y\,dy}{(x^2 + y^2)^n} = -\dfrac{1}{2}d\left(\dfrac{1}{(n-1)(x^2 + y^2)^{n-1}}\right)$ |
| $y\,dx - x\,dy$ | $1/y^2$ | $\dfrac{y\,dx - x\,dy}{y^2} = d\left(\dfrac{x}{y}\right)$ |
|  | $1/x^2$ | $\dfrac{y\,dx - x\,dy}{x^2} = -d\left(\dfrac{y}{x}\right)$ |
|  | $1/xy$ | $\dfrac{y\,dx - x\,dy}{xy} = d\left(\ln\dfrac{x}{y}\right)$ |
|  | $1/(x^2 + y^2)$ | $\dfrac{y\,dx - x\,dy}{x^2 + y^2} = d\left(\tan^{-1}\dfrac{x}{y}\right)$ |

Table 1–2 gives a number of special, but sometimes useful, formulas for producing integrating factors. The first formula, for example, will yield the integrating factor used in Example 3 of this section. These formulas are established in the exercises that follow.

## Table 1–2

| Given: $M\,dx + N\,dy = 0$ | |
|---|---|
| When | An integrating factor is |
| $\dfrac{1}{N}\left(\dfrac{\partial M}{\partial y} - \dfrac{\partial N}{\partial x}\right) = f(x)$ | $e^{\int f(x)\,dx}$ |
| $\dfrac{1}{M}\left(\dfrac{\partial M}{\partial y} - \dfrac{\partial N}{\partial x}\right) = -g(y)$ | $e^{\int g(y)\,dy}$ |
| $M$ and $N$ homogeneous of the same degree | $\dfrac{1}{Mx + Ny}$ |
| $M = yf(xy),\ N = xg(xy)$ $f(xy) \neq g(xy)$ | $\dfrac{1}{xy[f(xy) - g(xy)]}$ |

## EXERCISES

Use an integrating factor to find the general integral of the equations in Exercises 1 through 21.

1. $3x^4y^2\,dx + y\,dx + x\,dy = 0$

2. $y(y^2 + 1)\,dx + x(y^2 - 1)\,dy = 0$

3. $y\,dx + [y(x^2 + y^2) - x]\,dy = 0$

4. $\left(\dfrac{y}{x} + 2\right)dx + \left(\dfrac{x}{y} + 2\right)dy = 0$

5. $y(xy + 1)\,dx - x(xy - 1)\,dy = 0$

6. $(1 + xy)\,dx + x^2\,dy = 0$

7. $[x - y(x^2 + y^2)]\,dx + [y + x(x^2 + y^2)]\,dy = 0$

8. $y^2\,dx + (x^2 - xy - y^2)\,dy = 0$

9. $y(y^3 + 1)\,dx + x(y^3 - 2)\,dy = 0$

10. $(y^2 + 1)\,dx + y(x + y^2 - 1)\,dy = 0$

11. $(x^2 + y^2 + x)\,d_x + xy\,dy = 0$

12. $(x^4 + y^4)\,dx - xy^3\,dy = 0$

13. $(\sec x + y \tan x)\,dx + dy = 0$

14. $[2 \tan x + (x + y)\sec^2 x]\,dx + 2 \tan x\,dy = 0$

15. $y\,dx + x(1 - x^2y^2)\,dy = 0$

16. $y\,dx + x(1 - 3x^2y^2)\,dy = 0$

17. $x\,dy - y\,dx - (1 - x^2)\,dx = 0$

18. $(xy + \sqrt{1 + x^2y^2})\,dx + x^2\,dy = 0$

19. $(y^3 - 2x^2y)\,dx + (2xy^2 - x^3)\,dy = 0$

20. $x(x + y)\,dy - y^2\,dx = 0$

21. $\ ^{1}x^{2}dx + (x^2 - xy - y^2)\,dy = 0$

22. Let $M(x, y)$ and $N(x, y)$ be homogeneous of the same degree. Prove that $1/(Mx + Ny)$ is an integrating factor for $M\,dx + N\,dy = 0$.

23. a) Let $M$ and $N$ be continuously differentiable in a region $R$ of the $xy$-plane, and suppose that $N$ does not vanish anywhere in $R$. Prove that if

$$\frac{1}{N}\left(\frac{\partial M}{\partial y} - \frac{\partial N}{\partial x}\right) = f(x),$$

then $e^{\int f(x)\,dx}$ is an integrating factor for $M\,dx + N\,dy = 0$.

b) With hypotheses similar to those in (a), prove that if

$$\frac{1}{M}\left(\frac{\partial M}{\partial y} - \frac{\partial N}{\partial x}\right) = -g(y),$$

then $e^{\int g(y)\,dy}$ is an integrating factor for $M\,dx + N\,dy = 0$.

\* 24. Prove that $\mu = \mu(x, y)$ is an integrating factor for $M\,dx + N\,dy = 0$ if and only if $\mu$ satisfies the partial differential equation

$$M\frac{\partial \mu}{\partial y} - N\frac{\partial \mu}{\partial x} + \left(\frac{\partial M}{\partial y} - \frac{\partial N}{\partial x}\right)\mu = 0.$$

\* 25. a) Assuming that the necessary derivatives exist, prove that

$$\frac{1}{xy[f(xy) - g(xy)]}, \qquad f(xy) \neq g(xy),$$

is an integrating factor for the equation

$$yf(xy)\,dx + xg(xy)\,dy = 0.$$

b) Use the result in (a) to find the general integral of

$$y(x^2y^2 + 2)\,dx + x(2 - 2x^2y^2)\,dy = 0.$$

\* 26. Let $\mu = \mu(x, y)$ be an integrating factor for $M\,dx + N\,dy = 0$, and suppose that

$$dF = \mu M\,dx + \mu N\,dy.$$

Prove that every function of the form $\mu\phi(F)$, where $\phi$ is an arbitrary differentiable function of $F$, is also an integrating factor for $M\,dx + N\,dy = 0$, and hence conclude that every equation that admits one integrating factor actually admits infinitely many.

## 1–6  REDUCTION OF ORDER

In this section we shall consider two special forms of the general second-order equation

$$F(x, y, y', y'') = 0 \tag{1–41}$$

that can be solved by first-order methods.

**Dependent Variable Missing**   If $y$ does not appear explicitly in (1–41), then the substitution

$$y' = p$$

will reduce that equation to

$$F(x, p, p') = 0, \tag{1–42}$$

which is a first-order equation in $p$. If (1–42) can be solved for $p$, then $y$ can be found by integration.

**Example 1**   Solve the equation

$$2y'' - (y')^2 + 1 = 0.$$

*Solution.*   The substitution $y' = p$ reduces the equation to

$$2p' - p^2 + 1 = 0$$

or

$$\frac{2\,dp}{p^2 - 1} = dx.$$

Noting that

$$\frac{2}{p^2 - 1} = \frac{1}{p - 1} - \frac{1}{p + 1},$$

we have

$$\frac{dp}{p - 1} - \frac{dp}{p + 1} = dx,$$

which, when integrated, yields

$$\frac{p - 1}{p + 1} = c_1 e^x, \qquad c_1 \text{ arbitrary.}$$

We now solve for $p$ to obtain

$$p = \frac{1 + c_1 e^x}{1 - c_1 e^x} = 1 + \frac{2c_1 e^x}{1 - c_1 e^x}.$$

Thus

$$\frac{dy}{dx} = 1 + \frac{2c_1 e^x}{1 - c_1 e^x},$$

and

$$y = x - \ln(1 - c_1 e^x)^2 + c_2.$$

This technique can also be used with equations of order higher than two when the appropriate variables are missing.

**Example 2**   Solve the equation

$$xy''' - 2y'' = 0.$$

*Solution.*   If we make the substitution

$$y'' = q$$

(the use of the letters $p$ for $y'$ and $q$ for $y''$ is traditional), the equation becomes

$$xq' - 2q = 0$$

or

$$\frac{dq}{q} - 2\frac{dx}{x} = 0,$$

and we obtain

$$q = c_1 x^2, \qquad c_1 \text{ arbitrary.}$$

Thus

$$\frac{d^2y}{dx^2} = c_1 x^2,$$

and it follows that

$$y = c_1 x^4 + c_2 x + c_3,$$

where again $c_1$, $c_2$, and $c_3$ are arbitrary constants.

**Independent Variable Missing**  We now consider second-order equations of the form

$$F(y, y', y'') = 0 \qquad\qquad (1–43)$$

in which $x$ does not appear explicitly. Once again we make the substitution $y' = p$ and observe that

$$y'' = \frac{dp}{dx} = \frac{dp}{dy}\frac{dy}{dx} = p\frac{dp}{dy}.$$

Equation (1–43) can now be rewritten as

$$F\left(y, p, p\frac{dp}{dy}\right) = 0.$$

In this form it is a first-order equation in $y$ and $p$. The substitutions

$$y' = p \qquad \text{and} \qquad y'' = p\frac{dp}{dy}$$

are worth remembering. They appear frequently in applications.

**Example 3**  Solve the equation

$$y'' + k^2 y = 0.$$

*Solution.*  Using the substitution just suggested, we obtain

$$p\frac{dp}{dy} + k^2 y = 0,$$

which has

$$p^2 + k^2 y^2 = k^2 a^2,$$

where $a$ is an arbitrary constant, as its general integral. Thus

$$p = \pm k\sqrt{a^2 - y^2},$$

and

$$\frac{dy}{\sqrt{a^2 - y^2}} = \pm k\,dx.$$

A second integration yields

$$\sin^{-1}\frac{y}{a} = \pm kx + b.$$

It follows that

$$y = a \sin (kx + b).$$

In the exercises below, the reader is invited to show that this solution can also be written in the form

$$y = c_1 \sin kx + c_2 \cos kx,$$

where $c_1$ and $c_2$ are arbitrary constants. In this form it will play an important role in our later work.

### EXERCISES

Find the general integral of each of the following equations.

1. $xy'' = y'$
2. $yy'' + (y')^2 = 0$
3. $y''' - y'' = 1$
4. $y'' = 1 + (y')^2$
5. $y'' - k^2 y = 0$
6. $yy'' + (y')^3 = 0$
7. $yy'' = 2(y')^2 - 2y'$
8. $x^2 y'' = 2xy' + (y')^2$
9. Verify that $y = a \sin (kx + b)$, where $a$ and $b$ are arbitrary constants, can be written as

$$y = c_1 \sin kx + c_2 \cos kx,$$

where $c_1$ and $c_2$ are also arbitrary.

### 1–7  GEOMETRIC APPLICATIONS

Problems in geometry that involve slopes of curves often give rise to differential equations.

**Families of Plane Curves and Orthogonal Trajectories**  We have seen that the general solution of a first-order (ordinary) differential equation contains an arbitrary constant and therefore defines a one-parameter family of curves in the plane. Conversely, many one-parameter families of plane curves can be shown to be the integral curves of a first-order differential equation. Indeed, if

$$\Phi(x, y, c) = 0 \tag{1–44}$$

is an equation for such a family, and if we can eliminate $c$ between (1–44) and its derivative with respect to $x$, then we obtain a first-order differential equation

$$F(x, y, y') = 0 \tag{1–45}$$

which has (1–44) as its general integral. We call (1–45) a differential equation for the given one-parameter family, even though it occasionally has particular solutions that do not belong to that family. (See Chapter 11.)

**Example 1**  To find a differential equation for the family of parabolas

$$y = cx^2,$$

we differentiate,

$$y' = 2cx,$$

and then eliminate $c$ to obtain

$$y' = \frac{2y}{x} \quad \text{or} \quad xy' = 2y.$$

Whenever two families of plane curves have the property that each curve in either family is perpendicular, or **orthogonal**, to every curve in the other family, each is said to be a family of **orthogonal trajectories** of the other. Two curves are orthogonal at a point of intersection if and only if their slopes at the point are negative reciprocals of each other (Fig. 1–9). It follows that if the differential equation of a family of plane curves is

$$F(x, y, y') = 0,$$

then its orthogonal trajectories can be found by solving the equation

$$F\left(x, y, -\frac{1}{y'}\right) = 0.$$

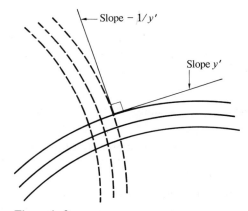

Slope $-1/y'$

Slope $y'$

**Figure 1–9**

**Example 2**   Find the orthogonal trajectories of the family of parabolas in Example 1.

*Solution.*   Since a differential equation for the family of parabolas is

$$xy' = 2y,$$

the orthogonal trajectories can be found by solving

$$x\left(-\frac{1}{y'}\right) = 2y.$$

This equation can be rewritten as

$$x\,dx + 2y\,dy = 0,$$

whose general integral is the family of ellipses

$$\frac{x^2}{2} + y^2 = c^2.$$

See Fig. 1–10.

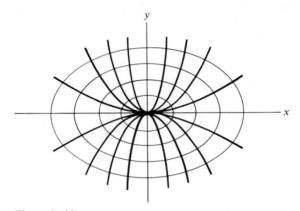

**Figure 1–10**

**Example 3**   Find the orthogonal trajectories of the family of circles through the points $(1, 0)$ and $(-1, 0)$.

*Solution.*   As Fig. 1–11 suggests, an equation for the family of circles is

$$x^2 + (y - c)^2 = 1 + c^2.$$

To find the orthogonal trajectories, we first differentiate and eliminate $c$. This gives

$$y' = -\frac{2xy}{1 - x^2 + y^2}.$$

(See Exercise 4.) Hence the orthogonal trajectories are the solution curves of the equation

$$y' = \frac{1 - x^2 + y^2}{2xy}, \tag{1–46}$$

together with the exceptional curve $x = 0$. We now rewrite (1–46) as

$$2xy\,dy - y^2\,dx = (1 - x^2)\,dx$$

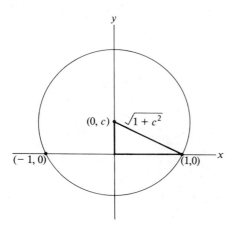

**Figure 1–11**

and observe that the left-hand side resembles the differential

$$\frac{2xy\,dy - y^2\,dx}{x^2} = d\left(\frac{y^2}{x}\right).$$

Thus $1/x^2$ is an integrating factor for (1–46), and the general integral of the equation is

$$\frac{y^2}{x} + 2c = -\frac{1}{x} - x$$

or

$$(x + c)^2 + y^2 = c^2 - 1.$$

Except for the $y$-axis, then, the orthogonal trajectories form a second one-parameter family of circles, but now with centers on the $x$-axis. See Fig. 1–12.

**Curves Defined by Differential Equations**   We conclude this section with two additional examples of curves that are obtained as solutions of differential equations.

    **Example 4**   Find all functions $y = y(x)$ with the property that a ray of light from the origin will be reflected by the graph of the function in the direction of the positive $y$-axis.

    *Solution.*   Since the incident and reflected rays form congruent angles with the tangent to the graph, the angle measures $\alpha$ and $\beta$ in Fig. 1–13 are equal. Hence

$$\tan \alpha = \tan \beta = \cot \phi.$$

Now, $\tan \theta = y/x$, $\tan \phi = dy/dx$, and $\alpha = \phi - \theta$. Thus

$$\tan \alpha = \frac{\tan \phi - \tan \theta}{1 + \tan \phi \tan \theta},$$

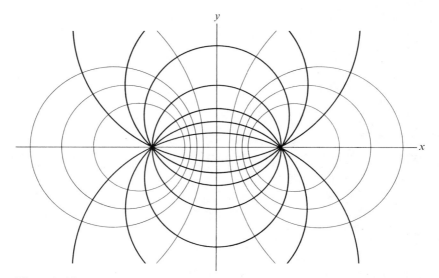

**Figure 1–12**

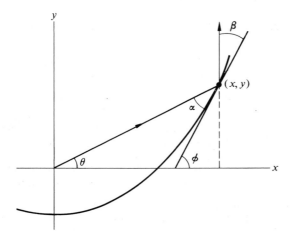

**Figure 1–13**

and the equality $\tan \alpha = \cot \phi$ yields

$$\frac{1}{y'} = \frac{y' - y/x}{1 + yy'/x}$$

or

$$\frac{1}{y'} = \frac{xy' - y}{x + yy'}.$$

It follows that

$$y' = \frac{y \pm \sqrt{x^2 + y^2}}{x}.$$

Since the right-hand side is homogeneous of degree zero, this equation can be solved by making the substitution $y = vx$, yielding the solutions

$$y = \frac{c}{2} x^2 - \frac{1}{2c}, \qquad c \neq 0.$$

(See Exercise 16.) The curves are parabolas opening upward about the $y$-axis with their foci at the origin.

**Example 5** Two skaters start on the $x$-axis, $A$ at the origin and $B$ at the point $(36, 0)$. Suppose that $A$ skates along the $y$-axis, that $B$ skates directly toward $A$ at all times, and that $B$ skates twice as fast as $A$. How far will $A$ travel before being caught by $B$?

*Solution.* Suppose that $A$ skates at a rate of $a$ units per second. Then at time $t$, with $t$ measured from the instant at which both start skating, $A$ will be at the point $(0, at)$. At the same time $B$ will be at the point $(x, y)$, whose coordinates, as we see from Fig. 1–14, satisfy the equation

$$\frac{dy}{dx} = \frac{y - at}{x}$$

or

$$xy' - y = -at.$$

To solve this equation for $y$, we first eliminate $t$ as follows. Differentiating with respect to $x$, we obtain

$$xy'' = -a \frac{dt}{dx}. \tag{1–47}$$

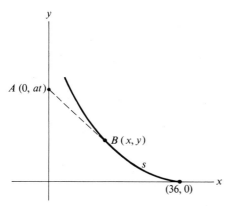

**Figure 1–14**

But since $ds/dt = 2a$,

$$\frac{dt}{dx} = \frac{dt}{ds}\frac{ds}{dx} = \frac{1}{2a}\frac{ds}{dx}.$$

Also,

$$\frac{ds}{dx} = -\sqrt{1 + \left(\frac{dy}{dx}\right)^2},$$

where the minus sign was chosen because $x$ decreases with increasing $s$. Thus (1–47) becomes

$$xy'' = \tfrac{1}{2}\sqrt{1 + (y')^2}$$

or, setting $y' = p$,

$$xp' = \tfrac{1}{2}\sqrt{1 + p^2}.$$

By separating variables we obtain

$$\frac{dp}{\sqrt{1 + p^2}} = \frac{1}{2}\frac{dx}{x}$$

and

$$\ln(p + \sqrt{1 + p^2}) = \tfrac{1}{2}\ln x + c.$$

To determine $c$, we observe that when $t = 0$, $x = 36$ and $p = y' = 0$. Thus $c = -\tfrac{1}{2}\ln 36$, and

$$p + \sqrt{1 + p^2} = \frac{x^{1/2}}{6}.$$

We now solve for $p = dy/dx$ to obtain

$$\frac{dy}{dx} = \frac{x^{1/2}}{12} - 3x^{-1/2},$$

from which it follows that

$$y = \frac{x^{3/2}}{18} - 6x^{1/2} + c.$$

Again the initial conditions enable us to determine $c$. We find that $c = 24$ and hence

$$y = \frac{x^{3/2}}{18} - 6x^{1/2} + 24.$$

Finally, by setting $x = 0$ we conclude that $A$ will skate 24 units before being caught by $B$.

## EXERCISES

Find a differential equation for each of the one-parameter families of curves in Exercises 1 through 10.

1. $y = ce^{cx}$

2. $\dfrac{x^2}{c^2} + \dfrac{y^2}{1 - c^2} = 1$

3. $(c^2 + 1)(y - cx) = 1$

4. $x^2 + (y - c)^2 = 1 + c^2$

5. The family of all lines through the point $(a, b)$.

6. The family of all lines whose $y$-intercept is twice their $x$-intercept.

7. The family of all lines tangent to the circle $x^2 + y^2 = 1$.

8. The family of all circles of radius $r$ whose centers lie on the line $y = x$.

9. The family of all circles through the origin whose centers are on the $x$-axis.

10. The family of all circles through the points $(1, 0)$ and $(0, 1)$.

Find the orthogonal trajectories of the families of curves in Exercises 11 through 15.

11. $\dfrac{x^2}{c^2} + \dfrac{y^2}{2c^2} = 1$

12. $xy = c$

13. $y = x - 1 + ce^{-x}$

14. $x(y^2 + 1) + xy^2 = c$

15. $(x + y)^2 = cx^2$

16. a) Verify that the general solution of

$$y' = \frac{y \pm \sqrt{x^2 + y^2}}{x}$$

is

$$y = \frac{c}{2} x^2 - \frac{1}{2c}, \qquad c \neq 0.$$

   b) Prove that these solutions are parabolas with their foci at the origin.

17. Find all the curves in the $xy$-plane with the property that the area of the triangle formed by the tangent to the curve, the $x$-axis, and the vertical line through the point of tangency is equal to the sum of the squares of the coordinates of the point of tangency.

18. Find all the curves in the $xy$-plane with the property that the area of the triangle formed by the tangent to the curve, the $x$-axis, and the normal to the curve is equal to the slope of the normal.

19. A curve lying above the $x$-axis has the property that the length of the arc joining any two points on the curve is equal to the area under the arc. What is the curve?

20. Suppose that the skaters in Example 5 skate at the same rate. How close will $B$ come to $A$?

21. Four bugs are at the corners of a square table whose corners have polar coordinates $(\pm 1, 0)$ and $(\pm 1, \pi/2)$. They begin to move at the same instant, each walking at the same rate toward the bug on its right. Find the path of each bug.

## 1–8 PHYSICAL APPLICATIONS

In this section we shall discuss several problems from physics which lead to first-order differential equations that we are now in a position to solve.

**Retarded Fall and Escape Velocity**  We begin by considering two problems involving motion in a straight line.

**Example 1**   An object of mass $m$ encounters air resistance that is directly proportional to its speed as it falls to earth. Find an equation for the motion of the object, given that it falls from rest.

*Solution.*   If we measure distance from the point where the object is released, with the positive direction downward, then the forces acting on the object are

$$mg, \text{ due to gravity, and}$$
$$-kv, \text{ due to air resistance,}$$

where $k$ is a positive constant and $v = dy/dt$ is the velocity of the object at time $t$. Newton's second law implies that

$$m \frac{dv}{dt} = mg - kv$$

or

$$\frac{dv}{dt} = g - Kv,$$

where $K = k/m$. We now separate variables,

$$\frac{dv}{g - Kv} = dt;$$

integrate,

$$-\frac{1}{K} \ln (g - Kv) = t + c;$$

and use the initial condition $v = 0$ when $t = 0$ to find that

$$c = -\frac{1}{K} \ln g.$$

Thus

$$\ln \frac{g - Kv}{g} = -Kt$$

and

$$v = \frac{g}{K} (1 - e^{-Kt}).$$

Note that as $t \to \infty$, $v$ approaches the constant value $g/K$. This limiting value is called the **terminal velocity** of the object.

To complete the description of the motion, we must solve

$$v = \frac{dy}{dt} = \frac{g}{K} (1 - e^{-Kt})$$

for $y$. An integration gives

$$y = \frac{g}{K} t + \frac{g}{K^2} e^{-Kt} + c,$$

and since our coordinate system was chosen so that $y = 0$ when $t = 0$, we have $c = -g/K^2$. Thus

$$y = \frac{g}{K^2}(Kt + e^{-Kt} - 1).$$

**Example 2**   Find the minimum velocity with which a rocket should be fired vertically upward in order to escape from the earth. Neglect air resistance.

*Solution.*   *Newton's law of gravitation* asserts that an object of mass $m$ is attracted toward the earth by a force of magnitude

$$\frac{km}{x^2},$$

where $x$ is the distance of the object from the center of the earth and $k$ is a positive constant. Thus, if we assume the rocket is given an instantaneous velocity $v_0$ at time $t_0$ and travels directly away from the earth thereafter, we have

$$m\frac{d^2x}{dt^2} = -\frac{km}{x^2}$$

or

$$\frac{d^2x}{dt^2} = -\frac{k}{x^2}.$$

To solve this equation, we substitute $x' = p$ and $x'' = p\, dp/dx$ to obtain

$$p\frac{dp}{dx} = -\frac{k}{x^2}.$$

Hence

$$\frac{p^2}{2} = \frac{k}{x} + c$$

or

$$\frac{v^2}{2} = \frac{k}{x} + c.$$

To find $c$, we observe that when $t = 0$, $v = v_0$ and $x = R$, where $R$ is the radius of the earth. (We assume that the rocket is fired from the earth's surface.) Thus

$$c = \frac{v_0{}^2}{2} - \frac{k}{R},$$

and we have

$$\frac{v^2}{2} = \frac{k}{x} + \frac{v_0{}^2}{2} - \frac{k}{R}.$$

To solve the problem at hand, we must now choose $v_0$ so that $v^2 > 0$ for all $x$. This implies that

$$v_0 > \sqrt{2k\left(\frac{1}{R} - \frac{1}{x}\right)},$$

from which we obtain

$$v_0 \geqslant \sqrt{\frac{2k}{R}}.$$

Finally, to determine $k$, we recall that at the surface of the earth the gravitational force on an object of mass $m$ has magnitude $mg$. Thus

$$mg = \frac{mk}{R^2},$$

and $k = gR^2$. Therefore, the **escape velocity** for the rocket is

$$v_0 = \sqrt{2gR} \approx 25{,}000 \text{ mph.}$$

**The Simple Pendulum**  In our next example we use the version of Newton's second law that equates the force acting on an object to the time rate of change in the momentum of the object (see p. 2).

**Example 3**  Describe the motion of a pendulum bob of mass $m$ swinging at the end of a light rigid rod of length $L$, as shown in Fig. 1–15. Neglect friction, air resistance, the mass of the rod, and the rotational inertia of the bob.

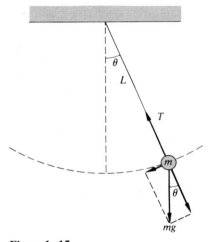

**Figure 1–15**

*Solution.*  The forces acting on the bob are the tension $T$ in the rod and the vertical force $mg$ due to gravity. The component of the gravitational force in the direction of the rod balances the tension, while the component perpendicular to the rod provides the force that causes the pendulum to oscillate. Since the magnitude of this force is $mg \sin \theta$, where $\theta$ is the angular displacement of the pendulum from the vertical, and since the momentum of the bob is

$$mv = mL\frac{d\theta}{dt},$$

we have

$$\frac{d}{dt}\left(mL\frac{d\theta}{dt}\right) = -mg\sin\theta$$

or

$$\frac{d^2\theta}{dt^2} = -\frac{g}{L}\sin\theta. \tag{1--48}$$

It turns out that this equation leads to an integral that cannot be evaluated in terms of elementary functions. However, if we assume that $\theta$ is small, and recall that

$$\sin\theta = \theta - \frac{\theta^3}{3!} + \frac{\theta^5}{5!} - \cdots,$$

we can replace $\sin\theta$ by $\theta$ in (1--48) without a substantial loss of accuracy. Then we have

$$\frac{d^2\theta}{dt^2} = -\frac{g}{L}\theta,$$

which, save for a difference in notation, is the equation that was solved in Example 3 of Section 1--6. Using the results of that example, we have

$$\theta = c_1\sin\sqrt{\frac{g}{L}}\,t + c_2\cos\sqrt{\frac{g}{L}}\,t.$$

Finally, if we assume that the pendulum began to oscillate by being released from rest at an angle $\alpha$ from the vertical, we find that

$$\theta = \alpha\cos\sqrt{\frac{g}{L}}\,t.$$

**The Hanging Cable**   As our final example in this section, we choose a problem that does not involve motion but still results in a differential equation.

**Example 4**   Find an equation for the curve assumed by a flexible cable of uniform density that is fastened at its two ends and allowed to hang under its own weight.

*Solution.*   We choose a coordinate system with the $y$-axis through the lowest point on the cable, as shown in Fig. 1--16. The weight of the portion $s$ of the cable between the lowest point and an arbitrary point $(x, y)$ induces a horizontal tension tension $T_0$ at the lowest point and a variable tension $T$ at $(x, y)$ which acts in the direction of the tangent to the curve. Since the cable is in equilibrium, the horizontal and vertical components of these forces must balance, and we have

$$T_0 = T\cos\theta, \qquad T\sin\theta = \rho s,$$

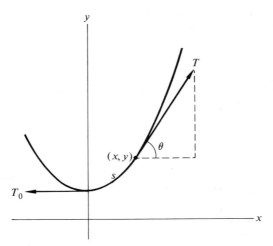

**Figure 1–16**

where $\rho$ is the density of the cable per unit length. Solving the first of these equations for $T$ and substituting in the second, we obtain

$$T_0 \tan \theta = \rho s.$$

But

$$\tan \theta = y' \qquad \text{and} \qquad s = \int_0^x \sqrt{1 + (y')^2} \, dt.$$

Thus

$$y' = a \int_0^x \sqrt{1 + (y')^2} \, dt,$$

where $a = \rho/T_0$. When this equation is differentiated with respect to $x$ we find that the shape of the cable is determined by the equation

$$y'' = a\sqrt{1 + (y')^2}.$$

Once again we use the method of reduction of order, setting $y' = p$ and $y'' = p'$, to obtain

$$p' = a\sqrt{1 + p^2}$$

or

$$\frac{dp}{\sqrt{1 + p^2}} = a \, dx.$$

Thus

$$\ln (p + \sqrt{1 + p^2}) = ax + c,$$

and since $p = 0$ when $x = 0$, we have $c = 0$. It follows that

$$p + \sqrt{1 + p^2} = e^{ax}$$

and

$$p = \tfrac{1}{2}(e^{ax} - e^{-ax}).$$

Integrating again, we obtain

$$y = \frac{1}{2a}(e^{ax} + e^{-ax}) + c.$$

We now adjust the scale on the $y$-axis so that $c = 0$ when $x = 0$ (i.e., we assume that $y(0) = 1/a$). When this is done,

$$y = \frac{1}{2a}(e^{ax} + e^{-ax}) = \frac{1}{a}\cosh ax.$$

As the reader may already know, this curve is called a **catenary**, the name being derived from the Latin word *catena*, for chain.

## EXERCISES

1. Suppose that a hole is bored from one side of the earth straight through to the other side, and that any object in the hole is attracted towards the center of the earth by a force that is directly proportional to its distance from the center.
   a) You stand at the edge of the hole and drop in a ball. Find an equation for the motion of the ball. (Neglect air resistance and friction.)
   b) When should you reach down to pick the ball up again?

2. Assume that evaporation causes a spherical raindrop to decrease in volume at a rate proportional to its surface area. How long will it take a raindrop of radius $r_0$ to evaporate completely?

3. An object of mass $m$ is released from rest and encounters air resistance proportional to the square of its speed as it falls to earth.
   a) Find the velocity of the object as a function of time.
   b) What is the terminal velocity of the object?
   c) Find the position of the object as a function of time.

4. A bullet fired vertically upward from the surface of the earth with initial velocity $v_0$ will encounter air resistance proportional to its speed.
   a) Find an equation for the motion of the bullet. Assume constant gravitational acceleration.
   *b) Show that it will take the bullet longer to fall back to the earth from its maximum height than to reach that height.

5. Find the shape of a flexible cable of negligible density that will support a bridge of uniform density $\rho$.

6. *Newton's law of cooling* asserts that the rate at which the temperature of an object changes is proportional to the temperature difference between it and its surroundings. Suppose that after 10 minutes of being surrounded by air whose temperature is 10 °C, an object cooled from 100 °C to 55 °C.
   a) How much longer would it take the object to cool to 19 °C?
   b) How long would it take the same object to cool from 500 °C to 250 °C when placed in a bath of ice water?

7. Water in an open bowl will flow out of a small hole in the bottom of the bowl with the

same speed it would have had if it had fallen freely from the water level to the hole. What shape should the bowl have if the water level is to decrease at a constant rate?

8. a) Use the method of reduction of order to show that the equation

$$\frac{d^2\theta}{dt^2} = -\frac{g}{L}\sin\theta,$$

describing the motion of a simple pendulum, can be reduced to

$$dt = -\sqrt{\frac{L}{2g}}\frac{d\theta}{\sqrt{\cos\theta - \cos\alpha}},$$

where $\alpha$ is the initial angular displacement of the pendulum from the vertical.

   b) Use the result in (a) to find a formula for the time it takes the pendulum to complete one full oscillation.

# The Elements
## of Linear Algebra

**2**

## 2–1 INTRODUCTION

In Section 1–6 we saw that the general solution of the second-order differential equation

$$y'' + k^2 y = 0 \tag{2–1}$$

can be written in the form

$$y = c_1 \sin kx + c_2 \cos kx, \tag{2–2}$$

where $c_1$ and $c_2$ are arbitrary constants. When $c_1 = 1$ and $c_2 = 0$, (2–2) yields the particular solution

$$y_1 = \sin kx;$$

and when $c_1 = 0$ and $c_2 = 1$, it yields

$$y_2 = \cos kx.$$

Furthermore, the general solution (2–2) can be recovered immediately from these particular solutions: just multiply each by a constant and add. We express this fact by saying that the general solution of $y'' + k^2 y = 0$ is a **linear combination** of the particular solutions $y_1 = \sin kx$ and $y_2 = \cos kx$.

A similar situation occurs with every differential equation of the form

$$y'' + a_1(x)y' + a_0(x)y = 0. \tag{2–3}$$

If $y_1 = y_1(x)$ and $y_2 = y_2(x)$ are solutions of this equation, then so is the linear combination

$$c_1 y_1 + c_2 y_2, \tag{2–4}$$

for if (2–4) is substituted for $y$ in (2–3), we obtain

$$(c_1 y_1 + c_2 y_2)'' + a_1(x)(c_1 y_1 + c_2 y_2)' + a_0(x)(c_1 y_1 + c_2 y_2)$$
$$= c_1 y_1'' + c_2 y_2'' + a_1(x)(c_1 y_1' + c_2 y_2') + a_0(x)(c_1 y_1 + c_2 y_2)$$
$$= c_1[y_1'' + a_1(x)y_1' + a_0(x)y_1] + c_2[y_2'' + a_1(x)y_2' + a_0(x)y_2]$$
$$= c_1 \cdot 0 + c_1 \cdot 0$$
$$= 0.$$

The importance of this seemingly innocent observation cannot be over-estimated. Indeed, much of the rest of our work with differential equations will be devoted to exploring its consequences. Among other things, it will lead to a comprehensive and highly satisfactory theory for solving equations like (2–3), a theory that stands in sharp contrast to the collection of unrelated techniques given in the preceding chapter. This theory appears as an application of the ideas developed in a branch of mathematics known as *linear algebra*, which, roughly speaking, can be described as a systematic exploration of situations where objects, like the solutions of Eq. (2–3), can be multiplied by real numbers and added to one another without losing their characteristic properties. As we shall see, examples of such situations abound in mathematics. In view of the importance of linear algebra in the study of differential equations, we shall devote the next two chapters to it. Then in Chapter 4 we shall return to the study of differential equations and use our newly gained algebraic knowledge to analyze equations like (2–3).

## 2–2 $\mathscr{R}^2$ AND $\mathscr{C}(I)$

The Cartesian plane of analytic geometry, denoted by $\mathscr{R}^2$, is one of the most familiar examples of what is known in mathematics as a **real vector space**. Each of its points, or **vectors**, is an ordered pair $(x_1, x_2)$ of real numbers whose individual entries, $x_1$ and $x_2$, are called the **components** of that vector. Geometrically, the vector $\mathbf{x} = (x_1, x_2)$ can be represented by an arrow drawn from the origin of the Cartesian plane to the point $(x_1, x_2)$, as shown in Fig. 2–1.

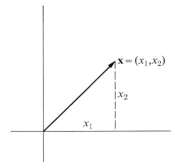

**Figure 2–1**

If $\mathbf{x} = (x_1, x_2)$ and $\mathbf{y} = (y_1, y_2)$ are any two vectors in $\mathcal{R}^2$, then by definition their **sum** is the vector

$$\mathbf{x} + \mathbf{y} = (x_1 + y_1, x_2 + y_2), \tag{2–5}$$

obtained by adding the corresponding components of $\mathbf{x}$ and $\mathbf{y}$. The graphical interpretation of this addition is the familiar "parallelogram law," which states that the vector $\mathbf{x} + \mathbf{y}$ is the diagonal of the parallogram determined by $\mathbf{x}$ and $\mathbf{y}$ (see Fig. 2–2). It follows at once from this definition that vector addition is both associative and commutative:

$$\mathbf{x} + (\mathbf{y} + \mathbf{z}) = (\mathbf{x} + \mathbf{y}) + \mathbf{z}, \tag{2–6}$$

$$\mathbf{x} + \mathbf{y} = \mathbf{y} + \mathbf{x}. \tag{2–7}$$

Moreover, if $\mathbf{0}$ denotes the vector $(0, 0)$, and $-\mathbf{x}$ the vector $(-x_1, -x_2)$, then

$$\mathbf{x} + \mathbf{0} = \mathbf{x}, \tag{2–8}$$

and

$$\mathbf{x} + (-\mathbf{x}) = \mathbf{0} \tag{2–9}$$

for every vector $\mathbf{x} = (x_1, x_2)$. Taken together, Eqs. (2–6) through (2–9) imply that vector addition behaves very much like the ordinary addition of arithmetic.

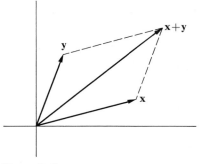

**Figure 2–2**

In addition to being able to add vectors in $\mathcal{R}^2$, we can also form the **product** of a real number $\alpha$ and a vector $\mathbf{x}$. The result, denoted by $\alpha\mathbf{x}$, is the *vector* each of whose components is $\alpha$ times the corresponding component of $\mathbf{x}$. Thus, if $\mathbf{x} = (x_1, x_2)$, then

$$\alpha\mathbf{x} = (\alpha x_1, \alpha x_2). \tag{2–10}$$

Geometrically, this vector can be viewed as $\mathbf{x}$ "magnified" by the factor $\alpha$, as illustrated in Fig. 2–3.

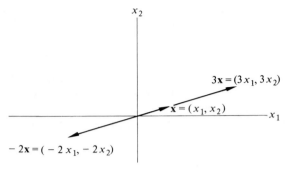

**Figure 2–3**

The principal algebraic properties of this multiplication are:

$$\alpha(\mathbf{x} + \mathbf{y}) = \alpha\mathbf{x} + \alpha\mathbf{y}, \tag{2–11}$$

$$(\alpha + \beta)\mathbf{x} = \alpha\mathbf{x} + \beta\mathbf{x}, \tag{2–12}$$

$$(\alpha\beta)\mathbf{x} = \alpha(\beta\mathbf{x}), \tag{2–13}$$

$$1\mathbf{x} = \mathbf{x}. \tag{2–14}$$

These equations follow easily from definitions (2–5) and (2–10), and save for (2–13), which we prove by way of illustration, their proofs are left as exercises.

To establish (2–13), let $\mathbf{x} = (x_1, x_2)$ be an arbitrary vector in $\mathcal{R}^2$, and let $\alpha$ and $\beta$ be real numbers. Then by repeated use of (2–10) we have

$$\begin{aligned}
(\alpha\beta)\mathbf{x} &= ((\alpha\beta)x_1, (\alpha\beta)x_2) \\
&= (\alpha(\beta x_1), \alpha(\beta x_2)) \\
&= \alpha(\beta x_1, \beta x_2) \\
&= \alpha(\beta(x_1, x_2)) \\
&= \alpha(\beta\mathbf{x}),
\end{aligned}$$

which is what we wished to show.

The reason for calling attention to Eqs. (2–11) through (2–14) is that they, together with (2–6) through (2–9), are precisely what make $\mathcal{R}^2$ a real vector space. Indeed, these equations are the axioms in the general definition of such a space, and once this definition has been given, the preceding discussion constitutes a verification of the fact that $\mathcal{R}^2$ is a real vector space. But before giving this definition, let us look at another example.

This time we consider the set $\mathscr{C}(I)$ consisting of all real-valued, continuous functions defined on an interval $I$ of the real line. For reasons that will soon become clear, we shall call any such function a vector, and we shall write it in boldface type. Thus $\mathbf{f}$ is a vector in $\mathscr{C}(I)$ if and only if $\mathbf{f}$ is a real-valued, continuous function on the interval $I$. Typical examples are sin $x$, cos $x$, and $e^x$, which are vectors in $\mathscr{C}(I)$ for any interval $I$.

At first sight it may seem that $\mathscr{C}(I)$ and $\mathscr{R}^2$ have nothing in common but the name "real vector space." But this is an instance where first impressions are misleading, for as we shall see, the two spaces are remarkably similar. The similarity arises from the fact that an addition and multiplication by real numbers can be defined in $\mathscr{C}(I)$ in such a way that these operations have exactly the same properties as the corresponding operations in $\mathscr{R}^2$.

Turning first to addition, let $\mathbf{f}$ and $\mathbf{g}$ be any two vectors in $\mathscr{C}(I)$. Then their sum, $\mathbf{f} + \mathbf{g}$, is defined to be the function (i.e., vector) whose value at each point $x$ in $I$ is the sum of the values of $\mathbf{f}$ and $\mathbf{g}$ at $x$. In other words,

$$(\mathbf{f} + \mathbf{g})(x) = \mathbf{f}(x) + \mathbf{g}(x) \tag{2–15}$$

(see Fig. 2–4). For example, if $\mathbf{f}$ and $\mathbf{g}$ are the functions sin $x$ and cos $x$, then their sum $\mathbf{f} + \mathbf{g}$ is the function sin $x$ + cos $x$. In particular, note that $\mathbf{f} + \mathbf{g}$ belongs to $\mathscr{C}(I)$ whenever $\mathbf{f}$ and $\mathbf{g}$ do.

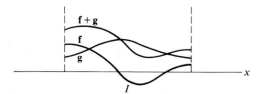

**Figure 2–4**

It is easy to verify that, apart from differences in notation and interpretation, Eqs. (2–6) through (2–9) remain valid in $\mathscr{C}(I)$. In fact, the equations

$$\mathbf{f} + (\mathbf{g} + \mathbf{h}) = (\mathbf{f} + \mathbf{g}) + \mathbf{h} \tag{2–16}$$

and

$$\mathbf{f} + \mathbf{g} = \mathbf{g} + \mathbf{f} \tag{2–17}$$

follow immediately from (2–15), while if $\mathbf{0}$ denotes the function whose value is zero at every point of $I$, then

$$\mathbf{f} = \mathbf{f} + \mathbf{0} \tag{2–18}$$

for every $\mathbf{f}$ in $\mathscr{C}(I)$ (Fig. 2–5). Finally, if $-\mathbf{f}$ is the function whose value at $x$ is $-\mathbf{f}(x)$ (i.e., $-\mathbf{f}$ is the reflection of $\mathbf{f}$ across $\mathbf{0}$, as shown in Fig. 2–6), then $\mathbf{f} + (-\mathbf{f})$ has the value zero at each point of $I$, and we have

$$\mathbf{f} + (-\mathbf{f}) = \mathbf{0}. \tag{2–19}$$

We have seen that the sum of two vectors in $\mathscr{R}^2$ is found by adding their corresponding components (Eq. 2–5). We can make a similar interpretation of

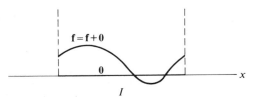

**Figure 2–5**

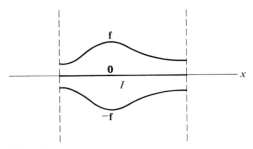

**Figure 2–6**

vector addition in the present example, as follows. If **f** is any vector in $\mathscr{C}(I)$, we agree to say that the "component" of **f** at the point $x$ is its functional value at $x$. Of course, every vector in $\mathscr{C}(I)$ then has infinitely many components, one for each $x$ in the interval $I$. But once we have accepted this fact, it becomes clear that Eq. (2–15) simply states that the sum of two vectors in $\mathscr{C}(I)$ is obtained by adding corresponding "components," just as in $\mathscr{R}^2$.

Next, if **f** is any vector in $\mathscr{C}(I)$ and $\alpha$ is an arbitrary real number, we define $\alpha\mathbf{f}$ by the equation

$$(\alpha\mathbf{f})(x) = \alpha\mathbf{f}(x). \tag{2–20}$$

In other words, $\alpha\mathbf{f}$ is the function whose value at $x$ is the product of the real numbers $\alpha$ and $\mathbf{f}(x)$. The similarity between this multiplication and the corresponding operation in $\mathscr{R}^2$ is clear, since $\alpha\mathbf{f}$ is also formed by multiplying each "component" of **f** by $\alpha$. (Figure 2–7 illustrates this multiplication for $\alpha = 2$.)

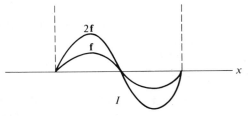

**Figure 2–7**

The analogy with $\mathscr{R}^2$ is now complete, for Eqs. (2–11) through (2–14) are also valid in $\mathscr{C}(I)$. We restate them here as

$$\alpha(\mathbf{f} + \mathbf{g}) = \alpha\mathbf{f} + \alpha\mathbf{g}, \tag{2–21}$$

$$(\alpha + \beta)\mathbf{f} = \alpha\mathbf{f} + \beta\mathbf{f}, \tag{2–22}$$

$$(\alpha\beta)\mathbf{f} = \alpha(\beta\mathbf{f}), \tag{2–23}$$

$$1\mathbf{f} = \mathbf{f}, \tag{2–24}$$

and leave their proofs as an exercise.

The space $\mathscr{C}(I)$ is much more than an idle example. A great deal of our later work will be devoted to a study of this and similar vector spaces.

### EXERCISES

In each of Exercises 1 through 5 compute the values of $\mathbf{x} + \mathbf{y}$ and $\alpha(\mathbf{x} + \mathbf{y})$ for the given vectors $\mathbf{x}$ and $\mathbf{y}$ in $\mathscr{R}^2$ and the real number $\alpha$. Illustrate each of your computations with a diagram.

1. $\mathbf{x} = (0, 2),$    $\mathbf{y} = (-1, 1),$    $\alpha = 3$
2. $\mathbf{x} = (\frac{1}{2}, 1),$    $\mathbf{y} = (1, -2),$    $\alpha = -2$
3. $\mathbf{x} = (-\frac{1}{2}, \frac{1}{3}),$    $\mathbf{y} = (-2, -1),$    $\alpha = -1$
4. $\mathbf{x} = (5, -2),$    $\mathbf{y} = (-3, 2),$    $\alpha = \frac{1}{2}$
5. $\mathbf{x} = (-5, -2),$    $\mathbf{y} = (-1, -1),$    $\alpha = -3$

In each of Exercises 6 through 10 compute the values of $\mathbf{f} + \mathbf{g}$ and $\alpha(\mathbf{f} + \mathbf{g})$ for the given vectors $\mathbf{f}$ and $\mathbf{g}$ in $\mathscr{C}(I)$ and the real number $\alpha$.

6. $\mathbf{f}(x) = 2x,$    $\mathbf{g}(x) = x^2 - x + 1,$    $\alpha = 2$
7. $\mathbf{f}(x) = \tan^2 x,$    $\mathbf{g}(x) = 1,$    $\alpha = -1$
8. $\mathbf{f}(x) = e^x,$    $\mathbf{g}(x) = e^{-x},$    $\alpha = \frac{1}{2}$
9. $\mathbf{f}(x) = \dfrac{x + 3}{x - 2},$    $\mathbf{g}(x) = -\dfrac{x - 2}{x + 3},$    $\alpha = \frac{1}{5}$
10. $\mathbf{f}(x) = \cos^2 x,$    $\mathbf{g}(x) = \sin^2 x,$    $\alpha = -3$

Solve each of the following vector equations for $\mathbf{x}$.

11. $2\mathbf{x} + 3(2, 1) = (0, 0)$
12. $4\mathbf{x} + 3\mathbf{x} = 2(-1, 3)$
13. $(\frac{1}{2} + \frac{2}{3})\mathbf{x} + 5(\frac{1}{6}(3, 4)) = \frac{1}{5}(2, -1)$
14. $\mathbf{x} - 3(2, 1) = 4\mathbf{x}$

Solve each of the following vector equations in $\mathscr{C}(I)$ for $\mathbf{f}$.

15. $2\mathbf{f} - 1 = \sec^2 x$
16. $\ln \mathbf{f} = x$
17. $e^{\mathbf{f}} = x + 2$
18. $2\mathbf{f} + e^{-x} = e^x$
19. Prove Eqs. (2–6) through (2–9) and (2–11) through (2–14), and illustrate each equation with a diagram.

## 2–3  REAL VECTOR SPACES

With the examples of the preceding section in mind we now give the definition of a real vector space.

> **Definition 2–1**   *A **real vector space** $\mathscr{V}$ is a collection of objects called **vectors**, together with an addition and a multiplication by real numbers which satisfy the following axioms.*
>
> **Axioms for addition**   *Given any pair of vectors* **x** *and* **y** *in* $\mathscr{V}$ *there exists a (unique) vector* **x** + **y** *in* $\mathscr{V}$ *called the **sum** of* **x** *and* **y**. *It is required that*
>
> i) *addition be **associative**,*
>
> $$\mathbf{x} + (\mathbf{y} + \mathbf{z}) = (\mathbf{x} + \mathbf{y}) + \mathbf{z};$$
>
> ii) *addition be **commutative**,*
>
> $$\mathbf{x} + \mathbf{y} = \mathbf{y} + \mathbf{x};$$
>
> iii) *there exist a vector* **0** *in* $\mathscr{V}$ *(called the **zero vector**, or just **zero**) such that*
>
> $$\mathbf{x} + \mathbf{0} = \mathbf{x}$$
>
> *for all* **x** *in* $\mathscr{V}$;
>
> iv) *for each* **x** *in* $\mathscr{V}$ *there exist a vector* $-\mathbf{x}$ *in* $\mathscr{V}$ *such that*
>
> $$\mathbf{x} + (-\mathbf{x}) = \mathbf{0}.$$
>
> **Axioms for scalar multiplication**   *Given any vector* **x** *in* $\mathscr{V}$ *and any real number* $\alpha$ *there exists a (unique) vector* $\alpha\mathbf{x}$ *in* $\mathscr{V}$ *called the **product**, or **scalar product**, of* $\alpha$ *and* **x**. *It is required that*
>
> v) $\alpha(\mathbf{x} + \mathbf{y}) = \alpha\mathbf{x} + \alpha\mathbf{y}$;
>
> vi) $(\alpha + \beta)\mathbf{x} = \alpha\mathbf{x} + \beta\mathbf{x}$;
>
> vii) $(\alpha\beta)\mathbf{x} = \alpha(\beta\mathbf{x})$;
>
> viii) $1\mathbf{x} = \mathbf{x}$.

Do not be discouraged by the length and formality of this definition; it looks much worse than it really is. The point here is simply that in order to deserve the name, a real vector space must have a number of elementary and eminently reasonable properties in common with $\mathscr{R}^2$. We have already seen that the function space $\mathscr{C}(I)$ has these properties, and now we give several additional examples before beginning the general study of vector spaces. We urge you to verify Axioms (i) through (viii) for each example.

**Example 1**   The real numbers, with the ordinary definitions of addition and multiplication, form a real vector space. The set of real numbers constitutes the collection of vectors and the set of scalars at one and the same time, so that Axioms (i) through (viii) are all familiar statements from arithmetic.

**Example 2**  Let $n$ be a fixed positive integer, and let $\mathscr{R}^n$ denote the totality of ordered $n$-tuples $(x_1, \ldots, x_n)$ of real numbers. If $\mathbf{x} = (x_1, \ldots, x_n)$ and $\mathbf{y} = (y_1, \ldots, y_n)$ are two such $n$-tuples, and $\alpha$ is a real number, set

$$\mathbf{x} + \mathbf{y} = (x_1 + y_1, \ldots, x_n + y_n) \tag{2-25}$$

and

$$\alpha\mathbf{x} = (\alpha x_1, \ldots, \alpha x_n). \tag{2-26}$$

Then $\mathscr{R}^n$, with this definition of vector addition and scalar multiplication, is a real vector space (the $n$-tuples being the vectors, and the real numbers being the scalars as always). Note that $\mathscr{R}^1$ is the vector space of Example 1 above, while $\mathscr{R}^2$ and $\mathscr{R}^3$ are the familiar spaces from analytic geometry.

**Example 3**  Let $\mathscr{P}$ denote the set of all polynomials in $x$ with real coefficients, and let addition of polynomials and multiplication by real numbers be defined as in high school algebra. Then $\mathscr{P}$ is a real vector space.

This completes our list of basic examples. Indeed, $\mathscr{R}^n$, $\mathscr{P}$, and $\mathscr{C}(I)$ are among the most important real vector spaces in mathematics, and most of our later work will be concerned with one or another of these spaces.

Before going on, we list a number of elementary consequences of Definition 2–1 that will often be used without explicit mention. Each of these statements is in effect a theorem about vector spaces and as such should be supported by a proof. We shall, however, omit the proofs, though all but the first are outlined in the exercises at the end of the section.

i)  Generalized associativity and commutativity of addition. Since vector addition is associative, we usually omit parentheses and write

$$\mathbf{x} + \mathbf{y} + \mathbf{z}$$

for the common value of $\mathbf{x} + (\mathbf{y} + \mathbf{z})$ and $(\mathbf{x} + \mathbf{y}) + \mathbf{z}$. The same remark applies to finite sums of the form

$$\mathbf{x}_1 + \mathbf{x}_2 + \cdots + \mathbf{x}_n;$$

they need not be festooned with parentheses to avoid ambiguity. Moreover, the commutativity of vector addition implies that the terms in such a sum can be rearranged at will without affecting its value.

ii)  If $\mathbf{x}$ is a vector and $\alpha$ a scalar, then

$$\alpha\mathbf{x} = \mathbf{0} \tag{2-27}$$

if and only if $\alpha = 0$ or $\mathbf{x} = \mathbf{0}$ (or both). This, of course, is exactly what one would expect, and the proof is perfectly straightforward. (See Exercise 11.) But do not confuse 0, the number zero, with $\mathbf{0}$, the zero vector. They are by no means the same, except in $\mathscr{R}^1$.

iii)  For every vector $\mathbf{x}$,

$$-\mathbf{x} = (-1)\mathbf{x}.$$

iv) In every vector space the zero vector is unique; it is the only vector with the property that

$$\mathbf{x} + \mathbf{0} = \mathbf{x} \qquad \text{for every } \mathbf{x}.$$

The fact that every vector in a vector space has an "additive inverse" allows us to introduce subtraction in the usual way. We define

$$\mathbf{x} - \mathbf{y}$$

to be an abbreviation for $\mathbf{x} + (-\mathbf{y})$ and then speak of *subtracting* $\mathbf{y}$ from $\mathbf{x}$.

Strictly speaking our insistence on using real numbers in the definition of a vector space is unnecessary. Complex numbers, for example, would do just as well, in which case we would have what is known as a complex vector space. Other types of scalars can also be used. But we shall have little occasion to use these more general vector spaces, and therefore in this book the term "vector space" will mean a *real* vector space as defined above.

As you continue to work with vector spaces, keep in mind that to specify a vector space you need to define three things:

1.  the collection of vectors,

2.  the way vectors are added, and

3.  the way vectors are multiplied by scalars.

Then you must verify Axioms (i) through (viii) for these three definitions. A change in even one of the definitions may result in an entirely different vector space or in no vector space at all.

### EXERCISES

1.  Prove that $\mathscr{R}^n$ is a real vector space.

2.  Find the value of $\alpha_1 \mathbf{x}_1 + \alpha_2 \mathbf{x}_2$ in $\mathscr{R}^4$ when
    a) $\mathbf{x}_1 = (2, 0, -1, 3),$ $\quad \mathbf{x}_2 = (1, -2, 2, 0),$ $\quad \alpha_1 = -1, \alpha_2 = 2;$
    b) $\mathbf{x}_1 = (1, 2, 3, 4),$ $\quad \mathbf{x}_2 = (0, 2, 1, -2),$ $\quad \alpha_1 = -2, \alpha_2 = \frac{1}{2};$
    c) $\mathbf{x}_1 = (-\frac{1}{2}, 2, 1, -3),$ $\quad \mathbf{x}_2 = (1, -2, \frac{1}{2}, 5),$ $\quad \alpha_1 = -2, \alpha_2 = \frac{1}{2}.$

3.  Prove that $\mathscr{P}$ is a real vector space. (Recall that the vectors in $\mathscr{P}$ may be written in the form

    $$a_0 + a_1 x + \cdots + a_n x^n,$$

    where $a_0, \ldots, a_n$ are real numbers.)

4.  Find the value of $\alpha_1 \mathbf{p}_1 + \alpha_2 \mathbf{p}_2$ in $\mathscr{P}$ when
    a) $\mathbf{p}_1(x) = x^2 - x + 1,$ $\quad \alpha_1 = 2,$
    $\mathbf{p}_2(x) = 3x^3 + 2x - 1,$ $\quad \alpha_2 = -1;$
    b) $\mathbf{p}_1(x) = 2x^4 - 4x^2 + 1,$ $\quad \alpha_1 = \frac{1}{2},$
    $\mathbf{p}_2(x) = -x^4 + 2x^3 + x^2 + 2,$ $\quad \alpha_2 = 1;$
    c) $\mathbf{p}_1(x) = \frac{1}{2}x^3 - 2x^2 + x - \frac{1}{5},$ $\quad \alpha_1 = 3,$
    $\mathbf{p}_2(x) = -x^2 + \frac{1}{2}x + \frac{1}{5},$ $\quad \alpha_2 = -2.$

5. Prove that the set of all solutions of the second-order differential equation

$$y'' + a_1(x)y' + a_0(x)y = 0$$

is a vector space if addition and scalar multiplication are defined as in $\mathscr{C}(I)$.

6. Let $\mathscr{V}$ be the set of all ordered pairs of real numbers. If $\mathbf{x} = (x_1, x_2)$ and $\mathbf{y} = (y_1, y_2)$ are any two elements in $\mathscr{V}$, define

$$\mathbf{x} + \mathbf{y} = (x_1 + y_1, 0)$$

and let scalar multiplication be defined as in $\mathscr{R}^2$. Is $\mathscr{V}$ a vector space? Why?

7. Repeat Exercise 6, this time with addition defined as in $\mathscr{R}^2$ and scalar multiplication changed to

$$\alpha\mathbf{x} = (\alpha x_1, 0).$$

8. With addition and scalar multiplication defined as in $\mathscr{C}(I)$, determine which of the following sets of functions is a vector space when $I$ is the closed interval $a \leqslant x \leqslant b$.
   a) All continuous functions $\mathbf{f}$ such that $\mathbf{f}(a) = \mathbf{f}(b)$.
   b) All continuous functions $\mathbf{f}$ such that $\mathbf{f}(a) = \mathbf{f}(b) = 1$.
   c) All functions that are zero on some subinterval of $[a, b]$.
   d) All functions that are zero on a fixed subinterval of $[a, b]$.
   e) All functions that are differentiable everywhere on $[a, b]$.

9. Let $\mathscr{R}^\infty$ be the set consisting of all infinite sequences of real numbers, with addition and scalar multiplication defined term-by-term:

$$\mathbf{x} + \mathbf{y} = (x_1 + y_1, x_2 + y_2, \ldots),$$
$$\alpha\mathbf{x} = (\alpha x_1, \alpha x_2, \ldots),$$

where $\mathbf{x} = (x_1, x_2, \ldots)$, $\mathbf{y} = (y_1, y_2, \ldots)$, and $\alpha$ is a real number. Prove that $\mathscr{R}^\infty$ is a real vector space.

\* 10. Let $\mathscr{R}^+$ be the set consisting of all *positive* real numbers, and define "addition" and "scalar multiplication" in $\mathscr{R}^+$ as follows. If $x$ and $y$ belong to $\mathscr{R}^+$, set

$$x + y = xy;$$

and if $x$ belongs to $\mathscr{R}^+$ and $\alpha$ is an arbitrary real number, set

$$\alpha x = x^\alpha.$$

Prove that $\mathscr{R}^+$ is a real vector space.

11. a) Use the equations

$$0\mathbf{x} = (0 + 0)\mathbf{x} = 0\mathbf{x} + 0\mathbf{x}$$

to prove that $0\mathbf{x} = \mathbf{0}$ for all $\mathbf{x}$.
   b) Give a proof like the one suggested in part (a) to show that

$$\alpha\mathbf{0} = \mathbf{0}$$

for all $\alpha$.
   c) Prove that if $\alpha\mathbf{x} = \mathbf{0}$, then $\alpha = 0$ or $\mathbf{x} = \mathbf{0}$. [*Hint*: Assume $\alpha \neq 0$ and multiply by $\alpha^{-1}$.]

12. Use the equations
$$0 = 0\mathbf{x} = (1 + (-1))\mathbf{x} = \mathbf{x} + (-1)\mathbf{x}$$
to prove that $(-1)\mathbf{x} = -\mathbf{x}$ for all $\mathbf{x}$.

13. Prove that the zero vector in a vector space is unique. [*Hint*: You know that $\mathbf{x} + \mathbf{0} = \mathbf{x}$ for all $\mathbf{x}$. Suppose $\mathbf{0}'$ to be a vector for which $\mathbf{x} + \mathbf{0}' = \mathbf{x}$ for all $\mathbf{x}$. Then show that $\mathbf{0}' = \mathbf{0}$.]

## 2–4 SUBSPACES

In this section we shall introduce the important notion of a subspace of a vector space.

*Definition 2–2*    *A subset $\mathscr{W}$ of a vector space $\mathscr{V}$ is a **subspace** of $\mathscr{V}$ if $\mathscr{W}$ itself is a vector space under the operations of addition and scalar multiplication defined in $\mathscr{V}$.*

*Remark.*   The term "subset" as used here and in similar contexts in the future means that every vector in $\mathscr{W}$ also belongs to $\mathscr{V}$. We shall always assume that there is at least one vector in $\mathscr{W}$, a fact which is sometimes expressed by saying that $\mathscr{W}$ is **nonempty**.

Before giving examples, let us consider the problem of determining when a given collection of vectors in $\mathscr{V}$ is a subspace of $\mathscr{V}$. One way of doing this, of course, is to verify that the set of vectors in question satisfies all of the requirements of Definition 2–1. However, the step-by-step verification of these axioms is both time-consuming and tedious, and it is usually much simpler to use the following criterion:

**Subspace criterion**    *A (nonempty) subset $\mathscr{W}$ of a vector space $\mathscr{V}$ is a subspace of $\mathscr{V}$ if every vector of the form*

$$\alpha_1 \mathbf{x}_1 + \alpha_2 \mathbf{x}_2 \tag{2–28}$$

*belongs to $\mathscr{W}$ whenever $\mathbf{x}_1$ and $\mathbf{x}_2$ belong to $\mathscr{W}$, and $\alpha_1$ and $\alpha_2$ are arbitrary scalars.*

*Proof.*   To prove this assertion we must show that $\mathscr{W}$ satisfies Definition 2–1. To this end, let $\mathbf{x}_1$ and $\mathbf{x}_2$ be any two vectors in $\mathscr{W}$. Then by (2–28), $1\mathbf{x}_1 + 1\mathbf{x}_2$ belongs to $\mathscr{W}$. But since $1\mathbf{x} = \mathbf{x}$ for all $\mathbf{x}$, it follows that $\mathbf{x}_1 + \mathbf{x}_2$ belongs to $\mathscr{W}$. Hence the sum of any vectors in $\mathscr{W}$ is again a vector in $\mathscr{W}$.

Next, set $\alpha_2 = 0$ in (2–28) and use (2–27) to conclude that all scalar multiples of a vector in $\mathscr{W}$ also belong to $\mathscr{W}$. This fact, together with the fact that $\mathscr{W}$ is nonempty, implies that the zero vector belongs to $\mathscr{W}$ since $\mathbf{0} = 0\mathbf{x}$ for any $\mathbf{x}$. Moreover, it also implies that $-\mathbf{x} = (-1)\mathbf{x}$ belongs to $\mathscr{W}$ whenever $\mathbf{x}$ does. Thus $\mathscr{W}$ satisfies Axioms (iii) and (iv) of Definition 2–1. Finally, the remaining axioms are automatically satisfied in $\mathscr{W}$ since they hold everywhere in $\mathscr{V}$ and $\mathscr{W}$ is contained in $\mathscr{V}$.  ∎

Note that Eq. (2–28) is just another way of saying

i)  the sum of any two vectors in $\mathscr{W}$ belongs to $\mathscr{W}$;

ii)  all scalar multiples of a vector in $\mathscr{W}$ belong to $\mathscr{W}$.

These two statements are usually described by saying that $\mathscr{W}$ is **closed** under addition and scalar multiplication. In these terms the subspace criterion reads as follows:

*A nonempty subset of a vector space is a subspace if it is closed under addition and scalar multiplication.*

**Example 1**  Every nonzero vector space $\mathscr{V}$ has at least two subspaces: (i) $\mathscr{V}$ itself, and (ii) the subspace $\mathcal{O}$ consisting of the zero vector by itself. The space $\mathcal{O}$ is called the **trivial subspace** of $\mathscr{V}$. A subspace of $\mathscr{V}$ that is distinct from $\mathscr{V}$ is called a **proper subspace** of $\mathscr{V}$.

**Example 2**  If $\mathscr{W}$ is the subset of $\mathscr{R}^3$ consisting of all the vectors whose third component is zero, then the sum of two such vectors and the product of such a vector and a real number also have their third components equal to zero. Thus $\mathscr{W}$ is closed under addition and scalar multiplication and hence is a subspace of $\mathscr{R}^3$. Geometrically, $\mathscr{W}$ is just the $xy$-plane in 3-space.

**Example 3**  Let $\mathscr{C}^{(1)}(I)$ denote the set of all continuously differentiable functions on an interval $I$. (See p. 8.) Then, since a differentiable function is continuous, $\mathscr{C}^{(1)}(I)$ is contained in $\mathscr{C}(I)$. Moreover, both the scalar multiple of a continuously differentiable function and the sum of two such functions are again continuously differentiable. Hence $\mathscr{C}^{(1)}(I)$ is closed under addition and scalar multiplication, and is thus a subspace of $\mathscr{C}(I)$. More generally, if $\mathscr{C}^{(n)}(I)$ denotes the set of all functions with continuous $n$th derivatives on $I$, then $\mathscr{C}^{(n)}(I)$ is a subspace of $\mathscr{C}^{(m)}(I)$ whenever $n \geqslant m$. This statement can be made to include $\mathscr{C}(I)$ by defining $\mathscr{C}^{(0)}(I)$ to be $\mathscr{C}(I)$.

**Example 4**  Let $\mathscr{P}_n$ be the set of all polynomials of the form

$$a_0 + a_1 x + \cdots + a_{n-1} x^{n-1},$$

where $a_0, \ldots, a_{n-1}$ are arbitrary real numbers and $n$ is a *fixed* positive integer. That is, $\mathscr{P}_n$ consists of all polynomials of degree *less than n*, together with zero polynomial. (By convention, the zero polynomial does not have a degree.) Then $\mathscr{P}_n$ is a subspace of $\mathscr{P}$.

Now that we know what a subspace is, it is natural to ask how one can construct subspaces in a vector space. To answer this question, we introduce the following definition.

**Definition 2–3**  *Every finite sum of the form*

$$\alpha_1 \mathbf{x}_1 + \cdots + \alpha_n \mathbf{x}_n,$$

*where $\alpha_1, \ldots, \alpha_n$ are scalars, is called a **linear combination** of the vectors $\mathbf{x}_1, \ldots, \mathbf{x}_n$.*

The connection between linear combinations of vectors and subspaces is all but immediate. Indeed, if $\mathscr{X}$ is any nonempty *subset* of vectors in $\mathscr{V}$, and if $\mathscr{S}(\mathscr{X})$ is the set of all (finite) linear combinations of the vectors in $\mathscr{X}$, then $\mathscr{S}(\mathscr{X})$ *is a subspace of $\mathscr{V}$ that contains* $\mathscr{X}$. For, if

$$\alpha_1 \mathbf{x}_1 + \cdots + \alpha_n \mathbf{x}_n$$

and

$$\beta_1 \mathbf{y}_1 + \cdots + \beta_m \mathbf{y}_m$$

are two linear combinations of vectors in $\mathscr{X}$, then their sum

$$\alpha_1 \mathbf{x}_1 + \cdots + \alpha_n \mathbf{x}_n + \beta_1 \mathbf{y}_1 + \cdots + \beta_m \mathbf{y}_m$$

is also such a linear combination, and $\mathscr{S}(\mathscr{X})$ is closed under addition. Since $\mathscr{S}(\mathscr{X})$ is clearly also closed under scalar multiplication, $\mathscr{S}(\mathscr{X})$ is a subspace of $\mathscr{V}$. Finally, the equation $\mathbf{x} = 1\mathbf{x}$ implies that each vector in $\mathscr{X}$ is a linear combination of vectors in $\mathscr{X}$ (namely $\mathbf{x}$ itself) and therefore belongs to $\mathscr{S}(\mathscr{X})$. Thus $\mathscr{X}$ is contained in $\mathscr{S}(\mathscr{X})$, and we have proved the following important theorem.

**Theorem 2–1**    *The set $\mathscr{S}(\mathscr{X})$ of all linear combinations of the vectors in a nonempty subset $\mathscr{X}$ of a vector space $\mathscr{V}$ is a subspace of $\mathscr{V}$ containing $\mathscr{X}$.*

$\mathscr{S}(\mathscr{X})$ is known as the subspace of $\mathscr{V}$ **spanned by** $\mathscr{X}$, a term that is meant to suggest that it is constructed from the vectors in $\mathscr{X}$ by addition and scalar multiplication. It is the *smallest* subspace of $\mathscr{V}$ that contains $\mathscr{X}$. (See Exercise 14.)

**Example 5**    If $\mathbf{x} = (x_1, x_2)$ is a nonzero vector in $\mathscr{R}^2$, then the subspace spanned by $\mathbf{x}$ consists of all the scalar multiples

$$\alpha \mathbf{x} = (\alpha x_1, \alpha x_2)$$

of $\mathbf{x}$. In terms of the standard Cartesian coordinates in $\mathscr{R}^2$ this set consists of all points $(x, y)$ such that

$$x = \alpha x_1, \qquad y = \alpha x_2.$$

There are just the parametric equations for the line through the origin in $\mathscr{R}^2$ determined by the point $(x_1, x_2)$. (See Fig. 2–8.)

In short, the subspace spanned by a single nonzero vector in $\mathscr{R}^2$ is the line through the origin determined by the vector. A similar result holds in $\mathscr{R}^3$ or, for that matter, in $\mathscr{R}^n$ for any $n$.

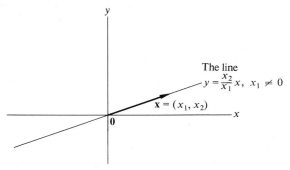

Figure 2–8

**Example 6**  In Section 2–1 we saw that the general solution of the second-order differential equation

$$y'' + k^2 y = 0 \tag{2–29}$$

consists of all linear combinations of the particular solutions

$$y_1 = \sin kx \quad \text{and} \quad y_2 = \cos kx.$$

In the language of linear algebra, we can say that the general solution of (2–29) is the subspace of $\mathscr{C}(-\infty, \infty)$ spanned by the vectors $\sin kx$ and $\cos kx$, where $\mathscr{C}(-\infty, \infty)$ is the space of continuous functions on the entire real line.

## EXERCISES

1. Determine which of the following subsets are subspaces of the indicated vector space. Give reasons for your answers.
   a) The set of all vectors in $\mathscr{R}^2$ of the form $\mathbf{x} = (1, x_2)$.
   b) The zero vector together with all vectors $\mathbf{x} = (x_1, x_2)$ in $\mathscr{R}^2$ for which $x_2/x_1$ has a constant value.
   c) The set of all vectors $\mathbf{x} = (x_1, x_2, x_3)$ in $\mathscr{R}^3$ for which $x_1 + x_2 + x_3 = 0$.
   d) The set of all vectors in $\mathscr{R}^3$ of the form $(x_1, x_2, x_1 + x_2)$.
   e) The set of all vectors $\mathbf{x} = (x_1, x_2, x_3)$ in $\mathscr{R}^3$ for which $x_1{}^2 + x_2{}^2 + x_3{}^2 = 1$.

2. Repeat Exercise 1 for the following subsets.
   a) The subset of $\mathscr{P}_n$ consisting of the zero polynomial and all polynomials of degree $n - 1$.
   b) The subset of $\mathscr{P}$ consisting of the zero polynomial and all polynomials of even degree.
   c) The subset of $\mathscr{P}$ consisting of the zero polynomial and all polynomials of degree 0.
   d) The subset of $\mathscr{P}_n$, $n > 1$, consisting of all polynomials which have $x$ as a factor.
   e) The subset of $\mathscr{P}$ consisting of all polynomials which have $x - 1$ as a factor.

3. Repeat Exercise 1 for the following subsets of $\mathscr{C}(I)$ when $I = [a, b]$.
   a) The set of all functions in $\mathscr{C}[a, b]$ which vanish at the point $x_0$ in $[a, b]$.
   b) The set of all nondecreasing functions in $\mathscr{C}[a, b]$.
   c) The set of all constant functions in $\mathscr{C}[a, b]$.
   d) The set of all functions $\mathbf{f}$ in $\mathscr{C}[a, b]$ such that $\mathbf{f}(a) = 1$.
   e) The set of all functions $\mathbf{f}$ in $\mathscr{C}[a, b]$ such that $\int_a^b \mathbf{f}(x)\, dx = 0$.

4. Prove that the only proper subspaces of $\mathscr{R}^3$ different from the trivial subspace are lines and planes through the origin.

5. Determine which of the following vectors belong to the subspace of $\mathscr{R}^3$ spanned by $(1, 2, 1)$ and $(2, 3, 4)$.
   a) $(4, 7, 6)$　　　　　　b) $(-\frac{1}{2}, -\frac{1}{2}, -\frac{3}{2})$　　　　　　c) $(\frac{1}{2}, 1, 1)$
   d) $(2, 9, 5)$　　　　　　e) $(2, 9, 4)$　　　　　　f) $(0, \frac{1}{3}, -\frac{2}{3})$

6. Determine which of the following vectors belong to the subspace of $\mathscr{R}^3$ spanned by $(1, -3, 2)$ and $(0, 4, 1)$.
   a) $(3, -1, 8)$　　　　　　b) $(2, -2, 1)$　　　　　　c) $(\frac{1}{2}, 1, \frac{3}{2})$
   d) $(2, -\frac{9}{2}, 0)$　　　　　　e) $(\frac{1}{3}, -1, \frac{2}{3})$　　　　　　f) $(\frac{2}{3}, 3, -\frac{4}{3})$

7. Determine which of the following polynomials belong to the subspace of $\mathscr{P}$ spanned by $x^3 + 2x^2 + 1$, $x^2 - 2$, and $x^3 + x$.
   a) $x^2 - x + 3$　　　　　　b) $x^2 - 2x + 1$　　　　　　c) $4x^3 - 3x + 5$
   d) $x^4 + 1$　　　　　　e) $-\frac{1}{2}x^3 + \frac{5}{2}x^2 - x - 1$　　f) $x - 5$

*See if linear combinations be written for an f*

8. Find the subspace of $\mathscr{R}^3$ spanned by each of the following sets of vectors.
   a) $(2, 1, 3), (-1, 2, 1)$          b) $(1, 0, 2), (2, 1, -2)$          c) $(-1, 1, 2), (0, 1, 0), (2, 4, 1)$

9. Find the subspace of $\mathscr{P}$ spanned by each of the following sets of vectors.
   a) $x^2, x(x + 1)$          b) $x + 1, x^2 - 1$          c) $1, x - 2, (x - 2)^2$

10. Let $\mathscr{X}$ be a nonempty subset of a vector space $\mathscr{V}$.
    a) Prove that $\mathscr{S}(\mathscr{S}(\mathscr{X})) = \mathscr{S}(\mathscr{X})$.
    b) Prove that $\mathscr{S}(\mathscr{X}) = \mathscr{X}$ if and only if $\mathscr{X}$ is a subspace of $\mathscr{V}$.

11. Prove that the vectors $(1, -1, 2)$ and $(3, 0, 1)$ span the same subspace of $\mathscr{R}^3$ as $(-1, -2, 3)$ and $(3, 3, -4)$.

12. Prove that the functions $\sin^2 x, \cos^2 x, \sin x \cos x$ span the same subspace of $\mathscr{C}(-\infty, \infty)$ as $1, \sin 2x, \cos 2x$.

13. Prove that the intersection $\mathscr{W}_1 \cap \mathscr{W}_2$ of two subspaces of a vector space $\mathscr{V}$ is a subspace of $\mathscr{V}$. (Recall that $\mathscr{W}_1 \cap \mathscr{W}_2$ consists of all vectors that belong to both $\mathscr{W}_1$ and $\mathscr{W}_2$.)

14. Let $\mathscr{X}$ be a nonempty subset of a vector space $\mathscr{V}$. Prove that $\mathscr{S}(\mathscr{X})$ is the intersection of all subspaces of $\mathscr{V}$ that contain $\mathscr{X}$.

## 2–5  LINEAR INDEPENDENCE AND DEPENDENCE

Now that we know how to construct subspaces in a vector space, we turn our attention to the question of efficiency. How can the subspace spanned by a finite (nonempty) set of vectors in $\mathscr{V}$ be described with as few vectors as possible? Consider $\mathscr{R}^3$, for instance, and the subspace $\mathscr{S}(\mathbf{x}_1, \mathbf{x}_2, \mathbf{x}_3)$ spanned by three nonzero, coplanar vectors, no two of which are collinear. It is clear that $\mathscr{S}(\mathbf{x}_1, \mathbf{x}_2, \mathbf{x}_3)$ is the plane through the origin in $\mathscr{R}^3$ containing these vectors (see Fig. 2–9), and it is equally clear that the given set contains more vectors than are needed to span this plane, since any two of them will do the job. Moreover, since it always takes at least two vectors to span a plane in $\mathscr{R}^3$, we can obtain a minimal subset of $\mathbf{x}_1, \mathbf{x}_2, \mathbf{x}_3$ that spans $\mathscr{S}(\mathbf{x}_1, \mathbf{x}_2, \mathbf{x}_3)$ by discarding any one of the given vectors.

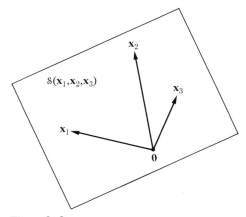

Figure 2–9

The purpose of this section is to show that it is always possible to reduce a finite set of vectors $x_1, \ldots, x_n$ to a minimal set that continues to span $\mathscr{S}(x_1, \ldots, x_n)$. But first we introduce some terminology.

**Definition 2–4**   *A vector $x$ is said to be **linearly dependent** on the vectors $x_1, \ldots, x_n$ if $x$ can be written as a linear combination of these vectors:*

$$x = \alpha_1 x_1 + \cdots + \alpha_n x_n.$$

*Otherwise, $x$ is said to be **linearly independent** of $x_1, \ldots, x_n$.*

Thus, if we denote the set $x_1, \ldots, x_n$ by $\mathscr{X}$, then $x$ is linearly dependent on $x_1, \ldots, x_n$ if and only if $x$ belongs to $\mathscr{S}(\mathscr{X})$, and $x$ is linearly independent of $x_1, \ldots, x_n$ if and only if $x$ does not belong to $\mathscr{S}(\mathscr{X})$. Note that each of the $x_i$, $1 \leqslant i \leqslant n$, is linearly dependent on $\mathscr{X}$, since each of these vectors belongs to $\mathscr{S}(\mathscr{X})$. Also since $0 = 0x$ for all $x$, *the zero vector is dependent on every (nonempty) set of vectors.*

**Example 1**   In $\mathscr{R}^3$ the vector $(1, 9, -10)$ is linearly dependent on $(2, 3, 1)$ and $(1, -1, 4)$ because

$$(1, 9, -10) = 2(2, 3, 1) - 3(1, -1, 4).$$

On the other hand, $(2, 1, -1)$ is linearly independent of $(1, 1, 0)$ and $(2, 0, 0)$, because no linear combination of these two vectors can have a nonzero third component.

**Example 2**   In $\mathscr{C}(-\infty, \infty)$, the vector space of real-valued continuous functions defined on the real line, the vector $\cos 2x$ is linearly dependent on $1$ and $\sin^2 x$ because

$$\cos 2x = 1 - 2 \sin^2 x.$$

It is convenient to extend the terminology of Definition 2–4 to include finite *sets* of vectors. We do this as follows.

**Definition 2–5**   *A finite set $\mathscr{X}$ of vectors is said to be **linearly independent** if no one of its members is linearly dependent on the other vectors in $\mathscr{X}$. Otherwise, $\mathscr{X}$ is said to be a **linearly dependent** set.*

Furthermore, for the sake of simplicity we shall often relax our terminology and say that the vectors $x_1, \ldots, x_n$ are linearly independent (dependent) when we mean that they form a linearly independent (dependent) set.

Since the concepts of linear independence and dependence will play an important role in the rest of our work, we need to be able to determine when a given set of vectors is linearly independent or dependent. The easiest method is to apply the following test.

**Test for linear independence**   *The vectors $x_1, \ldots, x_n$ are linearly independent if and only if the equation*

$$\alpha_1 x_1 + \cdots + \alpha_n x_n = 0$$

*implies that $\alpha_1 = \cdots = \alpha_n = 0$.*

*Proof.* Suppose that $\mathbf{x}_1, \ldots, \mathbf{x}_n$ are linearly independent and that

$$\alpha_1 \mathbf{x}_1 + \cdots + \alpha_n \mathbf{x}_n = \mathbf{0}.$$

We must show that all of the $\alpha_i$ are zero. Suppose that this were false. Then one of the $\alpha_i$, say $\alpha_1$, would be different from zero, and we could solve the preceding equation for $\mathbf{x}_1$ to obtain

$$\mathbf{x}_1 = -\frac{\alpha_2}{\alpha_1} \mathbf{x}_2 - \cdots - \frac{\alpha_n}{\alpha_1} \mathbf{x}_n.$$

But this would imply that $\mathbf{x}_1$ is linearly dependent on $\mathbf{x}_2, \ldots, \mathbf{x}_n$, contrary to our assumption. Hence $\alpha_1$ must be zero. By the same argument $\alpha_2 = \cdots = \alpha_n = 0$, as required.

Conversely, suppose that the equation

$$\alpha_1 \mathbf{x}_1 + \cdots + \alpha_n \mathbf{x}_n = 0$$

implies that $\alpha_1 = \cdots = \alpha_n = 0$. We must show that $\mathbf{x}_1, \ldots, \mathbf{x}_n$ are linearly independent. If this were not the case, then one of the $\mathbf{x}_i$, say $\mathbf{x}_1$, would be linearly dependent on $\mathbf{x}_2, \ldots, \mathbf{x}_n$, and we could find scalars $\alpha_2, \ldots, \alpha_n$ such that

$$\mathbf{x}_1 = \alpha_2 \mathbf{x}_2 + \cdots + \alpha_n \mathbf{x}_n,$$

or

$$-\mathbf{x}_1 + \alpha_2 \mathbf{x}_2 + \cdots + \alpha_n \mathbf{x}_n = \mathbf{0}.$$

But this is impossible, since the coefficient of $\mathbf{x}_1$ in this equation is $-1$. Hence the $\mathbf{x}_i$ must be linearly independent after all.  ∎

Naturally, since dependence is the negation of independence, there is a corresponding test for linear dependence. It reads as follows.

**Test for linear dependence**   *The vectors* $\mathbf{x}_1, \ldots, \mathbf{x}_n$ *are linearly dependent if and only if there exist scalars* $\alpha_1, \ldots, \alpha_n$, **not all zero**, *such that*

$$\alpha_1 \mathbf{x}_1 + \cdots + \alpha_n \mathbf{x}_n = \mathbf{0}.$$

**Example 3**   The vectors $\mathbf{x}_1 = (1, 3, -1, 2)$, $\mathbf{x}_2 = (2, 0, 1, 3)$, $\mathbf{x}_3 = (-1, 1, 0, 0)$ are linearly independent in $\mathscr{R}^4$ because the equation

$$\alpha_1 \mathbf{x}_1 + \alpha_2 \mathbf{x}_2 + \alpha_3 \mathbf{x}_3 = 0$$

implies that

$$\begin{aligned}
\alpha_1 + 2\alpha_2 - \alpha_3 &= 0, \\
3\alpha_1 \quad\quad\; + \alpha_3 &= 0, \\
-\alpha_1 + \alpha_2 \quad\quad\; &= 0, \\
2\alpha_1 + 3\alpha_2 \quad\quad\; &= 0,
\end{aligned}$$

from which it easily follows that $\alpha_1 = \alpha_2 = \alpha_3 = 0$.

Now let us return to the problem raised at the beginning of the section, and show how one can weed out extraneous vectors from any finite set $\mathbf{x}_1, \ldots, \mathbf{x}_n$ without disturbing $\mathscr{S}(\mathbf{x}_1, \ldots, \mathbf{x}_n)$. (We assume that the set contains at least one

nonzero vector.) The basic idea is obvious: just get rid of as many linearly dependent vectors as possible.

To accomplish this, we begin with the vector $\mathbf{x}_n$. If $\mathbf{x}_n$ is linearly dependent
on $\mathbf{x}_1, \ldots, \mathbf{x}_{n-1}$, then

$$\mathbf{x}_n = \alpha_1 \mathbf{x}_1 + \cdots + \alpha_{n-1} \mathbf{x}_{n-1},$$

and we can rewrite the expression

$$\mathbf{x} = \beta_1 \mathbf{x}_1 + \cdots + \beta_n \mathbf{x}_n$$

for an arbitrary vector in $\mathscr{S}(\mathbf{x}_1, \ldots, \mathbf{x}_n)$ in the form

$$\mathbf{x} = (\beta_1 + \alpha_1 \beta_n)\mathbf{x}_1 + \cdots + (\beta_{n-1} + \alpha_{n-1}\beta_n)\mathbf{x}_{n-1}.$$

This proves that $\mathbf{x}$ is already a linear combination of $\mathbf{x}_1, \ldots, \mathbf{x}_{n-1}$ and hence that
$\mathscr{S}(\mathbf{x}_1, \ldots, \mathbf{x}_{n-1}) = \mathscr{S}(\mathbf{x}_1, \ldots, \mathbf{x}_n)$. In this case we drop the vector $\mathbf{x}_n$ from the
set $\mathbf{x}_1, \ldots, \mathbf{x}_n$. If, on the other hand, $\mathbf{x}_n$ is *not* linearly dependent on $\mathbf{x}_1, \ldots, \mathbf{x}_{n-1}$,
we keep it.

If we repeat this procedure with each of the $\mathbf{x}_i$ in turn, dropping $\mathbf{x}_i$ if it is
linearly dependent on the remaining vectors in the (possibly modified) set, and
keeping it otherwise, it is clear that we will obtain a linearly independent subset
of $\mathbf{x}_1, \ldots, \mathbf{x}_n$ that spans the subspace $\mathscr{S}(\mathbf{x}_1, \ldots, \mathbf{x}_n)$. This is what we started out to
show. We have therefore proved the following theorem.

**Theorem 2–2**   *Every finite set of vectors $\mathscr{X}$ that contains at least one nonzero
vector contains a linearly independent subset which spans the subspace $\mathscr{S}(\mathscr{X})$.*

Note, however, that in general $\mathscr{X}$ may contain many such subsets. This was the
case in the example given at the beginning of this section, and we will encounter
other examples as we continue.

## EXERCISES

1. Show that the following sets of vectors in $\mathscr{R}^3$ are linearly independent.
   a) $(1, 0, 0), (0, 1, 0), (1, 1, 1)$       b) $(2, 1, 0), (0, 2, -1), (1, 1, 2)$

2. Show that the following sets of vectors in $\mathscr{R}^4$ are linearly independent.
   a) $(1, 1, 0, 0), (1, 0, 1, 0), (0, 1, 1, 0), (0, 1, 0, 1)$
   b) $(0, 0, 0, 1), (0, 0, 1, 1), (0, 1, 1, 1), (1, 1, 1, 1)$

3. Are the following sets of vectors linearly independent or dependent in $\mathscr{R}^3$?
   a) $(-2, 1, 3), (3, -2, -1), (-1, 0, 5)$
   b) $(-1, 3, 2), (3, 4, 0), (1, 4, 4)$
   c) $(2, -2, 1), (1, -3, 4), (-3, 1, 2)$

4. Are the following sets of vectors linearly independent or dependent in $\mathscr{P}_3$? [*Hint*: A
   polynomial is the zero polynomial if and only if all of its coefficients are zero.]
   a) $x + 1, x^2 - 2, x^2 - 3x - 5$
   b) $1, x + 1, (x + 1)^2$
   c) $x^2 + 3x - 2, x^2 + 4, -3x + 6$

5. Express each of the following vectors in $\mathscr{R}^3$ as a linear combination of $(4, -2, 1)$ and $(-3, 1, 2)$.

    a)  $(6, -4, 7)$                              b)  $2(10, -4, -3)$

6. Prove that the vectors $(a, b)$ and $(c, d)$ are linearly independent in $\mathscr{R}^2$ if and only if
$$ad - bc \neq 0.$$

7. Assume that $\mathbf{x}_1, \mathbf{x}_2$, and $\mathbf{x}_3$ are linearly independent in $\mathscr{V}$. Prove that $\mathbf{x}_1 + \mathbf{x}_2, \mathbf{x}_1 + \mathbf{x}_3$, and $\mathbf{x}_2 + \mathbf{x}_3$ are also linearly independent in $\mathscr{V}$.

8. a)  Show that every subset of a (finite) linearly independent set in a vector space is linearly independent.

    b)  Show that a (finite) set in a vector space is linearly dependent if it contains a linearly dependent subset.

## 2–6  LINEAR INDEPENDENCE IN VECTOR SPACES OF FUNCTIONS

In our study of differential equations it will often be necessary to test sets of solutions for linear independence, and though we already have one such test, we now present another that is much more convenient to use when working with differentiable functions.

*an if then not iff*

**Theorem 2–3** *Let* $y_1(x), \ldots, y_n(x)$ *be functions in* $\mathscr{C}^{(n-1)}(I)$, *and suppose that at some point* $x_0$ *in* $I$ *the vectors*

$$(y_1(x_0), y_1'(x_0), \ldots, y_1^{(n-1)}(x_0))$$
$$(y_2(x_0), y_2'(x_0), \ldots, y_2^{(n-1)}(x_0))$$
$$\cdot$$
$$\cdot$$
$$\cdot$$
$$(y_n(x_0), y_n'(x_0), \ldots, y_n^{(n-1)}(x_0))$$

*are linearly independent in* $\mathscr{R}^n$. *Then the functions* $y_1(x), \ldots, y_n(x)$ *are linearly independent in* $\mathscr{C}(I)$.

*doesn't prove dependence only existence theorem Prooving indep*

*Proof.*   We apply the test for linear independence, as follows. Suppose that

$$\alpha_1 y_1(x) + \alpha_2 y_2(x) + \cdots + \alpha_n y_n(x) = 0$$

for all $x$ in $I$. Then this identity, together with its first $n - 1$ derivatives, yields the system

$$\alpha_1 y_1(x) + \alpha_2 y_2(x) + \cdots + \alpha_n y_n(x) = 0$$
$$\alpha_1 y_1'(x) + \alpha_2 y_2'(x) + \cdots + \alpha_n y_n'(x) = 0$$
$$\cdot$$
$$\cdot$$
$$\cdot$$
$$\alpha_1 y_1^{(n-1)}(x) + \alpha_2 y_2^{(n-1)}(x) + \cdots + \alpha_n y_n^{(n-1)}(x) = 0,$$

which holds for all $x$ in $I$. We now set $x = x_0$ to obtain

$$\alpha_1 y_1(x_0) + \alpha_2 y_2(x_0) + \cdots + \alpha_n y_n(x_0) = 0$$
$$\alpha_1 y_1'(x_0) + \alpha_2 y_2'(x_0) + \cdots + \alpha_n y_n'(x_0) = 0$$
$$\vdots$$
$$\alpha_1 y_1^{(n-1)}(x_0) + \alpha_2 y_2^{(n-1)}(x_0) + \cdots + \alpha_n y_n^{(n-1)}(x_0) = 0,$$

which we rewrite in vector form as

$$\alpha_1 \begin{bmatrix} y_1(x_0) \\ y_1'(x_0) \\ \vdots \\ y_n^{(n-1)}(x_0) \end{bmatrix} + \alpha_2 \begin{bmatrix} y_2(x_0) \\ y_2'(x_0) \\ \vdots \\ y_2^{(n-1)}(x_0) \end{bmatrix} + \cdots + \alpha_n \begin{bmatrix} y_n(x_0) \\ y_n'(x_0) \\ \vdots \\ y_1^{(n-1)}(x_0) \end{bmatrix} = \begin{bmatrix} 0 \\ 0 \\ \vdots \\ 0 \end{bmatrix}.$$

(At this point it is more natural to display these vectors as columns rather than as rows.) By assumption the vectors in this equation are linearly independent in $\mathscr{R}^n$. Hence $\alpha_1 = \alpha_2 = \cdots = \alpha_n = 0$, and we are done. ∎

**Example 1**   Show that the functions

$$\sin x \quad \text{and} \quad \cos x$$

are linearly independent in $\mathscr{C}(-\infty, \infty)$.

*Solution.*   In this case the column vectors

$$\begin{bmatrix} y_i(x) \\ y_i'(x) \end{bmatrix}$$

are

$$\begin{bmatrix} \sin x \\ \cos x \end{bmatrix} \quad \text{and} \quad \begin{bmatrix} \cos x \\ -\sin x \end{bmatrix}.$$

When they are evaluated at $x = 0$, we obtain

$$\begin{bmatrix} 0 \\ 1 \end{bmatrix} \quad \text{and} \quad \begin{bmatrix} 1 \\ 0 \end{bmatrix},$$

which are linearly independent vectors in $\mathscr{R}^2$.

**Example 2**   Show that

$$e^x, \quad xe^x, \quad x^2 e^x$$

are linearly independent in $\mathscr{C}(-\infty, \infty)$.

*Solution.*  In this case the column vectors are

$$\begin{bmatrix} e^x \\ e^x \\ e^x \end{bmatrix}, \quad \begin{bmatrix} xe^x \\ (x+1)e^x \\ (x+2)e^x \end{bmatrix}, \quad \begin{bmatrix} x^2 e^x \\ x(x+2)e^x \\ x(x+4)e^x + 2e^x \end{bmatrix}.$$

Once again we set $x = 0$ to find

$$\begin{bmatrix} 1 \\ 1 \\ 1 \end{bmatrix}, \quad \begin{bmatrix} 0 \\ 1 \\ 2 \end{bmatrix}, \quad \begin{bmatrix} 0 \\ 0 \\ 2 \end{bmatrix}$$

which are easily seen to be linearly independent in $\mathscr{R}^3$.

It is natural to ask whether the negation of Theorem 2–3 will yield a test for linear dependence: if the vectors

$$(y_i(x), y_i'(x), \ldots, y_i^{(n-1)}(x)), \qquad 1 \leqslant i \leqslant n,$$

are linearly dependent in $\mathscr{R}^n$ *for all $x$ in $I$*, are the vectors $y_1(x), \ldots, y_n(x)$ then linearly dependent in $\mathscr{C}(I)$? It is not difficult to show that this *need not be true* for arbitrary functions in $\mathscr{C}^{(n-1)}(I)$. (See Exercise 9.) However, in Chapter 4 we will see that it *is* true whenever the $y_i(x)$ are solutions of an appropriate differential equation.

The preceding examples are not meant to suggest that the test for linear independence described by Theorem 2–3 should always be used when dealing with differentiable functions. Instead of producing column vectors automatically, it is often best to invent an argument that is suited to the particular functions in question. The following example illustrates such an argument, and another is suggested in the exercises at the end of the section.

**Example 3**  Prove that

$$e^x, \quad e^{2x}, \quad e^{3x}, \quad e^{4x}$$

are linearly independent in $\mathscr{C}(-\infty, \infty)$.

*Solution.*  In this case Theorem 2–3, with $x_0 = 0$, yields the vectors

$$\begin{bmatrix} 1 \\ 1 \\ 1 \\ 1 \end{bmatrix}, \quad \begin{bmatrix} 1 \\ 2 \\ 4 \\ 8 \end{bmatrix}, \quad \begin{bmatrix} 1 \\ 3 \\ 9 \\ 27 \end{bmatrix}, \quad \begin{bmatrix} 1 \\ 4 \\ 16 \\ 64 \end{bmatrix}.$$

The test for linear independence of Section 2–5 will certainly reveal them to be linearly independent in $\mathscr{R}^4$. But to apply this test, we must solve four linear equations in four unknowns, a tedious task at best. (In this regard, however, see p. 81.) As an alternative, we can argue as follows.

Suppose that

$$\alpha_1 e^x + \alpha_2 e^{2x} + \alpha_3 e^{3x} + \alpha_4 e^{4x} = 0 \tag{2-30}$$

for all $x$. Multiply by $e^{-x}$ to obtain

$$\alpha_1 + \alpha_2 e^x + \alpha_3 e^{2x} + \alpha_4 e^{3x} = 0.$$

Now differentiate to obtain the identity

$$\alpha_2 e^x + 2\alpha_3 e^{2x} + 3\alpha_4 e^{3x} = 0 \tag{2-31}$$

in which $\alpha_1$ no longer appears. Now repeat the maneuver: multiply by $e^{-x}$ and differentiate to obtain

$$2\alpha_3 e^x + 6\alpha_4 e^{2x} = 0. \tag{2-32}$$

One more repetition of this procedure yields

$$6\alpha_4 e^x = 0,$$

from which it follows that $\alpha_4 = 0$. Then in turn, (2–32) implies that $\alpha_3 = 0$, (2–31) that $\alpha_2 = 0$, and (2–30) that $\alpha_1 = 0$. The desired conclusion now follows from the test for linear independence.

This argument can be used to establish the linear independence of

$$e^{\alpha_1 x}, e^{\alpha_2 x}, \ldots, e^{\alpha_n x}$$

in $\mathscr{C}(-\infty, \infty)$ whenever the $\alpha_i$ are distinct real numbers. Just multiply by an appropriate power of $e^{-x}$ at each stage.

## EXERCISES

Prove that each of the following sets of functions is linearly independent in $\mathscr{C}(I)$ for the given interval $I$.

1. $\sin x$, $x \sin x$, $x^2 \sin x$, on $(-\infty, \infty)$
2. $x$, $x \ln x$, $x^2 \ln x$, on $(0, \infty)$
3. $1, x, x^2, \ldots, x^n$, on $(-\infty, \infty)$
4. $x^{1/2}$, $x^{1/3}$, on $(0, \infty)$
5. $e^{ax} \sin bx$, $e^{ax} \cos bx$, $b \neq 0$, on $(-\infty, \infty)$
6. Prove that

$$e^{\alpha_1 x}, e^{\alpha_2 x}, \ldots, e^{\alpha_n x}$$

   are linearly independent in $(-\infty, \infty)$ whenever the $\alpha_i$ are distinct real numbers. [*Hint*: See Example 3.]
7. Prove that $x^\alpha$, $x^\beta$, $x^\gamma$ are linearly independent in $\mathscr{C}(0, \infty)$ if and only if $\alpha$, $\beta$, $\gamma$ are distinct real numbers. [*Hint*: See Example 3.]
\* 8. Prove that any set of $n$ polynomials, one of each degree $0, 1, \ldots, n - 1$, is linearly independent in $\mathscr{P}_n$. [*Hint*: Use mathematical induction.]

9. Let $y_1(x) = x^3$ and $y_2(x) = |x^3|$.
   a) Show that $y_1$ and $y_2$ are linearly independent in $\mathscr{C}(-\infty, \infty)$.
   b) Show that the vectors

$$\begin{bmatrix} y_1(x) \\ y_1'(x) \end{bmatrix} \quad \text{and} \quad \begin{bmatrix} y_2(x) \\ y_2'(x) \end{bmatrix}$$

   are linearly dependent in $\mathscr{R}^2$ for all $x$, and therefore the negation of Theorem 2–3 does not yield a test for linear dependence.

* 10.  Prove that the functions

$$\sin x, \sin 2x, \ldots, \sin nx$$

are linearly independent in $\mathscr{C}(-\infty, \infty)$ for any positive integer $n$. [*Hint*: Let $\sum_{k=1}^{n} \alpha_k \sin kx = 0$; multiply by $\sin jx$, $1 \leqslant j \leqslant n$, and integrate from $-\pi$ to $\pi$.]

*11.  Repeat Exercise 10 for the functions

$$1, \sin x, \cos x, \sin 2x, \cos 2x, \ldots, \sin nx, \cos nx.$$

## 2–7  BASES

Linearly independent sets enjoy a special status in the study of vector spaces, and those linearly independent sets which span the entire space are particularly important. They therefore have a special name.

> **Definition 2–6**  *A finite linearly independent subset $\mathscr{B}$ of a vector space $\mathscr{V}$ is said to be a **basis** for $\mathscr{V}$ if $\mathscr{S}(\mathscr{B}) = \mathscr{V}$.*

The familiar vectors

$$\mathbf{i} = (1, 0, 0), \qquad \mathbf{j} = (0, 1, 0), \qquad \mathbf{k} = (0, 0, 1)$$

form a basis for $\mathscr{R}^3$, as we shall see in Example 1. For the moment we merely observe that every $\mathbf{x} = (x_1, x_2, x_3)$ in $\mathscr{R}^3$ can be written in *one and only one way* as a linear combination of these basis vectors, namely

$$\mathbf{x} = x_1\mathbf{i} + x_2\mathbf{j} + x_3\mathbf{k}.$$

It turns out that this property serves to characterize bases in vector spaces, as we now show.

> **Theorem 2–4**  *A set of vectors $\mathbf{e}_1, \ldots, \mathbf{e}_n$ is a basis for a vector space $\mathscr{V}$ if and only if every vector in $\mathscr{V}$ can be written **uniquely** as a linear combination of $\mathbf{e}_1, \ldots, \mathbf{e}_n$.*

*Proof.*  First, suppose that $\mathbf{e}_1, \ldots, \mathbf{e}_n$ is a basis for $\mathscr{V}$. Then the $\mathbf{e}_i$ span $\mathscr{V}$, and every vector in $\mathscr{V}$ can be written in *at least one* way as

$$\mathbf{x} = \alpha_1\mathbf{e}_1 + \cdots + \alpha_n\mathbf{e}_n. \tag{2–33}$$

To show that this is the only such expression possible, let

$$\mathbf{x} = \beta_1 \mathbf{e}_1 + \cdots + \beta_n \mathbf{e}_n \tag{2-34}$$

be another. Then by subtracting (2–34) from (2–33), we obtain

$$\mathbf{0} = (\alpha_1 - \beta_1)\mathbf{e}_1 + \cdots + (\alpha_n - \beta_n)\mathbf{e}_n. \tag{2-35}$$

But since $\mathbf{e}_1, \ldots, \mathbf{e}_n$ are a basis for $\mathscr{V}$, they are linearly independent. Hence by the test for linear independence, each of the coefficients in (2–35) is zero, and it follows that $\alpha_1 = \beta_1, \ldots, \alpha_n = \beta_n$, as desired.

Conversely, suppose that every vector in $\mathscr{V}$ can be written *uniquely* as a linear combination of $\mathbf{e}_1, \ldots, \mathbf{e}_n$. Then these vectors certainly span $\mathscr{V}$, and we need only prove their linear independence in order to show that they are a basis for $\mathscr{V}$. To accomplish this, we observe that $\mathbf{0} = 0\mathbf{e}_1 + \cdots + 0\mathbf{e}_n$. But our assumption concerning the uniqueness of such expressions implies that this is the *only* possible representation of $\mathbf{0}$ as a linear combination of $\mathbf{e}_1, \ldots, \mathbf{e}_n$. Thus, if $\alpha_1 \mathbf{e}_1 + \cdots + \alpha_n \mathbf{e}_n = \mathbf{0}$, we must have $\alpha_1 = \cdots = \alpha_n = 0$, and the test for linear independence now applies. ∎

**Example 1**   The vectors

$$\mathbf{e}_1 = (1, 0, \ldots, 0),$$
$$\mathbf{e}_2 = (0, 1, \ldots, 0),$$
$$\cdot$$
$$\cdot$$
$$\cdot$$
$$\mathbf{e}_n = (0, 0, \ldots, 1)$$

are a basis for $\mathscr{R}^n$, since $\mathbf{x} = x_1 \mathbf{e}_1 + \cdots + x_n \mathbf{e}_n$ is the only way of expressing the vector $\mathbf{x} = (x_1, \ldots, x_n)$ as a linear combination of $\mathbf{e}_1, \ldots, \mathbf{e}_n$. This particular basis is called the **standard basis** for $\mathscr{R}^n$.

**Example 2**   Again in $\mathscr{R}^n$ let

$$\mathbf{e}'_1 = (1, 0, \ldots, 0),$$
$$\mathbf{e}'_2 = (1, 1, \ldots, 0),$$
$$\cdot$$
$$\cdot$$
$$\cdot$$
$$\mathbf{e}'_n = (1, 1, \ldots, 1),$$

where in general $\mathbf{e}'_i$ is the $n$-tuple having ones in the first $i$ places and zeros thereafter. Then $\mathbf{e}'_1, \ldots, \mathbf{e}'_n$ is a basis for $\mathscr{R}^n$. To prove this, let $\mathbf{x} = (x_1, \ldots, x_n)$ be given, and let us attempt to find real numbers $\alpha_1, \ldots, \alpha_n$ such that $\mathbf{x} = \alpha_1 \mathbf{e}'_1 + \cdots + \alpha_n \mathbf{e}'_n$. For such an equality to hold we must have

$$(x_1, \ldots, x_n) = \alpha_1(1, 0, \ldots, 0) + \alpha_2(1, 1, \ldots, 0) + \cdots + \alpha_n(1, 1, \ldots, 1)$$
$$= (\alpha_1, 0, \ldots, 0) + (\alpha_2, \alpha_2, \ldots, 0) + \cdots + (\alpha_n, \alpha_n, \ldots, \alpha_n)$$
$$= (\alpha_1 + \alpha_2 + \cdots + \alpha_n, \quad \alpha_2 + \cdots + \alpha_n, \ldots, \alpha_n).$$

This yields the system of equations

$$\alpha_1 + \alpha_2 + \cdots + \alpha_n = x_1$$
$$\alpha_2 + \cdots + \alpha_n = x_2$$
$$\vdots$$
$$\alpha_n = x_n,$$

and it follows that

$$\alpha_1 \quad = x_1 - x_2,$$
$$\alpha_2 \quad = x_2 - x_3,$$
$$\vdots$$
$$\alpha_{n-1} = x_{n-1} - x_n,$$
$$\alpha_n \quad = x_n.$$

These equations show simultaneously that $\mathbf{x}$ can be written as a linear combination of $\mathbf{e}'_1, \ldots, \mathbf{e}'_n$ and that the coefficients of this relation are uniquely determined. Thus the $\mathbf{e}'_i$ are a basis for $\mathcal{R}^n$, as asserted.

**Example 3**   The polynomials $1, x, x^2, \ldots, x^{n-1}$ form a basis for the vector space $\mathcal{P}_n$ since each polynomial in this space can be written in one and only one way in the form $a_0 + a_1 x + \cdots + a_{n-1} x^{n-1}$.

**Example 4**   Let $\mathbf{p}_1, \ldots, \mathbf{p}_n$, be any finite set of polynomials in $\mathcal{P}$, and let $d$ be the maximum of the degrees of the $\mathbf{p}_i$. Then no linear combination of these polynomials is of degree greater than $d$, from which it follows that $\mathbf{p}_1, \ldots, \mathbf{p}_n$ is *not* a basis for $\mathcal{P}$, since $\mathcal{P}$ contains polynomials of arbitrarily high degree. *Thus $\mathcal{P}$ does not possess a basis* in the sense of Definition 2–6.

**EXERCISES**

1. Prove that each of the following sets of vectors is a basis for $\mathcal{R}^2$.
   a) $(1, 0), (0, -1)$
   b) $(\cos \theta, \sin \theta), (-\sin \theta, \cos \theta), \quad 0 \leqslant \theta \leqslant 2\pi, \theta$ fixed
   c) $(\alpha, 0), (0, \beta), \quad \alpha, \beta$ nonzero real numbers
   d) $(1, 1), (0, 1)$

2. Let $\alpha'$ and $\beta'$ be real numbers. Express each of the following vectors in $\mathcal{R}^2$ as a linear combination of the vectors in the various bases of Exercise 1:

$$\mathbf{i}, \quad \mathbf{j}, \quad \mathbf{i} + \mathbf{j}, \quad \alpha'\mathbf{i} + \beta'\mathbf{j}.$$

   [Recall that $\mathbf{i} = (1, 0)$ and $\mathbf{j} = (0, 1)$.]

3. a) Prove that $(2, 1, -2), (-1, 3, 1)$, and $(-2, 1, 3)$ are a basis for $\mathcal{R}^3$.
   b) Express $(4, -3, -3)$ as a linear combination of the basis in (a).

4. Prove that each of the following sets of vectors is a basis for $\mathcal{R}^4$.
   a) $(1, 0, 0, 0), (0, 1, 0, 0), (0, 0, 1, 0), (1, 1, 1, 1)$

b) $(1, 1, 0, 0), (0, 0, 1, 1), (-1, 0, 1, 1), (0, -1, 0, 1)$

c) $(2, -1, 0, 1), (1, 3, 2, 0), (0, -1, -1, 0), (-2, 1, 2, 1)$

d) $(1, -1, 2, 0), (1, 1, 2, 0), (3, 0, 0, 1), (2, 1, -1, 0)$

5. Express $(2, -2, 1, 3)$ as a linear combination of the vectors in each of the bases in Exercise 4.

6. Prove that the polynomials

$$1, \quad x, \quad \tfrac{3}{2}x^2 - \tfrac{1}{2}, \quad \tfrac{5}{2}x^3 - \tfrac{3}{2}x$$

are a basis for $\mathscr{P}_4$.

7. Express $x^2$ and $x^3$ as linear combinations of the basis vectors for $\mathscr{P}_4$ given in Exercise 6.

8. Complete Example 3 by proving that each polynomial in $\mathscr{P}_n$ can be written in one and only one way in the form $a_0 + a_1 x + \cdots + a_{n-1}x^{n-1}$. [*Hint*: Suppose both that

$$p(x) = a_0 + a_1 x + \cdots + a_{n-1}x^{n-1}$$

and that

$$p(x) = b_0 + b_1 x + \cdots + b_{n-1}x^{n-1}.$$

Then proceed as in the proof of Theorem 2–4.]

9. Show that the polynomials

$$1, x - a, (x - a)^2, \ldots, (x - a)^{n-1},$$

where $a$ is an arbitrary real number, are a basis for $\mathscr{P}_n$. [*Hint*: Consider the Taylor series expansion of a polynomial about the point $x = a$, and recall that if a function can be expanded in a convergent power series about $x = a$, say $f(x) = \sum_{k=0}^{\infty} a_k(x - a)^k$, then $a_k = f^{(k)}(a)/k!$.]

10. Let $\mathscr{X}$ be a finite linearly independent subset of a vector space $\mathscr{V}$, and suppose that every finite subset of $\mathscr{V}$ that *properly* contains $\mathscr{X}$ is linearly dependent. Prove that $\mathscr{X}$ is a basis for $\mathscr{V}$. [This shows that a basis is a *maximal* linearly independent set in a vector space.]

11. Let $\mathscr{X}$ be a finite subset of a vector space $\mathscr{V}$ which spans $\mathscr{V}$, and suppose that no *proper* subset of $\mathscr{X}$ spans $\mathscr{V}$. Prove that $\mathscr{X}$ is a basis for $\mathscr{V}$. [This shows that a basis is a *minimal* spanning set in a vector space.]

## 2–8 COORDINATE SYSTEMS

**Definition 2–7** *Let* $\mathbf{e}_1, \ldots, \mathbf{e}_n$ *be a basis for a vector space* $\mathscr{V}$ *and let*

$$\mathbf{x} = \alpha_1 \mathbf{e}_1 + \cdots + \alpha_n \mathbf{e}_n$$

*be the unique expression for* $\mathbf{x}$ *in terms of this basis (Theorem 2–4). Then the scalars* $\alpha_1, \ldots, \alpha_n$ *are called* **coordinates** *or* **components** *of* $\mathbf{x}$ *with respect to* $\mathbf{e}_1, \ldots, \mathbf{e}_n$, *and the basis vectors themselves are said to form a* **coordinate system** *for* $\mathscr{V}$. *Finally, the subspaces of* $\mathscr{V}$ *spanned by the individual* $\mathbf{e}_i$ *are called the* **coordinate axes** *of the given coordinate system.*

Thus a basis is a coordinate system, and the unique expression for a vector as a linear combination of basis vectors is nothing but a "decomposition" of the vector into its components along the various coordinate axes. In these terms, part of

Theorem 2–4 is the statement: *The coordinates of a vector are uniquely determined by the coordinate system.*

Nevertheless, we caution against expecting too much from a coordinate system, especially expecting coordinate axes to be mutually perpendicular. Strictly speaking, of course, the concept of perpendicularity in a real vector space has no meaning at this point in our discussion. But it is common knowledge that certain coordinate axes, such as the standard ones in $\mathscr{R}^n$, are mutually perpendicular. We want to emphasize the sometime nature of this phenomenon and to point out that "oblique" coordinate systems also exist. One example is the coordinate system $e'_1, \ldots, e'_n$ for $\mathscr{R}^n$ introduced in Example 2 of Section 2–7.

It is also worth mentioning explicitly that the coordinates of a vector change with a change of coordinate system. Failure to appreciate the implications of this statement often causes confusion for the unwary. For example, the vector $x = (4, 2)$ has coordinates 4, 2 with respect to the standard coordinate system $e_1 = (1, 0)$, $e_2 = (0, 1)$ in $\mathscr{R}^2$, since

$$(4, 2) = 4e_1 + 2e_2.$$

(Indeed, this is the reason for calling $e_1, e_2$ the *standard* coordinate system.) However, if we use the coordinate system $e'_1 = (1, 0)$, $e'_2 = (1, 1)$, then the coordinates of $x$ become 2, 2, because

$$(4, 2) = 2e'_1 + 2e'_2.$$

The vector in $\mathscr{R}^2$ having coordinates 4, 2 with respect to $e'_1, e'_2$ is the ordered pair $(6, 2)$. (See Figs. 2–10 and 2–11.)

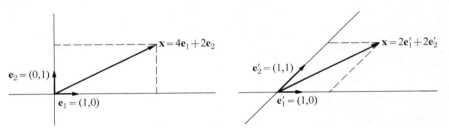

**Figure 2–10**

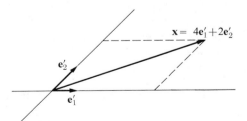

**Figure 2–11**

Whenever the components of vectors are known with respect to a basis it is possible to add the vectors and to multiply them by scalars just by adding and multiplying their components, as the next theorem shows.

**Theorem 2–5**   *Let $\mathbf{e}_1, \ldots, \mathbf{e}_n$ be any basis for a vector space $\mathcal{V}$. Then the sum of two vectors in $\mathcal{V}$ is found by adding their corresponding components, and the product of a vector and a scalar $\alpha$ is found by multiplying each component of the vector by $\alpha$.*

*Proof.*   If

$$\mathbf{x} = \alpha_1 \mathbf{e}_1 + \cdots + \alpha_n \mathbf{e}_n$$

and

$$\mathbf{y} = \beta_1 \mathbf{e}_1 + \cdots + \beta_n \mathbf{e}_n,$$

then it follows directly from the axioms defining a vector space that

$$\begin{aligned}
\mathbf{x} + \mathbf{y} &= (\alpha_1 \mathbf{e}_1 + \cdots + \alpha_n \mathbf{e}_n) + (\beta_1 \mathbf{e}_1 + \cdots + \beta_n \mathbf{e}_n) \\
&= (\alpha_1 \mathbf{e}_1 + \beta_1 \mathbf{e}_1) + \cdots + (\alpha_n \mathbf{e}_n + \beta_n \mathbf{e}_n) \\
&= (\alpha_1 + \beta_1) \mathbf{e}_1 + \cdots + (\alpha_n + \beta_n) \mathbf{e}_n.
\end{aligned}$$

Similarly,

$$\alpha \mathbf{x} = \alpha(\alpha_1 \mathbf{e}_1 + \cdots + \alpha_n \mathbf{e}_n) = (\alpha \alpha_1) \mathbf{e}_1 + \cdots + (\alpha \alpha_n) \mathbf{e}_n,$$

and the theorem is proved.   ∎

Theorem 2–5 places no restriction on the choice of the basis $\mathbf{e}_1, \ldots, \mathbf{e}_n$ and none on the nature of the vectors themselves. They may be $n$-tuples of real numbers, functions, or anything else, so long as they are vectors belonging to the same space and expressed, at the time of computation, in terms of the same basis.

Theorem 2–5 has many implications and certainly foreshadows the use of bases whenever numerical calculations with vectors are required. On the other hand, it turns out that whenever we are concerned with the general theory of vector spaces, bases are a distinct hindrance. This is because whenever we use a basis to prove a general statement about vector spaces, we must also show that the result is independent of the particular basis chosen to prove it. And it is usually as difficult to demonstrate this as it is to construct a coordinate-free proof of the original statement.

## EXERCISES

1.  Find the coordinates of each of the following vectors in $\mathcal{R}^3$ with respect to the basis $(1, 0, 0), (1, 1, 0), (1, 1, 1)$.

    a) $(0, 1, 0)$          b) $(-2, 1, 1)$          c) $(0, 0, 1)$
    d) $(-1, \frac{1}{2}, 1)$         e) $(4, -2, 2)$         f) $(1, 3, 2)$

2.  Repeat Exercise 1 with the basis $(1, 1, 0), (1, 0, 1), (0, 1, 1)$.

3.  Prove that the vectors $(2, 1, 0), (2, 1, 1), (2, 2, 1)$ form a basis for $\mathcal{R}^3$. Find the vectors in $\mathcal{R}^3$ which have the following coordinates with respect to this basis.

a) $1, 0, 0$       b) $-1, 2, 1$       c) $4, -5, 0$
d) $\frac{1}{2}, 2, 1$       e) $3, 1, -1$       f) $2, 2, 1$

4. Find the coordinates of the standard basis vectors in $\mathscr{R}^3$ with respect to the basis given in Exercise 3.

5. Find a basis in $\mathscr{R}^4$ with respect to which the vector $(-3, 1, 2, -1)$ has coordinates $1, 1, 1, 1$.

6. Does there exist a basis for $\mathscr{R}^2$ with respect to which an arbitrary vector $(x_1, x_2)$ has coordinates $2x_1$ and $3x_2$? [*Hint*: Suppose $(a_1, a_2)$ and $(b_1, b_2)$ to be such a basis. Write $(x_1, x_2)$ in terms of it, and see what this expression requires of the $a$'s and $b$'s.]

7. Find a basis for $\mathscr{R}^2$ with respect to which an arbitrary vector $(x_1, x_2)$ has coordinates $x_1$ and $x_1 + 2x_2$.

## 2–9  DIMENSION

Having observed that a basis in a vector space is nothing more than a coordinate system, we would find it disturbing, to say the least, if it turned out that the number of vectors in a basis for a vector space could change with the basis. This has not happened in any of the examples we have seen so far, but of course examples do not constitute a proof. Fortunately, the assurance we seek is an immediate consequence of the following theorem.

**Theorem 2–6**  *If $\mathscr{V}$ has a basis containing n vectors, then any set of $n + 1$ or more vectors in $\mathscr{V}$ is linearly dependent.**

*Proof.*  Let $\mathbf{e}_1, \ldots, \mathbf{e}_n$ be a basis for $\mathscr{V}$ and suppose, contrary to the assertion of the theorem, that $\mathscr{V}$ contains a linearly independent set $\mathbf{e}'_1, \ldots, \mathbf{e}'_m$ in which $m > n$. Express each of the $\mathbf{e}'_j$ as a linear combination of the $\mathbf{e}_i$, thereby obtaining the system of equations

$$
\begin{aligned}
\mathbf{e}'_1 &= \alpha_{11}\mathbf{e}_1 + \alpha_{21}\mathbf{e}_2 + \cdots + \alpha_{n1}\mathbf{e}_n, \\
\mathbf{e}'_2 &= \alpha_{12}\mathbf{e}_1 + \alpha_{22}\mathbf{e}_2 + \cdots + \alpha_{n2}\mathbf{e}_n, \\
&\;\;\vdots \\
\mathbf{e}'_m &= \alpha_{1m}\mathbf{e}_1 + \alpha_{2m}\mathbf{e}_2 + \cdots + \alpha_{nm}\mathbf{e}_n,
\end{aligned}
\tag{2–36}
$$

in which the $\alpha_{ij}$ are scalars. Since none of the $\mathbf{e}'_j$ is the zero vector, at least one of the $\alpha_{ij}$ must be different from zero in each of these equations. (Recall that the zero vector is linearly dependent on every vector in $\mathscr{V}$.) Thus, by relabeling the $\mathbf{e}_i$ if necessary, we may assume that $\alpha_{11} \neq 0$. This done, we solve the first equation for $\mathbf{e}_1$ and substitute the value obtained in the remaining $m - 1$ equations. This

---

* This result also applies to the trivial space consisting of just the zero vector, provided we agree that the empty set of vectors is a basis for this space. Notice that this convention forces us to view the empty set as a linearly independent set.

eliminates $\mathbf{e}_1$ from (2–36) and yields a system of equations of the form

$$\mathbf{e}'_2 = \beta_{22}\mathbf{e}_2 + \beta_{32}\mathbf{e}_3 + \cdots + \beta_{n2}\mathbf{e}_n + \beta_{12}\mathbf{e}'_1,$$
$$\mathbf{e}'_3 = \beta_{23}\mathbf{e}_2 + \beta_{33}\mathbf{e}_3 + \cdots + \beta_{n3}\mathbf{e}_n + \beta_{13}\mathbf{e}'_1,$$
$$\cdot$$
$$\cdot \qquad\qquad\qquad\qquad\qquad\qquad\qquad (2\text{–}37)$$
$$\cdot$$
$$\mathbf{e}'_m = \beta_{2m}\mathbf{e}_2 + \beta_{3m}\mathbf{e}_3 + \cdots + \beta_{nm}\mathbf{e}_n + \beta_{1m}\mathbf{e}'_1.$$

Focusing our attention on the first of these equations, we note that the linear independence of $\mathbf{e}'_1$ and $\mathbf{e}'_2$ implies that at least one of the coefficients $\beta_{22}, \beta_{32}, \ldots,$ $\beta_{n2}$ is different from zero. The $\mathbf{e}_i$ may now be relabeled again, if necessary, to make this nonzero coefficient $\beta_{22}$. Then a repetition of the argument just given, but this time applied to $\mathbf{e}_2$, will reduce (2–37) to the system

$$\mathbf{e}'_3 = \gamma_{33}\mathbf{e}_3 + \cdots + \gamma_{n3}\mathbf{e}_n + \gamma_{13}\mathbf{e}'_1 + \gamma_{23}\mathbf{e}'_2,$$
$$\cdot$$
$$\cdot$$
$$\cdot$$
$$\mathbf{e}'_m = \gamma_{3m}\mathbf{e}_3 + \cdots + \gamma_{nm}\mathbf{e}_n + \gamma_{1m}\mathbf{e}'_1 + \gamma_{2m}\mathbf{e}'_2.$$

Let us now speculate on the effect of our assumption that $m$ is greater than $n$. A moment's thought will reveal that by continuing the process of elimination begun above, we will eventually find ourselves confronted with a system of $m - n$ equations expressing each of the vectors $\mathbf{e}'_{n+1}, \ldots, \mathbf{e}'_m$ as a linear combination of $\mathbf{e}'_1, \ldots, \mathbf{e}'_n$. But this cannot be. Hence $m \leqslant n$ after all.   ∎

Now we are ready to prove that the number of vectors in every basis for a vector space is the same, a fact that we state as a theorem.

**Theorem 2–7**  *If $\mathscr{V}$ has a basis containing n vectors, then every basis for $\mathscr{V}$ contains n vectors.*

*Proof.*  If $\mathbf{e}_1, \ldots, \mathbf{e}_n$ and $\mathbf{e}'_1, \ldots, \mathbf{e}'_m$ are bases for $\mathscr{V}$, then Theorem 2–6 implies that $m \geqslant n$ and $m \leqslant n$. Hence $m = n$.   ∎

We have just seen that if a vector space $\mathscr{V}$ has a finite basis, then the number of vectors in that basis is an intrinsic property of $\mathscr{V}$. We call the number of basis elements the **dimension** of the space.

**Definition 2–8**  *A vector space is said to be **n-dimensional** if it has a basis containing n vectors, and **infinite dimensional** otherwise. We denote the fact that $\mathscr{V}$ is n-dimensional by writing*

$$\dim \mathscr{V} = n.$$

It follows from this definition and several of the examples given earlier that $\mathscr{R}^n$ and $\mathscr{P}_n$ are both $n$-dimensional, while $\mathscr{P}$ and the function space $\mathscr{C}^{(n)}(I)$ are infinite dimensional. In particular,

$\mathcal{R}^1$, the real line, is 1-dimensional,

$\mathcal{R}^2$, the Cartesian plane, is 2-dimensional, and

$\mathcal{R}^3$, the cartesian 3-space, is 3-dimensional,

just as our intuition demands. Also note that the convention adopted at the beginning of this section for the trivial space $\mathcal{O}$ implies that

$$\dim \mathcal{O} = 0.$$

Among the many consequences of Theorem 2–6 is a technique for constructing bases in a vector space. It goes as follows.

Let $\mathscr{V}$ be an $n$-dimensional vector space. If $n = 0$, the empty set is a basis for $\mathscr{V}$ and there is nothing to do. Otherwise, $n > 0$ and $\mathscr{V}$ contains nonzero vectors. Let $\mathbf{e}_1$ be one of them, and consider $\mathscr{S}(\mathbf{e}_1)$, the subspace spanned by $\mathbf{e}_1$. Either $\mathscr{S}(\mathbf{e}_1) = \mathscr{V}$, in which case $\mathbf{e}_1$ by itself is a basis for $\mathscr{V}$, or $\mathscr{S}(\mathbf{e}_1) \neq \mathscr{V}$. If the latter is true, then $\mathscr{V}$ contains a (nonzero) vector $\mathbf{e}_2$ that does not belong to $\mathscr{S}(\mathbf{e}_1)$.

Now consider the subspace $\mathscr{S}(\mathbf{e}_1, \mathbf{e}_2)$ spanned by the linearly independent vectors $\mathbf{e}_1$ and $\mathbf{e}_2$. Either $\mathscr{S}(\mathbf{e}_1, \mathbf{e}_2) = \mathscr{V}$, in which case $\mathbf{e}_1$ and $\mathbf{e}_2$ are a basis for $\mathscr{V}$, or there exists a (nonzero) vector $\mathbf{e}_3$ in $\mathscr{V}$ that does not belong to $\mathscr{S}(\mathbf{e}_1, \mathbf{e}_2)$. In the latter case, the vectors $\mathbf{e}_1, \mathbf{e}_2, \mathbf{e}_3$ are linearly independent, and we repeat the argument using the subspace $\mathscr{S}(\mathbf{e}_1, \mathbf{e}_2, \mathbf{e}_3)$.

Finally, since $\dim \mathscr{V} = n$ and since an $n$-dimensional vector space cannot contain more than $n$ linearly independent vectors (this is where Theorem 2–6 comes in), the process must end after $n$ steps with a basis for $\mathscr{V}$.

In addition to showing how a basis can be constructed in a finite-dimensional vector space, this argument establishes the fact that any linearly independent set of vectors in a finite-dimensional vector space that is not already a basis can be enlarged, or **extended** to a basis for the space. Since this result is often used in proving theorems about finite-dimensional vector spaces, we state it formally as a theorem.

**Theorem 2–8**    *Let $\mathbf{e}_1, \ldots, \mathbf{e}_m$ be linearly independent vectors in an $n$-dimensional vector space $\mathscr{V}$, and suppose that $m < n$. Then there exist $n - m$ vectors $\mathbf{e}_{m+1}, \ldots, \mathbf{e}_n$ in $\mathscr{V}$ such that*

$$\mathbf{e}_1, \ldots, \mathbf{e}_m, \mathbf{e}_{m+1}, \ldots, \mathbf{e}_n$$

*is a basis for $\mathscr{V}$.*

Finally, we observe that Theorem 2–6 also allows us to prove the reassuring fact that *every subspace $\mathscr{W}$ of a finite-dimensional vector space $\mathscr{V}$ is finite-dimensional, and*

$$\dim \mathscr{W} \leqslant \dim \mathscr{V}.$$

The proof is left as an exercise.

**EXERCISES**

1. What is the dimension of the subspace of $\mathcal{R}^3$ spanned by
   a) the vectors $(2, 1, -1), (3, 2, 1), (1, 0, -3)$?
   b) the vectors $(1, -1, 2), (0, 2, 1), (-1, 0, 1)$?

2. What is the dimension of the subspace of $\mathcal{R}^4$ spanned by
   a) the vectors $(1, 0, 2, -1), (3, -1, -2, 0), (1, -1, -6, 2), (0, 1, 8, -3)$?
   b) the vectors $(-\frac{1}{2}, \frac{1}{2}, 3, -1), (\frac{1}{2}, 0, 1, -\frac{1}{2}), (1, 1, 10, -4)$?

3. Let $\mathscr{W}$ be the set of all polynomials in $\mathscr{P}_n$ whose second derivative is zero; i.e., $\mathbf{p}(x)$ belongs to $\mathscr{W}$ if and only if $(d^2/dx^2)\mathbf{p}(x) = 0$. *show 2 closures basis is*
   a) Prove that $\mathscr{W}$ is a subspace of $\mathscr{P}_n$, and find a basis for $\mathscr{W}$. $\{1, x\}$
   b) Extend the basis for $\mathscr{W}$ found in (a) to a basis for $\mathscr{P}_n$.

4. Let $\mathscr{W}$ be the set of all polynomials $\mathbf{p}(x)$ in $\mathscr{P}_n$ such that

$$\mathbf{p}(1) = \mathbf{p}'(1) = 0.$$

   a) Prove that $\mathscr{W}$ is a subspace of $\mathscr{P}_n$, and find a basis for $\mathscr{W}$.
   b) Extend the basis for $\mathscr{W}$ found in (a) to a basis for $\mathscr{P}_n$.

5. a) Find the dimension of the subspace of $\mathscr{C}[-\pi, \pi]$ spanned by the vectors $1$, $\sin x$, $\sin^2 x$, $\cos^2 x$.
   b) Repeat part (a) for the vectors $\sin x \cos x$, $\sin 2x$, $\cos 2x$, $\sin^2 x$, $\cos^2 x$.

6. Prove that the vector space $\mathscr{C}[a, b]$ is infinite-dimensional.

7. What is the dimension of the subspace of all solutions (in $\mathscr{R}^3$) of the single linear equation

$$a_1 x_1 + a_2 x_2 + a_3 x_3 = 0?$$                                                   *2*

8. What is the dimension of the subspace of all solutions (in $\mathscr{R}^n$) of the single linear equation

$$a_1 x_1 + a_2 x_2 + \cdots + a_n x_n = 0?$$                         *n − 1*

9. Given the vectors

$$\mathbf{x}_1 = (2, 0, 1, 1) \qquad \text{and} \qquad \mathbf{x}_2 = (1, 1, 0, 3)$$

   in $\mathscr{R}^4$, find vectors $\mathbf{x}_3$ and $\mathbf{x}_4$ such that $\mathbf{x}_1, \mathbf{x}_2, \mathbf{x}_3, \mathbf{x}_4$ form a basis for $\mathscr{R}^4$.

10. Prove that every subspace $\mathscr{W}$ of a finite-dimensional vector space $\mathscr{V}$ is finite dimensional, with dim $\mathscr{W} \leqslant$ dim $\mathscr{V}$.

## 2–10  DETERMINANTS

In view of the results obtained in the preceding sections, it is clear that questions concerning linear independence and dependence of sets of vectors are of fundamental importance in linear algebra. We have already devised two tests for linear independence, one that can be applied in any vector space and another that is specifically tailored to $\mathscr{C}(I)$. In this section we shall discuss a third test, this time for sets of $n$ vectors in $\mathscr{R}^n$. The test is a numerical computation that is based on the vectors' components with respect to the standard basis $\mathbf{e}_1, \ldots, \mathbf{e}_n$.

It is not difficult to see that

$$a_1\mathbf{e}_1 + a_2\mathbf{e}_2 \quad \text{and} \quad b_1\mathbf{e}_1 + b_2\mathbf{e}_2$$

are linearly independent in $\mathscr{R}^2$ if and only if

$$a_1b_2 - a_2b_1 \neq 0 \tag{2-38}$$

(Exercise 6 of Section 2–5), and that

$$a_1\mathbf{e}_1 + a_2\mathbf{e}_2 + a_3\mathbf{e}_3, \quad b_1\mathbf{e}_1 + b_2\mathbf{e}_2 + b_3\mathbf{e}_3, \quad c_1\mathbf{e}_1 + c_2\mathbf{e}_2 + c_3\mathbf{e}_3$$

are linearly independent in $\mathscr{R}^3$ if and only if

$$a_1(b_2c_3 - b_3c_2) - a_2(b_1c_3 - b_3c_1) + a_3(b_1c_2 - b_2c_1) \neq 0. \tag{2-39}$$

The expressions on the left-hand side in (2–38) and (2–39) are examples of mathematical objects called "determinants." Determinants are extremely useful computational tools in dealing with a wide variety of problems in $\mathscr{R}^n$, and there is an extensive theory about them. A systematic exposition of this theory (both assertions and proofs) can be found in any standard text on linear algebra. We therefore omit it here. We shall, however, present those portions of this theory that will be of immediate value in our study of differential equations.

What is a determinant? The following definition provides the answer.

**Definition 2–9**    *A real-valued function*

$$\text{Det } (\mathbf{a}_1, \ldots, \mathbf{a}_n) \tag{2-38}$$

*defined on ordered n-tuples of vectors $\mathbf{a}_1, \ldots, \mathbf{a}_n$ of $\mathscr{R}^n$ is called an **$n \times n$ determinant**, or **determinant of order $n$**, if and only if it satisfies the following three conditions.*

1. *The function is linear in each of its variables, by which we mean that*

$$\text{Det } (\mathbf{a}_1, \ldots, \alpha\mathbf{a}_i + \beta\mathbf{a}_i', \ldots, \mathbf{a}_n) = \alpha \text{ Det } (\mathbf{a}_1, \ldots, \mathbf{a}_i, \ldots, \mathbf{a}_n)$$
$$+ \beta \text{ Det } (\mathbf{a}_1, \ldots, \mathbf{a}_i', \ldots, \mathbf{a}_n) \tag{2-40}$$

*for all scalars $\alpha$, $\beta$ and all vectors $\mathbf{a}_1, \ldots, \mathbf{a}_i, \mathbf{a}_i', \ldots, \mathbf{a}_n$ in $\mathscr{R}^n$.*

2. *If $\mathbf{a}_i = \mathbf{a}_j$ for $i \neq j$, then*

$$\text{Det } (\mathbf{a}_1, \ldots, \mathbf{a}_n) = 0. \tag{2-41}$$

3. *If $\mathbf{e}_1, \ldots, \mathbf{e}_n$ are the standard basis vectors in $\mathscr{R}^n$, taken in their natural order, then*

$$\text{Det } (\mathbf{e}_1, \ldots, \mathbf{e}_n) = 1. \tag{2-42}$$

Although it is far from obvious, these three conditions specify a **unique** function on $\mathscr{R}^n$ for each $n$. The following examples suggest why this is so.

**Example 1**  When $n = 1$ we are concerned with a real-valued function of a single vector $\mathbf{a} = a\mathbf{e}_1$ in $\mathscr{R}^1$, and the definition yields

$$\text{Det } (\mathbf{a}) = \text{Det } (a\mathbf{e}_1)$$
$$= a \text{ Det } (\mathbf{e}_1) \qquad \text{(by 2–40)}$$
$$= a. \qquad \text{(by 2–42)}$$

Hence in this case at least the determinant is unique.

**Example 2**  Let

$$\mathbf{a} = a_1\mathbf{e}_1 + a_2\mathbf{e}_2 \qquad \text{and} \qquad \mathbf{b} = b_1\mathbf{e}_1 + b_2\mathbf{e}_2$$

be vectors in $\mathscr{R}^2$. Then by using (2–40) twice, we have

$$\text{Det } (\mathbf{a}, \mathbf{b}) = \text{Det } (a_1\mathbf{e}_1 + a_2\mathbf{e}_2, b_1\mathbf{e}_1 + b_2\mathbf{e}_2)$$
$$= a_1 \text{ Det } (\mathbf{e}_1, b_1\mathbf{e}_1 + b_2\mathbf{e}_2) + a_2 \text{ Det } (\mathbf{e}_2, b_1\mathbf{e}_1 + b_2\mathbf{e}_2)$$
$$= a_1b_1 \text{ Det } (\mathbf{e}_1, \mathbf{e}_1) + a_1b_2 \text{ Det } (\mathbf{e}_1, \mathbf{e}_2) + a_2b_1 \text{ Det } (\mathbf{e}_2, \mathbf{e}_1) + a_2b_2 \text{ Det } (\mathbf{e}_2, \mathbf{e}_2).$$

But by (2–41)

$$\text{Det } (\mathbf{e}_1, \mathbf{e}_1) = \text{Det } (\mathbf{e}_2, \mathbf{e}_2) = 0,$$

and by (2–42)

$$\text{Det } (\mathbf{e}_1, \mathbf{e}_2) = 1.$$

Moreover, an easy argument based on Definition 2–9 reveals that

$$\text{Det } (\mathbf{e}_2, \mathbf{e}_1) = -1.$$

(See Exercise 6.) Hence

$$\text{Det } (\mathbf{a}, \mathbf{b}) = a_1b_2 - a_2b_1, \qquad (2\text{–}43)$$

which is precisely the expression in (2–38). This argument shows that *if* a determinant function exists on $\mathscr{R}^2$, it must be given by (2–43). We must still verify, however, that the function defined by this formula has the properties required of a determinant of order 2, a verification that also has been left as an exercise.

A similar argument applied to three vectors

$$\mathbf{a} = a_1\mathbf{e}_1 + a_2\mathbf{e}_2 + a_3\mathbf{e}_3,$$
$$\mathbf{b} = b_1e_1 + b_2\mathbf{e}_2 + b_3\mathbf{e}_3,$$
$$\mathbf{c} = c_1\mathbf{e}_1 + c_2\mathbf{e}_2 + c_3\mathbf{e}_3$$

would lead to

$$\text{Det } (\mathbf{a}, \mathbf{b}, \mathbf{c}) = a_1(b_2c_3 - b_3c_2) - a_2(b_1c_3 - b_3c_1) + a_3(b_1c_2 - b_2c_1), \qquad (2\text{–}44)$$

which is the expression in (2–39), and analogous formulas could be deduced for determinants of still higher orders.

Unfortunately, the simple Det $(\mathbf{a}, \mathbf{b}, \mathbf{c})$ notation does not display the standard-basis components of $\mathbf{a}$, $\mathbf{b}$ and $\mathbf{c}$ from which the value of the determinant has to be computed. We therefore shift to a notation that does:

$$\text{Det } (\mathbf{a}, \mathbf{b}, \mathbf{c}) = \begin{vmatrix} a_1 & a_2 & a_3 \\ b_1 & b_2 & b_3 \\ c_1 & c_2 & c_3 \end{vmatrix}.$$

This kind of notation is also used for determinants of other orders. Thus, if

$$\mathbf{a}_1 = a_{11}\mathbf{e}_1 + a_{12}\mathbf{e}_2 + \cdots + a_{1n}\mathbf{e}_n,$$
$$\mathbf{a}_2 = a_{21}\mathbf{e}_1 + a_{22}\mathbf{e}_2 + \cdots + a_{2n}\mathbf{e}_n,$$
$$\cdot$$
$$\cdot$$
$$\cdot$$
$$\mathbf{a}_n = a_{n1}\mathbf{e}_1 + a_{n2}\mathbf{e}_2 + \cdots + a_{nn}\mathbf{e}_n,$$

then the $n$th-order determinant Det $(\mathbf{a}_1, \ldots, \mathbf{a}_n)$ is denoted by

$$\begin{vmatrix} a_{11} & a_{12} \cdots a_{1n} \\ a_{21} & a_{22} \cdots a_{2n} \\ \cdot & \cdot & \cdot \\ \cdot & \cdot & \cdot \\ \cdot & \cdot & \cdot \\ a_{n1} & a_{n2} \cdots a_{nn} \end{vmatrix}$$

Our first important observation about determinants is that (2–44) can be rewritten in terms of $2 \times 2$ determinants:

$$\begin{vmatrix} a_1 & a_2 & a_3 \\ b_1 & b_2 & b_3 \\ c_1 & c_2 & c_3 \end{vmatrix} = a_1 \begin{vmatrix} b_2 & b_3 \\ c_2 & c_3 \end{vmatrix} - a_2 \begin{vmatrix} b_1 & b_3 \\ c_1 & c_3 \end{vmatrix} + a_3 \begin{vmatrix} b_1 & b_2 \\ c_1 & c_2 \end{vmatrix} \tag{2–45}$$

It can also be written as

$$\begin{vmatrix} a_1 & a_2 & a_3 \\ b_1 & b_2 & b_3 \\ c_1 & c_2 & c_3 \end{vmatrix} = a_1 \begin{vmatrix} b_2 & b_3 \\ c_2 & c_3 \end{vmatrix} - b_1 \begin{vmatrix} a_2 & a_3 \\ c_2 & c_3 \end{vmatrix} + c_1 \begin{vmatrix} a_2 & a_3 \\ b_2 & b_3 \end{vmatrix} \tag{2–46}$$

Still other possibilities exist.

Note that each of the $2 \times 2$ determinants appearing in these formulas can be obtained by deleting from the $3 \times 3$ determinant the row and column containing the coefficient of that $2 \times 2$ determinant. Also note that the coefficients in these formulas have alternating signs: $+$, $-$, $+$.

Analogous formulas can be established for determinants of all orders greater than one. For example, a fourth-order determinant can be evaluated by expressing it in terms of $3 \times 3$ determinants, each of which may then be evaluated by one of the formulas just given:

$$
\begin{vmatrix} a_1 & a_2 & a_3 & a_4 \\ b_1 & b_2 & b_3 & b_4 \\ c_1 & c_2 & c_3 & c_4 \\ d_1 & d_2 & d_3 & d_4 \end{vmatrix} = a_1 \begin{vmatrix} b_2 & b_3 & b_4 \\ c_2 & c_3 & c_4 \\ d_2 & d_3 & d_4 \end{vmatrix} - a_2 \begin{vmatrix} b_1 & b_3 & b_4 \\ c_1 & c_3 & c_4 \\ d_1 & d_3 & d_4 \end{vmatrix}
$$

$$
+ a_3 \begin{vmatrix} b_1 & b_2 & b_4 \\ c_1 & c_2 & c_4 \\ d_1 & d_2 & d_4 \end{vmatrix} - a_4 \begin{vmatrix} b_1 & b_2 & b_3 \\ c_1 & c_2 & c_3 \\ d_1 & d_2 & d_3 \end{vmatrix}.
$$

Again there is a variety of other formulas of this type.

**Example 3**   When the $2 \times 2$ determinants appearing in either (2–45) or (2–46) are evaluated, and the resulting expression is simplified, the formula for the $3 \times 3$ determinant becomes

$$
\begin{vmatrix} a_1 & a_2 & a_3 \\ b_1 & b_2 & b_3 \\ c_1 & c_2 & c_3 \end{vmatrix} = a_1 b_2 c_3 + a_2 b_3 c_1 + a_3 b_1 c_2 - a_3 b_2 c_1 - a_2 b_1 c_3 - a_1 b_3 c_2.
$$

Thus, for example,

$$
\begin{vmatrix} 2 & -1 & 3 \\ 0 & 1 & -6 \\ 4 & 2 & 7 \end{vmatrix} = 2 \cdot 1 \cdot 7 + (-1)(-6) \cdot 4 + 3 \cdot 0 \cdot 2 - 3 \ \ 1 \cdot 4 - (-1) \cdot 0 \cdot 7 - 2(-6) \cdot 2
$$

$$
= 14 + 24 - 12 + 24
$$

$$
= 50.
$$

The importance of determinants for our work stems from the following two theorems.

**Theorem 2–9**   *The vectors*

$$
a_{11}\mathbf{e}_1 + a_{12}\mathbf{e}_2 + \cdots + a_{1n}\mathbf{e}_n,
$$

.

.

.

$$
a_{n1}\mathbf{e}_1 + a_{n2}\mathbf{e}_2 + \cdots + a_{nn}\mathbf{e}_n
$$

*are linearly independent in $\mathscr{R}^n$ if and only if*

$$
\begin{vmatrix} a_{11} & a_{12} \cdots a_{1n} \\ \cdot & \cdot & \cdot \\ \cdot & \cdot & \cdot \\ \cdot & \cdot & \cdot \\ a_{n1} & a_{n2} \cdots a_{nn} \end{vmatrix} \neq 0.
$$

***Theorem 2–10***    *A system of n linear equations in n unknowns,*

$$a_{11}x_1 + a_{12}x_2 + \cdots + a_{1n}x_n = b_1,$$
$$a_{21}x_1 + a_{22}x_2 + \cdots + a_{2n}x_n = b_2,$$
$$\cdot$$
$$\cdot$$
$$\cdot$$
$$a_{n1}x_1 + a_{n2}x_2 + \cdots + a_{nn}x_n = b_n,$$

*has a unique solution if and only if the determinant formed from the coefficients of the system is different from zero; that is, if and only if*

$$\begin{vmatrix} a_{11} & a_{12} \cdots a_{1n} \\ a_{21} & a_{22} \cdots a_{2n} \\ \cdot & \cdot \quad\quad \cdot \\ \cdot & \cdot \quad\quad \cdot \\ \cdot & \cdot \quad\quad \cdot \\ a_{n1} & a_{n2} \cdots a_{nn} \end{vmatrix} \neq 0.$$

Proofs of both of these theorems can be found in any text on linear algebra.

**Example 4**    The vectors

$$\mathbf{a} = 2\mathbf{e}_1 - \mathbf{e}_2 + \mathbf{e}_3,$$
$$\mathbf{b} = \mathbf{e}_1 + 3\mathbf{e}_2 - \mathbf{e}_3,$$
$$\mathbf{c} = -\mathbf{e}_1 - \mathbf{e}_2 + 2\mathbf{e}_3$$

are linearly independent in $\mathscr{R}^3$ because

$$\begin{vmatrix} 2 & -1 & 1 \\ 1 & 3 & -1 \\ -1 & -1 & 2 \end{vmatrix} = 12 + 1 - 1 + 3 - 2 + 2 = 15.$$

Therefore, the system of equations

$$2x - y + z = b_1,$$
$$x + 3y - z = b_2,$$
$$-x - y + 2z = b_3$$

has a unique solution for every triple of real numbers $b_1$, $b_2$ and $b_3$.

**EXERCISES**

1. Evaluate the following determinants.

a) $\begin{vmatrix} 2 & 1 \\ 3 & -1 \end{vmatrix}$    b) $\begin{vmatrix} 5 & -10 \\ 1 & -2 \end{vmatrix}$

c) $\begin{vmatrix} 2 & 1 & -1 \\ 3 & -1 & 2 \\ 1 & 0 & 1 \end{vmatrix}$    d) $\begin{vmatrix} 4 & 2 & 1 \\ 1 & -1 & -3 \\ 2 & 4 & 7 \end{vmatrix}$

e)
$$\begin{vmatrix} 5 & 1 & -2 \\ 1 & 0 & 5 \\ -1 & 1 & 3 \end{vmatrix}$$
f)
$$\begin{vmatrix} 4 & 1 & -3 \\ 0 & 2 & 5 \\ 0 & 0 & 5 \end{vmatrix}$$

2. Evaluate:

a)
$$\begin{vmatrix} 1 & 3 & 0 & 5 \\ 2 & -1 & 4 & 1 \\ -2 & 3 & 1 & 2 \\ 3 & 5 & 1 & -1 \end{vmatrix}$$
b)
$$\begin{vmatrix} -2 & 2 & 1 & 3 \\ 0 & 1 & 2 & -1 \\ 4 & -3 & 1 & 3 \\ 2 & 1 & 1 & 5 \end{vmatrix}$$

3. Which of the following sets of vectors are linearly independent in $\mathscr{R}^3$?

a) $\mathbf{a} = \quad 2\mathbf{e}_1 - \quad \mathbf{e}_2 + 2\mathbf{e}_3$
   $\mathbf{b} = \quad -\mathbf{e}_1 + \quad \mathbf{e}_2 - 3\mathbf{e}_3$
   $\mathbf{c} = -2\mathbf{e}_1 + 2\mathbf{e}_2 - \quad \mathbf{e}_3$

b) $\mathbf{a} = \quad 4\mathbf{e}_1 - 2\mathbf{e}_2 + 3\mathbf{e}_3$
   $\mathbf{b} = \quad 2\mathbf{e}_1 - 8\mathbf{e}_2 + 7\mathbf{e}_3$
   $\mathbf{c} = \quad \mathbf{e}_1 + 3\mathbf{e}_2 - 2\mathbf{e}_3$

c) $\mathbf{a} = -3\mathbf{e}_1 + 5\mathbf{e}_2 + 2\mathbf{e}_3$
   $\mathbf{b} = \quad 4\mathbf{e}_1 - 3\mathbf{e}_2 - 3\mathbf{e}_3$
   $\mathbf{c} = \quad 2\mathbf{e}_1 + 7\mathbf{e}_2 + \quad \mathbf{e}_3$

d) $\mathbf{a} = -2\mathbf{e}_1 + 5\mathbf{e}_2 - 6\mathbf{e}_3$
   $\mathbf{b} = \quad 5\mathbf{e}_1 + \quad \mathbf{e}_2 - 3\mathbf{e}_3$
   $\mathbf{c} = -4\mathbf{e}_1 - \quad \mathbf{e}_2 + 6\mathbf{e}_3$

4. Which of the following sets of vectors are linearly independent in $\mathscr{R}^4$?

a) $\mathbf{a} = \quad 5\mathbf{e}_1 - \quad \mathbf{e}_2 - \quad \mathbf{e}_3 + 2\mathbf{e}_4$
   $\mathbf{b} = \quad \mathbf{e}_1 + 2\mathbf{e}_2 \qquad\quad + \quad \mathbf{e}_4$
   $\mathbf{c} = -3\mathbf{e}_1 \qquad\quad + 2\mathbf{e}_3 - 3\mathbf{e}_4$
   $\mathbf{d} = \qquad\qquad \mathbf{e}_2 - 4\mathbf{e}_3 + 5\mathbf{e}_4$

b) $\mathbf{a} = \quad 3\mathbf{e}_1 - 3\mathbf{e}_2 + \quad \mathbf{e}_3 + \quad \mathbf{e}_4$
   $\mathbf{b} = \quad -\mathbf{e}_1 + 4\mathbf{e}_2 - 3\mathbf{e}_3 + \quad \mathbf{e}_4$
   $\mathbf{c} = -2\mathbf{e}_1 - \quad \mathbf{e}_2 + \quad \mathbf{e}_3 - 2\mathbf{e}_4$
   $\mathbf{d} = \quad -\mathbf{e}_1 + 4\mathbf{e}_2 - 4\mathbf{e}_3 + \quad \mathbf{e}_4$

c) $\mathbf{a} = \quad 4\mathbf{e}_1 - 2\mathbf{e}_2 + 5\mathbf{e}_3 + \quad \mathbf{e}_4$
   $\mathbf{b} = \quad 2\mathbf{e}_1 + 3\mathbf{e}_2 - 3\mathbf{e}_3 - 2\mathbf{e}_4$
   $\mathbf{c} = \quad 2\mathbf{e}_1 - 5\mathbf{e}_2 + 8\mathbf{e}_3 + 2\mathbf{e}_4$
   $\mathbf{d} = \quad 4\mathbf{e}_1 - 2\mathbf{e}_2 + 5\mathbf{e}_3 - \quad \mathbf{e}_4$

5. Show that the value of a $3 \times 3$ determinant does not change if its rows and columns are interchanged; i.e., show that

$$\begin{vmatrix} a_1 & a_2 & a_3 \\ b_1 & b_2 & b_3 \\ c_1 & c_2 & c_3 \end{vmatrix} = \begin{vmatrix} a_1 & b_1 & c_1 \\ a_2 & b_2 & c_2 \\ a_3 & b_3 & c_3 \end{vmatrix}.$$

*Remark.* This result holds for determinants of any order.

6. Prove that

$$\text{Det}(\mathbf{a}_1, \ldots, \mathbf{a}_i, \ldots, \mathbf{a}_j, \ldots, \mathbf{a}_n) = -\text{Det}(\mathbf{a}_1, \ldots, \mathbf{a}_j, \ldots, \mathbf{a}_i, \ldots, \mathbf{a}_n).$$

[*Hint*: Consider $\text{Det}(\mathbf{a}_1, \ldots, \mathbf{a}_i + \mathbf{a}_j, \ldots, \mathbf{a}_i + \mathbf{a}_j, \ldots, \mathbf{a}_n)$, and use (2–40) and (2–41).]

7. Verify that the formula

$$\text{Det}(\mathbf{a}, \mathbf{b}) = a_1 b_2 - a_2 b_1$$

satisfies the conditions of Definition 2–9 when

$$\mathbf{a} = a_1 \mathbf{e}_1 + a_2 \mathbf{e}_2 \quad \text{and} \quad \mathbf{b} = b_1 \mathbf{e}_1 + b_2 \mathbf{e}_2.$$

# Linear Transformations and Linear Differential Operators  3

## 3–1 LINEAR TRANSFORMATIONS

Up to this point our study of real vector spaces has been a modest generalization of ideas implicit in analytic geometry. Although such notions as linear independence and dependence, subspaces, and bases may have been unfamiliar, they actually add little to the knowledge of vector spaces taught in elementary geometry. But as soon as these ideas are used to study functions defined on vector spaces, new and important things begin to happen.

The most important functions that arise in the study of vector spaces are known as linear transformations. They are defined as follows.

**Definition 3–1**  *A* **linear transformation**, *or* **linear operator**, *from a vector space $\mathscr{V}_1$ to a vector space $\mathscr{V}_2$ is a function A which associates with each vector* **x** *in $\mathscr{V}_1$ a unique vector A(**x**) in $\mathscr{V}_2$ in such a way that*

$$A(\mathbf{x}_1 + \mathbf{x}_2) = A(\mathbf{x}_1) + A(\mathbf{x}_2) \tag{3–1}$$

*and*

$$A(\alpha\mathbf{x}) = \alpha A(\mathbf{x}). \tag{3–2}$$

*This holds for all vectors* $\mathbf{x}_1$, $\mathbf{x}_2$, **x** *in $\mathscr{V}_1$ and for all scalars $\alpha$.*

In other words, a linear transformation is a function or mapping from one vector space to another which sends sums into sums and scalar products into scalar products. These requirements are sometimes paraphrased by saying that a linear transformation is "compatible" with the algebraic operations of vector addition and scalar multiplication. It is this compatibility which accounts for the importance of such functions in linear algebra.

One consequence of Definition 3–1 is that a linear transformation *A always* maps the zero vector of $\mathscr{V}_1$ onto the zero vector of $\mathscr{V}_2$:

$$A(\mathbf{0}) = \mathbf{0}. \tag{3–3}$$

Another is that $A$ maps linear combinations onto linear combinations:

$$A(\alpha_1\mathbf{x}_1 + \cdots + \alpha_n\mathbf{x}_n) = \alpha_1 A(\mathbf{x}_1) + \cdots + \alpha_n A(\mathbf{x}_n). \qquad (3\text{–}4)$$

The first of these assertions can be established by setting $\alpha = 0$ in (3–2); the second by repeated use of (3–1) and (3–2) in the obvious fashion. In particular, when $n = 2$, (3–4) becomes

$$A(\alpha_1\mathbf{x}_1 + \alpha_2\mathbf{x}_2) = \alpha_1 A(\mathbf{x}_1) + \alpha_2 A(\mathbf{x}_2). \qquad (3\text{–}5)$$

By itself, this equation can be (and often is) taken as the definition of a linear transformation, since (3–1) and (3–2) are satisfied if and only if (3–5) is. (See Exercise 17.) From time to time we shall use this fact to prove that a function is a linear transformation.

If $A$ is a linear transformation from $\mathscr{V}_1$ to $\mathscr{V}_2$, we write $A\colon \mathscr{V}_1 \to \mathscr{V}_2$ (read "$A$ maps $\mathscr{V}_1$ into $\mathscr{V}_2$") and say that $\mathscr{V}_1$ is the **domain** of $A$. We say $\mathscr{V}_2$ is the **codomain** of $A$, and if $\mathbf{x}$ is a vector of $\mathscr{V}_1$ we call the vector $A(\mathbf{x})$ the **image** of $\mathbf{x}$ under $A$. The set of all image vectors $A(\mathbf{x})$ is called variously the image of $\mathscr{V}_1$ under $A$, the image of $\mathscr{V}_1$ in $\mathscr{V}_2$, or simply the image of $A$, and is denoted by $\mathscr{I}(A)$. The set $\mathscr{I}(A)$ need not be all of $\mathscr{V}_2$, a possibility that is acknowledged by saying that $A$ maps $\mathscr{V}_1$ into $\mathscr{V}_2$(Fig. 3–1). However, if $\mathscr{I}(A)$ *is* $\mathscr{V}_2$, we say that $A$ maps $\mathscr{V}_1$ **onto** $\mathscr{V}_2$.

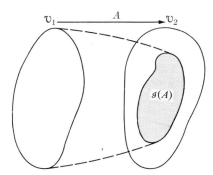

**Figure 3–1**

There is nothing in the definition of linear transformation to prevent $\mathscr{V}_1$ and $\mathscr{V}_2$ from being one and the same space. Indeed, this is one of the most fruitful settings in which to pursue the study of linear transformations. We sometimes alter our language when $\mathscr{V}_1$ and $\mathscr{V}_2$ are both the same space $\mathscr{V}$ by referring to $A$ as a linear transformation *on* $\mathscr{V}$, but this should not be construed as an assertion that $A$ maps $\mathscr{V}$ onto itself.

We conclude this section by giving a number of examples, several of which will figure prominently in our later work. For the most part we simply state the definitions of the functions in question and omit the verifications of linearity when they are routine.

**Example 1**   Let $\mathbf{x} = (x_1, x_2)$ be an arbitrary vector in $\mathscr{R}^2$ and set

$$A(\mathbf{x}) = (x_1, -x_2).$$

Geometrically, $A$ can be described as the linear transformation that maps $\mathscr{R}^2$ onto itself by reflection across the $x_1$-axis. (See Fig. 3–2 where, for generality, we show the effect of $A$ on an oblique coordinate system.)

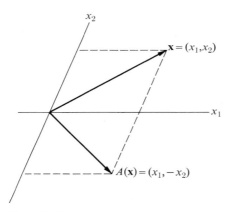

**Figure 3–2**

**Example 2**   Let $A$ be the mapping of $\mathscr{R}^2$ onto itself obtained by shearing the plane horizontally so that the $x_2$-axis is shifted through a 45° angle as shown in Fig. 3–3. Algebraically, $A$ is defined by the equation $A(x_1, x_2) = (x_1 + x_2, x_2)$ and is clearly linear.

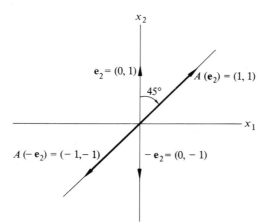

**Figure 3–3**

**Example 3**　Let $\mathscr{L}$ be any line through the origin in $\mathscr{R}^3$ and let $A$ be a fixed rotation about $\mathscr{L}$. Then, arguing geometrically, we can easily show that $A$ is a linear transformation of $\mathscr{R}^3$ onto itself.

**Example 4**　The mapping which sends each vector in $\mathscr{V}_1$ onto the zero vector in $\mathscr{V}_2$ is a linear transformation from $\mathscr{V}_1$ to $\mathscr{V}_2$ for all $\mathscr{V}_1$ and $\mathscr{V}_2$. It is called the **zero transformation** and is denoted by the symbol $O$, no matter which vector spaces are involved.

**Example 5**　A second linear transformation for which we reserve a special symbol is the **identity transformation** $I$ mapping a vector space $\mathscr{V}$ onto itself. The defining equation for $I$ is $I(\mathbf{x}) = \mathbf{x}$ for all $\mathbf{x}$ in $\mathscr{V}$. Its linearity is immediate.

**Example 6**　Consider the space $\mathscr{C}[a, b]$ of all real-valued, continuous functions on the interval $[a, b]$, and for each $\mathbf{f}$ in $\mathscr{C}[a, b]$ set

$$A(\mathbf{f}) = \int_a^x \mathbf{f}(t)\, dt, \qquad a \leqslant x \leqslant b.$$

Since $A(\mathbf{f})$ is continuous on $[a, b]$, $A$ can be viewed as a mapping of $\mathscr{C}[a, b]$ into itself. As such, it is linear since

$$A(\alpha_1 \mathbf{f}_1 + \alpha_2 \mathbf{f}_2) = \int_a^x [\alpha_1 \mathbf{f}_1(t) + \alpha_2 \mathbf{f}_2(t)]\, dt$$

$$= \int_a^x \alpha_1 \mathbf{f}_1(t)\, dt + \int_a^x \alpha_2 \mathbf{f}_2(t)\, dt$$

$$= \alpha_1 \int_a^x \mathbf{f}_1(t)\, dt + \alpha_2 \int_a^x \mathbf{f}_2(t)\, dt$$

$$= \alpha_1 A(\mathbf{f}_1) + \alpha_2 A(\mathbf{f}_2).$$

**Example 7**　For the same reasons as those just given, the mapping $A: \mathscr{C}[a, b] \to \mathscr{R}^1$ defined by

$$A(\mathbf{f}) = \int_a^b \mathbf{f}(x)\, dx$$

is also linear.

**Example 8**　Let $D$ denote the operation of differentiation on the space $\mathscr{C}^{(1)}(I)$; that is, $D(\mathbf{f}) = \mathbf{f}'$. Then the familiar identities

$$D(\mathbf{f}_1 + \mathbf{f}_2) = D(\mathbf{f}_1) + D(\mathbf{f}_2) \qquad \text{and} \qquad D(\alpha\mathbf{f}) = \alpha D(\mathbf{f})$$

imply that $D$ is a linear transformation from $\mathscr{C}^{(1)}(I)$ to $\mathscr{C}(I)$. More generally, the operation of taking the $n$th derivative is a linear transformation from the space of $n$-times continuously differentiable functions on an interval $I$ into the space $\mathscr{C}(I)$.

## EXERCISES

Prove that each of the following equations defines a linear transformation from $\mathscr{R}^2$ into (or onto) itself, and describe the effect of the transformation in geometric terms.

1.　$A(x_1, x_2) = -(x_1, x_2)$

2.　$A(x_1, x_2) = (2x_1, x_2)$

3.　$A(x_1, x_2) = 2(x_1, x_2/3)$

4.　$A(x_1, x_2) = 3(x_1, x_2)$

5.　$A(x_1, x_2) = \sqrt{2}(x_1 - x_2, x_1 + x_2)$

6.　$A(x_1, x_2) = (x_2, x_1)$

7.　$A(x_1, x_2) = -(x_2, x_1)$

8.　$A(x_1, x_2) = (x_1 + x_2, x_1 + x_2)$

9.　$A(x_1, x_2) = (0, 0)$

10.　$A(x_1, x_2) = (x_1 + x_2, 0)$

Which of the following equations defines a linear transformation on the space of polynomials $\mathscr{P}$?

11.  $A[\mathbf{p}(x)] = \mathbf{p}(x)^2$ 　　　　　　　12.  $A[\mathbf{p}(x)] = x(\mathbf{p}(x))$

13.  $A[\mathbf{p}(x)] = \mathbf{p}(x + 1) - \mathbf{p}(x)$ 　　14.  $A[\mathbf{p}(x)] = \mathbf{p}(x + 1) - \mathbf{p}(0)$

15.  $A[\mathbf{p}(x)] = \mathbf{p}''(x) - 2\mathbf{p}'(x)$ 　　　16.  $A[\mathbf{p}(x)] = \mathbf{p}(x^2)$

17.  Prove that $A: \mathscr{V}_1 \to \mathscr{V}_2$ is linear if and only if

$$A(\alpha_1 \mathbf{x}_1 + \alpha_2 \mathbf{x}_2) = \alpha_1 A(\mathbf{x}_1) + \alpha_2 A(\mathbf{x}_2)$$

for all $\mathbf{x}_1, \mathbf{x}_2$ in $\mathscr{V}_1$ and for all scalars $\alpha_1, \alpha_2$.

18.  Prove Eq. (3–4). [*Hint*: Use mathematical induction.]

19.  Let $\mathbf{e}_1, \ldots, \mathbf{e}_n$ be a basis for a finite-dimensional vector space $\mathscr{V}$, and for each index $i$, with $1 \leqslant i \leqslant n$, let $\eta_i$ be an arbitrary real number. Prove that the function $A: \mathscr{V} \to \mathscr{R}^1$ defined by

$$A(\alpha_1 \mathbf{e}_1 + \cdots + \alpha_n \mathbf{e}_n) = \alpha_1 \eta_1 + \cdots + \alpha_n \eta_n$$

for each vector $\alpha_1 \mathbf{e}_1 + \cdots + \alpha_n \mathbf{e}_n$ in $\mathscr{V}$ is a linear transformation.

20.  Let $\mathscr{V}_1$ be a finite-dimensional vector space with basis $\mathbf{e}_1, \ldots, \mathbf{e}_n$, let $\mathscr{V}_2$ be an arbitrary vector space, and let $\mathbf{y}_1, \ldots, \mathbf{y}_n$ be vectors in $\mathscr{V}_2$. For each $\mathbf{x} = \alpha_1 \mathbf{e}_1 + \cdots + \alpha_n \mathbf{e}_n$ in $\mathscr{V}_1$ set

$$A(\mathbf{x}) = \alpha_1 \mathbf{y}_1 + \cdots + \alpha_n \mathbf{y}_n.$$

Prove that $A$ is a linear transformation from $\mathscr{V}_1$ to $\mathscr{V}_2$.

21.  Find all the linear transformations that map $\mathscr{R}^1$ into (or onto) itself. [*Hint*: 1 is a basis for $\mathscr{R}^1$.]

## 3–2  ADDITION AND SCALAR MULTIPLICATION OF TRANSFORMATIONS

We begin the systematic study of linear transformations by describing several ways in which new transformations can be formed from old ones. Of these the simplest is the addition of two transformations, both of which map a given vector space $\mathscr{V}_1$ into the *same* space $\mathscr{V}_2$.

> **Definition 3–2**  Let $A$ and $B$ be linear transformations from $\mathscr{V}_1$ to $\mathscr{V}_2$. Then their **sum**, $A + B$, is the transformation from $\mathscr{V}_1$ to $\mathscr{V}_2$ defined for all $\mathbf{x}$ in $\mathscr{V}_1$ by the equation
>
> $$(A + B)(\mathbf{x}) = A(\mathbf{x}) + B(\mathbf{x}). \qquad (3–6)$$

This is of course just the familiar addition of functions now applied to linear transformations, and it is not difficult to show that $A + B$ is linear. Indeed, if $\mathbf{x}_1$ and $\mathbf{x}_2$ belong to $\mathscr{V}_1$ and $\alpha_1$ and $\alpha_2$ are scalars, then

$$\begin{aligned}
(A + B)(\alpha_1 \mathbf{x}_1 + \alpha_2 \mathbf{x}_2) &= A(\alpha_1 \mathbf{x}_1 + \alpha_2 \mathbf{x}_2) + B(\alpha_1 \mathbf{x}_1 + \alpha_2 \mathbf{x}_2) \\
&= \alpha_1 A(\mathbf{x}_1) + \alpha_2 A(\mathbf{x}_2) + \alpha_1 B(\mathbf{x}_1) + \alpha_2 B(\mathbf{x}_2) \\
&= \alpha_1 [A(\mathbf{x}_1) + B(\mathbf{x}_1)] + \alpha_2 [A(\mathbf{x}_2) + B(\mathbf{x}_2)] \\
&= \alpha_1 (A + B)(\mathbf{x}_1) + \alpha_2 (A + B)(\mathbf{x}_2).
\end{aligned}$$

Thus $A + B$ satisfies Eq. (3–5) and is therefore linear, as asserted.

**Example 1**    Let $D$ and $D^2$, respectively, denote the operations of taking first and second derivatives in $\mathscr{C}^{(2)}(I)$. Then the sum $D^2 + D$ is the linear transformation from $\mathscr{C}^{(2)}(I)$ to $\mathscr{C}(I)$ which sends each function $\mathbf{y}$ in $\mathscr{C}^{(2)}(I)$ onto the continuous function $\mathbf{y}'' + \mathbf{y}'$; that is,

$$(D^2 + D)(\mathbf{y}) = D^2\mathbf{y} + D\mathbf{y}.$$

**Example 2**    Let $\mathbf{K}(t)$ be a fixed function in $\mathscr{C}[a, b]$, let $I$ be the identity transformation on $\mathscr{C}[a, b]$, and let $A$ be the linear transformation mapping $\mathscr{C}[a, b]$ into itself given by

$$A(\mathbf{f}) = \int_a^x \mathbf{K}(t)\mathbf{f}(t)\, dt, \qquad a \leqslant x \leqslant b.$$

Then the sum $A + I$ is the linear transformation from $\mathscr{C}[a, b]$ into itself whose defining equation is

$$(A + I)(\mathbf{f}) = \int_a^x \mathbf{K}(t)\mathbf{f}(t)\, dt + \mathbf{f}.$$

The addition of linear transformations has a number of familiar and suggestive properties. In the first place,

$$A + (B + C) = (A + B) + C \tag{3-7}$$

and

$$A + B = B + A \tag{3-8}$$

whenever $A$, $B$, and $C$ are linear transformations from $\mathscr{V}_1$ to $\mathscr{V}_2$ (see Exercise 2). Secondly, the zero mapping from $\mathscr{V}_1$ to $\mathscr{V}_2$ defined in Example 4 of the preceding section acts as a "zero" for this addition since

$$A + O = O + A = A \tag{3-9}$$

for all $A : \mathscr{V}_1 \to \mathscr{V}_2$. Finally, if $A$ is any linear transformation from $\mathscr{V}_1$ to $\mathscr{V}_2$, and if we define $-A$ by the equation

$$(-A)(\mathbf{x}) = -A(\mathbf{x}) \tag{3-10}$$

for all $\mathbf{x}$ in $\mathscr{V}_1$, then we obtain a linear transformation from $\mathscr{V}_1$ to $\mathscr{V}_2$ with the property that

$$A + (-A) = (-A) + A = O. \tag{3-11}$$

In short, the addition of linear transformations from $\mathscr{V}_1$ to $\mathscr{V}_2$ satisfies all of the axioms postulated for addition in a vector space.

To complete this sequence of ideas, we now introduce a scalar multiplication on the set of linear transformations from $\mathscr{V}_1$ to $\mathscr{V}_2$.

**Definition 3–3**    *The **product** of a real number $\alpha$ and a linear transformation $A : \mathscr{V}_1 \to \mathscr{V}_2$ is the mapping $\alpha A$ from $\mathscr{V}_1$ to $\mathscr{V}_2$ defined by*

$$(\alpha A)(\mathbf{x}) = \alpha A(\mathbf{x}) \tag{3-12}$$

*for all $\mathbf{x}$ in $\mathscr{V}_1$. In other words, $\alpha A$ is the function whose value at $\mathbf{x}$ is computed by forming the scalar product of $\alpha$ and $A(\mathbf{x})$.*

We omit the proof that $\alpha A$ is linear, as well as the easy sequence of arguments required to show that the remaining axioms in the definition of a real vector space are now satisfied. These results are summarized in the following theorem.

**Theorem 3–1**    *The set of linear transformations from $\mathscr{V}_1$ to $\mathscr{V}_2$ is itself a real vector space under the definitions of addition and scalar multiplication given in this section.*

## EXERCISES

1. Cite the relevant axiom or definition needed to justify each step in the proof of the linearity of $A + B$.

2. Prove that addition of linear transformations is associative and commutative.

3. Prove that the mapping $-A$ defined in (3–10) is linear, and that
$$A + (-A) = (-A) + A = O.$$

4. a) Prove that $\alpha A$ as defined in the text is linear.
   b) Prove Theorem 3–1.

5. a) Let $\mathscr{X}$ be a nonempty subset of a vector space $\mathscr{V}_1$, and let $\mathscr{A}(\mathscr{X})$ denote the set of all linear transformations $A$ from $\mathscr{V}_1$ to $\mathscr{V}_2$ with the property that $A(\mathbf{x}) = \mathbf{0}$ for all $\mathbf{x}$ in $\mathscr{X}$. Prove that $\mathscr{A}(\mathscr{X})$ is a subspace of the space of all linear transformations from $\mathscr{V}_1$ to $\mathscr{V}_2$.
   b) What is $\mathscr{A}(\mathscr{X})$ when $\mathscr{X}$ consists of just the zero vector? when $\mathscr{X} = \mathscr{V}_1$?
   c) Prove that $\mathscr{A}(\mathscr{X}) = \mathscr{A}(\mathscr{S}(\mathscr{X}))$.

\* 6. Let $\mathscr{V}_1$ and $\mathscr{V}_2$ be finite-dimensional vector spaces with bases $\mathbf{e}_1, \ldots, \mathbf{e}_n$ and $\mathbf{e}'_1, \ldots, \mathbf{e}'_m$, respectively. For each pair of integers $i$ and $j$, with $1 \leqslant i \leqslant n$ and $1 \leqslant j \leqslant m$, define $A_{ij}: \mathscr{V}_1 \to \mathscr{V}_2$ by first defining $A_{ij}$ on the basis vectors of $\mathscr{V}_1$ according to the formula

$$A_{ij}(\mathbf{e}_k) = \begin{cases} \mathbf{e}'_j & \text{if } k = i, \\ \mathbf{0} & \text{if } k \neq i, \end{cases}$$

and then using (3–4) to obtain the value of $A_{ij}$ for each $\mathbf{x}$ in $\mathscr{V}_1$. (See Exercise 20, Section 3–1.)
   a) Prove that the $A_{ij}$ are linear transformations from $\mathscr{V}_1$ to $\mathscr{V}_2$ and that they are linearly independent in the vector space of all linear transformations from $\mathscr{V}_1$ to $\mathscr{V}_2$.
   b) Prove that the $A_{ij}$ span the space of linear transformations from $\mathscr{V}_1$ to $\mathscr{V}_2$, and hence deduce that this space is finite-dimensional with dimension $mn$. [*Hint:* Two linear transformations from $\mathscr{V}_1$ to $\mathscr{V}_2$ are identical if and only if they coincide on a basis for $\mathscr{V}_1$.]

## 3–3 PRODUCTS OF LINEAR TRANSFORMATIONS

In this section we introduce still a third way of constructing new linear transformations from old ones: composition or, as it is more commonly called, multiplication. The construction goes as follows.

Let $A$ and $B$ be linear transformations, and suppose that

$$A: \mathscr{V}_1 \to \mathscr{V}_2 \quad \text{and} \quad B: \mathscr{V}_2 \to \mathscr{V}_3.$$

Then for each $\mathbf{x}$ in $\mathscr{V}_1$, $A(\mathbf{x})$ is a vector in $\mathscr{V}_2$, and we can therefore apply $B$ to $A(\mathbf{x})$ to obtain the vector $B(A(\mathbf{x}))$ in $\mathscr{V}_3$ as suggested in Fig. 3–4. Thus $A$ and $B$ can be combined to produce a function from $\mathscr{V}_1$ to $\mathscr{V}_3$ which we will denote by $BA$ and call the *product* of $A$ and $B$ *in that order*, namely first $A$, then $B$. This is the content of the next definition.

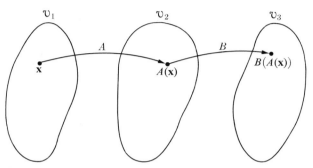

**Figure 3–4**

**Definition 3–3**   *If $A\colon \mathscr{V}_1 \to \mathscr{V}_2$ and $B\colon \mathscr{V}_2 \to \mathscr{V}_3$ are linear transformations, then their **product**, $BA$, is the mapping from $\mathscr{V}_1$ to $\mathscr{V}_3$ defined for all $\mathbf{x}$ in $\mathscr{V}_1$ by the equation*

$$BA(\mathbf{x}) = B(A(\mathbf{x})). \tag{3–13}$$

Notice that the product $BA$ exists only when the image of $A$ is contained in the domain of $B$, for only then does it make sense to speak of applying $B$ to the vector $A(\mathbf{x})$.

The crucial fact about products of linear transformations is that they too are linear.

**Theorem 3–2**   *The product of two linear transformations is itself a linear transformation.*

This theorem is an immediate consequence of Eq. (3–13), and we have left its proof as an exercise (Exercise 1).

At first sight it might seem more reasonable to denote the product of first $A$ then $B$ by $AB$ rather than $BA$. The explanation for not adopting this notation is simple. If it were used, (3–13) would have to be changed to read $AB(\mathbf{x}) = B(A(\mathbf{x}))$, and the writing of equations would then be an open invitation to error.

**Example 1**   Let $A\colon \mathscr{R}^3 \to \mathscr{R}^2$ be defined by

$$A(x_1, x_2, x_3) = (x_1, x_2),$$

and let $B\colon \mathscr{R}^2 \to \mathscr{R}^2$ be defined by

$$B(x_1, x_2) = (-x_1, -x_2).$$

Geometrically, $A$ projects $\mathcal{R}^3$ onto the $\mathbf{e}_1\mathbf{e}_2$- or $xy$-plane, and $B$ reflects $\mathcal{R}^2$ in the origin. Both $A$ and $B$ are linear, and $BA: \mathcal{R}^3 \to \mathcal{R}^2$ with

$$BA(x_1, x_2, x_3) = B(x_1, x_2) = (-x_1, -x_2).$$

On the other hand, the product $AB$ does not exist since $B(x)$ belongs to $\mathcal{R}^2$, while $A$ can be applied only to vectors in $\mathcal{R}^3$.

**Example 2**    Let $A$ and $B$ be the linear transformations mapping $\mathcal{R}^2$ onto itself defined by

$$A(x_1, x_2) = (-x_2, x_1), \qquad B(x_1, x_2) = (x_1, -x_2).$$

Geometrically, $A$ rotates $\mathcal{R}^2$ counterclockwise about the origin through 90°, and $B$ reflects $\mathcal{R}^2$ across the $\mathbf{e}_1$-axis. (See Fig. 3–5.) This time both products $BA$ and $AB$ are defined, but they are different transformations. Indeed,

$$BA(x_1, x_2) = B(A(x_1, x_2)) = B(-x_2, x_1) = (-x_2, -x_1),$$

while

$$AB(x_1, x_2) = A(B(x_1, x_2)) = A(x_1, -x_2) = (x_2, x_1),$$

so that $BA \neq AB$. For example, $BA(\mathbf{e}_1) = -\mathbf{e}_2$, while $AB(\mathbf{e}_1) = \mathbf{e}_2$. In short, *the multiplication of linear transformations is not commutative.*

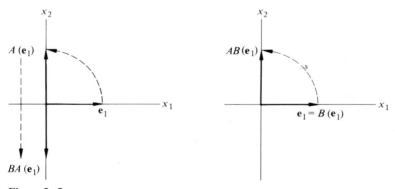

**Figure 3–5**

The preceding examples illustrated two ways in which the multiplication of linear transformations differs from "ordinary" multiplication. Why, then, call it multiplication at all? The answer lies in the following identities, which show that *most* of the properties usually associated with the term multiplication are still valid for products of linear transformations. Specifically, assuming that all of the indicated products are defined, we have

$$A(BC) = (AB)C, \tag{3-14}$$

$$(A_1 + A_2)B = A_1B + A_2B, \qquad A(B_1 + B_2) = AB_1 + AB_2, \tag{3-15}$$

$$(\alpha A)B = A(\alpha B) = \alpha(AB), \qquad \alpha \text{ a scalar,} \tag{3-16}$$

$$AI = A, \qquad IA = A, \qquad \text{where } I \text{ is the identity map.} \tag{3-17}$$

The first of these identities asserts that the multiplication of linear transformations is *associative*, the next two that it is *distributive* over addition, and the fourth that it *commutes* with the operation of scalar multiplication. Finally, (3–17) says that the identity transformation plays the same role in operator multiplication that the number one plays in arithmetic. Note, however, that two different identity maps are usually involved here. Strictly speaking, if $A: \mathscr{V}_1 \to \mathscr{V}_2$, then (3–17) ought to be written

$$AI_{\mathscr{V}_1} = A, \qquad I_{\mathscr{V}_2}A = A,$$

where $I_{\mathscr{V}_1}$ denotes the identity map on $\mathscr{V}_1$ and $I_{\mathscr{V}_2}$ the identity map on $\mathscr{V}_2$. But this notation is rarely used, since the meaning of the unidentified symbol $I$ is always clear from the context.

The proof of each of the preceding identities is a routine exercise in applying the definitions of the operations involved. For example, to establish (3–14), suppose that $C: \mathscr{V}_1 \to \mathscr{V}_2$, $B: \mathscr{V}_2 \to \mathscr{V}_3$, and $A: \mathscr{V}_3 \to \mathscr{V}_4$. Then each of the products $A(BC)$ and $(AB)C$ is a linear transformation from $\mathscr{V}_1$ to $\mathscr{V}_4$, and to prove their equality, we simply apply Definition 3–3 twice for each product:

$$[A(BC)](\mathbf{x}) = A(BC(\mathbf{x})) = A(B(C(\mathbf{x}))),$$
$$[(AB)C](\mathbf{x}) = AB(C(\mathbf{x})) = A(B(C(\mathbf{x})));$$

(3–14) now follows from the equality of the right-hand sides of these expressions. The remaining proofs have been left as exercises.

## EXERCISES

1. Prove Theorem 3–2.

2. Prove the distributivity formulas (3–15).

3. Find $AB$ and $BA$ for each of the following pairs of linear transformations on $\mathscr{R}^2$ or $\mathscr{R}^3$. Which of these pairs of transformations commute?
   a) $A(x_1, x_2) = (x_1, 0), B(x_1, x_2) = (0, x_2)$
   b) $A(x_1, x_2) = (x_1 + x_2, x_1 - x_2)$
       $B(x_1, x_2) = (x_1 - x_2, x_1 + x_2)$
   c) $A(x_1, x_2, x_3) = (x_2, x_1, x_1 + x_2 + x_3)$
       $B(x_1, x_2, x_3) = (x_2, x_1, -x_1 - x_2 - x_3)$
   d) $A(x_1, x_2, x_3) = (x_2, x_1 + x_3, x_1 + x_2 + x_3)$
       $B(x_1, x_2, x_3) = (x_1 + x_2 + x_3, 0, -x_1 - x_2 - x_3)$

4. Find $AB$ and $BA$ for each of the following pairs of linear transformations on the space of polynomials $\mathscr{P}$. What does each product do to the polynomial $x^2 - 2x + 1$?
   a) $A[\mathbf{p}(x)] = \mathbf{p}'(x)$      b) $A[\mathbf{p}(x)] = \mathbf{p}(x) + \mathbf{p}(0)$
       $B[\mathbf{p}(x)] = x\mathbf{p}(x)$        $B[\mathbf{p}(x)] = \mathbf{p}'(x)$
   c) $A[\mathbf{p}(x)] = \mathbf{p}(2x)$     d) $A[\mathbf{p}(x)] = \mathbf{p}(x + 1) - \mathbf{p}(x - 1)$
       $B[\mathbf{p}(x)] = \mathbf{p}'(x)$       $B[\mathbf{p}(x)] = \mathbf{p}(x - 1) - \mathbf{p}(x + 1)$

5  Let $D$ and $L$ denote the operations of differentiation and integration on $\mathscr{C}^{(2)}(I)$; that is,

$$Dy = \frac{d\mathbf{y}}{dx}, \qquad Ly = \int_a^x \mathbf{y}(t) \, dt.$$

Find $LD(\mathbf{y})$ and $DL(\mathbf{y})$ for any $\mathbf{y} = y(x)$ in $\mathscr{C}^{(2)}(I)$.

### 3-4  POLYNOMIALS IN A LINEAR TRANSFORMATION; LINEAR DIFFERENTIAL OPERATORS

If $A$ is a linear transformation on a vector space $\mathscr{V}$, by which we mean that $A \colon \mathscr{V} \to \mathscr{V}$, we can form repeated products of $A$ with itself to obtain the sequence of linear transformations on $\mathscr{V}$ known as the **powers** of $A$:

$$A^1 = A, \quad A^2 = AA, \quad A^3 = AA^2, \quad \ldots.$$

Moreover, it is customary to let $A^0$ denote the identity transformation on $\mathscr{V}$, by virtue of which $A^n$ is defined for all nonnegative integers $n$. It is clear that

$$A^m A^n = A^{m+n}, \qquad \text{and} \qquad (A^m)^n = A^{mn},$$

so that the usual rules for manipulating nonnegative exponents hold.

**Polynomials in $A$**   When the powers of $A \colon \mathscr{V} \to \mathscr{V}$ are combined by addition and scalar multiplication, we obtain polynomials in $A$. Thus, if

$$p(x) = a_n x^n + a_{n-1} x^{n-1} + \cdots + a_0$$

is a polynomial in $x$ with real coefficients, we define $p(A)$ to be the linear transformation on $\mathscr{V}$ obtained by substituting $A$ for $x$ in $p(x)$. Hence

$$p(A) = a_n A^n + a_{n-1} A^{n-1} + \cdots + a_0 I$$

or

$$p(A) = a_n A^n + a_{n-1} A^{n-1} + \cdots + a_0,$$

the factor $I$ being understood in the last term of this expression just as $x^0 = 1$ is understood in $p(x)$. It follows that if $\mathbf{x}$ is any vector in $\mathscr{V}$,

$$p(A)(\mathbf{x}) = a_n A^n(\mathbf{x}) + a_{n-1} A^{n-1}(\mathbf{x}) + \cdots + a_0 \mathbf{x}.$$

Multiplicatively, these polynomials obey all of the familiar rules of polynomial algebra, since the powers of $A$ commute with one another and are distributive over addition. In particular, *the multiplication of polynomials in a linear transformation is commutative*, since the identity

$$p(x)q(x) = q(x)p(x)$$

for ordinary polynomials implies that

$$p(A)q(A) = q(A)p(A).$$

This fact in turn allows us to factor polynomials in $A$ exactly as we do ordinary polynomials.

As an illustration, consider the space $\mathscr{C}^{(\infty)}(I)$ of all infinitely differentiable functions on $I$ and the differentiation operator $D$. Since $D$ maps $\mathscr{C}^{(\infty)}(I)$ into itself, we can construct polynomials in $D$, which are simply expressions of the form

$$a_n D^n + a_{n-1} D^{n-1} + \cdots + a_1 D + a_0, \tag{3–18}$$

where $a_0, \ldots, a_n$ are real numbers and $a_n \neq 0$. A linear transformation of this type is called an **$n$th-order constant coefficient linear differential operator**. Because of their algebraic properties, such operators play a special role in the study of differential equations, where they appear as linear transformations from $\mathscr{C}^{(n)}(I)$ to $\mathscr{C}(I)$ for any interval $I$.

**Example 1**   The polynomial
$$D^2 + D - 2$$

is a second-order constant coefficient linear differential operator, and if $y$ is any function in $\mathscr{C}^{(\infty)}(I)$ [or $\mathscr{C}^{(2)}(I)$], then

$$(D^2 + D - 2)y = D^2 y + Dy - 2y = y'' + y' - 2y.$$

This operator can also be written in factored form as $(D + 2)(D - 1)$ or $(D - 1)(D + 2)$, as the following computations show:

$$
\begin{aligned}
(D + 2)(D - 1)y &= (D + 2)[(D - 1)y] \\
&= (D + 2)(Dy - y) \\
&= D(Dy) - Dy + 2Dy - 2y \\
&= D^2 y + Dy - 2y \\
&= (D^2 + D - 2)y;
\end{aligned}
$$

thus
$$(D + 2)(D - 1) = D^2 + D - 2.$$

A similar calculation yields the equality
$$(D - 1)(D + 2) = D^2 + D - 2.$$

**Example 2**   Find the value of
$$(D - 1)(D + 2)y$$
when (a) $y = \sin x$, (b) $y = e^{-2x}$.

*Solution.*   a)  When $y = \sin x$,

$$
\begin{aligned}
(D - 1)(D + 2)y &= (D - 1)[(D + 2)\sin x] \\
&= (D - 1)(D \sin x + 2 \sin x) \\
&= (D - 1)(\cos x + 2 \sin x) \\
&= D(\cos x + 2 \sin x) - (\cos x + 2 \sin x) \\
&= \cos x - 3 \sin x.
\end{aligned}
$$

b) Similarly, when $y = e^{-2x}$,

$$(D - 1)(D + 2)y = (D - 1)[(D + 2)e^{-2x}]$$
$$= (D - 1)(De^{-2x} + 2e^{-2x})$$
$$= (D - 1)(-2e^{-2x} + 2e^{-2x})$$
$$= (D - 1)(0)$$
$$= 0.$$

The fact that $(D - 1)(D + 2)e^{-2x} = 0$ is described by saying that the operator $(D - 1)(D + 2)$ **annihilates** the function $y = e^{-2x}$. In other words, $y = e^{-2x}$ is a solution of the differential equation

$$y'' + y' - 2y = 0.$$

**Linear Differential Operators**  The coefficients in (3–18) need not be constants. Functions in $\mathscr{C}(I)$ can be used just as well, and the expression

$$a_n(x)D^n + a_{n-1}(x)D^{n-1} + \cdots + a_1(x)D + a_0(x), \qquad (3\text{--}19)$$

where $a_n(x)$ is not identically 0 on $I$, is called an **$n$th-order linear differential operator**. The linearity of $D$ and its powers implies that (3–19) is a linear transformation from $\mathscr{C}^{(n)}(I)$ to $\mathscr{C}(I)$. (See Exercise 2.) Typical examples of such operators are

$$xD + 1, \qquad D + x, \qquad x^2D^2 + 3xD - 1,$$

the first two of which are of first order, and the third of second order, on any interval of the real line. Operators like these can be multiplied, but, as the following example shows, *the multiplication is now noncommutative.*

**Example 3**  Find $AB$ and $BA$ given that

$$A = D + x \qquad \text{and} \qquad B = xD + 1.$$

*Solution.*  If $y$ belongs to $\mathscr{C}^{(2)}(I)$, then

$$AB(y) = (D + x)[(xD + 1)y]$$
$$= (D + x)(xDy + y)$$
$$= D(xDy + y) + x(xDy + y)$$
$$= D(xDy) + Dy + x^2Dy + xy,$$

and since $D(xDy) = xD^2y + Dy$, we have

$$AB(y) = xD^2y + 2Dy + x^2Dy + xy$$
$$= xD^2y + (x^2 + 2)Dy + xy$$
$$= [xD^2 + (x^2 + 2)D + x]y.$$

Hence

$$(D + x)(xD + 1) = xD^2 + (x^2 + 2)D + x,$$

and we see that the product *cannot* be computed by multiplying $D + x$ and $xD + 1$ as though they were ordinary polynomials in $x$ and $D$. This is one of the unpleasant consequences of a noncommutative multiplication, and it has a decisive effect on the study of differential equations.

Turning now to the product $BA$, we have

$$BA(y) = (xD + 1)[(D + x)y]$$
$$= (xD + 1)(Dy + xy)$$
$$= xD(Dy + xy) + Dy + xy$$
$$= xD^2y + xD(xy) + Dy + xy,$$

and since $D(xy) = xDy + y$, we have

$$BA(y) = xD^2y + x(xDy + y) + Dy + xy$$
$$= xD^2y + (x^2 + 1)Dy + 2xy$$
$$= [xD^2 + (x^2 + 1)D + 2x]y.$$

Hence

$$(xD + 1)(D + x) = xD^2 + (x^2 + 1)D + 2x,$$

and we see that

$$(D + x)(xD + 1) \neq (xD + 1)(D + x).$$

In view of the foregoing examples it is time to discuss the issue of vector notation for functions. All of the functions that we shall encounter in this text will belong to one of a number of vector spaces. Thus if we were inflexible in our use of boldface type, these functions would always be denoted by $\mathbf{f}$ or $\mathbf{f}(x)$. But the relentless use of boldface would produce such unsightly and confusing expressions as $x^n$, $\sin x$ and $d^2y/dx^2$. Such pedantry is pointless, so we shall use the symbols $f$ or $f(x)$ when we think the printed page or its reader would suffer from the use of boldface type.

**EXERCISES**

1. Write each of the following constant coefficient linear differential operators as a product of linear (i.e., first-order) factors.
   a) $D^2 - D - 12$      b) $D^3 + 6D^2 + 9D$
   c) $D^4 - 2D^2 + 1$      d) $D^3 + 4D^2 + D - 6$
   e) $D^2 + 2D - 5$ [*Hint*: Complete the square.]

2. Prove that (3–19) defines a linear transformation from $\mathscr{C}^{(n)}(I)$ to $\mathscr{C}(I)$.

3. Write each of the following products in the form $a_2(x)D^2 + a_1(x)D + a_0(x)$.
   a) $(xD + 1)^2$      b) $(2xD + 1)(D - 1)$
   c) $(D - 1)(2xD + 1)$      d) $(x^2D + 2x)(D - 2x)$
   e) $(D - 2x)(x^2D + 2x)$

4. a) Show that the operator $D - \alpha$, $\alpha$ a real number, annihilates the function $y = e^{\alpha x}$.

b) Show that the operator $(D - \alpha)^2$, $\alpha$ a real number, annihilates the functions $y = e^{\alpha x}$ and $y = xe^{\alpha x}$.

5. a) Show that the operator $(D^2 + \alpha^2)$, $\alpha$ a real number, annihilates the functions $y = \sin \alpha x$ and $y = \cos \alpha x$.

b) Show that the operator $(D^2 + \alpha^2)^2$, $\alpha$ a real number, annihilates the functions $y = \sin \alpha x$, $y = x \sin \alpha x$, $y = \cos \alpha x$, and $y = x \cos \alpha x$.

c) Show that the functions in (b) are linearly independent in $\mathscr{C}(I)$ for every interval $I$.

6. Show that each of the following operators annihilates the given function.

a) $D^2 - 2D + 2$,      $y = e^x \sin x$

b) $x^2 D^2 + 2xD - 6$,      $y = x^{-3}$

c) $(1 - x^2)D^2 - 2xD + 12$,      $y = \frac{5}{3}x^3 - x$

d) $x^2 D^2 + xD$,      $y = \ln |x|$,      $x \neq 0$

e) $x^2 D^2 + xD + (x^2 - \frac{1}{4})$,      $y = \sqrt{2/\pi x} \sin x$

7. Let $f$ and $g$ have derivatives of all orders. Prove that:

a) $D^2(fg) = f''g + 2f'g' + fg''$

b) $D^3(fg) = f'''g + 3f''g' + 3f'g'' + fg'''$

* c) $D^n(fg) = \displaystyle\sum_{k=0}^{n} \binom{n}{k}(D^{n-k}f)(D^k g)$, where

$$\binom{n}{k} = \frac{n!}{k!(n-k)!} = \frac{n(n-1)\cdots(n-k+1)}{k!}.$$

This result is known as **Leibnitz's rule**. (Use mathematical induction, and recall that $0! = 1$.)

8. Use the results of Exercise 7 to express each of the following linear differential operators in the form $a_n(x)D^n + \cdots + a_1(x)D + a_0(x)$.

a) $D^3(xD)$      b) $D^m(xD)$      c) $D^5(xD^2 + e^x)$

* 9. a) Prove that

$$(x^m D^m)x^k = k(k-1)\cdots(k - m + 1)x^k$$

for any real number $k$.

b) Show that if $a_0, a_1, a_2$ are constants, then

$$(a_2 x^2 D^2 + a_1 xD + a_0)x^k = [a_2 k(k-1) + a_1 k + a_0]x^k$$

for any real number $k$.

c) Show that $(xD)(x^3 D^3) = (x^3 D^3)(xD)$.

* 10. A linear differential operator is said to be an **equidimensional** or **Euler operator** if it can be written in the form

$$a_n x^n D^n + a_{n-1}x^{n-1}D^{n-1} + \cdots + a_1 xD + a_0,$$

where $a_0, \ldots, a_n$ are constants.

a) Find the value of $Lx^k$ when $L$ is equidimensional.

b) Prove that

$$(x^m D^m)(x^n D^n) = (x^n D^n)(x^m D^m)$$

for any pair of nonnegative integers $m$ and $n$, and hence deduce that the multiplication of equidimensional operators is commutative. [*Hint*: See Exercise 7.]

## 3–5 THE NULL SPACE AND IMAGE; INVERSES

Let $A$ be a linear transformation from $\mathscr{V}_1$ to $\mathscr{V}_2$, and let $\mathcal{N}(A)$ denote the set of all $\mathbf{x}$ in $\mathscr{V}_1$ such that $A(\mathbf{x}) = \mathbf{0}$. Then, as we have already observed, $\mathcal{N}(A)$ always contains the zero vector of $\mathscr{V}_1$. Actually we can say much more than this. For if $A(\mathbf{x}_1) = A(\mathbf{x}_2) = \mathbf{0}$, then

$$A(\alpha_1\mathbf{x}_1 + \alpha_2\mathbf{x}_2) = \alpha_1 A(\mathbf{x}_1) + \alpha_2 A(\mathbf{x}_2) = \mathbf{0}$$

for all scalars $\alpha_1, \alpha_2$, and it follows that $\mathcal{N}(A)$ *is a subspace of* $\mathscr{V}_1$. This subspace is called the **null space** or **kernel** of $A$ and is of fundamental importance in studying the behavior of $A$ on $\mathscr{V}_1$.

Of equal importance with the null space of $A$ is its image $\mathcal{I}(A)$, which, we recall, is the set of all $\mathbf{y}$ in $\mathscr{V}_2$ such that $\mathbf{y} = A(\mathbf{x})$ for some $\mathbf{x}$ in $\mathscr{V}_1$. It too is a subspace—this time of $\mathscr{V}_2$—since if $\mathbf{y}_1$ and $\mathbf{y}_2$ belong to $\mathcal{I}(A)$ with $\mathbf{y}_1 = A(\mathbf{x}_1)$, $\mathbf{y}_2 = A(\mathbf{x}_2)$, then

$$A(\alpha_1\mathbf{x}_1 + \alpha_2\mathbf{x}_2) = \alpha_1 A(\mathbf{x}_1) + \alpha_2 A(\mathbf{x}_2) = \alpha_1\mathbf{y}_1 + \alpha_2\mathbf{y}_2.$$

Hence $\alpha_1\mathbf{y}_1 + \alpha_2\mathbf{y}_2$ is also in the image of $A$, as required.

**Example 1**   Let $I: \mathscr{V} \to \mathscr{V}$ be the identity transformation. Then $\mathcal{N}(I) = \mathcal{O}$, the trivial subspace of $\mathscr{V}$, while $\mathcal{I}(I) = \mathscr{V}$.

**Example 2**   If $O: \mathscr{V}_1 \to \mathscr{V}_2$ is the zero transformation, then, by its very definition, $\mathcal{N}(O) = \mathscr{V}_1$ and $\mathcal{I}(O) = \mathcal{O}$.

**Example 3**   Let $D$ be the differentiation operator on the space of polynomials $\mathscr{P}_n$. Then the null space $D$ consists of all polynomials of degree zero together with the zero polynomial, while its image consists of the zero polynomial and all polynomials of degree $< n - 1$.

**Example 4**   Let $\mathscr{R}^\infty$ be the space of all infinite sequences $\{x_1, x_2, x_3, \ldots\}$ of real numbers, with addition and scalar multiplication defined termwise (see Exercise 9, Section 2–3), and let $A$ and $B$ be the linear transformations on $\mathscr{R}^\infty$ defined by

$$A\{x_1, x_2, x_3, \ldots\} = \{x_2, x_3, x_4, \ldots\},$$
$$B\{x_1, x_2, x_3, \ldots\} = \{0, x_1, x_2, \ldots\}.$$

Then $\mathcal{N}(A)$ is the subspace of $\mathscr{R}^\infty$ consisting of all sequences of the form $\{x_1, 0, 0, \ldots\}$, where $x_1$ is arbitrary, while $\mathcal{N}(B) = \mathcal{O}$. On the other hand, $\mathcal{I}(A) = \mathscr{R}^\infty$, while by definition $\mathcal{I}(B)$ consists of all sequences whose first entry is zero.

Our next example foreshadows the role null spaces will play in the study of linear differential equations.

**Example 5**   If $A: \mathscr{C}^{(2)}(-\infty, \infty) \to \mathscr{C}(-\infty, \infty)$ is the second-order linear differential operator $D^2 + 1$,

$$A(y) = \frac{d^2 y}{dx^2} + y,$$

then $\mathcal{N}(A)$ is simply the set of all solutions of the differential equation

$$\frac{d^2 y}{dx^2} + y = 0.$$

We have already observed several times that $\mathcal{N}(A)$ contains all functions of the form

$$y = c_1 \sin x + c_2 \cos x,$$

and in the next chapter we will show that these are the only functions in $\mathcal{N}(A)$.

More generally, if $L$ is the $n$th-order linear differential operator

$$a_n(x)D^n + a_{n-1}(x)D^{n-1} + \cdots + a_0(x),$$

where $a_0(x), \ldots, a_n(x)$ are continuous on an interval $I$, then

$$a_n(x)\frac{d^n y}{dx^n} + a_{n-1}(x)\frac{d^{n-1}y}{dx^{n-1}} + \cdots + a_0(x)y = 0 \qquad (3\text{--}20)$$

is known as an $n$th-order linear differential equation, and the task of solving (3–20) is identical with the task of describing the null space of $L$.

Now that we have introduced the null space and the image of a linear transformation, we will take a closer look at those transformations $A: \mathscr{V}_1 \to \mathscr{V}_2$ for which either

$$\text{(i) } \mathcal{N}(A) = \mathcal{O} \qquad \text{or} \qquad \text{(ii) } \mathcal{I}(A) = \mathscr{V}_2,$$

or both. The second of these equations asserts that $A$ maps $\mathscr{V}_1$ *onto* $\mathscr{V}_2$, and implies that for each $\mathbf{y}$ in $\mathscr{V}_2$ there exists at least one $\mathbf{x}$ in $\mathscr{V}_1$ such that $\mathbf{y} = A(\mathbf{x})$. The first, which says that the null space of $A$ contains only the zero vector, turns out to be equivalent to the assertion that $A$ is *one-to-one* in the sense of the following definition.

**Definition 3–4**    *A linear transformation $A: \mathscr{V}_1 \to \mathscr{V}_2$ is said to be* **one-to-one** *if and only if $A(\mathbf{x}_1) = A(\mathbf{x}_2)$ implies that $\mathbf{x}_1 = \mathbf{x}_2$.*

In other words, $A$ is one-to-one if and only if $A$ maps *distinct* vectors in $\mathscr{V}_1$ onto *distinct* vectors in $\mathscr{V}_2$; whence the name. (See Fig. 3–6.) This said, we prove

**Theorem 3–3**    *A linear transformation $A: \mathscr{V}_1 \to \mathscr{V}_2$ is one-to-one if and only if $\mathcal{N}(A) = \mathcal{O}$.*

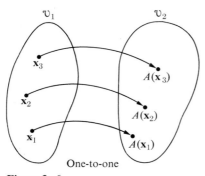
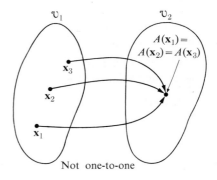

One-to-one                Not one-to-one

**Figure 3–6**

*Proof.*   Let $A$ be one-to-one, and suppose that $A(\mathbf{x}) = \mathbf{0}$. Then $A(\mathbf{x}) = A(\mathbf{0})$, and Definition 3–4 implies that $\mathbf{x} = \mathbf{0}$. Thus $\mathcal{N}(A) = \mathcal{O}$. Conversely, if $\mathcal{N}(A) = \mathcal{O}$ and $A(\mathbf{x}_1) = A(\mathbf{x}_2)$, then $A(\mathbf{x}_1) - A(\mathbf{x}_2) = \mathbf{0}$ or $A(\mathbf{x}_1 - \mathbf{x}_2) = \mathbf{0}$. Thus $\mathbf{x}_1 - \mathbf{x}_2 = \mathbf{0}$, and $\mathbf{x}_1 = \mathbf{x}_2$, as asserted.   ∎

Among the various transformations we have considered in this section, only $B: \mathcal{R}^\infty \to \mathcal{R}^\infty$ and $I$ were one-to-one, since only they had trivial null spaces. Additional examples of one-to-one transformations are: rotations of $\mathcal{R}^2$ about the origin, reflections across lines through the origin, etc. The reader should have no difficulty in augmenting this list indefinitely.

Linear transformations which are both one-to-one and onto are called **iso-morphisms**. They are also said to be **invertible** or **nonsingular**. They are of particular importance since, just as with ordinary one-to-one onto functions, they have inverses, and all of the standard facts concerning inverse functions hold for them. Indeed, if $A: \mathcal{V}_1 \to \mathcal{V}_2$ is one-to-one and onto, then *each* vector $\mathbf{y}$ in $\mathcal{V}_2$ is paired with a *unique* vector $\mathbf{x}$ in $\mathcal{V}_1$, and $A$ can therefore be used to define a function from $\mathcal{V}_2$ to $\mathcal{V}_1$. This function is called the **inverse** of $A$ and is denoted by $A^{-1}$ (read "$A$ inverse"). It can be described as *the* function from $\mathcal{V}_2$ to $\mathcal{V}_1$ such that for each $\mathbf{y}$ in $\mathcal{V}_2$

$$A^{-1}(\mathbf{y}) = \mathbf{x}, \qquad \text{where} \qquad A(\mathbf{x}) = \mathbf{y}. \qquad (3\text{–}21)$$

Loosely speaking, $A^{-1}$ is obtained from $A$ by reading the definition of $A$ from right to left, as suggested in Fig. 3–7. It is clear that $A^{-1}$ is a one-to-one map of $\mathcal{V}_2$ onto $\mathcal{V}_1$. Moreover, $A^{-1}$ is linear, since if $\mathbf{y}_1$ and $\mathbf{y}_2$ belong to $\mathcal{V}_2$ with $\mathbf{y}_1 = A(\mathbf{x}_1)$, $\mathbf{y}_2 = A(\mathbf{x}_2)$, then

$$A(\alpha_1\mathbf{x}_1 + \alpha_2\mathbf{x}_2) = \alpha_1\mathbf{y}_1 + \alpha_2\mathbf{y}_2,$$

and 3–21 implies that

$$A^{-1}(\alpha_1\mathbf{y}_1 + \alpha_2\mathbf{y}_2) = \alpha_1\mathbf{x}_1 + \alpha_2\mathbf{x}_2 = \alpha_1 A^{-1}(\mathbf{y}_1) + \alpha_2 A^{-1}(\mathbf{y}_2).$$

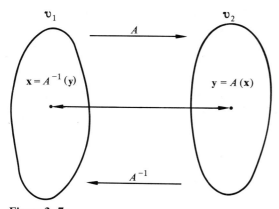

**Figure 3–7**

Having observed that $A^{-1}$ is one-to-one, onto, and linear, we see that it too is invertible, and if we simply parrot the construction given above, this time starting with $A^{-1}$, we find that $(A^{-1})^{-1} = A$. Finally, if we form the products $A^{-1}A$ and $AA^{-1}$, each of which is certainly defined, we find that both of them reduce to the identity transformation; that is,

$$A^{-1}A(\mathbf{x}) = \mathbf{x} \qquad \text{and} \qquad AA^{-1}(\mathbf{y}) = \mathbf{y}$$

for all $\mathbf{x}$ in $\mathscr{V}_1$ and all $\mathbf{y}$ in $\mathscr{V}_2$.* And with this we have proved the following theorem.

**Theorem 3–4**   *Every one-to-one linear transformation $A$ mapping $\mathscr{V}_1$ onto $\mathscr{V}_2$ has a unique inverse from $\mathscr{V}_2$ to $\mathscr{V}_1$ defined for all $\mathbf{y}$ in $\mathscr{V}_2$ by*

$$A^{-1}(\mathbf{y}) = \mathbf{x},$$

*where $A(\mathbf{x}) = \mathbf{y}$. $A^{-1}$ is also one-to-one, onto, and linear, with $(A^{-1})^{-1} = A$, and*

$$A^{-1}A = I_{v_1}, \qquad AA^{-1} = I_{v_2}. \tag{3–22}$$

It is an interesting and important fact that these last equations actually serve to characterize invertible linear transformations. When stated formally, this result reads as follows.

**Theorem 3–5**   *Let $A: \mathscr{V}_1 \to \mathscr{V}_2$ and $B: \mathscr{V}_2 \to \mathscr{V}_1$ be linear, and suppose that $BA$ and $AB$ are, respectively, the identity maps on $\mathscr{V}_1$ and $\mathscr{V}_2$. Then $A$ is one-to-one and onto, and $B = A^{-1}$.*

*Proof.*   Let $\mathbf{x}$ in $\mathscr{V}_1$ be such that $A(\mathbf{x}) = \mathbf{0}$. Then on the one hand,

$$B(A(\mathbf{x})) = B(\mathbf{0}) = \mathbf{0},$$

and on the other,

$$B(A(\mathbf{x})) = BA(\mathbf{x}) = I(\mathbf{x}) = \mathbf{x}.$$

Thus $\mathbf{x} = \mathbf{0}$, and $\mathscr{N}(A) = \mathcal{O}$.

Now let $\mathbf{y}$ be an arbitrary vector in $\mathscr{V}_2$. Then

$$\mathbf{y} = I(\mathbf{y}) = AB(\mathbf{y}) = A(B(\mathbf{y})),$$

and it follows that $\mathbf{y}$ is the image under $A$ of the vector $B(\mathbf{y})$ in $\mathscr{V}_1$. Thus $\mathscr{I}(A) = \mathscr{V}_2$, and we are done.   ∎

**Example 6**   If $A$ is any rotation of $\mathscr{R}^2$ about the origin through an angle $\theta$, then $A$ is invertible with $A^{-1}$ the rotation through $-\theta$, since $A^{-1}A = AA^{-1} = I$.

---

\*  These equations are the vector space analogs of such pairs of statements as

$$\sin^{-1}(\sin x) = x, \qquad \sin(\sin^{-1} x) = x, \qquad -\pi/2 \leqslant x \leqslant \pi/2,$$

and

$$e^{\ln x} = x, \qquad \ln(e^x) = x, \qquad x > 0,$$

which are familiar from calculus.

**Example 7** Let $A: \mathscr{R}^3 \to \mathscr{R}^3$ be defined by

$$A(x_1, x_2, x_3) = (x_1 + x_2, x_2, x_3).$$

Then $A$ is invertible, and

$$A^{-1}(x_1, x_2, x_3) = (x_1 - x_2, x_2, x_3),$$

since $A^{-1}A = AA^{-1} = I$.

Theorem 3–5 suggests a natural and valuable generalization of the notion of the inverse of a linear transformation $A: \mathscr{V}_1 \to \mathscr{V}_2$, namely a linear transformation $B: \mathscr{V}_2 \to \mathscr{V}_1$ such that

$$AB = I, \qquad BA \neq I.$$

The fact that such transformations exist can be seen by looking at Example 4. For the transformations $A$ and $B$ given there we have

$$AB\{x_1, x_2, x_3, \ldots\} = \{x_1, x_2, x_3, \ldots\},$$

and

$$BA\{x_1, x_2, x_3, \ldots\} = \{0, x_2, x_3, \ldots\}.$$

Transformations of this sort are encountered fairly often and are distinguished by name according to the following definition.

**Definition 3–5** *A linear transformation $B: \mathscr{V}_2 \to \mathscr{V}_1$ is said to be a* **right inverse** *of $A: \mathscr{V}_1 \to \mathscr{V}_2$ if the product $AB$ is the identity map on $\mathscr{V}_2$. Similarly, $B$ is said to be a* **left inverse** *of $A$ if $BA$ is the identity map of $\mathscr{V}_1$.*

*Remark.* If $B$ is a right (left) inverse of $A$, then $A$ is a left (right) inverse of $B$.

The example given a moment ago shows that a linear transformation may have a right or left inverse without having an inverse. However, it is easy to show that if $A$ has *both* a right inverse $B$ and a left inverse $C$, then $A$ is invertible, and $B = C = A^{-1}$. For then

$$AB = I, \qquad CA = I,$$

and

$$C(AB) = CI = C, \qquad (CA)B = IB = B.$$

But $C(AB) = (CA)B$. Thus $B = C$, and $AB = BA = I$. The assertion that $A$ is invertible with $B = C = A^{-1}$ now follows from Theorem 3–5.

**Example 8** Let $D$ and $L$ be, respectively, the differentiation and integration operators on $\mathscr{C}^{(1)}[a, b]$:

$$Dy = \frac{dy}{dx}, \qquad Ly = \int_a^x y(t)\, dt.$$

Then

$$LD(y) = \int_a^x y'(t)\, dt = y(x) - y(a),$$

while

$$DL(y) = \frac{d}{dx} \int_a^x y(t) \, dt = y(x),$$

and it follows that $DL = I$, while $LD \neq I$. In other words, the operation of integration on function spaces is a right inverse of differentiation but *not* a left inverse of differentiation.

## EXERCISES

1. Find the null space and, where applicable, the inverse of each of the following linear transformations on $\mathscr{R}^2$.
   a) $A(x_1, x_2) = 2(x_1, - x_2)$
   b) $A(x_1, x_2) = (x_2, 0)$
   c) $A(x_1, x_2) = (x_1 + x_2, x_1 + x_2)$
   d) $A(x_1, x_2) = (x_1 + x_2, x_1 - x_2)$

2. Repeat Exercise 1 for the following transformations on $\mathscr{R}^3$.
   a) $A(x_1, x_2, x_3) = (x_1 + x_2, x_2 + x_3, x_1)$
   b) $A(x_1, x_2, x_3) = (2x_1, -x_3, x_1 + x_3)$
   c) $A(x_1, x_2, x_3) = (x_2 + 2x_3, x_1 - x_2, 0)$
   d) $A(x_1, x_2, x_3) = (x_1 - x_2, x_1 + x_2 + x_3, x_2 + x_3)$

3. Repeat Exercise 1 for the following transformations on $\mathscr{P}$.
   a) $A(p) = \dfrac{d^2 p}{dx^2} - 2\dfrac{dp}{dx}$
   b) $A(p) = xp(x)$
   c) $A(p) = p(x) - p(0)$
   d) $A(p) = p(x)q(x)$, $q(x)$ a fixed nonconstant polynomial in $\mathscr{P}$

4. In the definition of $A^{-1}$ we insisted that $A$ be one-to-one. Why?

5. Let $A: \mathscr{R}^2 \to \mathscr{R}^2$ be defined by

   $$A(x_1, x_2) = (\alpha_1 x_1 + \alpha_2 x_2, \beta_1 x_1 + \beta_2 x_2),$$

   where $\alpha_1, \alpha_2, \beta_1, \beta_2$ are real numbers. Prove that $A$ is linear, and find a necessary and sufficient condition in terms of $\alpha_1, \alpha_2, \beta_1, \beta_2$ for $A$ to be invertible.

6. Let $A: \mathscr{V}_1 \to \mathscr{V}_2$ be a one-to-one linear transformation, and let $e_1, \ldots, e_n$ be linearly independent vectors in $\mathscr{V}_1$. Prove that $A(e_1), \ldots, A(e_n)$ are linearly independent in $\mathscr{V}_2$.

7. Let $A: \mathscr{V}_1 \to \mathscr{V}_2$ be linear, and suppose that

   $$\dim \mathscr{V}_1 = \dim \mathscr{V}_2 = n < \infty.$$

   Prove that $A$ is one-to-one if and only if it is onto. [*Hint*: See Exercise 6.]

8. Let $A$ and $B$ be invertible linear transformations mapping $\mathscr{V}$ onto itself. Prove that $AB$ and $BA$ are also invertible, and that

   $$(AB)^{-1} = B^{-1}A^{-1}, \qquad (BA)^{-1} = A^{-1}B^{-1}.$$

9. Let $A: \mathscr{V} \to \mathscr{V}$ be linear and suppose that $A^2 + I = A$. Prove that $A$ is invertible.

*10. Two vector spaces $\mathscr{V}_1$ and $\mathscr{V}_2$ are said to be *isomorphic* if and only if there exists an isomorphism $A: \mathscr{V}_1 \to \mathscr{V}_2$.
   a) Prove that two finite-dimensional spaces $\mathscr{V}_1$ and $\mathscr{V}_2$ are isomorphic if and only if $\dim \mathscr{V}_1 = \dim \mathscr{V}_2$.

b) Let $\mathscr{R}^+$ denote the space of Exercise 10 of Section 2–3. Prove that $\mathscr{R}^+$ and $\mathscr{R}^1$ are isomorphic by exhibiting an isomorphism $A: \mathscr{R}^+ \to \mathscr{R}^1$.

## 3–6  OPERATOR EQUATIONS

Much of the study of linear transformations is given over to devising methods for solving equations of the form

$$Ax = y, \tag{3–23}$$

where $\mathbf{y}$ is known, $\mathbf{x}$ is unknown, and $A$ is a linear transformation from $\mathscr{V}_1$ to $\mathscr{V}_2$. Such equations are known under the generic name of **operator equations** and will appear throughout this book as linear differential equations. In general the technique for solving an operator equation depends upon the operator involved as well as the underlying vector spaces. Nevertheless, there are a number of facts concerning such equations that can be proved without using anything other than the linearity of $A$, and we propose to get them on record here and now.

First, a vector $\mathbf{x}_0$ in $\mathscr{V}_1$ is said to be a **solution** of $(2–23)$ if $A(\mathbf{x}_0) = \mathbf{y}$, and the totality of such vectors is called the **solution set** of the equation. In the special case of a **homogeneous equation**

$$Ax = 0, \tag{3–24}$$

that is, an equation whose right-hand side is zero, this set is a *subspace* of $\mathscr{V}_1$.* It is then called the **solution space** of the equation. One of the most important properties of operator equations is that the problem of solving a **nonhomogeneous equation** $A\mathbf{x} = \mathbf{y}$ can all but be reduced to that of solving its **associated homogeneous equation** $A\mathbf{x} = \mathbf{0}$. In fact, if $\mathbf{x}_p$ is known to be a solution of $A\mathbf{x} = \mathbf{y}$, and if $\mathbf{x}_h$ is any solution whatever of $A\mathbf{x} = \mathbf{0}$, then $\mathbf{x}_p + \mathbf{x}_h$ is also a solution of $A\mathbf{x} = \mathbf{y}$, since

$$A(\mathbf{x}_p + \mathbf{x}_h) = A(\mathbf{x}_p) + A(\mathbf{x}_h)$$
$$= \mathbf{y} + \mathbf{0}$$
$$= \mathbf{y}.$$

Moreover, *every* solution $\mathbf{x}_0$ of $A\mathbf{x} = \mathbf{y}$ can be written in this form for a suitable $\mathbf{x}_h$, since from

$$A(\mathbf{x}_0 - \mathbf{x}_p) = A(\mathbf{x}_0) - A(\mathbf{x}_p)$$
$$= \mathbf{y} - \mathbf{y}$$
$$= \mathbf{0}$$

it follows that $\mathbf{x}_0 - \mathbf{x}_p = \mathbf{x}_h$ is a solution of $A\mathbf{x} = \mathbf{0}$, and hence that $\mathbf{x}_0 = \mathbf{x}_p + \mathbf{x}_h$, as asserted.

The solution $\mathbf{x}_p$ appearing in this argument is frequently called a **particular solution** of $A\mathbf{x} = \mathbf{y}$, and in these terms we can state the above result as follows.

---

\* This use of the word "homogeneous" is different from the one introduced in Section 1–3.

***Theorem 3–6***   *If $\mathbf{x}_p$ is a particular solution of $A\mathbf{x} = \mathbf{y}$, then the solution set of this equation consists of all vectors of the form $\mathbf{x}_p + \mathbf{x}_h$, where $\mathbf{x}_h$ is an arbitrary solution of the associated homogeneous equation $A\mathbf{x} = \mathbf{0}$.*

Geometrically this theorem asserts that the solution set of a nonhomogeneous operator equation can be obtained from the solution space of its associated homogeneous equation by *translating* that subspace by a particular solution $\mathbf{x}_p$ as suggested in Fig. 3–8. Algebraically it gives us a prescription for solving the operator equation $A\mathbf{x} = \mathbf{y}$: find *all* solutions of $A\mathbf{x} = \mathbf{0}$, *one* solution of $A\mathbf{x} = \mathbf{y}$, and add.

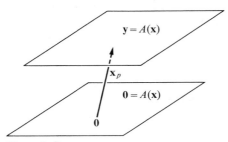

**Figure 3–8**

**Example 1**   Let $A: \mathscr{C}^{(2)}(-\infty, \infty) \to \mathscr{C}(-\infty, \infty)$ be the linear transformation $D^2 + I$ introduced in Example 5, Section 3–5. Then the operator equation

$$Ay = 1,$$

assumes the form

$$\frac{d^2y}{dx^2} + y = 1, \tag{3–25}$$

and its solution set consists of all functions in $\mathscr{C}^{(2)}(-\infty, \infty)$ which satisfy this equation on the entire real line. In this case it is obvious that $y = 1$ is one such function. Thus, to complete the solution of (3–25), it suffices to find *all* solutions of the homogeneous equation

$$\frac{d^2y}{dx^2} + y = 0.$$

In the next chapter we will show that the solution space of this equation is spanned by the functions $\sin x$ and $\cos x$. Once this fact has been established, Theorem 3–6 will imply that the solution set of (3–25) consists of all functions of the form

$$y = 1 + c_1 \sin x + c_2 \cos x,$$

where $c_1$ and $c_2$ are arbitrary constants.

**Example 2**   Let $A$ be a linear transformation from $\mathscr{R}^2$ to $\mathscr{R}^1$, let $\mathbf{e}_1$ and $\mathbf{e}_2$ be the standard basis vectors in $\mathscr{R}^2$, and let $A(\mathbf{e}_1) = \alpha_1$, $A(\mathbf{e}_2) = \alpha_2$, where $\alpha_1$ and $\alpha_2$ are real numbers. Then if $\mathbf{x} = x_1\mathbf{e}_1 + x_2\mathbf{e}_2$ is any vector in $\mathscr{R}^2$,

$$A(\mathbf{x}) = x_1A(\mathbf{e}_1) + x_2A(\mathbf{e}_2) = \alpha_1x_1 + \alpha_2x_2,$$

and the operator equation $A\mathbf{x} = \mathbf{0}$ is an abbreviated version of

$$\alpha_1 x_1 + \alpha_2 x_2 = 0. \tag{3–26}$$

Since (3–26) is the equation of the line through the origin in $\mathscr{R}^2$ with slope $-\alpha_1/\alpha_2$, the solution space of the equation $A\mathbf{x} = \mathbf{0}$ is just the set of points in $\mathscr{R}^2$ which comprise that line. In this case the solution set of the nonhomogeneous equation $A\mathbf{x} = \beta$, where $\beta$ is a real number, can be interpreted as a translation of the line described by (3–26) and shown in Fig. 3–9.

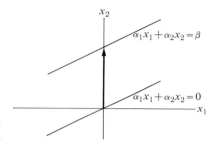

**Figure 3–9**

It goes without saying that in particular instances the solution set of $A\mathbf{x} = \mathbf{y}$ may be empty, in which case the equation has no solutions. In fact, one of the major problems in the study of operator equations (or arbitrary equations for that matter) is to determine conditions under which the equation will have solutions. This is the so-called **existence problem** for operator equations, and theorems which establish such conditions are called **existence theorems**.

Of equal or even greater importance is the problem of ascertaining when $A\mathbf{x} = \mathbf{y}$ admits *at most one* solution for any given $\mathbf{y}$ in $\mathscr{V}_2$. This problem is known as the **uniqueness problem** for operator equations, and can always be answered by examining the homogeneous equation $A\mathbf{x} = \mathbf{0}$ and using the following theorem.

> **Theorem 3–7**  *An operator equation $A\mathbf{x} = \mathbf{y}$ will have a unique solution (provided it has any solutions at all) if and only if its associated homogeneous equation $A\mathbf{x} = \mathbf{0}$ has no nonzero solutions, i.e., if and only if $\mathcal{N}(A) = \mathcal{O}$.*

The student should have no difficulty in convincing himself that this result is a consequence of Theorem 3–6 and the description of the solution set of $A\mathbf{x} = \mathbf{y}$ given there.

In the case where $A$ admits an inverse of one of the various types discussed in Section 3–5, the equation $A\mathbf{x} = \mathbf{y}$ can be solved immediately. If, for instance, $A$ is invertible, then from $A\mathbf{x} = \mathbf{y}$ we deduce that

$$A^{-1}(A\mathbf{x}) = A^{-1}\mathbf{y} \quad \text{or} \quad \mathbf{x} = A^{-1}\mathbf{y},$$

and the solution (which in this case must be unique) has been described in terms of $A^{-1}$. Similarly, if $B$ is either a right or left inverse of $A$, we find that the solution set of $A\mathbf{x} = \mathbf{y}$ is the set of all $\mathbf{x}$ in $\mathscr{V}_1$ such that $\mathbf{x} = B\mathbf{y}$. This technique for solving an operator equation is known as **inverting the operator**, and is used whenever an explicit formula for an inverse can be deduced from the definition of $A$. We shall meet it again in our study of differential equations.

### EXERCISES

1. Prove Theorem 3–7.

2. Let $A: \mathscr{R}^3 \to \mathscr{R}^2$ be the linear transformation defined by

$$A(\mathbf{e}_1) = (\alpha_1, \beta_1), \qquad A(\mathbf{e}_2) = (\alpha_2, \beta_2), \qquad A(\mathbf{e}_3) = (\alpha_3, \beta_3);$$

that is,

$$A(\mathbf{x}) = x_1 A(\mathbf{e}_1) + x_2 A(\mathbf{e}_2) + x_3 A(\mathbf{e}_3)$$

when $\mathbf{x} = x_1\mathbf{e}_1 + x_2\mathbf{e}_2 + x_3\mathbf{e}_3$.

a) Show that $\mathbf{x}$ is a solution of the homogeneous equation $A(\mathbf{x}) = \mathbf{0}$ if and only if the components of $\mathbf{x}$ satisfy the pair of linear equations

$$\alpha_1 x_1 + \alpha_2 x_2 + \alpha_3 x_3 = 0$$
$$\beta_1 x_1 + \beta_2 x_2 + \beta_3 x_3 = 0.$$

b) Describe the solution space of $A(\mathbf{x}) = \mathbf{0}$ geometrically. Does this equation ever have a unique solution?

c) Use the results of parts (a) and (b) to discuss the solutions of the equation $A(\mathbf{x}) = \mathbf{c}$, where $\mathbf{c}$ is a fixed vector in $\mathscr{R}^2$.

### *3–7 AN APPLICATION: FIRST-ORDER LINEAR DIFFERENCE EQUATIONS

As an application of the ideas developed in this chapter, we consider the **amortization problem** which arises when a debt must be paid in a number of equal monthly installments, with part of each installment being applied to the accrued interest on the debt and part to retire the principal. To fix our notation, let

> $A$ be the amount borrowed,
> $m$ be the amount of each monthly payment,
> $I$ be the monthly interest rate, and
> $p_n$ be the outstanding principal at the end of month $n$.

Then since interest in the amount $Ip_n$ is due at the end of the $n$th month, the difference

$$m - Ip_n$$

is applied toward retiring the principal, and we have

$$p_{n+1} = p_n - (m - Ip_n)$$

or

$$p_{n+1} - (1 + I)p_n = -m. \tag{3–27}$$

This equation and the initial condition

$$p_1 = A \tag{3-28}$$

constitute an initial-value problem involving a first-order constant-coefficient *linear difference equation.*

In order to analyze this problem, we introduce the vector space $\mathscr{R}^\infty$ of all infinite sequences

$$\mathbf{x} = (x_1, x_2, \ldots, x_n, \ldots)$$

of real numbers, where addition and scalar multiplication are defined term by term (see Exercise 9 of Section 2–3), and the difference operator $\Delta: \mathscr{R}^\infty \to \mathscr{R}^\infty$ is defined by

$$\Delta(x_1, x_2, \ldots, x_n, \ldots) = (x_2 - x_1, x_3 - x_2, \ldots, x_{n+1} - x_n, \ldots). \tag{3-29}$$

It is an easy exercise to show that $\Delta$ is linear, and since $\Delta$ maps $\mathscr{R}^\infty$ into itself, we can form the powers of $\Delta$:

$$\Delta^2, \Delta^3, \ldots, \Delta^n, \ldots,$$

which are also linear on $\mathscr{R}^\infty$.

> **Definition 3–6**  *An **nth-order constant-coefficient linear difference operator** is a polynomial of degree n in $\Delta$*
>
> $$a_n\Delta^n + a_{n-1}\Delta^{n-1} + \cdots + a_1\Delta + a_0, \tag{3-30}$$
>
> *where the $a_i$ are constants and $a_n \neq 0$. An equation of the form*
>
> $$L(\mathbf{x}) = \mathbf{c}, \tag{3-31}$$
>
> *where L is such an operator and $\mathbf{c}$ is a known vector in $\mathscr{R}^\infty$ is called an **nth-order constant-coefficient linear difference equation**.*

In particular, the most general first-order constant-coefficient linear difference equation is an equation of the form

$$a_1\,\Delta\mathbf{x} + a_0\mathbf{x} = \mathbf{c}, \tag{3-32}$$

where $a_1 \neq 0$, and $\mathbf{x}$ and $\mathbf{c}$ are vectors in $\mathscr{R}^\infty$. Since $a_1 \neq 0$, (3–32) can be rewritten in *normal form* as

$$\Delta\mathbf{x} + b\mathbf{x} = \mathbf{c}', \tag{3-33}$$

where $b = a_0/a_1$ and $\mathbf{c}' = (1/a_1)\mathbf{c}$. Now let

$$\mathbf{x} = (x_1, x_2, \ldots, x_n, \ldots)$$

and

$$\mathbf{c}' = (c_1, c_2, \ldots, c_n, \ldots).$$

Then (3–33) becomes

$$(x_2 + (b - 1)x_1, \; x_3 + (b - 1)x_2, \ldots, x_{n+1} + (b - 1)x_n, \ldots)$$
$$= (c_1, c_2, \ldots, c_n, \ldots),$$

which, by setting $a = b - 1$, we rewrite more simply as

$$(x_2 + ax_1, \; x_3 + ax_2, \ldots, x_{n+1} + ax_n, \ldots) = (c_1, c_2, \ldots, c_n, \ldots). \qquad (3\text{–}34)$$

Thus we have

> **Lemma 3–1**  *Every first-order constant-coefficient linear difference equation can be written in the form*
> $$T(\mathbf{x}) = \mathbf{c}, \qquad (3\text{–}35)$$
> *where $\mathbf{c}$ is a known vector in $\mathscr{R}^\infty$ and $T$ is the linear transformation defined on $\mathscr{R}^\infty$ by*
> $$T(x_1, x_2, \ldots, x_n, \ldots) = (x_2 + ax_1, \; x_3 + ax_2, \ldots, x_{n+1} + ax_n, \ldots),$$
> *where $a$ is a constant.*

In other words, the most general first-order constant-coefficient linear difference equation is equivalent to the sequence of equations

$$x_{n+1} + ax_n = c_n, \qquad n = 1, 2, \ldots, \qquad (3\text{–}36)$$

where $a$ and the $c_n$ are known constants. Save for notation, (3–27) is just such a sequence of equations.

To solve (3–35), we begin with the homogeneous equation

$$T(\mathbf{x}) = \mathbf{0}, \qquad (3\text{–}37)$$

which is equivalent to the sequence

$$x_{n+1} + ax_n = 0, \qquad n = 1, 2, \ldots. \qquad (3\text{–}38)$$

The solution of (3–38) is immediate. Starting with $n = 1$, we obtain, step-by-step,

$$x_2 = -ax_1$$
$$x_3 = -ax_2 = a^2 x_1$$
$$\vdots$$
$$x_n = (-a)^{n-1} x_1,$$

where $x_1$ is an arbitrary constant. It follows that the solution space of (3–37) is the one-dimensional subspace of $\mathscr{R}^\infty$ spanned by the vector

$$\mathbf{z}_1 = (1, -a, a^2, \ldots, (-a)^{n-1}, \ldots)$$

since all other solutions of $T(\mathbf{x}) = \mathbf{0}$ are scalar multiples of $\mathbf{z}_1$.

The solution of (3–36) is hardly more difficult. Once again we are free to choose $x_1$ as we wish, after which $x_2, x_3 \cdots$ are determined successively by the equations

$$x_2 = -ax_1 + c_1$$
$$x_3 = -ax_2 + c_2 = -a(-ax_1 + c_1) + c_2 = (-a)^2x_1 + (-a)c_1 + c_2$$
.
.
.

$$x_n = (-a)^n x_1 + \sum_{k=1}^{n} (-a)^{n-k} c_k. \tag{3–39}$$

Notice that when $x_1 = 0$, (3–39) yields the particular solution

$$\mathbf{w}_1 = \left(0, c_1, -ac_1 + c_2, \ldots, \sum_{k=1}^{n} (-a)^{n-k} c_k, \ldots\right)$$

of (3–36), while the first term in (3–39) is the $n$th component in an arbitrary solution of (3–38). Thus, as anticipated, every solution of the nonhomogeneous equation appears as the sum of a particular solution and an arbitrary solution of the associated homogeneous equation. Moreover, we have also proved that every initial-value problem

$$T(\mathbf{x}) = \mathbf{c}, \qquad x_1 = \alpha,$$

where $T$ and $\mathbf{c}$ are as before and $\alpha$ is a known constant, has a unique solution.

We now return to the amortization problem with which we began this section. We write the problem as

$$T(\mathbf{p}) = (-m, -m, \ldots), \qquad p_1 = A,$$

where

$$T(p_1, p_2, \ldots) = (p_2 - (1 + I)p_1, p_3 - (1 + I)p_2, \ldots).$$

A routine calculation reveals that

$$p_n = (A + m)(1 + I)^{n-1} - m \sum_{k=1}^{n} (1 + I)^{n-k}$$

(Exercise 1). But by the formula for the sum of the first $n$ terms in a geometric series,

$$1 + (1 + I) + (1 + I)^2 + \cdots + (1 + I)^{n-1} = \frac{1 - (1 + I)^n}{1 - (1 + I)} = \frac{(1 + I)^n - 1}{I}.$$

Thus

$$p_n = (A + m)(1 + I)^{n-1} + \frac{(1 + I)^n - 1}{I},$$

or

$$p_n = A(1 + I)^{n-1} - \frac{m[(1 + I)^{n-1} - 1]}{I}. \tag{3–40}$$

Given the values of $A$, $I$, and $m$, this formula can be used to determine the amount of the principal outstanding on the loan at the end of each month. It can also be used to determine the amount of the monthly payment, $m$, when $A$ and $I$ are known and the decision has been made to reduce the outstanding debt to zero at the end of $N$ months. This value is

$$m = \frac{AI}{1 - (1 + I)^{1-N}}. \tag{3–41}$$

### EXERCISES

1. Verify the formula

$$p_n = (A + m)(1 + I)^{n-1} - m \sum_{k=1}^{n} (1 + I)^{n-k}.$$

2. Verify that the solution of the initial-value problem

$$x_{n+1} - x_n = n + 1, \qquad x_1 = 2,$$

is $x_n = \frac{1}{2}(n^2 + n + 2)$.

3. Solve each of the following difference equations.
   a) $x_{n+1} + 2x_n = (-1)^n$     b) $x_{n+1} - 2x_n = 4$
   c) $x_{n+1} + x_n = n$           d) $x_{n+1} - x_n = n(n - 1)$

4. You want to borrow \$20,000 and are prepared to repay the loan in 25 years at the rate of \$140 per month. Will you be able to negotiate the loan if interest charges amount to 7% per year charged monthly on the unpaid balance?

# Linear
# Differential
# Equations

**4**

## 4–1 INTRODUCTION

An **$n$th-order linear differential equation** on an interval $I$ is an operator equation
of the form

$$Ly = h \tag{4-1}$$

where $h$ is continuous on $I$ and $L$ is an $n$th-order linear differential operator
defined on $I$. Such an equation is said to be **homogeneous** if $h$ is identically zero
on $I$, **nonhomogeneous** otherwise, and **normal** (on $I$) whenever the leading coefficient of $L$ does not vanish anywhere on $I$. A function $y = y(x)$ is a **solution** of
Eq. (4–1) if and only if it belongs to $\mathscr{C}^{(n)}(I)$ and satisfies the equation identically on $I$.
Thus an $n$th-order linear differential equation is an equation of the form

$$a_n(x)\frac{d^n y}{dx^n} + \cdots + a_1(x)\frac{dy}{dx} + a_0(x)y = h(x), \tag{4-2}$$

where the coefficients $a_0(x), \ldots, a_n(x)$ and right-hand side $h(x)$ are continuous on
an interval $I$ in which $a_n(x)$ is not identically zero. Typical examples are

$$\frac{d^2 y}{dx^2} + y = 0,$$

which is homogeneous, normal, and of order 2 on $(-\infty, \infty)$ or any of its subintervals, and

$$x^3 \frac{d^3 y}{dx^3} + x\frac{dy}{dx} = 3,$$

which is nonhomogeneous, normal, and of order 3 on $(0, \infty)$ and $(-\infty, 0)$ but
not normal on any interval containing the origin.

Given such an equation, we face the problem of finding all of its solutions on $I$,
provided solutions exist. They do. However, save for a few special types of equations, it is impossible to express them in closed form in terms of known functions.

113

Thus the systematic study of linear differential equations must be directed toward analyzing the general behavior of their solutions in the absence of specific techniques for exhibiting them. And it is here that the linearity of the equations becomes decisive. Because of it, the techniques of linear algebra can be used to obtain a reasonably complete description of the nature of the solutions.

To illustrate the role of linear algebra, let

$$Ly = 0 \tag{4-3}$$

be a *normal, homogeneous* linear differential equation of order $n$ on an interval $I$ of the $x$-axis. In this case the solution set of the equation is the *null space* of $L$ and hence is a *subspace* of $\mathscr{C}^{(n)}(I)$. It is called the **solution space** of the equation, and the task of solving (4–3) may be replaced by that of finding a *basis* for its solution space, provided the solution space is finite-dimensional. It is, and later in this chapter we shall prove the following fact:

*The solution space of any normal nth-order homogeneous linear differential equation is an n-dimensional subspace of $\mathscr{C}^{(n)}(I)$.*

(See Theorem 4–6.) Thus if $L$ is normal, and $y_1(x), \ldots, y_n(x)$ are *linearly independent* solutions of (4–3), then *every* solution of (4–3) is of the form

$$y(x) = c_1 y_1(x) + \cdots + c_n y_n(x) \tag{4-4}$$

for suitable real numbers $c_i$. Conversely, every function of this type is a solution of (4–3) whenever $y_1, \ldots, y_n$ are. For this reason (4–4), with the $c_i$ arbitrary, is called the **general solution** of (4–3), while any function obtained from the general solution by assigning definite values to the $c_i$ is called a **particular solution**.

These results can also be used to help solve a nonhomogeneous equation

$$Ly = h. \tag{4-5}$$

For, as we have seen, *the solution set of this equation can be found by adding **all** solutions of the associated homogeneous equation to any particular solution of* (4–5). This argument effectively reduces the problem of solving a nonhomogeneous equation to that of finding the general solution of its associated homogeneous equation. And in the next chapter we shall complete this reduction by giving a method by which a particular solution of (4–5) can always be found once the general solution of the associated homogeneous equation is known.

**Example**   Since $\sin x$ and $\cos x$ are linearly independent solutions of

$$y'' + y = 0$$

on $(-\infty, \infty)$, they are a basis for the solution space of the equation, and the general solution of the equation is

$$y = c_1 \sin x + c_2 \cos x,$$

where $c_1$ and $c_2$ are arbitrary constants. Therefore, since the function $x^2 + 1$ is easily seen to be a particular solution of

$$y'' + y = x^2 + 3,$$

the general solution of this nonhomogeneous equation is

$$y = x^2 + 1 + c_1 \sin x + c_2 \cos x.$$

Without theorems like the ones just cited there would be no guarantee that this expression included *every* solution of $y'' + y = x^2 + 3$.

## EXERCISES

1. a) Show that $e^{ax} \cos bx$ and $e^{ax} \sin bx$ are linearly independent solutions of

   $$(D^2 - 2aD + a^2 + b^2)y = 0, \qquad b \neq 0,$$

   on $(-\infty, \infty)$.
   b) What is the general solution of this equation?
   c) Find the particular solution of the equation in (a) that satisfies the "initial" conditions $y(0) = b$, $y'(0) = -a$.

2. a) Show that $e^{ax}$ and $xe^{ax}$ are linearly independent solutions of

   $$(D - a)^2 y = 0.$$

   b) Find the particular solution of this equation that satisfies the "initial" conditions $y(0) = 1$, $y'(0) = 2$.

3. a) Verify that $\sin^3 x$ and $\sin x - \tfrac{1}{3} \sin 3x$ are solutions of

   $$y'' + (\tan x - 2 \cot x)y' = 0$$

   on any interval where $\tan x$ and $\cot x$ are both defined. Are these solutions linearly independent?
   b) Find the general solution of this equation.

4. Show that $\tfrac{1}{9}x^3$ and $\tfrac{1}{9}(x^{3/2} + 1)^2$ are solutions of the <u>nonlinear</u> differential equation $(dy/dx)^2 - xy = 0$ on $(0, \infty)$. Is the sum of these functions a solution?

5. In each of the following show that the given functions span the solution space of the associated differential equation. Find a basis for the solution space in each case, and use it to obtain the general solution of the equation in question.
   a) $y'' - y = 0;$     $\sinh x, 2e^{-x}, -\cosh x$, on $(-\infty, \infty)$
   b) $x^2 y'' - 5xy' + 9y = 0;$     $2x^3 \ln x, x^3, x^3(2 \ln x - 1)$, on $(0, \infty)$
   c) $y'' + 4y = 0;$     $\sin 2x, -2 \cos 2x, -\cos (2x - 3)$, on $(-\infty, \infty)$
   d) $(1 - x^2)y'' - 2xy' + 2y = 0;$     $3x, \dfrac{x}{2} \ln \dfrac{1 + x}{1 - x} - 1, \dfrac{x}{2}$, on $(-1, 1)$

6. Let $L$ be a linear differential operator defined on an interval $I$, and suppose that $y_1$ and $y_2$ are, respectively, solutions of the equations

   $$Ly = h_1 \qquad \text{and} \qquad Ly = h_2.$$

   Show that $y_1 + y_2$ is a solution of $Ly = h_1 + h_2$.

7. Find a particular solution of

   $$a_n(x)y^{(n)} + a_{n-1}(x)y^{(n-1)} + \cdots + a_1(x)y' + a_0(x)y = h(x)$$

in each of the following special cases:

a) $h$ and $a_0$ are nonzero constants;

b) $h(x) = mx + b$, $a_0$ and $a_1$ are constants, and $a_0 \neq 0$;

c) $h(x) = mx$, $a_0 = 0$, and $a_1$ and $a_2$ are nonzero constants.

## 4–2 FIRST-ORDER LINEAR EQUATIONS

The techniques introduced in Chapter 1 can be used to solve the general first-order linear equation

$$a_1(x)\frac{dy}{dx} + a_0(x)y = h(x)$$

on an interval $I$ whenever $a_1(x) \neq 0$ for all $x$ in $I$. We begin by dividing the coefficients of the equation by $a_1(x)$ and rewriting it in normal form as

$$\frac{dy}{dx} + P(x)y = Q(x). \tag{4–6}$$

In the homogeneous case $Q(x) = 0$ on $I$, the variables can be separated to yield

$$\frac{dy}{y} = -P(x)\,dx,$$

and the general solution of the equation is

$$y_h = ce^{-\int P(x)\,dx},$$

where $c$ is an arbitrary constant.

To complete the task of solving (4–6), we must find a particular solution. We do this by looking for an integrating factor $\mu = \mu(x)$ *that is a function of $x$ alone* [see Exercise 23(a) of Section 1–5]. If such a $\mu$ exists, we must have

$$\frac{\partial}{\partial x}\,\mu(x) = \frac{\partial}{\partial y}\left[\mu(x)P(x)y - \mu(x)Q(x)\right].$$

But since

$$\frac{\partial}{\partial x}\,\mu(x) = \frac{d\mu}{dx}$$

and

$$\frac{\partial}{\partial y}\left[\mu(x)P(x)y - \mu(x)Q(x)\right] = \mu(x)P(x),$$

$\mu$ must satisfy the differential equation

$$\frac{d\mu}{dx} = \mu(x)P(x).$$

Thus

$$\frac{d\mu}{\mu} = P(x)\,dx,$$

and we find that

$$\mu = e^{\int P(x)\,dx}$$

is an integrating factor of (4–6). When both sides of (4–6) are multiplied by this factor, we obtain

$$e^{\int P(x)\,dx}\left(\frac{dy}{dx} + P(x)y\right) = Q(x)e^{\int P(x)\,dx},$$

which can be rewritten as

$$\frac{d}{dx}\left(e^{\int P(x)\,dx}y\right) = Q(x)e^{\int P(x)\,dx}.$$

We now integrate to obtain

$$e^{\int P(x)\,dx}y = \int Q(x)e^{\int P(x)\,dx}\,dx.$$

When we multiply both sides of this equation by $e^{-\int P(x)\,dx}$, we find that

$$y_p = e^{-\int P(x)\,dx}\int Q(x)e^{\int P(x)\,dx}\,dx$$

is a particular solution of (4–6). Since the general solution of (4–6) is obtained by forming the sum of $y_p$ and $y_h$, we have proved the following result.

**Theorem 4–1**    *If P and Q are continuous on an interval I of the x-axis, then the general solution of*

$$\frac{dy}{dx} + P(x)y = Q(x)$$

*on I is*

$$y = e^{-\int P(x)\,dx}\left(\int Q(x)e^{\int P(x)\,dx}\,dx + c\right), \qquad (4\text{–}7)$$

*where c is an arbitrary constant.*

Note that despite the welter of integrals, (4–7) contains only one arbitrary constant, namely $c$.

Since the technique that leads to this result is so simple, it is hardly worth the effort to memorize (4–7). Instead, *just multiply*

$$y' + P(x)y = Q(x)$$

by $e^{\int P(x)\,dx}$ *and integrate.*

**Example 1**    Find the general solution of

$$\frac{dy}{dx} + 2xy = x,$$

and the particular solution that satisfies the condition

$$y(0) = 0.$$

*Solution.*   We multiply by $e^{\int 2x \, dx} = e^{x^2}$ to obtain

$$\left(\frac{dy}{dx} + 2xy\right) e^{x^2} = xe^{x^2}$$

or

$$\frac{d}{dx}(ye^{x^2}) = xe^{x^2}.$$

Integrating, we have

$$ye^{x^2} = \int xe^{x^2} \, dx + c,$$

and it follows that

$$y = \left(\frac{e^{x^2}}{2} + c\right) e^{-x^2} = \frac{1}{2} + ce^{-x^2}$$

is the general solution of the given equation. Finally, setting $x = y = 0$ in this expression, we find that $c = -\frac{1}{2}$. Thus

$$y = \frac{1}{2}(1 - e^{-x^2})$$

is the particular solution that satisfies the condition $y(0) = 0$.

**Example 2**   Find the general solution of

$$x\frac{dy}{dx} + y = x. \qquad (4\text{--}8)$$

*Solution.*   Since the leading coefficient of this equation vanishes when $x = 0$, we can solve the equation only on $(0, \infty)$ and $(-\infty, 0)$. There however, it may be written as

$$\frac{dy}{dx} + \frac{1}{x}y = 1 \qquad (4\text{--}9)$$

and solved by introducing the integrating factor

$$e^{\int dx/x} = e^{\ln|x|} = |x|.$$

Upon multiplying (4–9) by $|x|$ and integrating, we obtain

$$y|x| = \int |x| \, dx + c = \begin{cases} \dfrac{x^2}{2} + c, & x > 0, \\[2mm] -\dfrac{x^2}{2} + c, & x < 0. \end{cases}$$

Thus

$$y = \begin{cases} \dfrac{c}{x} + \dfrac{x}{2}, & x > 0, \\[2mm] -\dfrac{c}{x} + \dfrac{x}{2}, & x < 0, \end{cases}$$

and since $c$ is arbitrary, we have

$$y = \frac{c}{x} + \frac{x}{2}.$$

We call attention to the fact that $x/2$ is the only solution of the given equation which is defined on the entire real line. Nevertheless, it is common practice to call $c/x + x/2$ the "general solution" of (4–8) without specifying the interval in question—a practice which is convenient but potentially misleading.

**Bernouilli's Equation**　In Chapter 1 we introduced the technique of making a change of variable to reduce a differential equation that cannot be solved to one whose solution can be obtained by known methods. We now use this technique to solve

$$\frac{dy}{dx} + P(x)y = Q(x)y^n, \tag{4–10}$$

where $n$ is an arbitrary real number. When $n = 0$ or 1 the equation is linear and can be solved as it stands. Otherwise, it is known as a **Bernouilli equation**, and we rewrite it as

$$y^{-n}\frac{dy}{dx} + P(x)y^{1-n} = Q(x). \tag{4–11}$$

[Note that when $n > 0$ we have suppressed the solution $y = 0$ in passing from (4–10) to (4–11).] Next, we make the change of variable

$$u = y^{1-n}.$$

Then

$$\frac{du}{dx} = (1 - n)y^{-n}\frac{dy}{dx},$$

and (4–11) becomes

$$\frac{1}{1 - n}\frac{du}{dx} + P(x)u = Q(x),$$

which is a normal first-order linear equation. We now solve this equation for $u$ and obtain the general solution of (4–10) as $y = u^{1/(1-n)}$.

**Example 3**　Solve the equation

$$\frac{dy}{dx} + y = (xy)^2. \tag{4–12}$$

*Solution.*　We rewrite the equation as

$$y^{-2}\frac{dy}{dx} + y^{-1} = x^2,$$

suppressing the solution $y = 0$ in the process, and make the change of variable

$$u = y^{-1}.$$

This gives

$$\frac{du}{dx} - u = -x^2,$$

from which we obtain

$$u = 2 + 2x + x^2 + ce^x$$

(see Exercise 17). Thus the solutions of (4–12) are

$$y = (2 + 2x + x^2 + ce^x)^{-1},$$

where $c$ is arbitrary, and $y = 0$.

## EXERCISES

Find the general solution of the equations in Exercises 1 through 15.

1. $xy' + 2y = 0$
2. $(\sin x)y' + (\cos x)y = 0$
3. $2y' + 3y = e^{-x}$
4. $3xy' - y = \ln x + 1$
5. $(x^2 + 1)y' - (1 - x)^2 y = xe^{-x}$
6. $(3x^2 + 1)y' - 2xy = 6x$
7. $(x \sin x)y' + (\sin x + x \cos x)y = xe^x$
8. $(x^2 + 1)y' + xy = (1 - 2x)\sqrt{x^2 + 1}$
9. $xy' + (1 - x^2)^{-1/2}y = (1 + \sqrt{1 - x^2})e^x$
10. $xy' + (2x + 1)^{-1/2}y = 1 + (2x + 1)^{1/2}$
11. $(1 + \sin x)y' + (2 \cos x)y = \tan x$
12. $yy' + xy^2 - x = 0$
13. $(x^2 + 1)y^{1/2}y' = xe^{3x/2} + (1 - x)^2 y^{3/2}$
14. $(xy^2)' = (xy)^3(x^2 + 1)$
* 15. $(x - 1)y' - 2y = \sqrt{(x^2 - 1)y}$
16. Verify the solution of $y' + y = (xy)^2$ given in Example 3.
17. Show that if $p(x)$ is a polynomial of degree $n$, then

$$y' + ay = p(x)$$

has a solution that is a polynomial of degree $n$ whenever $a \neq 0$.

18. Solve the integral equation

$$y(x) + \int_0^x y(t)\, dt = x.$$

[*Hint*: Differentiate.]

19. Solve the integral equation

$$y(x) + 2\int_0^x ty(t)\, dt = x^2.$$

[*Hint*: Differentiate.]

20. Discuss the behavior of the solutions of

$$y' + ay = e^{bx}$$

as $x \to \infty$, given that $a$ and $b$ are constants with $a \geqslant 0$.

## 4–3 APPLICATIONS $Skipped$

**Population Growth** In Chapter 1 we saw that when the rate of change of a population is proportional to the number of individuals in the population, the size of the population is described by the initial-value problem

$$\frac{dN}{dt} = kN, \qquad N(0) = N_0,$$

where $N = N(t)$ is the number of individuals in the population at time $t$, and $k$ is a factor of proportionality. If we assume that $k$ is a constant, the population then increases or decreases exponentially according to the formula

$$N = N_0 e^{kt}.$$

This model is open to criticism because prolonged exponential growth is unrealistic for most populations. Thus we must analyze the problem anew to obtain a more satisfactory model. One such analysis begins with the observation that populations change because of births and deaths, and with the assumption that at any instant the number of each is proportional to the total population at that instant. Thus if $B$ denotes the birth rate and $M$ the mortality rate for the population, then

$$\frac{dN}{dt} = (B - M)N. \tag{4–13}$$

If $B$ and $M$ were constants, this equation would reduce to $dN/dt = kN$, with $k = B - M$, but in general this is not the case. Equation (4–13) can therefore be used to construct a variety of models of population growth depending on the assumptions made about $B$ and $M$.

***The first model*** In many situations it is reasonable to assume that because of increasing pressure on such necessities as food and living space, the death rate increases in direct proportion to the size of the population, while the birth rate remains constant. This model might be used, for instance, to approximate the dynamics of economic growth in the face of finite resources. In this case

$$M = kN,$$

where $k$ is a positive constant, and (4–13) becomes

$$\frac{dN}{dt} = (B - kN)N$$

or

$$\frac{dN}{dt} - BN = -kN^2.$$

This is a Bernouilli equation, and a straightforward computation using the substitution $u = N^{-1}$ yields the general solution

$$N = \frac{B}{Bce^{-Bt} + k},$$

where $k$ is an arbitrary constant. Finally, if we set

$$N(0) = N_0,$$

we find that

$$N = \frac{BN_0}{(B - N_0k)e^{-Bt} + N_0k}.$$

(See Exercise 1.)

This result is more satisfactory than prolonged exponential growth, since now $N(t) \rightarrow B/k$ as $t \rightarrow \infty$ for any value of $N_0$ except 0. Indeed, if $B - N_0k > 0$, it is easy to show that $N'(t) > 0$ for all $t > 0$, in which case $N(t)$ approaches $B/k$ asymptotically from below. On the other hand, if $B - N_0k < 0$, then $N'(t) < 0$ for all $t > 0$ and $N(t)$ approaches $B/k$ asymptotically from above. Finally, if $B - N_0k = 0$, the population has the constant value $B/k$ for all $t$. Each of these possibilities is illustrated in Fig. 4–1.

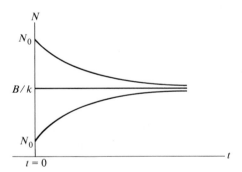

**Figure 4–1**

*The second model*   Here we suppose that the birth rate is proportional to the number of individuals present, while the death rate remains constant, so that (4–13) becomes

$$\frac{dN}{dt} = (kN - M)N$$

or

$$\frac{dN}{dt} + MN = kN^2.$$

It follows that

$$N = \frac{M}{Mce^{Mt} + k}.$$

(See Exercise 2.) The initial value $N(0) = N_0$ then yields the solution

$$N = \frac{MN_0}{(M - N_0k)e^{Mt} + N_0k}.$$

This situation is very different from the one portrayed by the first model. Three distinct types of solutions are possible depending on whether $N_0$ is greater than, equal to, or less than the *critical value* $M/k$. In fact,

(i)   $N(t) \to \infty$ *in a finite time* if $N_0 > M/k$;
(ii)  $N(t) = M/k$ for all $t$ if $N_0 = M/k$;
(iii) $N(t) \to 0$ as $t \to \infty$ if $N_0 < M/k$.

(See Exercise 2.) In the first case we have a population explosion, while in the last case the population faces eventual extinction (Fig. 4–2). Only in the second case is catastrophe averted.

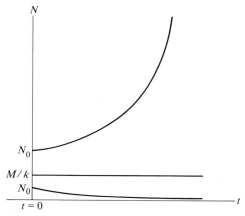

**Figure 4–2**

**Simple Electrical Circuits: The *RL*-Circuit**   The problem of describing the flow of current $I$ in a simple electrical circuit can also be phrased as an initial-value problem involving a differential equation. In this section we consider a simple *RL*-circuit of the type illustrated schematically in Fig. 4–3. The circuit consists of the following components.

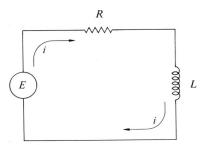

**Figure 4–3**

1. A source of *electromotive force* (emf), such as a battery or generator, of magnitude $E$ that serves to "push" current through the circuit.
2. A *resistance R* that opposes the flow of current with an emf of magnitude

$$E_R = IR.$$

   (This relation is known as *Ohm's law*.)
3. An *inductance L* that opposes any change in the flow of current with an emf of magnitude

$$E_L = L\frac{dI}{dt}.$$

Moreover, the components in this or any other electrical circuit obey *Kirchhoff's* (*second*) *law*:

*The algebraic sum of the electromotive forces around any closed loop in the circuit is zero.*

Thus for the circuit in Fig. 4–3,

$$E - E_R - E_L = 0,$$

or

$$L\frac{dI}{dt} + RI = E.$$

If we assume that the circuit was energized at time zero by closing the switch, the flow of current is obtained as the solution of the initial-value problem

$$L\frac{dI}{dt} + RI = E, \qquad I(0) = 0, \tag{4-14}$$

where $L$ and $R$ are constants.

We first solve this problem under the assumption that $E$ is a constant, $E_0$. In this case

$$I = \frac{E_0}{R}\left(1 - e^{-(R/L)t}\right), \tag{4-15}$$

and the current is the sum of two terms: a time-independent *steady-state* term

$$I_s = \frac{E_0}{R},$$

and a *transient* term

$$I_t = -\frac{E_0}{R} e^{-(R/L)t},$$

whose effect diminishes with time. (See Fig. 4–4.) Since $L$ appears only in the latter term, a simple *RL*-circuit operating under a constant impressed emf, or voltage, will eventually behave as if it were noninductive. The length of time for the transient term to become negligible is called the *delay time* of the circuit. It is a measure of the sensitivity of the circuit in responding to the constant voltage source $E_0$.

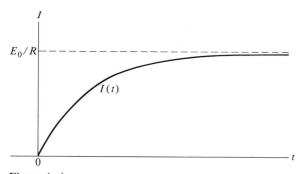

**Figure 4–4**

Let us now consider the more interesting case of an *RL*-circuit in which $E = E_0 \sin \omega t$, where $E_0$ and $\omega$ are positive constants. (This is the case of alternating rather than direct current.) Then (4–14) becomes

$$L\frac{dI}{dt} + RI = E_0 \sin \omega t, \qquad I(0) = 0,$$

and the solution has the rather complicated form

$$I = \frac{\omega E_0 L}{R^2 + \omega^2 L^2} e^{-(R/L)t} + \frac{E_0}{R^2 + \omega^2 L^2}(R \sin \omega t - \omega L \cos \omega t). \qquad (4-16)$$

(See Exercise 6.) In order to make the analysis of this solution easier, we introduce two auxiliary constants, the *steady-state impedance* $Z$ and the *phase angle* $\alpha$, defined by

$$R = Z \cos \alpha \qquad \text{and} \qquad \omega L = Z \sin \alpha,$$

where

$$Z = \sqrt{R^2 + \omega^2 L^2}.$$

Then (4–16) becomes

$$I = \frac{\omega E_0 L}{Z^2} e^{-(R/L)t} + \frac{E_0}{Z} \sin(\omega t - \alpha), \qquad (4\text{–}17)$$

and the solution is again the sum of a transient and a steady-state term. The steady-state term,

$$I_s = \frac{E_0}{Z} \sin(\omega t - \alpha), \qquad (4\text{–}18)$$

is sinusoidal, as we would expect, but it lags behind the impressed emf $E_0 \sin \omega t$ by the phase angle $\alpha$. Since

$$\sin \alpha = \frac{\omega L}{Z} \quad \text{and} \quad \cos \alpha = \frac{R}{Z},$$

$\alpha = 0$ in a purely resistive circuit ($L = 0$), and $\alpha = \pi/2$ in a purely inductive circuit ($R = 0$). In the first case the current and voltage are in phase; in the second case they are 90° out of phase. Notice that $Z$ plays the same role in (4–18) that the resistance $R$ plays in (4–15), the equation for the steady-state current in an $RL$-circuit operating under a constant emf. This analogy explains the term "impedance" used in this context.

In Fig. 4–5 we have sketched the graph of

$$I = I_s + I_t$$

for a typical $RL$-circuit. The diminishing effect of the transient term is clearly seen in the graph.

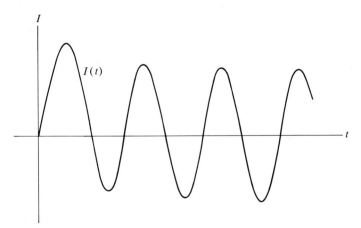

**Figure 4–5**

**One-Dimensional Neutron Transport**    As our final example, we present a simplified version of nuclear fission.

Suppose that a one-dimensional rod, situated as shown in Fig. 4–6, is bombarded by a beam of neutrons of intensity $I_0$ whose particles can move only in the $x$-direction (no scattering). On the average, in each unit length along the rod a certain fraction $p$ of the neutrons in the beam will interact with the atoms of the rod, and when an interaction occurs we suppose that the "triggering" neutron disappears and is replaced by two neutrons, one moving to the right and one moving to the left. The neutrons moving to the right constitute a reflected beam whose intensity we wish to determine. We shall assume that the speed of the neutrons in the beam is so great in comparison to the length of the rod that all reactions can be viewed as occurring instantaneously, without a time delay.

0                              $x - h$        $x$
**Figure 4–6**

For each unit of intensity of the incident beam, let $F(x)$ denote the intensity of the beam reflected from a rod of length $x$, let $I_0 = 1$, and consider the portion of the rod between $x - h$ and $x$, where $h$ is small. Since every neutron that interacts with the rod in this interval is replaced by a neutron moving to the left, the right-hand end of the portion of the rod from 0 to $x - h$ is struck by a beam of intensity one. This produces a reflected beam of intensity

$$F(x - h).$$

In addition, because of interactions in the interval between $x$ and $x - h$, a reflected beam of intensity

$$ph$$

will be created. Since any interactions that occur between the neutrons in either of these reflected beams will produce a replacement neutron still moving to the right, we have a contribution to $F(x)$ of magnitude

$$ph + F(x - h).$$

Next, since some of the neutrons in each of the reflected beams just described will interact with the rod, each beam will give rise to a secondary beam *moving to the left*, which in turn will produce additional contributions to the reflected beam when it strikes the right-hand end of the portion of the rod between 0 and $x - h$. The intensities of these secondary beams are, respectively,

$$ph\,F(x - h) \qquad \text{and} \qquad (ph)^2,$$

and they therefore will contribute

$$ph[F(x - h)]^2 \qquad \text{and} \qquad (ph)^2 F(x - h)$$

to the reflected beam. Repetitions of this argument yield a third or tertiary contribution, and so forth.

In order to make further progress with this problem, we now assume that $h$ is so small that terms involving powers of $h$ greater than one can be neglected. This allows us to discard the contribution $(ph)^2 F(x - h)$ to the secondary beam and *all* contributions thereafter, and results in the equation

$$F(x) = ph + F(x - h) + ph[F(x - h)]^2.$$

To solve this equation, we assume that $F$ is differentiable, and use the fact that $h$ is small to write

$$F(x - h) = F(x) - hF'(x).$$

A substitution then produces the equation

$$F = ph + F - hF' + ph(F - hF')^2$$

(we have not bothered to write the $x$'s). When terms involving $h^2$ are neglected, the last equation becomes simply

$$F' = p + pF^2. \tag{4–19}$$

This nonlinear equation is one of the simplest versions of the general *Ricatti equation*

$$y' = a_0(x) + a_1(x)y + a_2(x)y^2. \tag{4–20}$$

To solve (4–19), we rewrite it as

$$\frac{F'}{1 + F^2} = p$$

or

$$\frac{d}{dx}\left[ \tan^{-1} F(x) \right] = p,$$

and integrate. We then have

$$\tan^{-1} F(x) = px + c.$$

Finally, since $F(0) = 0$ (there is no reflection from a rod of length zero), $c = 0$, and

$$F(x) = \tan (px).$$

Note that when $x = \pi/2p$ the intensity of the reflected beam becomes infinite. This suggests the theoretical potential of such a device for producing power.

**EXERCISES**

1. a) Verify that the solution of

$$\frac{dN}{dt} - BN = -kN^2, \qquad N(0) = N_0,$$

is

$$N = \frac{BN_0}{(B - N_0k)e^{-Bt} + N_0k}.$$

b) Prove that $N(t) \to B/k$ as $t \to \infty$ for all $N_0 \neq 0$.

2. a) Verify that the solution of

$$\frac{dN}{dt} + MN = kN^2, \qquad N(0) = N_0,$$

is

$$N = \frac{MN_0}{(M - N_0k)e^{Mt} + N_0k}.$$

b) Show that $N(t) \to \infty$ in a finite time if $N_0 > M/k$, and $N(t) \to 0$ as $t \to \infty$ if $N_0 < M/k$.

3. If a certain business were operated on a nonprofit basis, its value $v(t)$ at time $t$ would increase at a rate proportional to that value. The managers of the business decide, however, to reduce its value at a rate equal to a fixed fraction of the value to obtain a profit $p(t)$ from the business.
   a) Find a pair of differential equations that can be used to compute $v(t)$ and $p(t)$.
   b) Find $v$ as a function of time given that $v(0) = v_0$.
   c) Find $p$ as a function of time given that $p(0) = 0$.

4. The natural rate of increase of bacteria in a colony is proportional to the number present, and without interference the size of the colony will double in $3 \ln 2$ days. Suppose, however, that $N_1$ bacteria are removed from the colony every day.
   a) Show that when the bacteria are removed daily at noon the size of the colony will gradually increase if
   $$N_1 < N_0(e^{1/3} - 1),$$
   where $N_0 = N(0)$, but that it will decrease if $N_1 > N_0(e^{1/3} - 1)$.
   b) Show that when the bacteria are removed at a uniform rate throughout the day, the size of the colony will gradually increase if
   $$N_1 < \frac{N_0}{3}$$
   and will decrease otherwise.

5. In a certain chemical reaction substance $A$ is converted into substance $B$ at a rate proportional to the amount of $A$ present, and $B$ in turn is converted into substance $C$ at a rate proportional to the amount of $B$ present. Find $B(t)$, the amount of $B$ present at time $t$, if the conversion rates are $k_1$ and $k_2$, respectively, and if the reaction began with $A_0$ units of substance $A$.

6. Verify Eqs. (4–15) and (4–16).

7. Find the current flow in a simple $RL$-circuit under a damped sinusoidal emf $E = E_0e^{-at} \sin bt$, where $E_0$, $a$, and $b$ are constants, and $a > 0$. Assume that $I(0) = I_0$.

8. Verify that when $F = F(x)$ is differentiable and when terms of degree greater than one in $h$ are neglected the equation

$$F(x) = ph + F(x - h) + ph[F(x - h)]^2$$

can be rewritten as

$$F'(x) = p + p[F(x)]^2.$$

* 9. Show that when second-order effects are taken into consideration the intensity of the reflected neutron beam in the example discussed in the text must satisfy the differential equation

$$\frac{h^2}{2} F'' - (1 + 2ph)F' - (p + p^2hF + pF^2 + p^2hF^3) = 0.$$

[*Hint*: Consider the first three terms in the Taylor series expansion of $F$ about $x$.]

10.  a) Prove that the change of variable $v = y'/y$ will reduce the second-order homogeneous linear differential equation

$$y'' + a_1(x)\, y' + a_0(x)y = 0 \qquad (4-21a)$$

to the Riccati equation

$$v' + v^2 + a_1(x)v + a_0(x) = 0. \qquad (4-21b)$$

Hence deduce that the problem of solving (4–21a) is equivalent to that of solving the simultaneous pair of first-order equations

$$\frac{dy}{dx} = vy, \qquad \frac{dv}{dx} = -v^2 - a_1(x)v - a_0(x). \qquad (4-21c)$$

Equation (4–21b) is called the *Riccati equation associated with* (4–21a).

  b) What conditions ought one to impose on (4–21c) to correspond to the conditions $y(0) = y_0$, $y'(0) = y_1$ on (4–21a)?

  c) Prove that every Riccati equation (4–20) in which $a_2(x) \neq 0$, can be converted to a second-order homogeneous linear differential equation by making the change of variable $y = -v'/(a_2v)$.

11. Find the Riccati equation associated with $y'' - y = 0$. Solve this equation and hence find the general solution of $y'' - y = 0$.

## 4–4  EXISTENCE AND UNIQUENESS OF SOLUTIONS: INITIAL-VALUE PROBLEMS

In Section 4–2 we saw that every normal first-order linear differential equation

$$y' + P(x)y = Q(x) \qquad (4-22)$$

defined on an interval $I$ has solutions on $I$. In fact, it has infinitely many, one for each value of $c$ in the expression

$$y = \left[ c + \int Q(x)e^{\int P(x)\,dx}\, dx \right] e^{-\int P(x)\,dx}.$$

Geometrically this fact implies that the general solution of (4–22) is a one-parameter family of plane curves whose graphs traverse the vertical strip of the $xy$-plane determined by $I$, as suggested in Fig. 4–7. Furthermore, it is an easy matter to show that if $(x_0, y_0)$ is any point in this region, then (4–22) admits at least one solution whose graph passes through $(x_0, y_0)$. (See Exercise 2.) These results

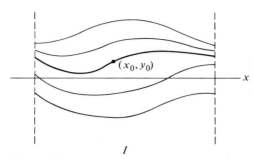

**Figure 4–7**

constitute an existence theorem for normal first-order linear differential equations, and we restate them here in slightly different language as Theorem 4–2.

**Theorem 4–2**    *Let L be a normal first-order linear differential operator defined on an interval I, and let $x_0$ be any point in I. Then for every real number $y_0$ the initial-value problem*

$$Ly = h, \qquad y(x_0) = y_0$$

*has **at least one** solution.*

Next, we propose to show that this solution is *unique*. To this end let $y_1$ and $y_2$ be solutions of

$$y' + P(x)y = Q(x), \qquad y(x_0) = y_0. \tag{4–23}$$

Substituting $y_1 - y_2$ for $y$ in these equations, we find that

$$[y_1 - y_2]' + P(x)[y_1 - y_2] = [y_1' + P(x)y_1] - [y_2' + P(x)y_2]$$
$$= Q(x) - Q(x) = 0$$

and

$$(y_1 - y_2)(x_0) = y_1(x_0) - y_2(x_0) = y_0 - y_0 = 0.$$

Thus $y_1 - y_2$ is a solution of the (homogeneous) initial-value problem

$$y' + P(x)y = 0, \qquad y(x_0) = 0. \tag{4–24}$$

But the general solution of $y' + P(x)y = 0$ is

$$y = ce^{f(x)},$$

where $f(x) = -\int P(x)\,dx$ and $c$ is an arbitrary constant; and since $e^{f(x)}$ does not vanish for any $x$, the requirement that $y(x_0) = 0$ forces us to set $c = 0$. Hence (4–24) has $y = 0$ as its only solution and therefore $y_1 - y_2 = 0$. This implies that $y_1 = y_2$, which is just what we wanted to show. When stated formally this result reads as follows.

**Theorem 4–3**    *Every initial-value problem involving a normal first-order linear differential operator has **at most one** solution.*

∴ it has exactly one

Theorems 4–2 and 4–3 assert that an initial-value problem for a normal first-order differential equation has *precisely one* solution and therefore imply that the general solution of a normal first-order linear differential equation is a one-parameter family of curves whose graphs *fill* a vertical strip of the plane. The initial condition $y(x_0) = y_0$ has the effect of selecting from the members of this family *the* solution whose graph passes through the point $(x_0, y_0)$.

The general theory of linear differential equations can properly be said to begin with the theorems that generalize these results to $n$th-order equations. In the special case treated above, the theorems were proved by simply exhibiting the solution. But in the general case this procedure is impossible, and as a result the corresponding existence and uniqueness theorems are more difficult to prove. For this reason we have deferred the proofs to a later chapter (Chapter 12). Here we simply state the results. First, the existence theorem:

**Theorem 4–4**    *Let L be a normal nth-order linear differential operator defined on an interval I, and let $x_0$ be any point in I. Then if $y_0, y_1, \ldots, y_{n-1}$ are arbitrary real numbers, the **initial-value problem***

$$Ly = h,$$
$$y(x_0) = y_0, \quad y'(x_0) = y_1, \ldots, y^{(n-1)}(x_0) = y_{n-1}$$

*has **at least one** solution on I.*

Next the uniqueness theorem:

**Theorem 4–5**    *Every initial-value problem involving a normal nth-order linear differential operator has **exactly one** solution.*

*Remark.*   Since the general solution of a normal $n$th-order linear differential equation involves $n$ arbitrary constants, it is not surprising that an initial-value problem for such an equation must impose $n$ additional conditions on the general solution to fix the values of these constants. Following the pattern in the first-order case, we have done this by selecting a point $x_0$ in $I$ (the *initial point*) and then requiring that the solution satisfy the $n$ *initial conditions*

$$y(x_0) = y_0,$$
$$y'(x_0) = y_1,$$
$$\vdots$$
$$y^{(n-1)}(x_0) = y_{n-1},$$

(4–25)

where $y_0, \ldots, y_{n-1}$ are arbitrary real numbers. Note that these conditions can indeed be imposed on $y$ since any solution of an $n$th-order linear differential equation belongs to $\mathscr{C}^{(n)}(I)$ for some $I$, and hence has derivatives up to and including those of order $n$ at each point in $I$.

In later sections we shall explore some of the deeper consequences of these theorems, and for the moment we mention in passing a useful fact that is an immediate consequence of them.

**Corollary 4–1**  *If L is a normal nth-order linear differential operator defined on I, then the initial-value problem*

$$Ly = 0,$$

$$y(x_0) = y'(x_0) = \cdots = y^{(n-1)}(x_0) = 0,$$

*has y = 0 as its only solution.*

*Proof.*  There is really nothing to do since $y = 0$ is clearly a solution of the given problem, and by Theorem 4–5 it is the only one.  ∎

Finally, since the equation $Ly = h$ is an operator equation of the type discussed in Section 3–6, it follows that if $L$ were one-to-one we could, in theory at least, solve this equation by finding an inverse for $L$. However, as we have seen, linear differential operators are *not* one-to-one, so in the absence of additional restrictions on the domain of $L$ it is impossible to use the method of inverting the operator to solve the given equation. This is precisely where the initial conditions enter the problem. Indeed, the uniqueness theorem stated above can be read as asserting that a complete set of initial conditions of the type displayed in (4–25) provides just such a restriction on $L$. In other words, if we let the domain of $L$ consist of only those functions in $\mathscr{C}^{(n)}(I)$ which satisfy such a set of initial conditions, then $L$ becomes one-to-one and hence has a unique inverse $G$. Hence the task of solving an initial-value problem involving $L$ comes down to finding an explicit formula for the inverse operator $G$, since once $G$ is known the problem

$$Ly = h,$$

$$y(x_0) = y_0, \quad y'(x_0) = y_1, \quad \cdots, \quad y^{(n-1)}(x_0) = y_{n-1},$$

can be solved by computing $G[h]$. This is just what we shall do in the next chapter when we turn our attention to linear differential operators with constant coefficients. But first we shall continue our study of the behavior of the solutions of an arbitrary (normal) linear differential equation.

## EXERCISES

1. Solve each of the following initial-value problems and determine the domain of the solution.
   a)  $xy' + 2y = 0, \quad y(1) = -1$
   b)  $(\sin x)y' + (\cos x)y = 0, \quad y(3\pi/4) = 2$
   c)  $2y' + 3y = e^{-x}, \quad y(-3) = -3$
   d)  $(1 - x^2)y' - 2xy = 0; \quad y(0) = 1$
   e)  $(1 + \sin x)y' + (\cot x)y = \cos x, \quad y(\pi/2) = 1$

f) $x^2 y' + xy/\sqrt{1 - x^2} = 1 + \sqrt{1 - x^2}, \quad y(\tfrac{1}{2}) = 0$

g) $(x \sin x) y' + (\sin x + x \cos x) y = e^x/x, \quad y(-\tfrac{1}{2}) = 0$

2. Let $P$ and $Q$ be continuous on an interval $I$, and let $x_0$ be a point in $I$.

   a) Show that the general solution of

   $$y' + P(x)y = Q(x)$$

   can be written as

   $$y = e^{-\int_{x_0}^x P(t)\, dt} \left( \int_{x_0}^x Q(t) e^{\int_{x_0}^t P(u)\, du}\, dt + c \right),$$

   where $x_0$ and $x$ are in $I$, and $c$ is an arbitrary constant.

   b) Use the result in (a) to find the particular solution of $y' + P(x)y = Q(x)$ that satisfies the initial condition $y(x_0) = y_0$.

3. Prove that the graph of a nontrivial solution of a *homogeneous* normal first-order linear differential equation cannot intersect the x-axis. (Recall that the trivial solution of such an equation is the solution $y = 0$.)

4. Prove that two distinct solutions of a normal first-order linear differential equation cannot intersect.

5. Give an example to show that the conclusion of Theorem 4–4 fails when the equation is not normal on $I$.

6. a) Find an inverse $G$ for the linear differential operator $D$ such that $G[h](x_0) = y_0$.

   b) Find an inverse $G$ for the linear differential operator $D - k$, where $k$ is a constant, such that the condition $G[h](x_0) = y_0$.

\* 7. Solve the initial-value problem

   $$y' + ay = \int_0^b y(t)\, dt, \qquad y(0) = c,$$

   where $a$, $b$, and $c$ are constants. [*Hint*: Set $\int_0^b y(t)\, dy = k$, solve for $y$ in terms of $k$, and then determine $k$.]

8. Let $P, Q_1$, and $Q_2$ be continuous for all $x \geq 0$, and let $y_1$ be a solution of

   $$y' + P(x)y = Q_1(x), \qquad y(0) = c_1,$$

   and $y_2$ a solution of

   $$y' + P(x)y = Q_2(x), \qquad y(0) = c_2.$$

   a) Show that if $Q_1 = Q_2$ and $c_1 > c_2$, then $y_1(x) > y_2(x)$ for all $x \geq 0$.

   b) Show that if $Q_1(x) > Q_2(x)$ for all $x \geq 0$ and $c_1 = c_2$, then $y_1(x) > y_2(x)$.

## 4–5 DIMENSION OF THE SOLUTION SPACE

In this section we shall use the existence and uniqueness theorems stated in the preceding section to give an elegant proof of the fact that the dimension of the solution space of every *normal* homogeneous linear differential equation is equal to the order of the equation. Before we begin, however, we note that this result need not hold for an equation whose leading coefficient vanishes somewhere in the interval under consideration. (An example is provided by the equation $xy' + y = 0$ on any interval containing the origin.) Thus the assumption of normality is essential in the following discussion.

**Theorem 4–6**   *The solution space of any normal nth-order homogeneous linear differential equation*

$$a_n(x)\frac{d^n y}{dx^n} + \cdots + a_0(x)y = 0 \qquad (4\text{--}26)$$

*defined on an interval $I$ is an n-dimensional subspace of $\mathscr{C}^{(n)}(I)$.*

*Proof.* Let $x_0$ be a fixed point in $I$. Then by Theorem 4–4 this equation admits solutions $y_1(x), \ldots, y_n(x)$ which satisfy the initial conditions

$$
\begin{aligned}
y_1(x_0) &= 1, & y_1'(x_0) &= 0, \ldots, y_1^{(n-1)}(x_0) = 0 \\
y_2(x_0) &= 0, & y_2'(x_0) &= 1, \ldots, y_2^{(n-1)}(x_0) = 0 \\
&\;\cdot \\
&\;\cdot \qquad\qquad\qquad\qquad\qquad\qquad\qquad (4\text{--}27) \\
&\;\cdot \\
y_n(x_0) &= 0, & y_n'(x_0) &= 0, \ldots, y_n^{(n-1)}(x_0) = 1.
\end{aligned}
$$

In other words, $y_1(x), \ldots, y_n(x)$ have the property that the vectors

$$(y_i(x_0), y_i'(x_0), \ldots, y_i^{(n-1)}(x_0)), \qquad i = 1, \ldots, n,$$

are the standard basis vectors in $\mathscr{R}^n$. This choice of solutions is illustrated in Fig. 4–8 for a second-order equation, in which case

$$(y_1(x_0), y_1'(x_0)) = (1, 0), \qquad (y_2(x_0), y_2'(x_0)) = (0, 1).$$

We assert that these solutions are a *basis* for the solution space of (4–26).

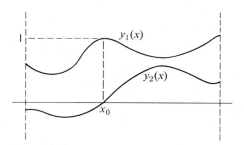

**Figure 4–8**

In the first place, the linear independence of

$$y_1(x), \ldots, y_n(x)$$

in $\mathscr{C}^{(n)}(I)$ follows at once from Theorem 2–3. Thus we need only show that *every* solution of (4–26) can be written as a linear combination of $y_1(x), \ldots, y_n(x)$. To this end let $y(x)$ be an arbitrary solution of the equation and suppose that

$$y(x_0) = a_0, \quad y'(x_0) = a_1, \ldots, y^{(n-1)}(x_0) = a_{n-1}.$$

Then by Theorem 4–5 we know that $y(x)$ is *the* solution of (4–26) which satisfies these particular initial conditions. But by (4–27) we see that the function

$$a_0 y_1(x) + a_1 y_2(x) + \cdots + a_{n-1} y_n(x)$$

also satisfies this initial-value problem. Hence

$$y(x) = a_0 y_1(x) + a_1 y_2(x) + \cdots + a_{n-1} y_n(x),$$

and we are done.  ∎

Notice that the particular numerical values used to fix the solutions $y_1(x), \ldots, y_n(x)$ did not play an essential role in the argument just given. The success of the proof depended only on the linear independence of the vectors

$$(y_i(x_0), y_i'(x_0), \ldots, y_i^{(n-1)}(x_0)), \qquad i = 1, \ldots, n,$$

in $\mathcal{R}^n$, and the choice made in (4–27) merely served to simplify our computations.

**Example 1**   Since the second-order equation

$$\frac{d^2 y}{dx^2} - y = 0 \tag{4–28}$$

is normal on the entire $x$-axis, its solution space is a 2-dimensional subspace of $\mathscr{C}(-\infty, \infty)$. By substitution in the equation we can easily show that

$$y_1(x) = \tfrac{1}{2}(e^x + e^{-x}) = \cosh x,$$
$$y_2(x) = \tfrac{1}{2}(e^x - e^{-x}) = \sinh x$$

are solutions of (4–28) on $(-\infty, \infty)$. Moreover, the vectors

$$(y_1(0), y_1'(0)) = (1, 0),$$
$$(y_2(0), y_2'(0)) = (0, 1)$$

are linearly independent. Thus, by Theorem 4–6, $\cosh x$ and $\sinh x$ are a basis for the solution space of this equation, and it follows that the general solution of (4–28) is

$$y = c_1 \cosh x + c_2 \sinh x,$$

where $c_1$ and $c_2$ are arbitrary constants.

**Example 2**   The functions

$$y_1(x) = e^x, \qquad y_2(x) = e^{-x}$$

provide a second pair of solutions of Eq. (4–28). In this case

$$y_1(0) = 1, \qquad y_1'(0) = 1,$$
$$y_2(0) = 1, \qquad y_2'(0) = -1,$$

and since the vectors $(1, 1)$ and $(1, -1)$ are linearly independent in $\mathcal{R}^2$, $e^x$ and $e^{-x}$ also form a basis for the solution space of the equation. Hence the general solution of (4–28) may also be written

$$y = c_1 e^x + c_2 e^{-x},$$

which, of course, is just a variant of the solution obtained in Example 1.

**Example 3** The functions

$$y_1(x) = \sin 2x, \qquad y_2(x) = \cos 2x$$

are solutions of the normal second-order equation

$$\frac{d^2 y}{dx^2} + 4y = 0 \tag{4–29}$$

on $(-\infty, \infty)$. Furthermore,

$$y_1(0) = 0, \qquad y_1'(0) = 2,$$

while

$$y_2(0) = 1, \qquad y_2'(0) = 0,$$

and since $(0, 2)$ and $(1, 0)$ are linearly independent vectors in $\mathcal{R}^2$, $\sin 2x$ and $\cos 2x$ are linearly independent in $\mathcal{C}(-\infty, \infty)$. Hence these functions are a basis for the solution space of (4–29), and the general solution of that equation is

$$y = c_1 \sin 2x + c_2 \cos 2x.$$

## EXERCISES

1. Show that the given functions form a basis for the solution space of each of the following differential equations, and find a basis for the solution space of the equation that satisfies initial conditions of the form (4–27) at the point $x_0$.
   a) $y'' - 2y' + y = 0$; $e^x, xe^x$; $x_0 = 0$
   b) $y''' - y'' - 2y' = 0$; $1, 2e^{2x}, e^{-x}$; $x_0 = 0$
   c) $y''' - y' = 0$; $1, e^x, e^{-x}$; $x_0 = 0$
   d) $x^2 y'' - 2y = 0, x > 0$; $3x^2, 1/x$; $x_0 = 1$
   e) $x^3 y''' + 2x^2 y'' - xy' + y = 0, x > 0$; $x, 1/x, x \ln x$; $x_0 = 1$

2. Repeat Exercise 1 for: $y^{(iv)} + 2y'' + y = 0$; $\sin x, \cos x, x \sin x, x \cos x$; $x_0 = 0$.

3. Use the equation $xy' + y = 0$ to show that the conditions of Theorem 4–6 need not hold when the hypothesis of normality is not satisfied.

4. Prove that distinct solutions of a normal second-order linear differential equation never have a point of mutual tangency.

5. Suppose that $y_1(x), \ldots, y_n(x)$ are solutions of a normal homogeneous linear differential equation of order $m \geq n$ on an interval $I$. Prove that if $y_1(x), \ldots, y_n(x)$ are linearly independent in $\mathcal{C}(I)$, then they are linearly independent in $\mathcal{C}(J)$ for any subinterval $J$ of $I$. Give an example to show that this conclusion need not hold if $y_1(x), \ldots, y_n(x)$ do not satisfy such an equation.

### 4–6 THE WRONSKIAN

In Chapter 2 we proved that $y_1, \ldots, y_n$ are linearly independent functions in $\mathscr{C}^{(n-1)}(I)$ whenever there exists a point $x_0$ in $I$ at which the $n$ vectors

$$(y_i(x_0), y_i'(x_0), \ldots, y_i^{(n-1)}(x_0)), \qquad i = 1, \ldots, n, \qquad (4\text{–}30)$$

are linearly independent in $\mathscr{R}^n$. For our present purposes this result can be stated more conveniently in terms of a determinant.

Let $y_1, \ldots, y_n$ be arbitrary functions in $\mathscr{C}^{(n-1)}(I)$, and for each $x$ in $I$ consider the determinant

$$\begin{vmatrix} y_1(x) & y_2(x) & \cdots & y_n(x) \\ y_1'(x) & y_2'(x) & \cdots & y_n'(x) \\ \vdots & & & \vdots \\ y_1^{(n-1)}(x) & y_2^{(n-1)}(x) \cdots & y_n^{(n-1)}(x) \end{vmatrix}. \qquad (4\text{–}31)$$

*Wronskian*

This determinant defines a real-valued function on the interval $I$ known as the **Wronskian** of $y_1, \ldots, y_n$. We will denote the Wronskian by $W[y_1, \ldots, y_n]$, to indicate its dependence on $y_1, \ldots, y_n$. Its value at $x$ will be denoted by $W[y_1, \ldots, y_n](x)$ or $W[y_1(x), \ldots, y_n(x)]$. In short, the Wronskian of $y_1, \ldots, y_n$ is the real-valued function whose defining equation is

$$W[y_1, \ldots, y_n](x) = \begin{vmatrix} y_1(x) & y_2(x) & \cdots & y_n(x) \\ y_1'(x) & y_2'(x) & \cdots & y_n'(x) \\ \vdots & & & \vdots \\ y_1^{(n-1)}(x) & y_2^{(n-1)}(x) \cdots & y_n^{(n-1)}(x) \end{vmatrix}. \qquad (4\text{–}32)$$

For example,

$$W[x, \sin x] = \begin{vmatrix} x & \sin x \\ 1 & \cos x \end{vmatrix} = x \cos x - \sin x,$$

and

$$W[x, 2x] = \begin{vmatrix} x & 2x \\ 1 & 2 \end{vmatrix} = 0.$$

Since an $n \times n$ determinant is nonzero if and only if its columns are linearly independent when viewed as vectors in $\mathscr{R}^n$ (see Section 2–10), the Wronskian of $y_1, \ldots, y_n$ is different from zero at a point $x_0$ if and only if the columns of (4–32) are linearly independent when $x = x_0$. But for each $x_0$ in $I$ the columns of (4–32) are none other than the vectors in (4–30). Therefore we have the following theorem.

**Theorem 4–7**  *The functions $y_1, \ldots, y_n$ are linearly independent in $\mathscr{C}^{(n-1)}(I)$, and hence also in $\mathscr{C}(I)$, whenever their Wronskian is **not** identically zero on $I$.*

**Example 1**   Since

$$W[e^x, e^{-x}] = \begin{vmatrix} e^x & e^{-x} \\ e^x & -e^{-x} \end{vmatrix} = -2,$$

the functions $e^x$ and $e^{-x}$ are linearly independent in $\mathscr{C}(I)$ for any interval $I$.

**Example 2**   The functions $x$, $x^{1/2}$, $x^{3/2}$ are linearly independent in $\mathscr{C}(I)$ for any sub-interval $I$ of the positive $x$-axis, since

$$W[x, x^{1/2}, x^{3/2}] = \begin{vmatrix} x & x^{1/2} & x^{3/2} \\ 1 & \frac{1}{2}x^{-1/2} & \frac{3}{2}x^{1/2} \\ 0 & -\frac{1}{4}x^{-3/2} & \frac{3}{4}x^{-1/2} \end{vmatrix} = -\frac{1}{4}.$$

More generally, $x^\alpha$, $x^\beta$, $x^\gamma$ are linearly independent in $\mathscr{C}(I)$, where $I$ is as above, if and only if $\alpha$, $\beta$, $\gamma$ are distinct real numbers (see Exercise 6 of Section 2–6).

**Example 3**   The functions $x^3$ and $|x|^3$ are linearly independent in $\mathscr{C}(-\infty, \infty)$, for if $c_1 x^3 + c_2 |x|^3 \equiv 0$, then

$$c_1(1)^3 + c_2|1|^3 = 0,$$
$$c_1(-1)^3 + c_2|-1|^3 = 0,$$

and $c_1 = c_2 = 0$. On the other hand, the Wronskian of $x^3$ and $|x|^3$ is identically zero on $(-\infty, \infty)$, since if $x \geqslant 0$, then

$$W[x^3, |x|^3] = \begin{vmatrix} x^3 & x^3 \\ 3x^2 & 3x^2 \end{vmatrix} = 0,$$

while if $x < 0$, then

$$W[x^3, |x|^3] = \begin{vmatrix} x^3 & -x^3 \\ 3x^2 & -3x^2 \end{vmatrix} = 0.$$

This shows that the converse of Theorem 4–7 is *false*. One cannot deduce the linear *dependence* of a set of functions in $\mathscr{C}(I)$ from the fact that their Wronskian vanishes identically on $I$. *Warning* (See Fig. 4–9.)

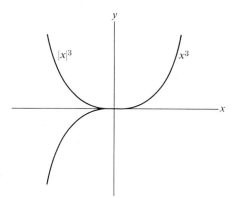

**Figure 4–9**

Example 3 notwithstanding, it *is* true that the Wronskian of a linearly dependent set of functions in $\mathscr{C}(I)$ vanishes identically on $I$, provided, of course, that
the Wronskian exists in the first place. Hence, instead of abandoning the search
for a converse to Theorem 4–7, we shall weaken our requirements and ask whether
it is possible to impose additional conditions on a set of functions which, together
with the vanishing of their Wronskian on $I$, will imply linear dependence. This
can in fact be done: *we simply require that the functions be solutions of a homogeneous linear differential equation.*

**Theorem 4–8**    *Let* $y_1, \ldots, y_n$ *be solutions of a normal nth-order homogeneous
linear differential equation*

$$a_n(x)\frac{d^n y}{dx^n} + \cdots + a_0(x)y = 0 \tag{4–33}$$

*on an interval* $I$, *and suppose that* $W[y_1, \ldots, y_n]$ *is identically zero on* $I$. *Then*
$y_1, \ldots, y_n$ *are linearly dependent in* $\mathscr{C}(I)$.

*Proof.*    Let $x_0$ be any point in $I$, and consider the system of linear equations

$$
\begin{aligned}
c_1 y_1(x_0) + \cdots + c_n y_n(x_0) &= 0 \\
c_1 y_1'(x_0) + \cdots + c_n y_n'(x_0) &= 0 \\
&\quad\vdots \\
c_1 y_1^{(n-1)}(x_0) + \cdots + c_n y_n^{(n-1)}(x_0) &= 0,
\end{aligned}
\tag{4–34}
$$

in the unknowns $c_1, \ldots, c_n$. Since the Wronskian of $y_1, \ldots, y_n$ vanishes identically
on $I$, the determinant of (4–34) is zero, and the system has a *nontrivial* solution
$(\bar{c}_1, \ldots, \bar{c}_n)$. (See Section 2–10.) Thus the function

$$y(x) = \sum_{i=1}^{n} \bar{c}_i y_i(x)$$

is a solution of the initial-value problem consisting of (4–33) and the initial
conditions
$$y(x_0) = 0, \quad y'(x_0) = 0, \quad \ldots, \quad y^{(n-1)}(x_0) = 0.$$

Hence, by Corollary 4–1,
$$\bar{c}_1 y_1(x) + \cdots + \bar{c}_n y_n(x) = 0$$

for all $x$ in $I$, and the linear dependence of $y_1, \ldots, y_n$ now follows from the fact
that the $\bar{c}_i$ are not all zero.    ∎

Actually we have established a result that is considerably stronger than the
one advertised. For the preceding proof made use only of the fact that the Wronskian of $y_1, \ldots, y_n$ vanished *at a single point* in $I$, so that the conclusion will remain

valid under this less restrictive hypothesis. Combined with Theorem 4–7, this observation yields

**Theorem 4–9** *A set of n solutions of a normal nth-order homogeneous linear differential equation is linearly independent in $\mathscr{C}(I)$, hence is a basis for the solution space of the equation, if and only if its Wronskian **never** vanishes on I.*

**Example 4** By substitution we can show that $\sin^3 x$ and $1/\sin^2 x$ are solutions of

$$\frac{d^2y}{dx^2} + \tan x \frac{dy}{dx} - 6(\cot^2 x)y = 0$$

on any interval $I$ in which $\tan x$ and $\cot x$ are both defined. Moreover,

$$W\left[\sin^3 x, \frac{1}{\sin^2 x}\right] = \begin{vmatrix} \sin^3 x & \dfrac{1}{\sin^2 x} \\ 3\sin^2 x \cos x & -\dfrac{2\cos x}{\sin^3 x} \end{vmatrix} = -5\cos x.$$

Since $\cos x$ is never zero on an interval $I$ on which $\tan x$ is defined, Theorem 4–9 implies that $\sin^3 x$ and $1/\sin^2 x$ are linearly independent in $\mathscr{C}(I)$ and that the general solution of the given equation is

$$y = c_1 \sin^3 x + \frac{c_2}{\sin^2 x}.$$

(This result can also be obtained directly from Theorem 4–7.)

**Example 5** The functions

$$y_1(x) = \sin^3 x, \qquad y_2(x) = \sin x - \tfrac{1}{3}\sin 3x$$

are solutions of

$$\frac{d^2y}{dx^2} + (\tan x - 2\cot x)\frac{dy}{dx} = 0 \qquad\qquad (4\text{–}35)$$

on any interval $I$ in which $\tan x$ and $\cot x$ are defined. But

$$W[y_1(x), y_2(x)] = \begin{vmatrix} \sin^3 x & \sin x - \tfrac{1}{3}\sin 3x \\ 3\sin^2 x \cos x & \cos x - \cos 3x \end{vmatrix}$$

$$= \sin^3 x(\cos x - \cos 3x) - 3\sin^2 x \cos x(\sin x - \tfrac{1}{3}\sin 3x)$$

$$= \sin^2 x(\sin 3x \cos x - \sin x \cos 3x) - \sin^2 x(2\sin x \cos x)$$

$$= \sin^2 x \sin 2x - \sin^2 x \sin 2x = 0.$$

Hence $y_1$ and $y_2$ are linearly dependent in $\mathscr{C}(I)$ and do *not* form a basis for the solution space of (4–35). In fact, $\sin^3 x = \tfrac{3}{4}\sin x - \tfrac{1}{4}\sin 3x$. However, any constant $c$ is a solution of (4–35), and since $c$ and $\sin^3 x$ are linearly independent in $\mathscr{C}(I)$ when $c \neq 0$, the general solution of the equation is

$$y = c_1 + c_2 \sin^3 x.$$

Of course, this expression may also be written as

$$y = c_1 + c_2(\sin x - \tfrac{1}{3}\sin 3x).$$

## EXERCISES

By computing Wronskians show that each of the following sets of functions is linearly independent in $\mathscr{C}(I)$ for the indicated interval $I$.

1. $1, e^{-x}, 2e^{2x}$, on any interval $I$

2. $e^x, \sin 2x$, on any interval $I$

3. $1, x, x^2, \ldots, x^n$, on any interval $I$

4. $\ln x, x \ln x$, on $(0, \infty)$

5. $x^{1/2}, x^{1/3}$, on $(0, \infty)$

6. $e^{ax} \sin bx, e^{ax} \cos bx$ $(b \neq 0)$, on any interval $I$

7. $e^x, e^x \sin x$, on any interval $I$

8. $e^{-x}, xe^{-x}, x^2 e^{-x}$, on any interval $I$

9. $1, \sin^2 x, 1 - \cos x$, on any interval $I$

10. $\ln \dfrac{x-1}{x+1}, 1$, on $(-\infty, -1)$

11. $\sqrt{1-x^2}, x$, on $(-1, 1)$

12. $\sin (x/2), \cos^2 x$, on any interval $I$

13. Show that $x^\alpha, x^\beta, x^\gamma$ are linearly independent in $\mathscr{C}(0, \infty)$ if and only if they are linearly independent in $\mathscr{C}(I)$ for *every* subinterval $I$ of $(0, \infty)$. [*Hint*: First establish the following assertions, and then use Theorem 4–9:

   a) $x^\alpha, x^\beta, x^\gamma$ satisfy the linear differential equation

$$x^3 y''' + a_2 x^2 y'' + a_1 xy' + a_0 y = 0,$$

   where

$$a_2 = 3 - \alpha - \beta - \gamma,$$
$$a_1 = 1 - \alpha - \beta - \gamma + \alpha\beta + \alpha\gamma + \beta\gamma,$$
$$a_0 = -\alpha\beta\gamma.$$

   b) $W(x^\alpha, x^\beta, x^\gamma) = x^{\alpha+\beta+\gamma-3} \begin{vmatrix} 1 & 1 & 1 \\ \alpha & \beta & \gamma \\ \alpha(\alpha-1) & \beta(\beta-1) & \gamma(\gamma-1) \end{vmatrix}$,

   and hence $W(x^\alpha, x^\beta, x^\gamma)$ either vanishes nowhere in $(0, \infty)$ or vanishes identically.]

14. Let $f$ belong to $\mathscr{C}^{(1)}[a, b]$ and suppose that $f$ is not the zero function. By computing their Wronskian show that $f(x)$ and $xf(x)$ are linearly independent in $\mathscr{C}^{(1)}[a, b]$. (Also see Exercise 15.)

15. Show that $f(x)$ and $xf(x)$ are linearly independent in $\mathscr{C}[a, b]$ if $f$ is continuous and not identically zero on $[a, b]$.

16. Suppose that $f$ is an odd function in $\mathscr{C}^{(1)}[-a, a]$ (that is, $f(-x) = -f(x)$) and that $f(0) = f'(0) = 0$. Show that

$$W[f(x), |f(x)|] = 0$$

   for all $x$ in $[-a, a]$, but that $f$ and $|f|$ are linearly independent in $\mathscr{C}^{(1)}[-a, a]$ unless $f$ is identically zero. Compare this result with Example 3 in the text.

17. Let $f$ and $g$ be any two functions in $\mathscr{C}^{(1)}(I)$ and suppose that $g$ never vanishes in $I$. Prove

that if $W[f(x), g(x)] \equiv 0$ on $I$, then $f$ and $g$ are linearly dependent in $\mathscr{C}^{(1)}(I)$. [Hint: Calculate $d/dx(f(x)/g(x))$.]

*18. Let $f$ and $g$ be any two functions in $\mathscr{C}^{(1)}(I)$ which have only finitely many zeros in $I$ and no common zeros. Prove that if $W[f(x), g(x)] \equiv 0$ and $I$, then $f$ and $g$ are linearly dependent in $\mathscr{C}^{(1)}(I)$. [Hint: Apply the result of Exercise 17 to the finite number of subintervals of $I$ on which $f$ or $g$ never vanishes.]

*19. a) Show that

$$W[e^{a_1 x}, \ldots, e^{a_n x}] = e^{(a_1 + \cdots + a_n)x} \begin{vmatrix} 1 & 1 & \cdots & 1 \\ a_1 & a_2 & \cdots & a_n \\ a_1^2 & a_2^2 & \cdots & a_n^2 \\ \cdot & \cdot & & \cdot \\ \cdot & \cdot & & \cdot \\ \cdot & \cdot & & \cdot \\ a_1^{n-1} & a_2^{n-1} & \cdots & a_n^{n-1} \end{vmatrix}.$$

b) The determinant appearing in part (a) is known as a *Vandermonde* determinant. Show that it is zero if and only if $a_i = a_j$ for some pair of indices $i, j$ with $i \neq j$. [Hint: Expand the determinant by cofactors of the 1st column to obtain a polynomial in $a_1$. Is $a_2$ a root of this polynomial?]

*20. Prove that if $u_1, \ldots, u_n$ are a basis for the solution space of the linear differential equation

$$y^{(n)} + a_{n-1}(x)y^{(n-1)} + \cdots + a_0(x)y = 0,$$

then the $a_i(x)$, $0 \leqslant i \leqslant n - 1$, are *uniquely determined* by $u_1, \ldots, u_n$. [Hint: For each index $i$, let $u_{i,x_0}^{(j)}$ be *the* solution of the given equation that satisfies the initial conditions $u_{i,x_0}^{(j)}(x_0) = 0$, $0 \leqslant j \leqslant n - 1$, $j \neq i$, and $u_{i,x_0}^{(i)}(x_0) = -1$. Then $a_i(x_0) = u_{i,x_0}^{(n)}(x_0)$.]

21. Let $u_1$ and $u_2$ be linearly independent solutions of the normal second-order linear differential equation

$$y'' + a_1(x)y' + a_0(x)y = 0.$$

Express the coefficients $a_0(x)$, $a_1(x)$ in terms of $u_1$ and $u_2$. [Hint: Let $y$ be an arbitrary solution of the equation, and consider the Wronskian of $y$, $u_1$, $u_2$.]

22. Generalize the result of the preceding exercise to an $n$th-order equation

$$y^{(n)} + a_{n-1}(x)y^{(n-1)} + \cdots + a_0(x)y = 0.$$

23. For each pair of functions listed below, use the results of Exercise 21 to find a homogeneous second-order linear differential equation whose solution space has the functions as a basis.

a) $x, xe^x$
b) $x, x^2$
c) $\sin x, \cos x$
d) $x, \sin x$
e) $x, \ln x$

## 4–7 ABEL'S FORMULA

According to Theorem 4–8, the Wronskian of a set of solutions of a normal homogeneous linear differential equation vanishes either identically or not at all. This fact can also be deduced from our next theorem, which gives an explicit formula, called **Abel's formula**, for the Wronskian in this case.

> **Theorem 4–10**  Let $y_1, \ldots, y_n$ be solutions on an interval $I$ of the normal nth-order equation
>
> $$\frac{d^n y}{dx^n} + a_{n-1}(x)\frac{d^{n-1}y}{dx^{n-1}} + \cdots + a_0(x)y = 0. \qquad (4\text{–}36)$$
>
> Then
>
> $$W[y_1, \ldots, y_n](x) = ce^{-\int a_{n-1}(x)\,dx} \qquad (4\text{–}37)$$
>
> for an appropriate constant $c$.

*Proof.*    To avoid using general properties of determinants, we shall prove (4–37) only for $n = 2$. The general proof is identical to the one we give here except that it uses a special formula for the derivative of an nth-order determinant (see Exercise 9). Thus suppose that $y_1$ and $y_2$ are solutions of

$$\frac{d^2 y}{dx^2} + a_1(x)\frac{dy}{dx} + a_0(x)y = 0.$$

Then a straightforward calculation gives

$$\frac{d}{dx} W[y_1, y_2](x) = \frac{d}{dx} \begin{vmatrix} y_1(x) & y_2(x) \\ y_1'(x) & y_2'(x) \end{vmatrix}$$

$$= \frac{d}{dx}\left[ y_1(x)y_2'(x) - y_2(x)y_1'(x) \right]$$

$$= y_1(x)y_2''(x) - y_1''(x)y_2(x)$$

$$= y_1(x)\left[ -a_1(x)y_2'(x) - a_0(x)y_2(x) \right]$$

$$\qquad\qquad - y_2(x)\left[ -a_1(x)y_1'(x) - a_0(x)y_1(x) \right]$$

$$= -a_1(x)\left[ y_1(x)y_2'(x) - y_2(x)y_1'(x) \right]$$

$$= -a_1(x)W[y_1(x), y_2(x)].$$

Thus $W[y_1, y_2](x)$ is differentiable on $I$, and satisfies the first-order linear differential equation

$$\frac{dy}{dx} + a_1(x)y = 0.$$

But the general solution of this equation is

$$y = ce^{-\int a_1(x)\,dx}.$$

Hence

$$W[y_1, y_2](x) = ce^{-\int a_1(x)\,dx}$$

for some value of $c$.  ∎

The determination of the appropriate value of $c$ for Abel's formula (4–37) depends, of course, on one's choice of the constant of integration in $\int a_{n-1}(x)\,dx$. But once the choice has been made, the process is not at all mysterious. For example, if we choose a point $x_0$ of $I$, and use $\int_{x_0}^x a_{n-1}(x)\,dx$ for $\int a_{n-1}(x)\,dx$, then

$c$ is determined by the equation

$$W[y_1, \ldots, y_n](x) = ce^{-\int_{x_0}^{x} a_{n-1}(x)\,dx}.$$

If we set $x = x_0$ we see that $c = W[y_1, \ldots, y_n](x_0)$, which we abbreviate as $W(x_0)$ to get

$$W[y_1, \ldots, y_n](x) = W(x_0)e^{-\int_{x_0}^{x} a_{n-1}(x)\,dx}. \tag{4–38}$$

This formula shows that the Wronskian of any basis for the solution space of a normal homogeneous linear differential equation is determined up to a multiplicative constant by the equation itself, and does not depend on the particular basis used to compute it. This observation will be important later on.

**Example 1**   Since the equation

$$x\frac{d^2y}{dx^2} + \frac{dy}{dx} + xy = 0 \qquad \text{divide by } x \text{ other}$$

is normal on $(0, \infty)$, the Wronskian of any two solutions $y_1$, $y_2$ of this equation must be of the form

$$y'' + \frac{1}{x}y + ty = 0$$

$$W[y_1, y_2](x) = ce^{-\int(dx/x)} = \frac{c}{x}.$$

If, in addition, $y_1$ and $y_2$ satisfy the initial conditions

$$y_1(x_0) = a_0, \qquad y_1'(x_0) = a_1,$$
$$y_2(x_0) = b_0, \qquad y_2'(x_0) = b_1,$$

at some point $x_0 > 0$, then

$$c = x_0\begin{vmatrix} a_0 & b_0 \\ a_1 & b_1 \end{vmatrix} = x_0(a_0b_1 - a_1b_0),$$

and

$$W[y_1, y_2](x) = \frac{x_0(a_0b_1 - a_1b_0)}{x}.$$

$$W(y_1, y_2) = W(x_0) \cdot e^{-\int_{x_0}^{x} \frac{dt}{t}}$$
$$= W(x_0) \cdot e^{-\ln x - \ln x_0}$$
$$= W(x_0)\, e^{\ln \frac{1}{x}} \cdot e^{\ln x_0}$$
$$= W(x_0) \cdot \frac{1}{x_0} \cdot e^{\ln \frac{1}{x}}$$
$$= \frac{x_0\, W(x_0)}{x}$$

Our first substantial application of Abel's formula will be to find the general solution of a normal second-order homogeneous linear differential equation when a nonvanishing solution of the equation is already known. Specifically, let $y_1$ be a solution of

$$y'' + a_1(x)y' + a_0(x)y = 0 \tag{4–39}$$

that does not vanish at any point of $I$. Since every solution $y_2$ of (4–39) satisfies the equation

$$W[y_1, y_2](x) = ce^{-\int a_1(x)\,dx}$$

on $I$, a solution $y_2$ can be found by solving the nonhomogeneous *first-order* equation

$$y_1(x)\frac{dy_2}{dx} - y_1'(x)y_2 = ce^{-\int a_1(x)\,dx}.$$

By (4–7) the general solution of this equation on any interval on which $y_1 \neq 0$ is

$$y_2 = cy_1(x) \int \frac{e^{-\int a_1(x)\, dx}}{y_1(x)^2}\, dx + ky_1(x),$$

where $k$ is an arbitrary constant. In particular, this formula for $y_2$ is valid on the interval $I$, and when $c = 1$ and $k = 0$, we get

$$y_2(x) = y_1(x) \int \frac{e^{-\int a_1(x)\, dx}}{y_1(x)^2}\, dx.$$

For a general solution of (4–39), of course, we need a $y_2$ that is linearly independent of $y_1$, but a calculation of $W[y_1, y_2]$ shows this to be the case. (See Exercise 10.) Thus we have

> **Theorem 4–11**  *If $y_1$ is a solution of*
>
> $$y'' + a_1(x)y' + a_0(x)y = 0 \qquad (4\text{–}40)$$
>
> *on an interval $I$, and if $y_1$ does not vanish at any point of $I$, then*
>
> $$y_2(x) = y_1(x) \int \frac{e^{-\int a_1(x)\, dx}}{y_1(x)^2}\, dx \qquad (4\text{–}41)$$
>
> *is a solution on $I$. Moreover, $y_1$ and $y_2$ are linearly independent and the general solution of (4–40) is*
>
> $$y = c_1 y_1 + c_2 y_2,$$
>
> *where $c_1$ and $c_2$ are arbitrary constants.*

**Example 2**   By substitution we find that $x^2$ is a solution on $(0, \infty)$ of the second-order equation

$$x^2 y'' + x^3 y' - 2(1 + x^2)y = 0. \qquad (4\text{–}42)$$

Hence a second linearly independent solution in $\mathscr{C}(0, \infty)$ can be found by solving the first-order equation

$$x^2 y' - 2xy = e^{-x^2/2}.$$

From Formula (4–41) we obtain

$$y_2 = x^2 \int \frac{e^{-\int x\, dx}}{x^4}\, dx = x^2 \int x^{-4} e^{-x^2/2}\, dx,$$

and the general solution of (4–42) on $(0, \infty)$ is

$$y = x^2 \left( c_1 + c_2 \int x^{-4} e^{-x^2/2}\, dx \right).$$

(The solution is left in integral form because $\int x^{-4} e^{-x^2/2}\, dx$ cannot be expressed in terms of elementary functions.)

**Example 3**    The function $y_1 = 1$ is a solution of

$$y'' + (\tan x - 2 \cot x)y' = 0 \tag{4–43}$$

on any interval in which $\tan x$ and $\cot x$ are both defined. From (4–41) we obtain a second solution

$$\begin{aligned}
y_2(x) &= \int e^{-\int(\tan x - 2\cot x)\,dx}\,dx \\
&= \int e^{\ln(\cos x \, \sin^2 x)}\,dx \\
&= \int \sin^2 x \cos x\,dx \\
&= \tfrac{1}{3}\sin^3 x.
\end{aligned}$$

Thus the general solution of (4–43) is

$$y = c_1 + c_2 \sin^3 x,$$

which agrees with the result obtained at the end of the last section.

## EXERCISES

1. For each of the following differential equations find the Wronskian of the solutions $y_1, y_2$ that satisfy the given initial conditions.

   a) $x^2 y'' + xy' + (x^2 + 1)y = 0;$    $y_1(1) = 0,\ y_1'(1) = 1,\ y_2(1) = y_2'(1) = 1$

   b) $(1 - x^2)y'' - 2xy' + n(n + 1)y = 0$ (n a positive integer);    $y_1(0) = y_1'(0) = 2,$
   $y_2(0) = 1,\ y_2'(0) = -1$

   c) $x^2 y'' - 3xy' + y = 0;$    $y_1(-1) = y_1'(-1) = 2,\ y_2(-1) = 0,\ y_2'(-1) = -1$

   d) $y'' + 2xy = 0;$    $y_1(0) = y_1'(0) = 1,\ y_2(0) = 1,\ y_2'(0) = 0$

   e) $y'' - (\sin x)y' + 3(\tan x)y = 0;$    $y_1(0) = 1,\ y_1'(0) = 0,\ y_2(0) = 0,\ y_2'(0) = 1$

   f) $\sqrt{1 + x^3}\,y'' - x^2 y' + y = 0;$    $y_1(1) = 1,\ y_1'(1) = 0,\ y_2(1) = -1,\ y_2'(1) = 1$

In Exercises 2 through 8, a solution $y_1$ of each differential equation is given. Find a second, linearly independent solution by the method of this section.

2. $y'' - 4y' + 4y = 0;$     $y_1(x) = e^{2x}$

3. $y'' - 2ay' + a^2 y = 0;$     $y_1(x) = e^{ax}$

4. $3xy'' - y' = 0;$     $y_1(x) = 1$

5. $y'' + (\tan x)y' - 6(\cot^2 x)y = 0;$     $y_1(x) = \sin^3 x$

6. $(1 - x^2)y'' - 2xy' = 0;$     $y_1(x) = 1$

7. $(1 - x^2)y'' - 2xy' + 2y = 0;$     $y_1(x) = x$

8. $2xy'' - (e^x)y' = 0;$     $y_1(x) = 1$

9. a) Prove that if the functions $f_{ij}$ are in $\mathscr{C}^{(1)}(I),\ 1 \leqslant i, j \leqslant 2$, then so is the function $F$ defined by

   $$F(x) = \begin{vmatrix} f_{11}(x) & f_{12}(x) \\ f_{21}(x) & f_{22}(x) \end{vmatrix},$$

   and that

   $$\frac{d}{dx}\begin{vmatrix} f_{11}(x) & f_{12}(x) \\ f_{21}(x) & f_{22}(x) \end{vmatrix} = \begin{vmatrix} f_{11}'(x) & f_{12}'(x) \\ f_{21}'(x) & f_{22}'(x) \end{vmatrix} + \begin{vmatrix} f_{11}(x) & f_{12}(x) \\ f_{21}'(x) & f_{22}'(x) \end{vmatrix}.$$

\*     b) Generalize the result of part (a) to $n$th-order determinants, and show in particular that

$$\frac{d}{dx}\begin{vmatrix} f_{11}(x) & f_{12}(x) & \cdots & f_{1n}(x) \\ f_{21}(x) & f_{22}(x) & \cdots & f_{2n}(x) \\ \cdot & \cdot & & \cdot \\ \cdot & \cdot & & \cdot \\ \cdot & \cdot & & \cdot \\ f_{n1}(x) & f_{n2}(x) & \cdots & f_{nn}(x) \end{vmatrix}$$

can be expressed as the sum of $n$ determinants, the $i$th of which is obtained from $|f_{ij}(x)|$ by differentiating the functions in the $i$th column.

10.  Show that the functions

$$y_1 = y_1(x), \qquad y_2 = y_1(x) \int \frac{e^{-\int a_1(x)\,dx}}{y_1(x)^2}\,dx$$

described in Theorem 4–11 are linearly independent in $\mathscr{C}(I)$. (You may prefer to rewrite $y_2$ as

$$y_2 = y_1(x) \int_{x_0}^x \frac{e^{-\int_{x_0}^t a_1(u)\,du}}{y_1(t)^2}\,dt.)$$

## \*4–8  THE SOLUTIONS OF $y'' + y = 0$

By now it should be obvious that the theorems we have in hand are powerful tools for studying linear differential equations. What is not so obvious is that they can also be used to obtain detailed information about the solutions of an individual equation. Before going any further we propose to illustrate this aspect of our results. Since the method we wish to illustrate is applied almost exclusively in rather complicated situations, we have chosen to introduce it by means of an example which, though somewhat artificial, has the merit of absolute clarity. In the process we will undoubtedly give the impression that we are resolutely shutting our eyes to the obvious (which, in fact, is what we will be doing), but the technique in question and the spirit underlying its application are of considerable importance.

The problem we set for ourselves is to study the solutions of the second-order equation

$$y'' + y = 0 \tag{4–44}$$

without using any information other than what is provided by the equation itself and the general theorems we have proved so far.

In the first place, this equation is normal on $(-\infty, \infty)$. Thus we can apply Theorem 4–6 to assert that its solution space is spanned by two linearly independent functions, $C(x)$ and $S(x)$, which are defined for all $x$ and satisfy the initial conditions

$$\begin{aligned} C(0) &= 1, & C'(0) &= 0, \\ S(0) &= 0, & S'(0) &= 1. \end{aligned} \tag{4–45}$$

Moreover, from the identities

$$C''(x) + C(x) = 0, \qquad S''(x) + S(x) = 0, \tag{4–46}$$

which are valid for all $x$, we conclude that both $C(x)$ and $S(x)$ are infinitely differentiable on $(-\infty, \infty)$, and that all of their derivatives are also solutions of (4–44). For example, the identity $C''(x) + C(x) = 0$ implies that $C''(x)$ is differentiable, since $C(x)$ is, and that $C'''(x) + C'(x) = 0$. But this is just (4–44) again with $C'(x)$ in place of $y$. $S'(x)$ is treated similarly, and the argument can be repeated to establish the assertion for still higher derivatives. In particular,

$$C'''(x) = -C'(x), \qquad S'''(x) = -S'(x),$$
$$C^{(iv)}(x) = C(x), \qquad S^{(iv)}(x) = S(x),$$

and the derivatives of $C(x)$ and $S(x)$ repeat in cycles of four. Finally, (4–45) and (4–46) imply that

$$C'(0) = 0, \qquad C''(0) = -1,$$
$$S'(0) = 1, \qquad S''(0) = 0,$$

and it follows that $C'(x)$ is *the* solution of (4–44) which satisfies the initial conditions $C'(0) = 0, C''(0) = -1$, while $S'(x)$ is *the* solution which satisfies $S'(0) = 1, S''(0) = 0$. Thus

$$C'(x) = -S(x), \qquad S'(x) = C(x),$$

and we have evaluated *all* of the derivatives of $C(x)$ and $S(x)$.

Using these results, we now prove the identity

$$S(x)^2 + C(x)^2 = 1. \tag{4–47}$$

Indeed, since

$$\frac{d}{dx}[S(x)^2 + C(x)^2] = 2S(x)S'(x) + 2C(x)C'(x) = 0,$$

it follows that

$$S(x)^2 + C(x)^2 = k,$$

where $k$ is a constant, and setting $x = 0$ yields $k = 1$. Among other things, (4–47) implies that $|C(x)| \leq 1, |S(x)| \leq 1$ for all $x$.

Next, we establish the addition formulas

$$C(a + x) = C(a)C(x) - S(a)S(x),$$
$$S(a + x) = S(a)C(x) + C(a)S(x), \tag{4–48}$$

where $a$ is an arbitrary real number. We begin by observing that

$$\frac{d^2}{dx^2} C(a + x) = -C(a + x) \qquad \text{and} \qquad \frac{d^2}{dx^2} S(a + x) = -S(a + x).$$

Thus $C(a + x)$ and $S(a + x)$ are solutions of (4–44) and as such must be linear combinations of $C(x)$ and $S(x)$; that is,

$$C(a + x) = \alpha_1 C(x) + \alpha_2 S(x),$$
$$S(a + x) = \beta_1 C(x) + \beta_2 S(x), \tag{4–49}$$

for suitable constants $\alpha_1, \alpha_2, \beta_1, \beta_2$. To obtain the values of these constants we set $x = 0$ in (4–49) and the identities obtained from it by differentiation. This gives $\alpha_1 = C(a)$, $\beta_1 = S(a)$, $\alpha_2 = -S(a)$, $\beta_2 = C(a)$, and (4–48) is proved.

In much the same fashion we find that

$$C(-x) = C(x), \qquad S(-x) = -S(x), \tag{4–50}$$

and conclude that the graph of $C(x)$ is symmetric about the $y$-axis, while that of $S(x)$ is symmetric about the origin (Exercise 1).

At this point we could derive those long and all too familiar lists of trigonometric identities involving $C(x)$ and $S(x)$. However, it is much more instructive to prove that these functions are periodic with period $2\pi$. Here we begin by *defining* $\pi/2$ to be the smallest positive real number such that $C(x) = 0$. (The proof that such a number exists has been left as Exercise 2.) Then $C(x)$ is positive on the interval $(0, \pi/2)$, and since $S'(x) = C(x)$, we conclude that $S(x)$ is *increasing* on that interval. But $S(0) = 0$, and hence $S(x)$ is also positive on $(0, \pi/2)$. Thus by (4–47), $S(\pi/2) = 1$, and the addition formulas now give

$$C(\pi) = -1, \qquad C(3\pi/2) = 0, \qquad C(2\pi) = 1,$$
$$S(\pi) = 0, \qquad S(3\pi/2) = -1, \qquad S(2\pi) = 0.$$

Hence

$$C(x + 2\pi) = C(x)C(2\pi) - S(x)S(2\pi) = C(x),$$
$$S(x + 2\pi) = S(x)C(2\pi) + C(x)S(2\pi) = S(x),$$

and the periodicity of $C(x)$ and $S(x)$ is established.

To complete the discussion, we must show that $2\pi$ is the *smallest* period for each of these functions. For $C(x)$ the argument goes as follows. From (4–48) we obtain

$$C(x + \pi/2) = -S(x),$$
$$C(x + \pi) = -S(x),$$
$$C(x + 3\pi/2) = S(x),$$

and it follows that $C(x)$ is negative on the interval $(\pi/2, 3\pi/2)$ and positive on $(3\pi/2, 2\pi)$. For similar reasons $S(x)$ is positive on $(0, \pi)$ and negative on $(\pi, 2\pi)$. But since $C'(x) = -S(x)$, $C(x)$ is decreasing on $(0, \pi)$ and increasing on $(\pi, 2\pi)$. This fact, together with $C(3\pi/2) = 0$ and $C(2\pi) = 1$, implies that $2\pi$ is the smallest positive real number such that $C(x) = 1$, and we are done.

## EXERCISES

1. Establish (4–50) by showing that $C(-x)$, $S(-x)$ are solutions of (4–44) and expressing them in terms of the basis $C(x)$, $S(x)$ for the solution space.

\* 2. Show that there is a least positive real number $\alpha$ such that $C(\alpha) = 0$. [*Hint*: Assume the contrary; then argue that
   a) $C(x) > 0$ for $0 < x < \infty$,    b) $S(x) > 0$ for $0 < x < \infty$,
   c) $C(x)$ is strictly decreasing and concave downwards on $(0, \infty)$,
      and derive a contradiction from (a) and (c).\* Having established that $C(x)$ has positive zeros, consider the greatest lower bound $\alpha$ of the set of positive zeros of $C(x)$.]

3. Let $E(x)$ be the unique solution on $(-\infty, \infty)$ of the initial-value problem $y' - y = 0$, $y(0) = 1$. Establish the following properties of $E(x)$.
   a) $E(x)$ has derivations of all orders, and $E^{(n)}(x) = E(x)$.
   b) $E(x) > 0$ on $(0, \infty)$. [*Hint*: Otherwise let $x_0$ be the smallest positive real number for which $E(x_0) = 0$, and derive a contradiction by applying the mean-value theorem. (Why would such a smallest number exist?)]
   c) $E(x)$ is strictly increasing and concave upwards on $(0, \infty)$.
   d) $E(a + x) = E(a)E(x)$ for all real numbers $a$, $x$.
   e) $E(-x) = 1/E(x)$ for every real number $x$. [*Hint*: Apply (d).]
   f) $0 < E(x) < 1$ on $(-\infty, 0)$.
   g) $\lim_{x \to \infty} E(x) = \infty$. [*Hint*: Use (c).]
   h) $\lim_{x \to -\infty} E(x) = 0$.
   i) Set $E(1) = e$, and show that $E(n) = e^n$.

4. a) Prove that every solution of a homogeneous linear differential equation with constant coefficients has derivatives of all orders at every point on the $x$-axis.
   b) Generalize this result to homogeneous equations with variable coefficients and to nonhomogeneous equations.

---

\*  To establish these facts rigorously, the intermediate-value and mean-value theorems from calculus must be applied. You may prefer to give an intuitive argument based on a consideration of the graphs of $C(x)$ and $S(x)$.

# Equations with Constant Coefficients

**5**

## 5–1 INTRODUCTION

Linear differential equations of the form

$$a_n y^{(n)} + a_{n-1} y^{(n-1)} + \cdots + a_0 y = h(x) \tag{5–1}$$

in which $a_0, \ldots, a_n$ are constants and $a_n \neq 0$, are in many respects the simplest of all differential equations. For one thing, they can be discussed *entirely* within the context of linear algebra, and form the only substantial class of equations of order greater than one which can be explicitly solved. This, plus the fact that such equations arise in a surprisingly wide variety of physical problems, accounts for the special place they occupy in the theory of linear differential equations.

We shall begin the discussion of this chapter by considering the homogeneous version of Eq. (5–1), which can be written in normal form as

$$(D^n + a_{n-1} D^{n-1} + \cdots + a_0) y = 0, \tag{5–2}$$

or as

$$Ly = 0, \tag{5–3}$$

where $L$ is the linear differential operator $D^n + a_{n-1} D^{n-1} + \cdots + a_0$. Algebraically such constant-coefficient operators behave *exactly* as if they were ordinary polynomials in $D$, and can therefore be factored according to the rules of elementary algebra. In particular, *every linear differential operator with constant coefficients can be expressed as a product of constant-coefficient operators of degrees one and two* (Exercise 2). As we shall see, this fact essentially reduces the task of solving (5–2) to the second-order case where complete results can be obtained with relative ease.

This done, we shall take up the problem of finding a particular solution of $Ly = h$ given the general solution of the associated homogeneous equation $Ly = 0$. Here the restriction on the coefficients of $L$ will be dropped and much more far-reaching results obtained. The language of operator theory and the ideas of linear algebra will dominate this portion of our discussion and furnish just that

*lemma - a proof whose use is to prove a more*
*general statement or theory and has no*
*general use itself*

measure of insight needed to make it intelligible. Finally, we shall end the chapter with a number of special results and applications, among which is a derivation, due originally to Newton, of Kepler's laws of planetary motion.

**EXERCISES**

1. a) Prove that the product of two complex numbers $a + bi$ and $c + di$ is real if and only if either

 i) $a + bi = 0$ or $c + di = 0$, or
 ii) $a = kc$ and $b = -kd$, for some real number $k \neq 0$.

 [*Hint*: Recall that $a + bi$ is real if and only if $b = 0$, and that the product $(a + bi)(c + di)$ is computed by using the distributive law and the rule $i^2 = -1$.]

 b) Let $P(x)$ be a polynomial with *real* coefficients, and suppose that $P(x)$ has $a + bi$, $b > 0$, as a root; that is, $P(a + bi) = 0$. Prove that $a - bi$ is also a root of $P(x)$.

2. a) Let $P(x)$ be a polynomial of degree $n$, $n > 0$, with real coefficients. Use the fact that $P(x)$ has *exactly* $n$ roots in the complex number system to prove that $P(x)$ can be factored into a product of linear and quadratic factors with real coefficients. [*Hint*: See Exercise 1(b).]

 b) Use the result obtained in part (a) to deduce that every polynomial of odd degree with real coefficients has at least one real root.

3. Find the second-degree polynomial that has $a + bi$ and $a - bi$, $b > 0$, as roots.

4. Write each of the following linear differential operators as a product of first- and second-order operators with real coefficients that cannot be factored further.

 a) $D^2 - 2$                 b) $D^2 - \sqrt{3}D - 6$
 c) $D^3 + 4D^2 + 5D + 2$      d) $D^3 - D^2 + D - 1$
 e) $D^3 - D^2 - D + 1$        f) $D^4 - 5D^2 + 4$
 g) $D^4 - 4D^3 + D^2 + 6D$     h) $D^4 + 1$

## 5–2 HOMOGENEOUS EQUATIONS OF ORDER TWO

We have already pointed out that the technique for solving constant-coefficient linear differential equations depends on the commutativity of the operator multiplication involved. To make this dependence explicit and at the same time phrase it in the form best suited to our immediate needs, we begin with a lemma.

 **Lemma 5–1** *If $L_1, \ldots, L_n$ are constant-coefficient linear differential operators, then the null space of each is included in the null space of their product.*

 *Proof.* To prove this assertion, we must show that $(L_1 \cdots L_n)y = 0$ whenever $L_i y = 0$. But this follows immediately because we are free to change the order in which the operators are applied:

$$(L_1 \cdots L_n)y = (L_1 \cdots L_{i-1}L_{i+1} \cdots L_n L_i)y$$
$$= (L_1 \cdots L_{i-1}L_{i+1} \cdots L_n)(L_i y)$$
$$= (L_1 \cdots L_{i-1}L_{i+1} \cdots L_n)0$$
$$= 0. \quad \blacksquare$$

**Example 1**   The second-order equation

$$(D^2 - 4)y = 0 \tag{5-4}$$

can be rewritten as

$$(D + 2)(D - 2)y = 0.$$

Hence $e^{2x}$ and $e^{-2x}$ are solutions of (5-4), since they are the solutions of the first-order equations $(D - 2)y = 0$ and $(D + 2)y = 0$. Furthermore, an easy computation reveals that these functions are linearly independent in $\mathscr{C}(-\infty, \infty)$. It follows from Theorem 4-6 that the general solution of (5-4) is

$$y = c_1 e^{2x} + c_2 e^{-2x},$$

where $c_1$ and $c_2$ are arbitrary constants.

This simple example suggests that we attempt to solve the general second-order equation

$$(D^2 + a_1 D + a_0)y = 0 \tag{5-5}$$

by decomposing the operator $D^2 + a_1 D + a_0$ into linear factors. To this end we first find the roots $\alpha_1$, $\alpha_2$ of the quadratic equation

$$m^2 + a_1 m + a_0 = 0, \qquad \leftarrow \text{Characteristic equation} \tag{5-6}$$
$$\text{if } D^2 + a, D + a_0\, y = 0$$

known as the **characteristic equation** of (5-5), and then rewrite (5-5) as

$$(D - \alpha_1)(D - \alpha_2)y = 0. \tag{5-7}$$

This done, the argument falls into cases that depend on the nature of $\alpha_1$ and $\alpha_2$.

CASE 1.   $\alpha_1$ *and* $\alpha_2$ *real and unequal.*  Here the reasoning used in Example 1 carries over without change; the functions $e^{\alpha_1 x}$ and $e^{\alpha_2 x}$ are linearly independent solutions of (5-7), and

$$y = c_1 e^{\alpha_1 x} + c_2 e^{\alpha_2 x}$$

is the general solution.

CASE 2.   $\alpha_1 = \alpha_2 = \alpha$.  In this case (5-7) becomes

$$(D - \alpha)^2 y = 0, \tag{5-8}$$

and our earlier argument yields just one solution of the equation, namely $e^{\alpha x}$. But we can apply the method of Section 4-7 to find a second, linearly independent solution by solving the first-order equation

$$W[e^{\alpha x}, y(x)] = e^{2\alpha x}.$$

Since

$$W[e^{\alpha x}, y(x)] = \begin{vmatrix} e^{\alpha x} & y(x) \\ \alpha e^{\alpha x} & y'(x) \end{vmatrix} = e^{\alpha x}[y'(x) - \alpha y(x)],$$

we have $y' - \alpha y = e^{\alpha x}$, and Formula (4–7) yields

$$y = xe^{\alpha x}.$$

Hence the general solution of (5–8) is   $y = c_1 e^{\alpha z} + c_2 x e^{\alpha x}$

$$y = (c_1 + c_2 x)e^{\alpha x}.$$

CASE 3.   $\alpha_1$ *and* $\alpha_2$ *complex.* Here $\alpha_1 = a + bi$, $\alpha_2 = a - bi$, where $a$ and $b$ are real, and $b \neq 0$. Fortunately $e^{\alpha_1 x}$ and $e^{\alpha_2 x}$ continue to make sense when $\alpha_1$ and $\alpha_2$ are complex, and if we yield to the impulse to carry out the steps of Case 1, albeit only formally, we are led to the expression

*although*

$$\begin{aligned} y &= c_1 e^{\alpha_1 x} + c_2 e^{\alpha_2 x} \\ &= c_1 e^{(a+bi)x} + c_2 e^{(a-bi)x} \\ &= e^{ax}(c_1 e^{ibx} + c_2 e^{-ibx}) \end{aligned}$$

as a candidate for the general solution of (5–7). Things will look somewhat better at this point if we invoke *Euler's formula* (Exercise 34)

$$e^{ix} = \cos x + i \sin x$$

to rewrite the preceding expression as

$$\begin{aligned} y &= e^{ax}[c_1(\cos bx + i \sin bx) + c_2(\cos bx - i \sin bx)] \\ &= e^{ax}[(c_1 + c_2)\cos bx + i(c_1 - c_2)\sin bx] \\ &= c_3 e^{ax} \cos bx + c_4 e^{ax} \sin bx, \end{aligned}$$

for we thereby return to real-valued functions of a real variable. Thus a purely formal investigation leads us to ask if the functions $e^{ax} \cos bx$ and $e^{ax} \sin bx$ could be a basis for the solution space of (5–7). That is, are these two functions solutions of (5–7) and are they linearly independent in $\mathscr{C}(-\infty, \infty)$? The answer is yes, the verification is easy, and we leave the details as an exercise.

Since these three cases include all the possible combinations of $\alpha_1$ and $\alpha_2$, we have completed the task of solving the general second-order homogeneous linear differential equation with constant coefficients. For convenience we summarize these results as follows.

**Theorem 5–1**   *To solve a second-order homogeneous linear differential equation of the form*

$$(D^2 + a_1 D + a_0)y = 0,$$

*first find the roots* $\alpha_1$ *and* $\alpha_2$ *of the characteristic equation*

$$m^2 + a_1 m + a_0 = 0.$$

*The general solution of the given equation can then be expressed in terms of* $\alpha_1$ *and* $\alpha_2$ *and arbitrary constants* $c_1$ *and* $c_2$ *as follows:*

*conditions approach*

| $\alpha_1, \alpha_2$ | General solution |
|---|---|
| Real, $\alpha_1 \neq \alpha_2$ | $c_1 e^{\alpha_1 x} + c_2 e^{\alpha_2 x}$ |
| Real, $\alpha_1 = \alpha_2 = \alpha$ | $(c_1 + c_2 x)e^{\alpha x}$ |
| Complex, $\alpha_1 = a + bi$<br>$\quad\quad\quad \alpha_2 = a - bi$ | $e^{ax}(c_1 \cos bx + c_2 \sin bx)$ |

**Example 2**   Find the general solution of

$$\text{a)} \quad y'' - 2\sqrt{2}y' + 2y = 0;$$
$$\text{b)} \quad y'' - 2y' + 5y = 0.$$

*Solution.*   a)  The characteristic equation is

$$m^2 - 2\sqrt{2}m + 2 = 0,$$

and factors as

$$(m - \sqrt{2})^2 = 0.$$

The only root of the characteristic equation is $m = \sqrt{2}$, so the general solution of the equation is

$$y = (c_1 + c_2 x)e^{\sqrt{2}x}.$$

b)  In this case the characteristic equation is

$$m^2 - 2m + 5 = 0,$$

and cannot be written as a product of linear factors with real coefficients. (Its discriminant is negative.) By completing the square or applying the quadratic formula, we find the roots to be

$$m = 1 \pm 2i.$$

Hence the general solution of the differential equation is

$$y = e^x(c_1 \sin 2x + c_2 \cos 2x).$$

**EXERCISES**

Find the general solution of each of the following differential equations.

1. $y'' + y' - 2y = 0$
2. $3y'' - 5y' + 2y = 0$
3. $8y'' + 14y' - 15y = 0$
4. $y'' - 2y' = 0$
5. $y'' + 4y = 0$
6. $3y'' + 2y = 0$
7. $y'' + 4y' + 8y = 0$
8. $4y'' - 4y' + 3y = 0$
9. $y'' - 2y' + 2y = 0$
10. $9y'' - 12y' + 4y = 0$
11. $y'' + 2y' + 4y = 0$
12. $2y'' - 2\sqrt{2}y' + y = 0$

13. $2y'' - 5\sqrt{3}y' + 6y = 0$                14. $9y'' + 6y' + y = 0$

15. $64y'' - 48y' + 17y = 0$

In Exercises 16 through 25 find the solutions of the given initial-value problems.

16. $2y'' - y' - 3y = 0$;   $y(0) = 2, y'(0) = -\frac{7}{2}$

17. $y'' - 8y' + 16y = 0$;   $y(0) = \frac{1}{2}, y'(0) = -\frac{1}{3}$

18. $4y'' - 12y' + 9y = 0$;   $y(0) = 1, y'(0) = \frac{7}{2}$

19. $y'' + 2y = 0$;   $y(0) = 2, y'(0) = 2\sqrt{2}$

20. $4y'' - 4y' + 5y = 0$;   $y(0) = \frac{1}{2}, y'(0) = 1$

21. $y'' + 4y' + 13y = 0$;   $y(0) = 0, y'(0) = -2$

22. $9y'' - 3y' - 2y = 0$;   $y(0) = 3, y'(0) = 1$

23. $y'' - 2\sqrt{5}y' + 5y = 0$;   $y(0) = 0, y'(0) = 3$

24. $16y'' + 8y' + 5y = 0$;   $y(0) = 4, y'(0) = -1$

25. $y'' - \sqrt{2}y' + y = 0$;   $y(0) = \sqrt{2}, y'(0) = 0$

26. Prove that $e^{\alpha_1 x}$ and $e^{\alpha_2 x}$ are linearly independent in $\mathscr{C}(-\infty, \infty)$ whenever $\alpha_1$ and $\alpha_2$ are distinct real numbers.

27. Verify that $xe^{\alpha x}$ is a solution of the second-order equation $(D - \alpha)^2 y = 0$, and prove that this function and $e^{\alpha x}$ are linearly independent in $\mathscr{C}(-\infty, \infty)$.

28. Verify that $e^{ax} \cos bx$ and $e^{ax} \sin bx$ are linearly independent solutions of the equation

$$(D - \alpha_1)(D - \alpha_2)y = 0$$

when $\alpha_1 = a + bi$ and $\alpha_2 = a - bi, b \neq 0$.

29. Find a constant-coefficient linear differential equation whose general solution is:
    a) $(c_1 + c_2 x)e^{-3x}$                b) $c_1 e^x \sin 2x + c_2 e^x \cos 2x$
    c) $(c_1 + c_2 x)e^{-2x} + 1$             d) $c_1 e^{-x} + c_2 e^{-3x} + x + 4$
    e) $c_1 \sin 3x + c_2 \cos 3x + x/3$

30. For each of the following functions find a second-order linear differential equation with constant coefficients that has the given function as a particular solution.
    a) $x(1 + e^x)$                           b) $4 \sin x \cos x$
    c) $(1 + 2e^x)e^{2x} + 6x + 5$            d) $\cos x(1 - 4 \sin^2 x)$
    e) $e^{3x} + e^{2x} + xe^{3x}$

31. a) Show that the general solution of the second-order equation

$$[D^2 - 2aD + (a^2 + b^2)]y = 0$$

can be written in the form

$$y = c_1 e^{ax} \cos (bx + c_2),$$

where $c_1$ and $c_2$ are arbitrary constants. This form is frequently called the *phase-amplitude* form of the solution.

   b) Write the general solution of $(D^2 + 4)y = 0$ in phase-amplitude form.

32. If $L = (D - \alpha)^2$, $\alpha$ real, show that

$$Le^{kx} = (k - \alpha)^2 e^{kx}.$$

Differentiate both sides of this identity with respect to $k$ to prove that

$$Lxe^{kx} = (k - \alpha)[2e^{kx} + x(k - \alpha)e^{kx}],$$

and then show that $xe^{\alpha x}$ is a solution of $Ly = 0$.

33. a) Find the solution of
$$(D^2 - 2D + 26)y = 0$$
whose graph passes through the point $(0, 1)$ with slope 2.

   b) Solve the problem of part (a) again, this time writing the general solution in the form
$$y = c_1 e^{(a + bi)x} + c_2 e^{(a - bi)x}.$$

   Evaluate $c_1$ and $c_2$ formally, then use Euler's formula to show that the resulting solution can be transformed into the solution found in (a).

*34. When $z$ is a complex number, the function $e^z$ is defined by the infinite series

$$e^z = 1 + z + \frac{z^2}{2!} + \cdots + \frac{z^n}{n!} + \cdots,$$

and it can be shown that this series converges absolutely for all values of $z$. Set $z = ix$ in this series, and use the fact that $i^2 = -1$ to prove Euler's formula. [*Hint*: Since the series is absolutely convergent for all $z$, its terms may be rearranged at will.]

## 5–3 HOMOGENEOUS EQUATIONS OF ARBITRARY ORDER

The technique for solving homogeneous constant-coefficient linear differential equations is now all but complete. Thus, to solve

$$(D^4 - 2D^3 + 2D^2 - 2D + 1)y = 0, \tag{5–9}$$

we first decompose the operator into linear and quadratic factors, as suggested in Section 5–1, to obtain the equivalent equation

$$(D - 1)^2(D^2 + 1)y = 0.$$

Then we invoke Lemma 5–1 to assert that the solution space of each of the second-order equations $(D - 1)^2 y = 0$ and $(D^2 + 1)y = 0$ is contained in the solution space of (5–9). Thus $e^x$, $xe^x$, $\sin x$, and $\cos x$ are solutions of (5–9), and since these functions are linearly independent in $\mathscr{C}(-\infty, \infty)$, the general solution of the equation is

$$y = (c_1 + c_2 x)e^x + c_3 \sin x + c_4 \cos x.$$

This, in brief, is how all homogeneous constant-coefficient equations are solved, and save for the difficulty occasioned by equations such as

$$(D - 1)^4 y = 0 \qquad \text{and} \qquad (D^2 + 1)^2 y = 0,$$

whose characteristic equations do not have enough distinct roots to produce the required number of linearly independent solutions, we are done. On the other hand, our experience with the equation $(D - \alpha)^2 y = 0$ suggests that the missing solutions for the equations just mentioned are, respectively, $x^2 e^x$, $x^3 e^x$, and $x \sin x$, $x \cos x$, as indeed they are. In fact, there is a general pattern that makes the formulation of such "missing" solutions an easy matter.

To see what the pattern is, we go directly to the general homogeneous constant-coefficient linear equation

$$(D^n + a_{n-1}D^{n-1} + \cdots + a_0)y = 0. \tag{5-10}$$

We decompose the operator into powers of linear and quadratic factors, the linear factors being determined by the real roots of the *characteristic equation*

$$m^n + a_{n-1}m^{n-1} + \cdots + a_0 = 0, \tag{5-11}$$

and the quadratic factors by the complex roots. In this factorization a typical power of a linear factor has the form $(D - \alpha)^r$, where $\alpha$ is a real root of (5–11) of multiplicity $r$. Similarly, a typical power of a quadratic factor has the form $[D^2 - 2aD + (a^2 + b^2)]^r$, corresponding to the pair of complex roots $a \pm bi$, $b \neq 0$. According to Lemma 5–1, all we have to do to find the solutions of (5–10) is to find the null spaces of these factors. The following lemma is the key to this last step.

**Lemma 5–2**  *If $r$ is a positive integer and $\alpha$, $a$, and $b$ are real numbers with $b \neq 0$, then*

i) $(D - \alpha)^r(x^{r-1}e^{\alpha x}) = 0$, *and*

ii) $(D^2 - 2aD + a^2 + b^2)^r(x^{r-1}e^{ax}\sin bx) = 0$,
   $(D^2 - 2aD + a^2 + b^2)^r(x^{r-1}e^{ax}\cos bx) = 0$.

*Proof.*  We shall establish (i), and leave (ii) as an exercise. The proof is by induction on $r$.

The result clearly holds when $r = 1$, and if we now assume that

$$(D - \alpha)^r(x^{r-1}e^{\alpha x}) = 0$$

has been proved for $r \geqslant 1$, we have

$$\begin{aligned}
(D - \alpha)^{r+1}(x^r e^{\alpha x}) &= (D - \alpha)^r(D - \alpha)(x^r e^{\alpha x}) \\
&= (D - \alpha)^r(rx^{r-1}e^{\alpha x} + \alpha x^r e^{\alpha x} - \alpha x^r e^{\alpha x}) \\
&= (D - \alpha)^r(rx^{r-1}e^{\alpha x}) \\
&= r(D - \alpha)^r(x^{r-1}e^{\alpha x}) \\
&= 0.
\end{aligned}$$

Thus the result also holds for $r + 1$, and the proof is complete. ∎

Of course, if $(D - \alpha)^r$ annihilates $x^{r-1}e^{\alpha x}$, then it also annihilates the functions $x^{r-2}e^{\alpha x}$, $x^{r-3}e^{\alpha x}, \ldots, e^{\alpha x}$, because $(D - \alpha)^r x^{r-2}e^{\alpha x} = (D - \alpha)(D - \alpha)^{r-1}x^{r-2}e^{\alpha x}$, which by the lemma is $(D - \alpha)(0) = 0$, and so on. Thus, if $x^{r-1}e^{\alpha x}$ belongs to the null space of $(D - \alpha)^r$, then so do the other $r - 1$ functions. Moreover, the $r$ functions taken together are linearly independent in $\mathscr{C}(-\infty, \infty)$ (see Example 5 and Exercise 24), so that they form a basis for the null space of $(D - \alpha)^r$. Furthermore, if $\alpha \neq \beta$, these functions are linearly independent of the corresponding functions for any $(D - \beta)^s$. (See Exercise 28.)

At this point it should not come as a surprise that the null spaces of the factors $(D^2 - 2aD + a^2 + b^2)^r$ are spanned by powers of $x$ times $e^{ax} \sin bx$ and $e^{ax} \cos bx$. These functions are linearly independent of each other, and the functions in the null space of a power of one quadratic factor are linearly independent of the functions in the null space of any other. Finally, all of these "quadratic generators" are linearly independent of the (nonzero) functions in the null spaces of the linear factors. (See Exercise 25.)

Granting all this, we can now state the theorem that tells us how to find the general solution of *every* homogeneous constant-coefficient linear differential equation.

**Theorem 5–2**   *Let*

$$y^{(n)} + a_{n-1}y^{(n-1)} + \cdots + a_0 y = 0 \tag{5-12}$$

*be an nth-order homogeneous linear differential equation with constant real coefficients, let $\alpha_1, \ldots, \alpha_k$ be the distinct roots of its characteristic polynomial, and suppose that*

$$m^n + a_{n-1}m^{n-1} + \cdots + a_0 = (m - \alpha_1)^{r_1}(m - \alpha_2)^{r_2} \cdots (m - \alpha_k)^{r_k}.$$

i) *If $\alpha_i$ is real, then the functions*

$$e^{\alpha_i x}, \quad xe^{\alpha_i x}, \quad \ldots, \quad x^{r_i-1}e^{\alpha_i x}$$

*are linearly independent solutions of (5-12).*

ii) *If $\alpha_i = a + bi$ ($a, b$ real, $b \neq 0$), then the functions*

$$e^{ax} \sin bx, \quad xe^{ax} \sin bx, \quad \ldots, \quad x^{r_i-1} e^{ax} \sin bx$$
$$e^{ax} \cos bx, \quad xe^{ax} \cos bx, \quad \ldots, \quad x^{r_i-1} e^{ax} \cos bx$$

*are linearly independent solutions of (5-12).*

iii) *Every solution of (5-12) can be written uniquely as a linear combination of the solutions in (i) and (ii).*

**Example 1**   Find the general solution of

$$(D^3 + 1)y = 0. \tag{5-13}$$

*Solution.*   Since

$$D^3 + 1 = (D + 1)(D^2 - D + 1),$$

the roots of the characteristic equation are

$$-1, \quad \frac{1}{2} + \frac{\sqrt{3}}{2} i, \quad \text{and} \quad \frac{1}{2} - \frac{\sqrt{3}}{2} i.$$

Thus the general solution of (5–13) is

$$y = c_1 e^{-x} + e^{x/2}\left(c_2 \sin \frac{\sqrt{3}}{2}x + c_3 \cos \frac{\sqrt{3}}{2}x\right).$$

**Example 2**   Solve the equation

$$y^{(7)} - 2y^{(5)} + y^{(3)} = 0.$$

*Solution.*   In operator notation this equation becomes

$$(D^7 - 2D^5 + D^3)y = 0,$$

and since

$$D^7 - 2D^5 + D^3 = D^3(D - 1)^2(D + 1)^2,$$

the general solution of the equation is

$$y = c_1 + c_2 x + c_3 x^2 + (c_4 + c_5 x)e^x + (c_6 + c_7 x)e^{-x}.$$

**Example 3**   Find a linear differential equation that has $e^{2x}$ and $xe^{-3x}$ among its solutions.

*Solution.*   We must find a linear differential operator $L$ with the property that

$$Le^{2x} = 0 \quad\text{and}\quad Lxe^{-3x} = 0;$$

in other words, $L$ must annihilate $e^{2x}$ and $xe^{-3x}$ (see p. 96). Since any operator that contains the factors $D - 2$ and $(D + 3)^2$ will have this property,

$$(D - 2)(D + 3)^2 y = 0$$

or

$$y''' + 4y'' - 3y' - 18y = 0$$

is an equation of the required type. There are, of course, infinitely many others.

**Example 4**   Find a constant-coefficient linear differential operator $L$ which, when applied to the equation

$$(D^2 + 1)(D - 1)y = e^x + 2 - 7x \sin x,$$

produces the homogeneous equation

$$L(D^2 + 1)(D - 1)y = 0.$$

*Solution.*   We seek an operator that will annihilate $e^x$, 2, and $7x \sin x$. According to Theorem 5–2, it will suffice to set

$$L = D(D - 1)(D^2 + 1)^2.$$

**Example 5**   As a final example we shall prove that the functions

$$e^x, \quad xe^x, \quad x^2 e^x,$$
$$e^{-2x}, \quad xe^{-2x}$$

from which the general solution of

$$(D - 1)^3(D + 2)^2 y = 0$$

is constructed are linearly independent in $\mathscr{C}(-\infty, \infty)$.

We observe first that the Wronskian is far too cumbersome to use when so many functions are involved, so we appeal directly to the definition of linear independence.

Suppose that

$$(c_1 + c_2 x + c_3 x^2)e^x + (c_4 + c_5 x)e^{-2x} \equiv 0. \qquad (5\text{–}14)$$

Then

$$(c_1 + c_2 x + c_3 x^2)e^{3x} + c_4 + c_5 x \equiv 0,$$

and if we apply the operator $(D - 3)^3$ to this expression, the first three terms will be annihilated, leaving

$$(D - 3)^3(c_4 + c_5 x) \equiv 0.$$

A routine computation reduces this identity to

$$c_5 - c_4 - c_5 x \equiv 0,$$

from which we immediately obtain $c_4 = c_5 = 0$. This allows us to rewrite (5–14) as

$$c_1 + c_2 x + c_3 x^2 \equiv 0,$$

which in turn implies that $c_1 = c_2 = c_3 = 0$, thus completing the proof.

## EXERCISES

Find the general solution of each of the following differential equations.

1. $y''' + 3y'' - y' - 3y = 0$
2. $y''' + 5y'' - 8y' - 12y = 0$
3. $4y''' + 12y'' + 9y' = 0$
4. $y''' + 6y'' + 13y' = 0$
5. $2y''' + y'' - 8y' - 4y = 0$
6. $y''' + 3y'' + y' + 3y = 0$
7. $y^{(iv)} - y'' = 0$
8. $y^{(iv)} - 8y'' + 16y = 0$
9. $y^{(iv)} + 18y'' + 81y = 0$
10. $4y^{(iv)} - 8y''' - y'' + 2y' = 0$
11. $y^{(iv)} + y''' + y'' = 0$
12. $y^{(iv)} = 0$
13. $y^{(iv)} - 4y''' + 6y'' - 4y' + y = 0$
14. $y^{(v)} + 2y''' + y' = 0$
15. $y^{(iv)} + y = 0$

In Exercises 16 through 23 find a linear differential operator that annihilates the given function.

16. $3 + 4x - 2e^{-2x}$
17. $x(2x + 1)\sin x$
18. $(x^3 + 2x - 1)e^x \cos 3x$
19. $x^2 e^{(x+1)}$
20. $x^2 \sin x \cos x$
21. $xe^x \sin^2 x$
22. $x \sin(x + \pi/4)$
23. $(xe^x + 1)^3$

24. Prove that for any real number $\alpha$ and any positive integer $n$, the functions

$$e^{\alpha x}, \quad xe^{\alpha x}, \quad \ldots, \quad x^{n-1}e^{\alpha x}$$

are linearly independent in $\mathscr{C}(-\infty, \infty)$. [*Hint:* A polynomial of positive degree has only a finite number of roots.]

25. Prove that the functions

$$e^{2x}, \quad xe^{2x}, \quad e^{2x}\sin x, \quad e^{2x}\cos x$$

are linearly independent in $\mathscr{C}(-\infty, \infty)$.

26. a) Let $P(x)$ be a polynomial with real coefficients, and let $L = P(D)$ be the associated constant-coefficient linear differential operator. Prove that

$$Le^{\alpha x} = P(\alpha)e^{\alpha x}.$$

  b) Use the result in (a) to show that if $L$ is a constant-coefficient linear differential operator, then $Le^{\alpha x} = 0$ if and only if $L$ has $D - \alpha$ as a factor.

27. Prove that

$$(D - \alpha)^k x^m e^{\alpha x} = \begin{cases} 0 & \text{if } k > m, \\ m(m - 1) \cdots (m - k + 1)x^{m-k}e^{\alpha x} & \text{if } k \leqslant m. \end{cases}$$

In particular, deduce that

$$(D - \alpha)^m x^m e^{\alpha x} = m!e^{\alpha x}.$$

28. Prove that if $\alpha$ and $\beta$ are distinct real numbers, and $r$ and $s$ are positive integers, then the functions

$$e^{\alpha x}, \quad xe^{\alpha x}, \quad \ldots, \quad x^{r-1}e^{\alpha x},$$
$$e^{\beta x}, \quad xe^{\beta x}, \quad \ldots, \quad x^{s-1}e^{\beta x}$$

are linearly independent in $\mathscr{C}(-\infty, \infty)$.

## 5–4 NONHOMOGENEOUS EQUATIONS: VARIATION OF PARAMETERS AND GREEN'S FUNCTIONS

In Section 4–1 we saw that the general solution of the nonhomogeneous linear differential equation

$$a_n(x)\frac{d^n y}{dx^n} + \cdots + a_0(x)y = h(x) \qquad (5\text{--}15)$$

can be written in the form

$$y = y_p + y_h,$$

where $y_p$ is any *particular solution* of (5–15) and $y_h$ is the *general solution* of the associated homogeneous equation

$$a_n(x)\frac{d^n y}{dx^n} + \cdots + a_0(x)y = 0. \qquad (5\text{--}16)$$

We now turn our attention to the problem of finding such a particular solution *under the assumption that $y_h$ is known.* In this portion of our discussion we will be able to drop the requirement that the coefficients of the equation be constants, in spite of the fact that thus far we know how to find $y_h$ only in the constant-coefficient case. We shall, however, continue to assume that (5–15) is normal on some interval

$I$ (i.e., that $a_n(x) \neq 0$ for all $x$ in $I$). Accordingly, we can divide by the leading co-efficient and rewrite the equation in normal form as

$$\frac{d^n y}{dx^n} + a_{n-1}(x)\frac{d^{n-1}y}{dx^{n-1}} + \cdots + a_0(x)y = h(x). \tag{5-17}$$

In the language of linear operators, the problem of finding a particular solution of (5–17) consists of finding *some one* function in $\mathscr{C}^{(n)}(I)$ that satisfies the equation

$$Ly = h,$$

where $L = D^n + a_{n-1}(x)D^{n-1} + \cdots + a_0(x)$. And this, as we saw in Section 4–4, is equivalent to the problem of finding a (*right*) *inverse* $G$ for $L$, that is, a linear transformation

$$G: \mathscr{C}(I) \to \mathscr{C}^{(n)}(I)$$

such that

$$L(G[h]) = h$$

for all $h$ in $\mathscr{C}(I)$. The existence of such inverses is guaranteed by the fact that (5–17) has solutions for every $h$ in $\mathscr{C}(I)$ (see Theorem 4–4); the only problem is how to go about finding one. Conceptually the process is easy.

As described in Section 4–4, we first impose a "complete" set of initial conditions on Eq. (5–17) in order to guarantee that the equation has a unique solution. This procedure has the effect of restricting the domain of $L$ in such a way that $L$ becomes one-to-one and therefore has an inverse. Since the particular solution obtained is immaterial, we choose the simplest of all possible initial conditions, namely

$$y(x_0) = 0, \quad y'(x_0) = 0, \quad \ldots, \quad y^{(n-1)}(x_0) = 0,$$

where $x_0$ is an arbitrary but fixed point in $I$. This done, we "invert" the operator $L$ to obtain a formula for $G$. Here, of course, is the heart of the problem, and except for the fact that we would expect $G$ to be an integral operator (recall the fundamental theorem of calculus), there is little in the way of past experience to guide us. Surprisingly, however, the solution is relatively straightforward.

We begin by considering operators of order two, in which case (5–17) becomes

$$y'' + a_1(x)y' + a_0(x)y = h(x). \tag{5-18}$$

We seek a particular solution $y_p$ of this equation such that

$$y_p(x_0) = 0, \quad y_p'(x_0) = 0, \tag{5-19}$$

where $x_0$ is a point in $I$. We assume that the general solution

$$y_h = c_1 y_1(x) + c_2 y_2(x) \tag{5-20}$$

of the homogeneous equation associated with (5–18) is already known. The construction of $y_p$ is begun by making the unjustified but not unreasonable assumption that any particular solution of (5–18) ought to be related to the expression for $y_h$.

We therefore attempt to alter the latter in such a way that it becomes a solution of the given equation. One way of doing this is to allow the *parameters* $c_1$ and $c_2$ in (5–20) to vary with $x$ in the hope that a solution of (5–18) of the form

$$y_p = c_1(x)y_1(x) + c_2(x)y_2(x) \tag{5–21}$$

will emerge.

If (5–21) is substituted in (5–18), and the notation simplified by suppressing mention of the variable $x$, we obtain

$$c_1(y_1'' + a_1 y_1' + a_0 y_1) + c_2(y_2'' + a_1 y_2' + a_0 y_2) + (c_1' y_1 + c_2' y_2)'$$
$$+ a_1(c_1' y_1 + c_2' y_2) + (c_1' y_1' + c_2' y_2') = h. \tag{5–22}$$

Moreover, since $y_1$ and $y_2$ are solutions of the homogeneous equation $y'' + a_1 y' + a_0 y = 0$, the first two terms in (5–22) vanish, and we have

$$(c_1' y_1 + c_2' y_2)' + a_1(c_1' y_1 + c_2' y_2) + (c_1' y_1' + c_2' y_2') = h.$$

This identity, which must hold if (5–21) is to be a solution of (5–18), will be satisfied if $c_1$ and $c_2$ can be chosen so that

$$\begin{aligned} c_1'(x)y_1(x) + c_2'(x)y_2(x) &= 0, \\ c_1'(x)y_1'(x) + c_2'(x)y_2'(x) &= h(x), \end{aligned} \tag{5–23}$$

for all $x$ in $I$. Thus it remains to show that these equations serve to determine $c_1(x)$ and $c_2(x)$, and that this can be done in such a way that the function

$$y_p = c_1(x)y_1(x) + c_2(x)y_2(x)$$

satisfies the initial conditions given in (5–19).

Now, for each $x$ in $I$, (5–23) may be viewed as a pair of linear equations in the unknowns $c_1'(x)$ and $c_2'(x)$. As such, the determinant of its coefficients is

$$\begin{vmatrix} y_1(x) & y_2(x) \\ y_1'(x) & y_2'(x) \end{vmatrix},$$

which we recognize as the Wronskian of the linearly independent solutions $y_1(x)$ and $y_2(x)$ of the homogeneous equation associated with (5–18). According to Theorem 4–8, this determinant is a continuous function of $x$ that *never vanishes* on $I$. Hence by Theorem 2–10 the system (5–23) has a unique solution for $c_1'(x)$ and $c_2'(x)$, and once this solution is known, $c_1(x)$ and $c_2(x)$ can be found by integration. In the next paragraph we give the explicit formulation. As you will see there, the required initial conditions are satisfied by choosing suitable limits of integration, completing the determination of $y_p$. For obvious reasons, this method of allowing the parameters $c_1$ and $c_2$ to vary in order to determine the solution $y_p$ is called **variation of parameters**.

Starting with (5–23), we easily obtain

$$c_1'(x) = -\frac{h(x)y_2(x)}{W[y_1(x), y_2(x)]}, \qquad c_2'(x) = \frac{h(x)y_1(x)}{W[y_1(x), y_2(x)]}.$$

Thus

$$c_1(x) = -\int_{x_0}^x \frac{h(t)y_2(t)}{W[y_1(t), y_2(t)]}\, dt, \qquad c_2(x) = \int_{x_0}^x \frac{h(t)y_1(t)}{W[y_1(t), y_2(t)]}\, dt, \quad (5\text{–}24)$$

and if these values are substituted in (5–21) and the terms are combined, we will find that $y_p$ can be written in integral form as

$$y_p(x) = \int_{x_0}^x \frac{y_2(x)y_1(t) - y_1(x)y_2(t)}{W[y_1(t), y_2(t)]}\, h(t)\, dt. \qquad (5\text{–}25)$$

In finding this formula for $y_p$ we have also found a formula for the inverse that we were seeking for the linear differential operator

$$L = D^2 + a_1(x)D + a_0(x).$$

For if $h$ is any function in $\mathscr{C}(I)$, and if we set

$$G[h] = \int_{x_0}^x K(x, t)h(t)\, dt, \qquad (5\text{–}26)$$

where

$$K(x, t) = \frac{y_2(x)y_1(t) - y_1(x)y_2(t)}{W[y_1(t), y_2(t)]}, \qquad (5\text{–}27)$$

then it is not difficult to show that $G$ maps $\mathscr{C}(I)$ to $\mathscr{C}^{(2)}(I)$, acts as a right inverse for $L$, and has the further property that $G[h](x_0) = G[h]'(x_0) = 0$. (See Exercise 29.) Notice that the function $K(x, t)$ is independent of the particular choice of $x_0$ in the interval $I$. In the next section we shall see that it is also independent of the basis chosen for the solution space of $Ly = 0$. Thus $K(x, t)$ is completely determined by the operator $L$. It is called the **Green's function for $L$ for initial-value problems** or, more simply the **Green's function for $L$**.

**Example 1**    Find the general solution of the second-order equation

$$y'' + y = \tan x. \qquad (5\text{–}28)$$

*Solution 1.*  Since the general solution of the associated homogeneous equation is

$$y_h = c_1 \sin x + c_2 \cos x,$$

we seek a particular solution of (5–28) of the form

$$y_p = c_1(x) \sin x + c_2(x) \cos x,$$

where $c_1(x)$ and $c_2(x)$ are determined from the pair of equations

$$c_1'(x) \sin x + c_2'(x) \cos x = 0,$$
$$c_1'(x) \cos x - c_2'(x) \sin x = \tan x$$

(see Eq. 5–23). We then have

$$c_1'(x) = \sin x, \qquad c_2'(x) = -\sin x \tan x,$$

and it follows that

$$c_1(x) = \int \sin x \, dx = -\cos x,$$

$$c_2(x) = -\int \frac{\sin^2 x}{\cos x} \, dx = -\int (\sec x - \cos x) \, dx$$

$$= -\ln |\sec x + \tan x| + \sin x.$$

(It is customary to omit the constants of integration at this point since we are only seeking a particular solution of the given equation.) Thus

$$y_p = -\cos x \sin x + (\sin x - \ln |\sec x + \tan x|) \cos x$$

$$= -\cos x \ln |\sec x + \tan x|,$$

and the general solution of (5–28) is

$$y = -\cos x \ln |\sec x + \tan x| + c_1 \sin x + c_2 \cos x.$$

*Solution 2.*   Alternatively, we can solve (5–28) by finding the Green's function $K(x, t)$ for the operator $L = D^2 + 1$, and then using (5–26). According to (5–27),

$$K(x, t) = \frac{\sin x \cos t - \cos x \sin t}{W[\cos t, \sin t]} = \sin (x - t),$$

and if we set $x_0 = 0$ in (5–26), the expression

$$G[h] = \int_0^x \sin (x - t) h(t) \, dt$$

defines a right inverse for $L$. Notice that we are now in a position to find a particular solution of *any* linear differential equation involving the operator $D^2 + 1$, a fact that vividly illustrates the economy of using Green's functions. We now set $h(t) = \tan t$ to obtain

$$y_p(x) = \int_0^x \sin (x - t) \tan t \, dt$$

$$= \int_0^x (\sin x \cos t - \cos x \sin t) \tan t \, dt$$

$$= \sin x \int_0^x \sin t \, dt - \cos x \int_0^x \frac{\sin^2 t}{\cos t} \, dt$$

$$= \sin x - \cos x \ln |\sec x + \tan x|.$$

From this we obtain the same general solution that was found previously.

**Example 2**   In Section 5–8 we shall show that the general solution of

$$xy'' + y' = 0$$

on $(0, \infty)$ and $(-\infty, 0)$ is $c_1 + c_2 \ln |x|$. Hence the nonhomogeneous equation

$$xy'' + y' = x + 1 \tag{5–29}$$

has a particular solution of the form $y_p = c_1(x) + c_2(x) \ln |x|$, where $c_1(x)$ and $c_2(x)$ are determined from the equations

$$c_1'(x) + c_2'(x) \ln |x| = 0,$$

$$c_2'(x) \cdot \frac{1}{x} = \frac{x+1}{x}.$$

[Here $h(x) = (x+1)/x$ since we must divide by $x$ to put (5–29) in normal form.] Thus

$$c_1'(x) = -(x+1) \ln |x|, \qquad c_2'(x) = x + 1,$$

and

$$c_1(x) = \frac{x^2}{4} + x - \left( \frac{x^2}{2} + x \right) \ln |x|,$$

$$c_2(x) = \frac{x^2}{2} + x.$$

Accordingly,

$$y_p = \frac{x^2}{4} + x - \left( \frac{x^2}{2} + x \right) \ln |x| + \left( \frac{x^2}{2} + x \right) \ln |x| = \frac{x^2}{4} + x,$$

and the general solution of (5–29) is

$$y = \frac{x^2}{4} + x + c_1 + c_2 \ln |x|.$$

*1–15*

*1. find $y_1$ & $y_2$        3 plug in for        4 plug in for*

**EXERCISES**  *2. form $W[y_1, y_2]$        $c_1'$, $c_2'$ find        $y_p = c_1 y_1 + c_2 y_2$*

Find the general solution of each of the following differential equations. *5 $y = y_p + c_1 y_1 + c_2 y_2$*

1. $(D^2 + 1)y = \dfrac{1}{\cos x}$

2. $(D^2 - D - 2)y = e^{-x} \sin x$

3. $(D^2 + 4D + 4)y = xe^{2x}$

4. $(D^2 + 3D - 4)y = x^2 e^x$

5. $(4D^2 + 4D + 1)y = xe^{-x/2} \sin x$

6. $(D^2 + 4)y = \dfrac{e^{2x}}{2}$

7. $(D^2 + 10D - 12)y = \dfrac{(e^{2x} + 1)^2}{e^{2x}}$

8. $(D + 3)^2 y = (x + 1)e^x$

9. $(D^2 - 2D + 2)y = e^{2x} \sin x$

10. $(4D^2 - 8D + 5)y = e^x \tan^2 \dfrac{x}{2}$

For each of the following differential equations verify that the given expression is the general solution of the associated homogeneous equation and then find a particular solution of the equation.

11. $x^2 y'' - 2xy' + 2y = x^3 \ln x, \ x > 0; \ y_h = c_1 x + c_2 x^2$

12. $x^2 y'' - xy' + y = x(x + 1); \ y_h = (c_1 + c_2 \ln |x|)x$

13. $(\sin 4x)y'' - 4(\cos^2 2x)y' = \tan x; \ y_h = c_1 + c_2 \cos 2x$

14. $xy'' - (1 + 2x^2)y' = x^5 e^{x^2}; \ y_h = c_1 + c_2 e^{x^2}$

15. $(1 - x^2)y'' - 2xy' = 2x, -1 < x < 1; y_h = c_1 + c_2 \ln \dfrac{1 + x}{1 - x}$

In each of the following exercises find a Green's function for the given linear differential operator.

16. $D^2 + 3$                                  17. $D^2 - D - 2$

18. $D^2 + 4D + 4$                          19. $4D^2 - 8D + 5$

20. $D^2 + 3D - 4$                          21. $x^2D^2 - 2xD + 2$

22. $xD^2 - (1 + 2x^2)D$                 23. $(1 - x^2)D^2 - 2xD$

\* 24. Solve the initial-value problem

$$(D^2 + 2aD + b^2)y = \sin \omega t, \qquad y(0) = y'(0) = 0,$$

where $a, b, \omega$ are (real) constants and $a < b$. Consider separately the cases (i) $a = 0$, $b = \omega$, and (ii) $a \neq 0$ or $b \neq \omega$, and sketch the solution curve in each case.

\* 25. Find a Green's function for the operator in Exercise 24 and use this function to obtain the desired solution as $G[\sin \omega t]$.

26. The second-order equation

$$(D - 1)(xD + 3)y = e^x$$

may be solved by setting $(xD + 3)y = u$ and then successively solving the first-order equations

$$(D - 1)u = e^x \qquad \text{and} \qquad (xD + 3)y = u.$$

Use this technique to show that the general solution of this equation is

$$y = \frac{c_1}{x^3} + \frac{c_2 e^x}{x^3}(x^2 - 2x + 2) + e^x.$$

27. Use the technique introduced in the preceding exercise to show once more that $xe^{\alpha x}$ is a solution of $(D - \alpha)^2 y = 0$.

\* 28. Let $K(x, t)$ denote the Green's function for initial-value problems involving the operator $L = D^2 + a_1(x)D + a_0(x)$, and assume that $L$ is defined on an interval $I$ of the $x$-axis.
   a) What is the domain of $K(x, t)$ in the $xt$-plane?
   b) Prove that $K(x, x) = 0$ and $K_x(x, x) = 1$ for all $x$ in $I$. [*Note*: $K_x$ denotes the partial derivative of $K(x, t)$ with respect to $x$.]
   c) Show that for each fixed $t$ in $I$ the function $\varphi(x) = K(x, t)$ is a solution on $I$ of the initial-value problem $Ly = 0; \varphi(t) = 0, \varphi'(t) = 1$.
   d) Use the results of parts (b) and (c) to deduce that $K(x, t)$ is independent of the particular basis $y_1(x)$ and $y_2(x)$ chosen for the solution space of the homogeneous equation $Ly = 0$.

29. With $K(x, t)$ as in the preceding exercise, show that the function $y_p$ defined by

$$y_p(x) = \int_{x_0}^x K(x, t)h(t)\, dt$$

satisfies the initial conditions $y_p(x_0) = y_p'(x_0) = 0$ for all $h$ in $\mathscr{C}(I)$. [*Hint*: Use Leibnitz's formula for differentiating integrals, namely,

$$\frac{d}{dx} \int_{a(x)}^{b(x)} f(x, t)\, dt = \int_{a(x)}^{b(x)} \frac{\partial f}{\partial x}(x, t) + f(x, b(x))b'(x) - f(x, a(x))a'(x).]$$

* 30. Find the Green's function for initial-value problems on $(0, \infty)$ for the operator

$$L = D^2 + \frac{1}{x}D + \left(1 - \frac{p^2}{x^2}\right),$$

where $p$ is a nonnegative real number.

## *5–5 VARIATION OF PARAMETERS AND GREEN'S FUNCTIONS (continued)

The method of variation of parameters can easily be extended to apply to equations of arbitrary order. We begin with a normal equation

$$y^{(n)} + a_{n-1}(x)y^{(n-1)} + \cdots + a_0(x)y = h(x) \tag{5–30}$$

defined on an interval $I$, and again assume that the general solution

$$y_h = c_1 y_1(x) + \cdots + c_n y_n(x) \tag{5–31}$$

of the associated homogeneous equation is known. Then, following the argument given in the second-order case, we seek a particular solution of the form

$$y_p = c_1(x)y_1(x) + \cdots + c_n(x)y_n(x), \tag{5–32}$$

where, in addition to the requirement that $y_p$ satisfy (5–30), we impose the following $n - 1$ conditions on the unknown functions $c_1(x), \ldots, c_n(x)$:

$$c_1' y_1 + \cdots + c_n' y_n = 0,$$
$$c_1' y_1' + \cdots + c_n' y_n' = 0,$$
$$\vdots \tag{5–33}$$
$$c_1' y_1^{(n-2)} + \cdots + c_n' y_n^{(n-2)} = 0$$

for all $x$ in $I$. If the expression for $y_p$ is now substituted in (5–30) and the above conditions are used, we obtain the additional equation

$$c_1' y_1^{(n-1)} + \cdots + c_n' y_n^{(n-1)} = h(x). \tag{5–34}$$

For each $x$ in $I$, (5–33) together with (5–34) may be viewed as a system of $n$ linear equations in the unknowns $c_1', \ldots, c_n'$ whose determinant is $W[y_1(x), \ldots, y_n(x)]$. Our earlier reasoning still applies, and we can obtain a particular solution of (5–30) by solving this system for $c_1', \ldots, c_n'$, integrating, and then substituting the resulting functions in (5–32).

In fact, if $V_k(x)$ denotes the determinant obtained from $W[y_1(x), \ldots, y_n(x)]$ by replacing its $k$th column with

$$\begin{bmatrix} 0 \\ \cdot \\ \cdot \\ \cdot \\ 0 \\ 1 \end{bmatrix},$$

a straightforward computation gives

$$c'_k(x) = \frac{V_k(x)h(x)}{W[y_1(x), \ldots, y_n(x)]}. \tag{5–35}$$

(See Exercise 17.) Hence, just as in the second-order case, the particular solution can be written in integral form as

$$y_p(x) = \int_{x_0}^x \frac{y_1(x)V_1(t) + \cdots + y_n(x)V_n(t)}{W[y_1(t), \ldots, y_n(t)]} h(t) \, dt, \tag{5–36}$$

where $x_0$ is any point in $I$. This formula is usually written as

$$y_p(x) = \int_{x_0}^x K(x, t) h(t) \, dt, \tag{5–37}$$

where

$$K(x, t) = \frac{y_1(x)V_1(t) + \cdots + y_n(x)V_n(t)}{W[y_1(t), \ldots, y_n(t)]}, \tag{5–38}$$

or, for those who prefer determinants,

$$K(x, t) = \frac{\begin{vmatrix} y_1(t) & \cdots & y_n(t) \\ y'_1(t) & \cdots & y'_n(t) \\ \cdot & & \cdot \\ \cdot & & \cdot \\ \cdot & & \cdot \\ y_1^{(n-2)}(t) & \cdots & y_n^{(n-2)}(t) \\ y_1(x) & \cdots & y_n(x) \end{vmatrix}}{\begin{vmatrix} y_1(t) & \cdots & y_n(t) \\ y'_1(t) & \cdots & y'_n(t) \\ \cdot & & \cdot \\ \cdot & & \cdot \\ \cdot & & \cdot \\ y_1^{(n-2)}(t) & \cdots & y_n^{(n-2)}(t) \\ y_1^{(n-1)}(t) & \cdots & y_n^{(n-1)}(t) \end{vmatrix}}. \tag{5–39}$$

The function $K(x, t)$ defined here is called the **Green's function for the operator** $L = D^n + a_{n-1}(x)D^{n-1} + \cdots + a_0(x)$ (for initial-value problems in the interval

*I*), and the expression

$$G[h] = \int_{x_0}^{x} K(x, t)h(t)\, dt \tag{5-40}$$

defines a right-inverse $G: \mathscr{C}(I) \to \mathscr{C}^{(n)}(I)$ for the operator $L$. In fact, $G$ is *the* inverse of $L$ that satisfies the initial conditions

$$G(h)(x_0) = G(h)'(x_0) = \cdots = G(h)^{(n-1)}(x_0) = 0 \tag{5-41}$$

for all $h$ in $\mathscr{C}(I)$.

**Example**   Find a particular solution $y_p$ for the equation

$$3y''' + 5y'' - 2y' = r(x), \tag{5-42}$$

where $r(x)$ is continuous on $(-\infty, \infty)$.

*Solution.*   Here the general solution of the associated homogeneous equation is $c_1 + c_2 e^{-2x} + c_3 e^{x/3}$. Hence

$$y_p = c_1(x) + c_2(x)e^{-2x} + c_3(x)e^{x/3},$$

where $c_1(x)$, $c_2(x)$, $c_3(x)$ satisfy the identities

$$c_1'(x) + c_2'(x)e^{-2x} + c_3'(x)e^{x/3} = 0,$$
$$-2c_2'(x)e^{-2x} + \tfrac{1}{3}c_3'(x)e^{x/3} = 0,$$
$$4c_2'(x)e^{-2x} + \tfrac{1}{9}c_3'(x)e^{x/3} = \frac{r(x)}{3}.$$

[To obtain the last of these identities, we had to divide (5–42) by 3 to put it in normal form.] Thus

$$c_1'(x) = -\frac{r(x)}{2},$$
$$c_2'(x) = \tfrac{1}{14}e^{2x}r(x),$$
$$c_3'(x) = \tfrac{3}{7}e^{-x/3}r(x),$$

and it follows that

$$y_p = -\frac{1}{2}\int r(x)\, dx + \frac{e^{-2x}}{14}\int e^{2x}r(x)\, dx + \frac{3e^{x/3}}{7}\int e^{-x/3}r(x)\, dx$$

$$= \int \left[-\tfrac{1}{2} + \tfrac{1}{14}e^{-2(x-t)} + \tfrac{3}{7}e^{(x-t)/3}\right]r(t)\, dt.$$

Alternatively, we could have computed the Green's function $K(x, t)$ for the (normal) operator $D^3 + \tfrac{5}{3}D^2 - \tfrac{2}{3}D$, and used (5–37) to express $y_p$ as an integral involving $K(x, t)$. Starting with the basis $1, e^{-2x}, e^{x/3}$ for the solution space of the associated homogeneous equation, we would then obtain

$$
K(x, t) = \frac{
\begin{vmatrix}
1 & e^{-2t} & e^{t/3} \\
0 & -2e^{-2t} & \frac{1}{3}e^{t/3} \\
1 & e^{-2x} & e^{x/3}
\end{vmatrix}
}{
\begin{vmatrix}
1 & e^{-2t} & e^{t/3} \\
0 & -2e^{-2t} & \frac{1}{3}e^{t/3} \\
0 & 4e^{-2t} & \frac{1}{9}e^{t/3}
\end{vmatrix}
} = \frac{\frac{7}{3}e^{-5t/3} - \frac{1}{3}e^{-2x}e^{t/3} - 2e^{x/3}e^{-2t}}{-\frac{2}{9}e^{-5t/3} - \frac{4}{3}e^{-5t/3}}
$$

$$
= -\frac{3}{2} + \frac{3}{14}e^{-2(x-t)} + \frac{9}{7}e^{(x-t)/3}.
$$

Thus

$$
y_p = G\left[\frac{r(x)}{3}\right] = \int_{x_0}^{x} \left[-\tfrac{1}{2} + \tfrac{1}{14}e^{-2(x-t)} + \tfrac{3}{7}e^{(x-t)/3}\right]r(t)\,dt,
$$

which agrees with the result obtained previously.

The remainder of this section will be devoted to taking a closer look at Green's functions for initial-value problems and to establishing some of their more important properties. Throughout the discussion we shall assume that

$$
L = D^n + a_{n-1}(x)D^{n-1} + \cdots + a_0(x)
$$

is a fixed linear differential operator on $\mathscr{C}^{(n)}(I)$ and that $K(x, t)$ is the Green's function for $L$. It then follows from the very way $K(x, t)$ was constructed (see Exercises 18 through 21) that:

1. $K(x, t)$ is defined throughout the region $R$ of the $xt$-plane consisting of all points $(x, t)$ with $x$ and $t$ in $I$ (see Fig. 5–1);
2. $K(x, t)$ and $\partial K/\partial x$, $\partial^2 K/\partial x^2, \ldots, \partial^n K/\partial x^n$ are continuous everywhere in $R$;

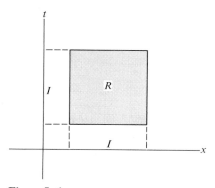

**Figure 5–1**

3. For *every* $x_0$ in $I$ and *every* $h$ in $\mathscr{C}(I)$, the function

$$y(x) = \int_{x_0}^x K(x, t)h(t)\, dt$$

is a solution on $I$ of the initial-value problem

$$Ly = h,$$

$$y(x_0) = y'(x_0) = \cdots = y^{(n-1)}(x_0) = 0.$$

These properties are actually sufficient to characterize the Green's function for initial-value problems involving $L$ in the sense that $K(x, t)$ is the *unique* function defined in $R$ which satisfies Properties 1, 2, and 3. This assertion will be proved as Theorem 5–4 and is mentioned here only to motivate the following definition.

**Definition 5–1**   *A function $H(x, t)$ is said to be a **Green's function for initial-value problems involving the linear differential operator $L$** if and only if $H(x, t)$ enjoys Properties 1, 2, and 3 just listed.*

This said, we proceed at once to give an alternative and for our purposes much more useful description of a Green's function for $L$. For convenience, we shall denote the various derivatives $\partial H/\partial x$, $\partial^2 H/\partial x^2$, ... in the following argument by $H_1, H_2, \ldots$. With this notation in effect, we have

**Theorem 5–3**   *Let $H(x, t)$ be defined throughout the region $R$ of Fig. 5–1, and suppose that $H$ and its partial derivatives $H_1, H_2, \ldots, H_n$ are continuous everywhere in $R$. Then $H(x, t)$ is a Green's function for the linear differential operator*

$$L = D^n + a_{n-1}(x)D^{n-1} + \cdots + a_0(x)$$

*if and only if the following identities are satisfied throughout $R$:*

$$H(x, x) \equiv 0,$$
$$H_1(x, x) \equiv 0,$$
$$\cdot$$
$$\cdot \tag{5-43}$$
$$\cdot$$
$$H_{n-2}(x, x) \equiv 0,$$
$$H_{n-1}(x, x) \equiv 1,$$

*and*

$$H_n(x, t) + a_{n-1}(x)H_{n-1}(x, t) + \cdots + a_0(x)H(x, t) \equiv 0. \tag{5-44}$$

*Proof.*   (You may find it helpful as you read this proof to write out the corresponding argument for the case where $n = 2$.) First assume that $H(x, t)$ is a Green's function for $L$. Then by Property 3 above the function

$$y(x) = \int_{x_0}^x H(x, t)h(t)\, dt \tag{5-45}$$

is a solution of the initial-value problem

$$Ly = h; \qquad y(x_0) = y'(x_0) = \cdots = y^{(n-1)}(x_0) = 0,$$

for every $x_0$ in $I$ and every $h$ in $\mathscr{C}(I)$. We now differentiate (5–45) with the help of Leibnitz's formula* to obtain

$$y'(x) = \int_{x_0}^{x} H_1(x, t)h(t)\, dt + H(x, x)h(x),$$

which when $x = x_0$ reduces to

$$H(x_0, x_0)h(x_0) = 0$$

(recall that $y'(x_0) = 0$). But by assumption this expression is valid for all $h$ in $\mathscr{C}(I)$ and hence, in particular, for $h \equiv 1$. Thus

$$H(x_0, x_0) = 0,$$

and since $x_0$ can be chosen arbitrarily in $I$, we have

$$H(x, x) \equiv 0,$$

and

$$y'(x) = \int_{x_0}^{x} H_1(x, t)h(t)\, dt. \qquad (5\text{–}47)$$

We now repeat the argument, starting with (5–47), to get, first

$$y''(x) = \int_{x_0}^{x} H_2(x, t)h(t)\, dt + H_1(x, x)h(x),$$

then

$$H_1(x, x) \equiv 0,$$

and finally

$$y''(x) = \int_{x_0}^{x} H_2(x, t)h(t)\, dt.$$

Continuing in this fashion, we eventually arrive at

$$H_{n-2}(x, x) \equiv 0,$$

and

$$y^{(n-1)}(x) = \int_{x_0}^{x} H_{n-1}(x, t)h(t)\, dt.$$

Differentiating once more, we obtain

$$y^{(n)}(x) = \int_{x_0}^{x} H_n(x, t)h(t)\, dt + H_{n-1}(x, x)h(x);$$

---

\* Leibnitz's formula is

$$\frac{d}{dx} \int_{a(x)}^{b(x)} f(x, t)\, dt = \int_{a(x)}^{b(x)} \frac{\partial f}{\partial x}(x, t)\, dt + f(x, b(x))b'(x) - f(x, a(x))a'(x). \qquad (5\text{–}46)$$

whence

$$y^{(n)}(x_0) = H(x_0, x_0)h(x_0). \tag{5-48}$$

But since $y(x)$ is a solution of $Ly = h$, it follows that

$$y^{(n)}(x) + a_{n-1}(x)y^{(n-1)}(x) + \cdots + a_0(x)y(x) \equiv h(x),$$

and the initial conditions in effect imply that

$$y^{(n)}(x_0) = h(x_0).$$

This, together with (5–48) and the fact that $h$ and $x_0$ are still arbitrary, implies that

$$H_{n-1}(x, x) \equiv 1$$

and

$$y^{(n)}(x) = \int_{x_0}^x H_n(x, t)h(t) \, dt + h(x).$$

Thus, in particular, we have established the several identities listed in (5–43).

To establish (5–44), we substitute the formulas obtained above for $y, y', \ldots,$ $y^{(n-1)}$ in $Ly = h$. After the various terms are collected we have

$$\int_{x_0}^x [H_n(x, t) + a_{n-1}(x)H_{n-1}(x, t) + \cdots + a_0(x)H(x, t)]h(t) \, dt \equiv 0. \tag{5-49}$$

Finally, the fact that this expression holds for all $x_0$ in $I$ and all $h$ in $\mathscr{C}(I)$ allows us to conclude that the bracketed portion of the integrand is identically zero (see Exercise 23). With this the first part of the proof is complete.

As for the remainder, the argument needed to show that (5–43) and (5–44) imply that $H(x, t)$ is a Green's function for $L$ is an elementary computation, and has therefore been relegated to the exercises (see Exercise 24). ∎

Among other things, the preceding theorem asserts that for *any* fixed $t_0$ in the interval $I$, the function

$$k(x) = H(x, t_0)$$

is a solution of the initial-value problem

$$Ly = 0; \qquad y(t_0) = y'(t_0) = \cdots = y^{(n-2)}(t_0) = 0, \; y^{(n-1)}(t_0) = 1.$$

But as we know, the solution of this problem is unique. Thus the values of $H(x, t)$ are *uniquely determined* by the operator $L$ on the line segment consisting of those points $(x, t)$ in $R$ with $t = t_0$ (see Fig. 5–2). However, $t_0$ can be chosen arbitrarily in $I$. We therefore have our main result.

**Theorem 5–4** *The Green's function for initial-value problems on $I$ involving a normal linear differential operator $L$ is **uniquely** determined by $L$, and hence must coincide with the function $K(x, t)$ defined by (5–38) or (5–39). In particular, $K(x, t)$ is independent of the basis for the solution space of $Ly = 0$ used in computing it.*

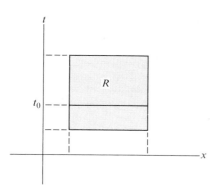

**Figure 5–2**

Everything that has been said up to this point in our discussion of Green's functions applies to arbitrary (normal) linear differential operators. As might be expected, much more precise information can be given for operators with constant coefficients, so we conclude this section with a theorem that describes the Green's functions obtained for this special case.

*Theorem 5–5*    *The Green's function for a constant-coefficient linear differential operator $L$ can be written in the form $k(x - t)$, where $k(x)$ is the solution on $(-\infty, \infty)$ of the initial-value problem*

$$Ly = 0; \qquad y(0) = y'(0) = \cdots = y^{(n-2)}(0) = 0, \; y^{(n-1)}(0) = 1.$$

*Proof.*   The function $H(x, t) = k(x - t)$ clearly satisfies the identities listed in (5–43), and because the coefficients of $L$ are constant the equality $Lk = 0$ implies (5–44). Hence by Theorem 5–4 the function $k(x - t)$ is *the* Green's function for $L$.  ∎

### EXERCISES

Use the method of variation of parameters without employing Formulas (5–37) through (5–39) to find the general solution of each of the following differential equations.

1. $y''' - y'' - y' + y = 4xe^x$             2. $y''' - y' = \sin x$

3. $y''' - 2y'' = 4(x + 1)$                   4. $y''' - 3y'' - y' + 3y = 1 + e^x$

5. $y''' - 7y' + 6y = 2 \sin x$               6. $y''' - 3y' - 2y = 9e^{-x}$

7. $y''' - y' = \sin x$                       8. $y''' + y'' + y' + y = 2(\sin x + \cos x)$

9. $y^{(iv)} - y'' = 2xe^x$                   10. $y^{(iv)} - y = x^2 + 1$

In Exercises 11 through 16 compute the Green's function $K(x, t)$ for the given operator (a) by using Formula (5–38) or (5–39) and (b) by applying Theorem 5–5.

11. $D^2(D - 1)$                              12. $D(D^2 - 4)$

13. $D^3 - 6D^2 + 11D - 6$                    14. $D^3 + \frac{5}{2}D^2 - \frac{3}{2}$

15. $D^2(D^2 - 1)$                           16. $D^4 - 1$

17. Verify Formula (5–35).

Exercises 18 through 21 concern properties of the function $K(x, t)$ defined by Formula (5–38) or (5–39). Establish each of them.

18. When $n = 2$, Formula (5–39) reduces to (5–27) of the preceding section.

19. $K(x, t)$ is defined and has continuous derivatives through order $n$ in the region $R$ described in the text.

20. The partial derivatives with respect to $x$ of $K(x, t)$ satisfy the identities

$$K(x, x) \equiv K_1(x, x) \equiv \cdots \equiv K_{n-2}(x, x) \equiv 0, \qquad K_{n-1}(x, x) \equiv 1,$$

in the interval $I$. Here

$$K_i(x, t) = \frac{\partial^i}{\partial x^i} K(x, t).$$

21. For each $x_0$ in $I$ and each $h$ in $\mathscr{C}(I)$ the function

$$y(x) = \int_{x_0}^{x} K(x, t)h(t)\, dt$$

satisfies the initial conditions $y(x_0) = \cdots = y^{(n-1)}(x_0) = 0$ and the equation $Ly = h(x)$. [*Hint:* Use Leibnitz's formula and the result of Exercise 20.]

22. If $f$ is continuous on the interval $[a, b]$ and if

$$\int_a^b f(x)g(x)\, dx = 0$$

for *every* $g$ in $\mathscr{C}[a, b]$, then $f \equiv 0$ on $[a, b]$. [*Hint:* Assume that $f(x_0) \neq 0$ and use the continuity of $f$ to obtain an interval $(x_0 - \delta, x_0 + \delta)$ in which $|f(x)| \geq |f(x_0)|/2$. Then find a function $g$ in $\mathscr{C}(I)$ for which the integral above is different from zero.]

23. Use the result of Exercise 22 to prove the assertion made in the text concerning the bracketed term in (5–49).

24. Let $H(x, t)$ satisfy the hypotheses of Theorem 5–3, and the identities given in (5–43) and (5–44). Prove that for every $x_0$ in $I$ and every $h$ in $\mathscr{C}(I)$ the function

$$y(x) = \int_{x_0}^{x} H(x, t)h(t)\, dt$$

satisfies the initial-value problem

$$Ly = 0; \qquad y(x_0) = y'(x_0) = \cdots = y^{(n-1)}(x_0) = 0.$$

[*Hint:* Use Leibnitz's formula.]

25. Show that the function $H(x, t) = k(x - t)$ of Theorem 5–5 satisfies (5–43) and (5–44).

## 5–6  REDUCTION IN ORDER

In Section 4–7 we used Abel's formula and the technique of *reduction in order* to find the general solution of a (normal) second-order homogeneous linear differential equation when one nontrivial solution of the equation was already known. We can now extend this technique to find the general solution of a second-order *non*homogeneous equation when we know a nontrivial solution of the associated homogeneous equation. In doing so we shall illustrate one of the remarkable properties of linear differential equations, namely that it is sometimes possible to find all the solutions of a nonhomogeneous equation $Ly = h$ without knowing all the solutions of the associated homogeneous equation $Ly = 0$.

The key to extending the technique of reduction in order to solve nonhomogeneous equations is variation of parameters, which proves to be a powerful tool in this new context.

**Example 1** Find the general solution of

$$x^2 y'' + 2xy' - 2y = x \qquad (5–50)$$

on the interval $(0, \infty)$.

*Solution.* The associated homogeneous equation

$$x^2 y'' + 2xy' - 2y = 0$$

clearly has $y_p = x$ as a particular solution. In fact, any constant times $x$ is also a solution of the homogeneous equation, and our earlier experience with variation of parameters suggests that we seek solutions of the nonhomogeneous equation (5–50) of the form

$$y = xc(x).$$

Then

$$y' = xc'(x) + c(x),$$
$$y'' = xc''(x) + 2c'(x),$$

and substituting in (5–50), we obtain

$$x^2(xc'' + 2c') + 2x(xc' + c) - 2xc = x,$$

which reduces to

$$x^3 c'' + 4x^2 c' = x.$$

Thus (5–50) will have a solution of the form $y = xc(x)$ if $c$ can be chosen to satisfy

$$c'' + \frac{4}{x} c' = x^{-2}, \qquad x \neq 0.$$

But this equation is *first-order linear* in $c'$, and can therefore be solved by the technique introduced in Section 4–2. An easy computation reveals that it has $x^4$ as an integrating factor and that

$$c' = \frac{1}{3x} + k_1 x^{-4},$$

where $k_1$ is an arbitrary constant. A second integration yields

$$c = \tfrac{1}{3} \ln x + k_1 x^{-3} + k_2,$$

and it follows that for all constants $k_1$ and $k_2$

$$y = \frac{x}{3} \ln x + k_1 x^{-2} + k_2 x \tag{5-51}$$

is a solution of (5–50). In fact, (5–51) is the general solution of the equation, as can easily be deduced from the linear independence of the functions $x^{-2}$ and $x$ in $\mathscr{C}(0, \infty)$.

*Remark.*  Had this technique been applied to finding the general solution of the homogeneous equation

$$x^2 y'' + 2xy' - 2y = 0,$$

it would have yielded

$$y = k_1 x^{-2} + k_2 x, \qquad x > 0.$$

The same result would have been obtained with the earlier technique of reduction in order based on Abel's formula. Indeed, it is not difficult to show that in the homogeneous case the two methods are variants of each other. (See Exercise 14.)

We now turn to the general case of an arbitrary second-order linear equation

$$a_2(x)y'' + a_1(x)y' + a_0(x)y = h(x) \tag{5-52}$$

which is normal on an interval $I$. Suppose that $y_p$ is a nontrivial solution of the associated homogeneous equation. Then if we set

$$y = y_p c(x),$$

we will have

$$y' = y_p c'(x) + y_p' c(x),$$
$$y'' = y_p c''(x) + 2y_p' c'(x) + y_p'' c(x),$$

and substitution in (5–52) will yield

$$a_2(y_p c'' + 2y_p' c' + y_p'' c) + a_1(y_p c' + y_p' c) + a_0 y_p c = h$$

or

$$a_2 y_p c'' + (2a_2 y_p' + a_1 y_p)c' + (a_2 y_p'' + a_1 y_p' + a_0 y_p)c = h.$$

But by assumption $a_2 y_p'' + a_1 y_p' + a_0 y_p = 0$. Hence the last equation reduces to

$$a_2 y_p c'' + (2a_2 y_p' + a_1 y_p)c' = h,$$

which is first-order linear in $c'$ and can therefore be solved. Once $c'$ has been found, $c$ can be obtained by integration, and the general solution of (5–52) can then be expressed as the product of $c$ and the known solution $y_p$. (See Exercise 11 for details.)

**Example 2**  Find the general solution of

$$xy'' + (1 - x)y' - y = e^x \tag{5-53}$$

for $x > 0$.

*Solution.* In this case the associated homogeneous equation is

$$xy'' + (1 - x)y' - y = 0,$$

and the solution $y_p = e^x$ is easily found by inspection. (The phrase "found by inspection," which is often encountered in the study of differential equations, is just a dodge to hide the fact that the process was one of trial and error.) Hence the general solution of (5–53) can be written in the form

$$y = c(x)e^x,$$

and a routine computation reveals that $c'$ must satisfy the equation

$$xc'' + (x + 1)c' = 1.$$

Rewriting this equation as

$$c'' + \left(1 + \frac{1}{x}\right)c' = \frac{1}{x},$$

we find that it has

$$e^{\int (1 + 1/x)\, dx} = xe^x$$

as an integrating factor, and that its general solution is

$$c' = \frac{e^{-x}}{x}(k_1 + e^x) = k_1 \frac{e^{-x}}{x} + \frac{1}{x}.$$

Hence

$$c = \ln x + k_1 \int \frac{e^{-x}}{x}\, dx + k_2,$$

and the general solution of (5–53) is

$$y = e^x \ln x + k_1 e^x \int \frac{e^{-x}}{x}\, dx + k_2 e^x.$$

## EXERCISES

Use the given solution of the associated homogeneous equation to find the general solution of the equations in Exercises 1 through 10.

1. $y'' + xy' = 3x; \qquad y_p = 1$

2. $xy'' - y' = 0, x > 0; \qquad y_p = 1$

3. $y'' - \frac{2}{x^2}y = 0, x > 0; \qquad y_p = x^2$

4. $xy'' - (x + 2)y' + 2y = x^3 + x, x > 0; \qquad y_p = e^x$

5. $y'' - 4xy' + (4x^2 - 2)y = 0; \qquad y_p = e^{x^2}$

6. $xy'' + (2 + x)y' + y = e^{-x}, x > 0; \qquad y_p = 1/x$

7. $4x^2 y'' - 8xy' + 9y = 0, x > 0; \qquad y_p = x^{3/2}$

8. $x^2y'' + xy' - y = \dfrac{\ln x}{x}, x > 0; \qquad y_p = x$

9. $xy'' + (x - 1)y' - y = 0, x > 0; \qquad y_p = e^{-x}$

10. $y'' - 2xy' + 2y = 1, x > 0; \qquad y_p = x$

11. The substitution $y = y_p c(x)$ transforms the second-order equation (5–52) into an equation that is first-order in $c'$. Once this equation has been solved for $c'$, the function $c$ can be obtained by integration. Prove that $y_p c(x)$ is the general solution of (5–52). [*Hint*: Let $c'(x) = k_1 f(x)$, where $k_1$ is an arbitrary constant. Then $c(x) = k_1 \int f(x)\, dx + k_2$, and $y_p c(x) = k_1 y_p \int f(x)\, dx + k_2 y_p$. Compute the Wronskian of $y_p \int f(x)\, dx$ and $y_p$ and conclude that these functions are linearly independent on $I$, the interval of normality of (5–52).]

12. The equation

$$(1 - x^2)y'' - 2xy' + 2y = 0$$

has $y_p = x$ as a particular solution. Use the method of variation of parameters to find a second solution on the interval $(-1, 1)$ that is linearly independent of $y_p$.

13. a) Verify that $y_p = x^{-1/2} \sin x$ is a solution of

$$x^2y'' + xy' + (x^2 - \tfrac{1}{4})y = 0$$

for $x > 0$.

b) Use the method of variation of parameters to find a second solution of the equation that is linearly independent of $y_p$.

14. Use the method of reduction in order and the solution

$$y = e^{-\int P(x)\, dx}$$

of the homogeneous equation

$$y' + P(x)y = 0$$

to find the general solution of

$$y' + P(x)y = Q(x).$$

15. Let $y_1$ be a nontrivial solution of the normal second-order equation

$$y'' + a_1(x)y' + a_0(x)y = 0.$$

Use the method of variation of parameters to find the general solution of this equation and compare the result to the solution obtained by means of Abel's formula.

* 16. Let $y_1$ and $y_2$ be linearly independent solutions of the third-order equation

$$y''' + a_2(x)y'' + a_1(x)y' + a_0(x)y = 0$$

on an interval $I$.

a) Show that the method of variation of parameters can be used with the solution $y_1$ to reduce the order of the third-order equation by one.

b) Show that $(y_2/y_1)'$ is a solution of the second-order equation obtained in (a) on any subinterval of $I$ on which $y_1$ has no zeros, and hence deduce that any normal (homogeneous) third-order linear differential equation can be solved on such a subinterval as soon as two linearly independent solutions are known.

17. Use the technique outlined in Exercise 16 to find the general solution of

$$x^3y''' - 3xy' + 3y = 0, \qquad x > 0,$$

given that $y_1 = x$ and $y_2 = x^3$ are solutions.

## 5–7 THE METHOD OF UNDETERMINED COEFFICIENTS

The method of variation of parameters enables us to find a particular solution of a nonhomogeneous linear differential equation whenever the general solution of its associated homogeneous equation is known. However, it is not always the most efficient way of producing such a solution, and for certain equations the work involved can be considerably reduced. For instance, it would be pointless to use variation of parameters to find a particular solution of $(D^2 - D + 5)y = 3$ since the solution $y_p = \frac{3}{5}$ is immediately evident. And even with an equation such as

$$(D^2 + 3)y = e^x$$

it is obvious that a solution of the form

$$y_p = Ae^x$$

must exist for a suitable value of $A$. In fact, by substituting $Ae^x$ in the equation we obtain

$$Ae^x + 3Ae^x = e^x,$$

from which it follows that $A = \frac{1}{4}$, and $y_p = e^x/4$.

This method, in which a particular solution is known up to certain undetermined constants, and the values of these constants are found by using the differential equation, is known as the **method of undetermined coefficients**. It is clear that the method depends for its success on our ability to recognize the form of a particular solution, and for this reason lacks the generality possessed by the method of variation of parameters. Nevertheless, it can be used often enough to merit attention.

One type of equation for which this method always works is a *constant-coefficient* equation

$$Ly = h \tag{5–54}$$

in which $h$ itself is a solution of a linear differential equation with constant coefficients. For then we can find a linear differential operator $L_1$ which annihilates $h$ (i.e., is such that $L_1h = 0$), and it follows that every solution of (5–54) is also a solution of the homogeneous equation

$$L_1Ly = 0. \tag{5–55}$$

Thus we can obtain a particular solution for (5–54) by appropriately determining the constants in the general solution of (5–55). A few examples will suffice to illustrate this technique.

**Example 1**   Since $D^3$ annihilates the right-hand side of the equation

$$(D^2 + 1)y = 3x^2 + 4, \tag{5-56}$$

a particular solution of (5–56) can be found among the solutions of the homogeneous equation

$$D^3(D^2 + 1)y = 0. \tag{5-57}$$

In other words, (5–56) has a particular solution of the form

$$y_p = c_1 + c_2x + c_3x^2 + c_4 \sin x + c_5 \cos x \tag{5-58}$$

for suitably chosen values of $c_1, \ldots, c_5$. In fact, we can say even more than this if we observe that $c_4 \sin x + c_5 \cos x$ is the general solution of the homogeneous equation $(D^2 + 1)y = 0$. For then it is clear that the last two terms in (5–58) will be annihilated when substituted in (5–56), and so, instead of dragging them through our computations only to see them disappear in the process, we can begin by setting

$$y_p = c_1 + c_2x + c_3x^2.$$

Substituting this expression in (5–56), we obtain

$$2c_3 + c_1 + c_2x + c_3x^2 = 3x^2 + 4,$$

from which it follows that

$$c_1 + 2c_3 = 4, \qquad c_2 = 0, \qquad c_3 = 3.$$

Thus $c_1 = -2, c_2 = 0, c_3 = 3$, and

$$y_p = 3x^2 - 2.$$

**Example 2**   To find a particular solution of

$$(D^2 - 4D + 4)y = 2e^{2x} + \cos x, \tag{5-59}$$

we apply the operator $(D - 2)(D^2 + 1)$ to the equation and obtain

$$(D - 2)^3(D^2 + 1)y = 0.$$

The general solution of the preceding equation is

$$y = c_1e^{2x} + c_2xe^{2x} + c_3x^2e^{2x} + c_4 \sin x + c_5 \cos x,$$

and since the first two terms are annihilated by the operator $D^2 - 4D + 4$, we look for a particular solution of (5–59) of the form

$$y_p = c_3x^2e^{2x} + c_4 \sin x + c_5 \cos x.$$

In this case

$$y_p' = 2c_3xe^{2x} + 2c_3x^2e^{2x} + c_4 \cos x - c_5 \sin x,$$
$$y_p'' = 2c_3e^{2x} + 8c_3xe^{2x} + 4c_3x^2e^{2x} - c_4 \sin x - c_5 \cos x,$$

and substitution in (5–59) yields

$$2c_3e^{2x} + (3c_4 + 4c_5) \sin x + (3c_5 - 4c_4) \cos x = 2e^{2x} + \cos x.$$

Hence

$$2c_3 = 2, \qquad 3c_4 + 4c_5 = 0, \qquad 3c_5 - 4c_4 = 1,$$

and

$$c_3 = 1, \qquad c_4 = -\tfrac{4}{25}, \qquad c_5 = \tfrac{3}{25}.$$

Accordingly,

$$y_p = x^2 e^{2x} - \tfrac{4}{25} \sin x + \tfrac{3}{25} \cos x.$$

## EXERCISES

Use the method of undetermined coefficients to find a particular solution for each of the following differential equations.

1. $D(D + 1)y = 2x + 3e^x$
2. $D(D + 1)y = 2 + e^{-x}$
3. $D(D - 1)y = \sin x$
4. $(D^2 + 1)y = 3 \cos x$
5. $(D^2 + 4D + 2)y = xe^{-2x}$
6. $(D^2 + D - 6)y = 6(x + 1)$
7. $(6D^2 + 2D - 1)y = 7x(x + 1)e^x$
8. $(D^2 - 5D + 6)y = -2 + 36x^2 + e^x$
9. $(D^2 - 4D + 5)y = (x + 1)^3$
10. $(D^2 - 4D + 8)y = e^{2x}(1 + \sin 2x)$
11. $(D^2 + 6D + 10)y = x^4 + 2x^2 + 2$
12. $(D^2 - D + \tfrac{1}{4})y = xe^{x/2}$
13. $D(D^2 - 2D + 10)y = 3xe^x$
14. $(D^3 + 3D^2 + 3D + 1)y = x^4 + 4x^3 + 10x^2 + 20x + 1$
15. $(D^3 - D^2 - D + 1)y = 2(x + 2e^{-x})$
16. $(D^3 - 3D - 2)y = e^x(1 + xe^x)$
17. $(D^3 + D - 1)y = \sin x + \cos x$
18. $D^2(D^2 + 1)y = 1 + 2xe^x$
19. $D(D^2 - 1)(D - 2)y = x^2 + 2x + 3 - 2e^x$
20. $(D^4 + 5D^2 + 4)y = 2 \cos x$

Give the form of a particular solution for each of the following equations. The coefficients need not be evaluated.

21. $(D^2 - 4D + 4)y = x(2e^{2x} + x \sin x)$
22. $(D^2 + 2D + 2)y = x^2 - 3xe^{-2x} \cos 5x$
23. $(D^2 + 1)^3(D - 1)y = 3e^{-x} + 5x^2 \cos x$
24. $D^2(D^4 - 4D^3 + 6D^2 - 4D + 1)y = (x^2 + 1)(1 - e^x)$
25. $(D^8 - 2D^4 + 1)y = (2x - 1) \cosh x + x^3 \sin x$
26. $(D^3 - 1)(D^2 + D - 2)y = e^{x/2} \sin \sqrt{3}x - x \cos \sqrt{3}x$
27. $(D^3 - 1)^3 y = (2x + 1)^2 e^x + \dfrac{x \sin x}{2}$
28. $(D^2 - 2D + 1)(D^2 - 4)^2 y = x \sinh x + \cosh 2x$
29. $[(4D^2 - 4D + 5)(D^2 + 2D + 1)]^2 y = x(1 + e^{x/2} \sin x - x \cos x)$
30. $D(D^2 - 4)^5 y = (x + 1)^2[(x + 1) + \sinh 2x]$

## 5-8  THE EULER EQUATION

A linear differential equation of the form

$$x^n \frac{d^n y}{dx^n} + a_{n-1} x^{n-1} \frac{d^{n-1} y}{dx^{n-1}} + \cdots + a_1 x \frac{dy}{dx} + a_0 y = 0, \qquad (5\text{--}60)$$

where $a_0, \ldots, a_{n-1}$ are constants, is called a (homogeneous) **Euler equation** of order $n$. This equation is defined on the entire $x$-axis, but is normal only on intervals that do not contain the point $x = 0$. It is one of the relatively few equations with variable coefficients that can be solved in closed form in terms of elementary functions, and it is important because its solutions are to some extent typical of those of a large class of linear differential equations whose leading coefficients vanish at the origin.

Equation (5–60) is solved by converting it into a linear equation with constant coefficients through the change of variable

$$u = \ln x$$

when $x > 0$. (On the interval $(-\infty, 0)$ the change of variable $u = \ln(-x)$ must be used.) We begin by using the chain rule to express the various derivatives $dy/dx$, $d^2 y/dx^2, \ldots$ in terms of the variable $u$:

$$\frac{dy}{dx} = \frac{dy}{du} \frac{du}{dx} = \frac{1}{x} \frac{dy}{du},$$

$$\frac{d^2 y}{dx^2} = \frac{d}{dx}\left(\frac{1}{x} \frac{dy}{du}\right) = \frac{1}{x} \frac{d}{dx}\left(\frac{dy}{du}\right) - \frac{1}{x^2} \frac{dy}{du}$$

$$= \frac{1}{x} \frac{d^2 y}{du^2} \frac{du}{dx} - \frac{1}{x^2} \frac{dy}{du} = \frac{1}{x^2}\left(\frac{d^2 y}{du^2} - \frac{dy}{du}\right),$$

and so on. These formulas follow a pattern, which becomes apparent when they are written in operator notation, where $D, D^2, \ldots$ are used to denote differentiation *with respect to $u$*:

$$\frac{dy}{dx} = \frac{1}{x} Dy,$$

$$\frac{d^2 y}{dx^2} = \frac{1}{x^2} D(D - 1)y,$$

$$\cdot$$
$$\cdot$$
$$\cdot$$

$$\frac{d^n y}{dx^n} = \frac{1}{x^n} D(D - 1) \cdots (D - n + 1)y \qquad (5\text{--}61)$$

(Exercise 17). Hence (5–60) becomes

$$[D(D - 1) \cdots (D - n + 1) + a_{n-1}D(D - 1) \cdots (D - n + 2)$$
$$+ \cdots + a_2 D(D - 1) + a_1 D + a_0]y = 0,$$

which, as asserted, is a constant-coefficient equation in $D = d/du$.

In particular, when $n = 2$, Eq. (5–60) becomes

$$x^2 \frac{d^2 y}{dx^2} + a_1 x \frac{dy}{dx} + a_0 y = 0, \qquad (5\text{--}62)$$

and the change of variable* $u = \ln |x|$ yields the equation

$$[D(D - 1) + a_1 D + a_0]y = 0.$$

Thus the general solution $y = y(x)$ of (5–62) is determined by the roots $\alpha_1$, $\alpha_2$ of the equation

$$m(m - 1) + a_1 m + a_0 = 0,$$

known as the **indicial equation** associated with (5–62). An easy computation shows that

i) if $\alpha_1$ and $\alpha_2$ are real, $\alpha_1 \neq \alpha_2$, then

$$y = c_1 |x|^{\alpha_1} + c_2 |x|^{\alpha_2};$$

ii) if $\alpha_1 = \alpha_2 = \alpha$, then

$$y = |x|^{\alpha}(c_1 + c_2 \ln |x|);$$

iii) if $\alpha_1 = a + bi, \alpha_2 = a - bi, b \neq 0$, then

$$y = |x|^a [c_1 \sin (b \ln |x|) + c_2 \cos (b \ln |x|)].$$

These results will be used in Chapter 7.

**Example 1**   Find the solution of

$$x^2 y'' + 2xy' - 6y = 0 \qquad (5\text{--}63)$$

whose graph passes through the point (1, 1) with slope zero.

*Solution.*   The indicial equation associated with (5–63) is

$$m(m - 1) + 2m - 6 = 0$$

or

$$m^2 + m - 6 = 0,$$

and has 2 and $-3$ as roots. Thus the general solution of (5–63) on $(0, \infty)$ is

$$y = c_1 x^2 + c_2 x^{-3},$$

---

* If we use the absolute value at this point, our solutions will be valid on both the positive and negative $x$-axis.

and since the initial conditions

$$y(1) = 1, \qquad y'(1) = 0$$

imply that $c_1 = \frac{3}{5}$ and $c_2 = \frac{2}{5}$, the required solution is

$$y = \tfrac{3}{5}x^2 + \tfrac{2}{5}x^{-3}.$$

**Example 2**    Earlier in this chapter (Example 2 of Section 5–4) we said that the general solution of

$$xy'' + y' = 0 \tag{5–64}$$

is

$$y = c_1 + c_2 \ln |x|. \tag{5–65}$$

To prove this assertion, we multiply (5–64) by $x$ to obtain the second-order Euler equation

$$x^2 y'' + xy' = 0,$$

whose associated indicial equation is

$$m(m - 1) + m = 0$$

or

$$m^2 = 0.$$

From this (5–65) follows at once.

The general solution of a nonhomogeneous Euler equation can be obtained by variation of parameters as in Example 2 of Section 5–4, or by computing the Green's function of the operator involved. But here, too, the problem can be reduced to one involving a constant-coefficient equation, and it is usually easier to solve in this form.

**Example 3**    Find the general solution of

$$x^2 y'' + xy' = x + x^2 \tag{5–66}$$

on $(0, \infty)$.

*Solution.* If $u = \ln x$, $x > 0$, then $x = e^u$, $x^2 = e^{2u}$, and the transformed version of (5–66) is

$$D^2 y = e^u + e^{2u}. \tag{5–67}$$

The general solution of $D^2 y = 0$ is $c_1 + c_2 u$, and a particular solution of (5–67) can be found by using the method of undetermined coefficients with the expression $Ae^u + Be^{2u}$. An easy computation yields $A = 1$, $B = \frac{1}{4}$. Thus the general solution of (5–66) on $(0, \infty)$ is

$$y = x + \frac{x^2}{4} + c_1 + c_2 \ln x.$$

**EXERCISES**

Find the general solution of the equations in Exercises 1 through 11.

1. $x^2 y'' + 2xy' - 2y = 0$
2. $4x^2 y'' - 8xy' + 9y = 0$
3. $x^2 y'' + xy' + 9y = 0$
4. $x^2 y'' - 3xy' + 7y = 0$

5. $x^2 y'' + x y' - p^2 y = 0,$     $p$ a constant

6. $2x^2 y'' + x y' - y = 0$

7. $x^3 y''' - 2x^2 y'' - 17xy' - 7y = 0$

8. $x^3 y''' - 3x^2 y'' + 6xy' - 6y = 0$

9. $x^3 y''' + 4x^2 y'' - 2y = 0$

10. $x^3 y''' + 4x^2 y'' - 8xy' + 8y = 0$

11. $x^4 y^{(iv)} + 6x^3 y''' + 7x^2 y'' + xy' - y = 0$

In Exercises 12 through 17, find a particular solution of each equation on $(0, \infty)$.

12. $x^2 y'' + x y' - 9y = x^3 + 1$

13. $x^2 y'' + x y' + 9y = \sin(\ln x^3)$

14. $x^2 y'' + 4xy' + 2y = 2 \ln x$

15. $x^3 y''' + 4x^2 y'' + xy' + y = x$

16. $x^2 y''' + x y'' + 4y' = 1 + \cos(2 \ln x)$

17. Prove (5–61) by mathematical induction.

18. Compute the Green's function for the operator

$$x^2 D^2 + a_1 x D + a_0$$

on $\mathscr{C}^{(2)}(0, \infty)$.

## 5–9  CIRCUITS CONTAINING A CAPACITOR

In Section 4–3 we saw that the flow of current $I$ in a simple (one-loop) electrical circuit containing a resistance $R$, an inductance $L$, and a source of electromotive force $E$ is described by the initial-value problem

$$L \frac{dI}{dt} + RI = E, \qquad I(0) = 0.$$

We now consider one-loop circuits that contain a capacitor of capacitance $C$ which has the ability to store charge $Q$. The electrical behavior of a capacitor is such that as it gathers charge it resists further charge accumulation with an emf of magnitude

$$E_C = \frac{1}{C} Q.$$

Thus by Kirchhoff's second law (p. 124) the flow of current in a circuit of the type shown in Fig. 5–3 must satisfy the differential equation

$$L \frac{dI}{dt} + RI + \frac{1}{C} Q = E. \tag{5–68}$$

As it stands, this equation involves two time-dependent variables, $I$ and $Q$. However, since current is defined to be the time rate of change of charge, $I$ and $Q$ are related by the equation

$$I = \frac{dQ}{dt}.$$

Thus $dI/dt = d^2 Q/dt^2$, and (5–68) can be rewritten in terms of $Q$ as

$$L \frac{d^2 Q}{dt^2} + R \frac{dQ}{dt} + \frac{1}{C} Q = E. \tag{5–69}$$

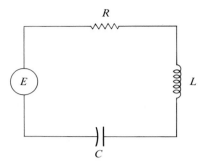

**Figure 5–3**

This equation admits a remarkably varied collection of solutions depending on the values of $L$, $R$, and $C$, the nature of the impressed emf, and the initial conditions imposed on the circuit. We shall explore a few of the many possibilities, leaving still others as exercises.

**The $LC$-Circuit**   We first consider a circuit with a resistance small enough to be ignored, in which case (5–69) becomes

$$L \frac{d^2Q}{dt^2} + \frac{1}{C} Q = E. \tag{5–70}$$

The circuit can then be analyzed as follows.

   ***The homogeneous equation***   The homogeneous equation associated with (5–70) is

$$L \frac{d^2Q}{dt^2} + \frac{1}{C} Q = 0,$$

the characteristic equation is

$$Lm^2 + \frac{1}{C} = 0,$$

and the characteristic roots are

$$m = \pm \frac{1}{\sqrt{LC}} i.$$

Thus the general solution of the homogeneous equation is

$$Q = c_1 \sin \frac{t}{\sqrt{LC}} + c_2 \cos \frac{t}{\sqrt{LC}},$$

and the corresponding current $I = dQ/dt$ is

$$I = \frac{c_1}{\sqrt{LC}} \cos \frac{t}{\sqrt{LC}} - \frac{c_2}{\sqrt{LC}} \sin \frac{t}{\sqrt{LC}}.$$

***The initial conditions***   We now assume that the circuit was closed at time $t = 0$ with an initial charge of $Q_0$ on the capacitor and an initial current of zero:

$$Q(0) = Q_0, \qquad I(0) = 0.$$

It follows that $c_1 = 0$ and $c_2 = Q_0$, and therefore we have

$$Q = Q_0 \cos \frac{t}{\sqrt{LC}}, \qquad I = -\frac{Q_0}{\sqrt{LC}} \sin \frac{t}{\sqrt{LC}}. \qquad (5\text{--}71)$$

Thus the flow of current in an $LC$-circuit with a charged capacitor but without an impressed emf is sinusoidal.

***The impressed emf***   We consider two cases:

1.  $E = E_0$, *a constant.* Here (5–70) clearly has

    $$Q = CE_0$$

    as a particular solution, and the behavior of the circuit is given by the expressions

    $$Q = Q_0 \cos \frac{t}{\sqrt{LC}} + CE_0$$

    and

    $$I = -\frac{Q_0}{\sqrt{LC}} \sin \frac{t}{\sqrt{LC}}.$$

    Note that in this case the impressed emf has no effect on the flow of current in the circuit.

2.  $E = E_0 \sin \omega t$, *where $E_0$ and $\omega$ are positive constants.* In this case we use the method of undetermined coefficients and seek a particular solution of

    $$L\frac{d^2Q}{dt^2} + \frac{1}{C} Q = E_0 \sin \omega t$$

    in the form

    $$Q_p = A \sin \omega t + B \cos \omega t.$$

    A routine computation yields the equations

    $$\left(\frac{1}{C} - L\omega^2\right) A = E_0 \qquad \left(\frac{1}{C} - L\omega^2\right) B = 0,$$

    and it follows that if $\omega \neq 1/\sqrt{LC}$, then

    $$A = \frac{CE_0}{1 - CL\omega^2} \qquad \text{and} \qquad B = 0.$$

    Thus

    $$Q_p = \frac{CE_0}{1 - CL\omega^2} \sin \omega t,$$

with a corresponding contribution to the current of

$$I_p = \frac{CE\omega}{1 - CL\omega^2} \cos \omega t.$$

Finally, the flow of current in the circuit is

$$I = -\frac{Q_0}{\sqrt{LC}} \sin \frac{t}{\sqrt{LC}} + \frac{CE_0\omega}{1 - CL\omega^2} \cos \omega t.$$

The case $\omega = 1/\sqrt{LC}$ is left as an exercise.

**The *RLC*-Circuit**    We now consider a circuit that contains all three components, $R$, $L$, and $C$.

   ***The homogeneous equation***    In this case the homogeneous equation is

$$L \frac{d^2Q}{dt^2} + R \frac{dQ}{dt} + \frac{1}{C} Q = 0,$$

and its characteristic equation,

$$Lm^2 + Rm + \frac{1}{C} = 0,$$

has as its roots

$$m_1 = -\frac{R}{2L} + \sqrt{\left(\frac{R}{2L}\right)^2 - \frac{1}{LC}}, \qquad m_2 = -\frac{R}{2L} - \sqrt{\left(\frac{R}{2L}\right)^2 - \frac{1}{LC}}.$$

The general solution depends on the nature of $m_1$ and $m_2$ as follows.

1. $(R/2L)^2 - 1/LC > 0$. Here $m_1$ and $m_2$ are real and distinct, and

$$Q_h = c_1 e^{m_1 t} + c_2 e^{m_2 t}$$

where $c_1$ and $c_2$ are arbitrary constants. The corresponding current then is

$$I_h = c_1 m_1 e^{m_1 t} + c_2 m_2 e^{m_2 t}.$$

Moreover, since $\sqrt{(R/2L)^2 - 1/LC} < R/2L$, both $m_1$ and $m_2$ are negative and $Q_h$ and $I_h$ are transients (diminish with time).

2. $(R/2L)^2 - 1/LC = 0$. In this case $m_1 = m_2 = -R/2L$, and

$$Q_h = (c_1 + c_2 t)e^{-(R/2L)t},$$

$$I_h = \left[\left(c_2 - \frac{R}{2L} c_1\right) - \frac{R}{2L} c_2 t\right] e^{-(R/2L)t}.$$

(5–72)

Again the charge and current are transients.

3. $(R/2L)^2 - 1/LC < 0$. Here $m_1$ and $m_2$ are complex conjugates and can be

written as

$$m_1 = -\frac{R}{2L} + \omega_0 i, \qquad m_2 = -\frac{R}{2L} - \omega_0 i,$$

where

$$\omega_0 = \sqrt{\frac{1}{LC} - \left(\frac{R}{2L}\right)^2}.$$

It follows that

$$Q_h = e^{-(R/2L)t}(c_1 \sin \omega_0 t + c_2 \cos \omega_0 t),$$

and

$$I_h = e^{-(R/2L)t}\left[\left(\omega_0 c_1 - \frac{R}{2L} c_2\right) \cos \omega_0 t - \left(\omega_0 c_2 + \frac{R}{2L} c_1\right) \sin \omega_0 t\right]$$

Both are transients in this case as well.

*The initial conditions*    Again we impose the initial conditions

$$Q(0) = Q_0, \qquad I(0) = 0,$$

which describe a circuit in which the flow of current is due to a discharging capacitor.

1. *$m_1$ and $m_2$ are real and distinct.* This case is left as an exercise.
2. *$m_1 = m_2 = -R/2L$.* Setting $t = 0$ in (5–72), we find that

$$c_1 = Q_0 \qquad \text{and} \qquad c_2 = \frac{Q_0 R}{2L}.$$

Hence

$$Q_h = Q_0 \left(1 + \frac{Rt}{2L}\right) e^{-(R/2L)t}$$

and

$$I_h = -Q_0 \left(\frac{R}{2L}\right)^2 te^{-(R/2L)t}.$$

Thus if $Q_0$ is negative, $I_h$ increases from 0 when $t = 0$ to a maximum value of

$$-\frac{Q_0 R}{2L} e^{-1}$$

when $t = 2L/R$, and decreases asymptotically to zero thereafter. (See Fig. 5–4.) In this case the circuit is said to be *critically damped*.

3. *$m_1$ and $m_2$ are complex conjugates.* This is the most interesting case of the three. An easy computation reveals that

$$c_1 = \frac{R}{2L\omega_0} Q_0 \qquad \text{and} \qquad c_2 = Q_0.$$

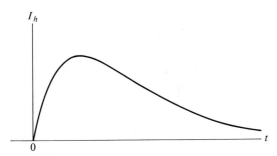

**Figure 5–4**

It follows that

$$Q_h = Q_0 e^{-(R/2L)t} \left( \frac{R}{2L\omega_0} \sin \omega_0 t + \cos \omega_0 t \right)$$

and

$$I_h = -Q_0 \left( \omega_0 + \frac{R^2}{4L^2\omega_0} \right) e^{-(R/2L)t} \sin \omega_0 t.$$

The second of these equations can be simplified to read

$$I_h = -\frac{Q_0}{LC\omega_0} e^{-(R/2L)t} \sin \omega_0 t.$$

The current now oscillates, but its magnitude decreases exponentially, as shown in Fig. 5–5. The time $T$ for the current to complete one "cycle" is known as the *natural period* of the circuit, and its reciprocal $v = 1/T$ is called the *natural frequency*. Clearly

$$T = \frac{2\pi}{\omega_0} = \frac{2\pi}{\sqrt{1/LC - (R/2L)^2}}$$

and

$$v = \frac{1}{2\pi} \sqrt{\frac{1}{LC} - \left( \frac{R}{2L} \right)^2}.$$

In 1888 the German physicist Hertz used a circuit of this type to produce the first recorded instances of electromagnetic radiation.

**The Impressed emf**   We now consider the case where the circuit is operating under a sinusoidal emf

$$E = E_0 \sin \omega t,$$

where $E_0$ and $\omega$ are positive constants. In this case it is easy to see that (5–69) has a steady-state solution of the form

$$Q_s = A \sin \omega t + B \cos \omega t.$$

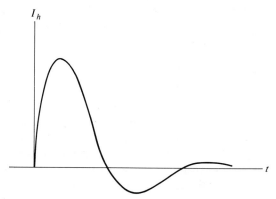

**Figure 5–5**

Using the method of undetermined coefficients, we find that

$$A = -\frac{E_0[\omega L - (1/\omega C)]}{\omega[R^2 + (\omega L - (1/\omega C))^2]} = -\frac{E_0\gamma}{\omega Z^2}$$

$$B = -\frac{E_0 R}{\omega[R^2 + (\omega L - (1/\omega C))^2]} = -\frac{E_0 R}{\omega Z^2},$$

where

$$\gamma = \omega L - \frac{1}{\omega C} \quad \text{and} \quad Z^2 = R^2 + \left(\omega L - \frac{1}{\omega C}\right)^2.$$

Thus

$$Q_s = -\frac{E_0\gamma}{\omega Z^2}\sin \omega t - \frac{E_0 R}{\omega Z^2}\cos \omega t$$

or, more simply,

$$Q_s = -\frac{E_0}{\omega Z}\cos (\omega t - \alpha),$$

where $\sin \alpha = \gamma/Z$ and $\cos \alpha = R/Z$. Finally, the steady-state current in the circuit is

$$I_s = \frac{E_0}{Z}\sin (\omega t - \alpha).$$

As in the case of an $RL$-circuit operating under a sinusoidal emf, $\alpha$ and $Z$ are called the *phase angle* and *steady-state impedance* of the circuit.

*Resonance*    Since the steady-state impedance

$$Z = \sqrt{R^2 + \left(\omega L - \frac{1}{\omega C}\right)^2}$$

in an $RLC$-circuit operating under an impressed emf of magnitude $E_0 \sin \omega t$ is a function of $\omega$, the maximum value $E_0/Z$ of the steady-state current $I_s$ is also a function of $\omega$. If $L$ and $C$ are fixed, then $E/Z$ is at a maximum when $\omega L - 1/\omega C$ vanishes, i.e., when $\omega = 1/\sqrt{LC}$. For this value of $\omega$,

$$\frac{E_0}{Z} = \frac{E_0}{R}.$$

Furthermore, this maximum increases with decreasing $R$. Physically, these observations imply that the circuit will respond quite differently to different frequencies of the input voltage (emf). The more $\omega$ differs from $1/\sqrt{LC}$, the larger will be the impedance and the smaller the response of the circuit. Thus, by minimizing the resistance $R$, we can make an $RLC$-circuit into a highly selective instrument that will discriminate sharply against inputs whose frequencies differ from $1/\sqrt{LC}$. (See Fig. 5–6.) The frequency

$$v = 1/\sqrt{LC}$$

is known as the *natural* or *resonating frequency* of the circuit. Such circuits are extremely useful as tuning circuits and as filters in electronic equipment.

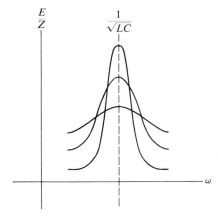

**Figure 5–6**

**EXERCISES**

1. *The RC-Circuit.* The differential equation governing the behavior of a simple non-inductive circuit containing a resistance $R$, a capacitance $C$, and a source of emf $E$ is

$$R\frac{dQ}{dt} + \frac{1}{C}Q = E.$$

a) Find the charge $Q$ and current $I$ in such a circuit as functions of time given that

$$E = E_0 \quad \text{(a constant)} \quad \text{and} \quad Q(0) = 0.$$

b) Find $Q$ and $I$ in such a circuit when

$$E = 0 \quad \text{and} \quad Q(0) = Q_0.$$

  c) Find the steady-state charge and current in an $RC$-circuit operating under an impressed emf of $E = E_0 \sin \omega t$, where $E_0$ and $\omega$ are positive constants.

  d) Discuss the way in which the steady-state current varies with $\omega$.

2. Determine the steady-state behavior of an $LC$-circuit operating under an impressed emf of $E = E_0 \sin t / \sqrt{LC}$.

3. Find the charge and current in an $RLC$-circuit in which $Q(0) = Q_0$, $I(0) = 0$, $E = 0$ and the roots of the characteristic equation are real and distinct.

## 5-10 PLANETARY MOTION: KEPLER'S LAWS AND NEWTON'S LAW OF GRAVITATION

The major intellectual triumph of seventeenth-century physics and mathematics was the discovery of Kepler's laws of planetary motion and their subsequent derivation from Newton's second law of motion and law of gravitation. The beauty and elegance of this derivation will stand for all time as one of the most remarkable achievements of the human mind. And the derivation itself illustrates far better than any other example the power of differential equations as an analytical tool for describing the physical universe.

Briefly, the problem is to describe the motion of a planet about a sun under the influence of the gravitational attraction between them. If we suppose this mutual attraction to be the only force acting on the planet, then an immediate consequence of Newton's second law

$$\mathbf{F} = m\mathbf{a}$$

is that the motion of the planet will take place in a plane. At each instant the deviation from the straight line in which the planet would move if there were no force acting on it (i.e., the line of the planet's velocity vector) is caused entirely by the force $\mathbf{F}$. Since $\mathbf{F}$ always points toward the sun, the planet will stay in the plane determined by its velocity vector and the center of the sun's mass.

Here, for the first time in our work, it will be natural to use polar coordinates. Thus we begin by translating some familiar facts into polar-coordinate form.

**A. The Conic Sections in Polar Coordinates**   Among all the various definitions of the conic sections one of the most satisfactory (and economical) asserts that a conic is the locus described by a point $P$ as it moves in the plane in such a way that the ratio of its distance from a fixed point $F$ to its distance from a fixed line $l$ is a constant. This ratio, which must be positive, is known as the *eccentricity* of the conic, and is denoted by $e$. The point $F$ and line $l$ are called the *focus* and *directrix* of the conic.

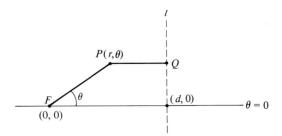

**Figure 5–7**

To derive an equation for an arbitrary conic, we place $F$ at the pole or origin of the coordinate system and take the ray from $F$ perpendicular to $l$ as the polar axis. Then if $P$ is an arbitrary point on the locus, we see from Fig. 5–7 that the defining equation for the conic is

$$\frac{PF}{PQ} = e. \tag{5–73}$$

Now let $P$ be the point $(r, \theta)$, and let $d$ be the distance from $F$ to $l$. Then

$$PF = r, \qquad PQ = d - r \cos \theta,$$

and (5–73) becomes

$$\frac{r}{d - r \cos \theta} = e.$$

Solving for $r$, we have

$$r = \frac{ed}{1 + e \cos \theta}, \tag{5–74}$$

which is the equation we shall use later in the section. The loci described by this equation fall into three classes, as follows.

  i) $0 < e < 1$. The locus is an ellipse.

 ii) $e = 1$. The locus is a parabola.

iii) $e > 1$. The locus is a hyperbola.

Each of these possibilities is illustrated in Fig. 5–8.*

**B. Newton's Second Law in Polar Coordinates**   In Chapter 1 we introduced Newton's second law of motion

$$\mathbf{F} = m\mathbf{a},$$

---

\* The circle $r = a$ is missing from this list. It can be obtained as the limiting case of (5–74) where $e \to 0$ and $d$ increases so that the product $ed$ has the constant value $a$. In the limit the directrix becomes "the line at infinity."

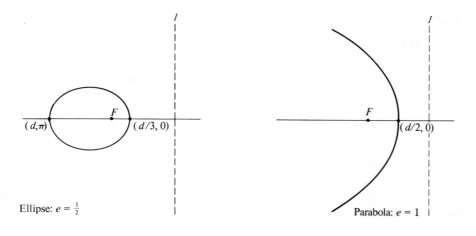

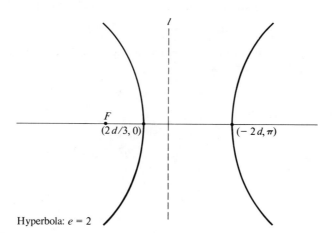

**Figure 5–8**

relating the force **F** acting on a moving object of constant mass $m$ to the acceleration **a** experienced by the object. This equation can also be written

$$\mathbf{F} = m\,\frac{d^2\mathbf{r}}{dt^2}, \tag{5-75}$$

where **r** is the directed distance of the object from the origin. In the case of motion in a plane this equation can be rewritten

$$F_x\mathbf{i} + F_y\mathbf{j} = m\left(\frac{d^2x}{dt^2}\,\mathbf{i} + \frac{d^2y}{dt^2}\,\mathbf{j}\right), \tag{5-76}$$

where

i) **i** and **j** are the standard basis vectors in $\mathcal{R}^2$,

ii) $F_x$ and $F_y$ are the $x$- and $y$-components of **F**, and

iii) $d^2x/dt^2$ and $d^2y/dt^2$ are the $x$- and $y$-components of the acceleration.

Equation (5–76) is equivalent to the pair of scalar equations

$$F_x = m\frac{d^2x}{dt^2}, \qquad F_y = m\frac{d^2y}{dt^2}.$$

When working in polar coordinates we resolve **F** into the components of magnitude $F_r$ and $F_\theta$, the first of which is in the direction of **r**, and the second perpendicular to **r**, as shown in Fig. 5–9. To find a corresponding resolution of **a**, we observe that

$$\mathbf{u}_r = \cos\theta\,\mathbf{i} + \sin\theta\,\mathbf{j}$$

is a *unit* vector (i.e., a vector of length one) in the direction of **r**, and

$$\mathbf{u}_\theta = -\sin\theta\,\mathbf{i} + \cos\theta\,\mathbf{j}$$

is a unit vector perpendicular to $\mathbf{u}_r$, oriented as shown in Fig. 5–10. (See Exercise 1.) In terms of $\mathbf{u}_r$ and $\mathbf{u}_\theta$ we write **a** as

$$\mathbf{a} = a_r\mathbf{u}_r + a_\theta\mathbf{u}_\theta.$$

Since **a** can also be written as

$$\mathbf{a} = \frac{d^2x}{dt^2}\mathbf{i} + \frac{d^2y}{dt^2}\mathbf{j},$$

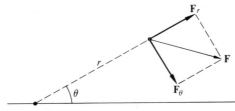

**Figure 5–9**

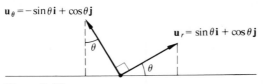

$\mathbf{u}_\theta = -\sin\theta\,\mathbf{i} + \cos\theta\,\mathbf{j}$

$\mathbf{u}_r = \sin\theta\,\mathbf{i} + \cos\theta\,\mathbf{j}$

**Figure 5–10**

it follows that

$$a_r = \frac{d^2x}{dt^2} \cos\theta + \frac{d^2y}{dt^2} \sin\theta,$$

$$a_\theta = -\frac{d^2x}{dt^2} \sin\theta + \frac{d^2y}{dt^2} \cos\theta.$$

(See Exercise 2(a).) Finally, if we use the relations

$$x = r\cos\theta, \qquad y = r\sin\theta,$$

to express $d^2x/dt^2$ and $d^2y/dt^2$ in polar coordinates, and substitute in the preceding equations, we find that

$$a_r = \frac{d^2r}{dt^2} - r\left(\frac{d\theta}{dt}\right)^2$$

$$a_\theta = 2\frac{dr}{dt}\frac{d\theta}{dt} + r\frac{d^2\theta}{dt^2}. \tag{5-77}$$

(See Exercise 2(b).) From (5-77) we can obtain at once the scalar form of Newton's second law in polar coordinates:

$$F_r = m\left[\frac{d^2r}{dt^2} - r\left(\frac{d\theta}{dt}\right)^2\right],$$

$$F_\theta = m\left[2\frac{dr}{dt}\frac{d\theta}{dt} + r\frac{d^2\theta}{dt^2}\right]. \tag{5-78}$$

**C. Central Forces and Kepler's Second Law**   We now turn to the problem of describing the motion of an object under a so-called *central force*, that is, a force that acts only in the radial direction. (To a first approximation this will describe the motion of the planets, because the dominant force acting on each planet is the gravitational attraction between it and the sun.) In this case $F_\theta = 0$, and Newton's second law implies that

$$2\frac{dr}{dt}\frac{d\theta}{dt} + r\frac{d^2\theta}{dt^2} = 0.$$

Multiplying this equation by $r$, we have

$$\left(2r\frac{dr}{dt}\right)\frac{d\theta}{dt} + r^2\frac{d^2\theta}{dt^2} = 0$$

or

$$\frac{d}{dt}\left(r^2\frac{d\theta}{dt}\right) = 0.$$

It follows that

$$r^2\frac{d\theta}{dt} = h, \tag{5-79}$$

where $h$ is a constant. Moreover, if we agree to measure $\theta$ so that $\theta$ increases as $t$ increases, then $h$ will be positive.

Equation (5–79) has a simple interpretation in terms of area. Indeed, if we let $A\big|_{t_0}^{t_1}$ denote the area swept out by the radius vector from the origin to the moving object between times $t_0$ and $t_1$, then

$$A\big|_{t_0}^{t} = \frac{1}{2}\int_{t_0}^{t_1} r^2 \, d\theta,$$

and it follows from (5–79) that

$$A\big|_{t_0}^{t_1} = \frac{1}{2}\int_{t_0}^{t_1} h \, dt = \tfrac{1}{2}h(t_1 - t_0).$$

Thus for a particular object moving under the action of a central force, $A\big|_{t_0}^{t_1}$ depends only on $t_0$ and $t_1$. This is *Kepler's second law*:

*The radius vector from the sun to a planet sweeps out equal areas in equal times.*

This remarkable fact is illustrated in Fig. 5–11.

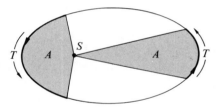

**Figure 5–11**

**D. Newton's Law of Gravitation, Kepler's First and Third Laws**   One of the most surprising facts about the preceding analysis is that it is valid for *all* motion under the action of a central force, regardless of the nature of the force. However, in order to derive an equation of motion we need specific information about the force itself.

For planets this information is provided by Newton's famous **law of gravitation**:

*Any two objects in the universe attract one another with a force that is directly proportional to the product of their masses and inversely proportional to the square of the distance between them.*

Thus two objects of masses $m_1$ and $m_2$ that are a distance $r$ apart attract one another with a force of magnitude

$$F = \Gamma\frac{m_1 m_2}{r^2},$$

where $\Gamma$ is a constant known as *Newton's gravitational constant*.

To continue our analysis of planetary motion, we let $M$ be the mass of the sun and $m$ the mass of the planet under consideration. Then the magnitude of the radial force acting on the planet is

$$F_r = \Gamma \frac{Mm}{r^2},$$

and (5–78) yields the fundamental equation

$$-\Gamma \frac{M}{r^2} = \frac{d^2 r}{dt^2} - r \left( \frac{d\theta}{dt} \right)^2, \tag{5–80}$$

where the minus sign on the left was introduced because the force is directed toward the pole.

To solve this equation, we recall that

$$\frac{d\theta}{dt} = \frac{h}{r^2}$$

(Eq. 5–79), and make the change of variable

$$u = \frac{1}{r}.$$

Then

$$\frac{du}{dt} = -\frac{1}{r^2} \frac{dr}{dt},$$

and it follows that

$$\frac{dr}{dt} = -r^2 \frac{du}{dt}$$

$$= -r^2 \frac{du}{d\theta} \frac{d\theta}{dt}$$

$$= -h \frac{du}{d\theta},$$

and

$$\frac{d^2 r}{dt^2} = -h \frac{d}{dt} \left( \frac{du}{d\theta} \right) = -h \frac{d^2 u}{d\theta^2} \frac{d\theta}{dt} = -h^2 u^2 \frac{d^2 u}{d\theta^2}.$$

Thus (5–80) becomes

$$-\Gamma M u^2 = -h^2 u^2 \frac{d^2 u}{d\theta^2} - h^2 u^3,$$

which simplifies to

$$\frac{d^2 u}{d\theta^2} + u = \frac{\Gamma M}{h^2}$$

and has

$$u = c_1 \sin \theta + c_2 \cos \theta + \frac{\Gamma M}{h^2}$$

as its general solution. We now agree to choose the polar axis so that $r$ is a minimum when $\theta = 0$. This choice implies that $dr/dt = 0$ and $d^2r/dt^2 > 0$ when $\theta = 0$, so

$$\frac{du}{d\theta} = 0 \quad \text{and} \quad \frac{d^2u}{d\theta^2} < 0$$

when $\theta = 0$, and an easy calculation reveals that $c_1 = 0$ and $c_2 > 0$. Thus

$$u = A \cos \theta + \frac{\Gamma M}{h^2},$$

where $A$ is a positive constant. Since $r = 1/u$, we obtain

$$r = \frac{h^2/\Gamma M}{1 + (Ah^2/\Gamma M) \cos \theta}.$$

This is the polar equation of a conic with eccentricity

$$e = \frac{Ah^2}{\Gamma M} \quad \text{and} \quad d = \frac{1}{A}.$$

Therefore, the path of a planet describes either an ellipse, a parabola, or a hyperbola. The fact that the planets remain in orbit about the sun implies that each orbit is an ellipse, and we have thus established *Kepler's first law*:

*The orbit of a planet is an ellipse with the sun at one focus.*

Finally, *Kepler's third law* states the relationship between the period $P$ of the planet's revolution in orbit, and the length $L$ of the major axis of the orbital ellipse:

$$P^2 = \frac{\pi^2}{2\Gamma M} L^3. \tag{5--81}$$

The proof is left as an exercise. Notice that the constant $\pi^2/2\Gamma M$ appearing in this relation is independent of the particular planet under consideration.

As we said at the beginning of this section, the derivation of Kepler's laws that we have just given (a derivation which is essentially Newton's own), is one of the most remarkable arguments in all of mathematical physics. Perhaps its most conspicuous feature is its seeming simplicity: it requires no mathematics more complicated than elementary calculus and the solution of two very simple differential equations. The reason for this simplicity lies hidden in Newton's second law of motion and law of gravitation, the two most powerful tools for describing the physical universe that have ever been forged. Some appreciation for their power can be gained from the fact that without them, and using only observational data, it took Kepler twenty years to discover his three laws of planetary motion. With Newton's laws, deriving them is but an afternoon's work.

## EXERCISES

1. In Euclidean geometry the *length* or *norm* of a vector

$$\mathbf{x} = x_1\mathbf{e}_1 + \cdots + x_n\mathbf{e}_n$$

in $\mathcal{R}^n$ is defined to be

$$\|\mathbf{x}\| = \sqrt{x_1^2 + \cdots + x_n^2},$$

where $\mathbf{e}_1, \ldots, \mathbf{e}_n$ are the standard basis vectors. (In $\mathcal{R}^2$ or $\mathcal{R}^3$ this definition reduces to the usual length of the vector from the origin to the point $(x_1, x_2)$ or $(x_1, x_2, x_3)$ obtained by the Pythagorean theorem.)

a) Verify that $\mathbf{u}_r = \cos\theta\,\mathbf{i} + \sin\theta\,\mathbf{j}$ is a unit vector in the plane.

b) Show that the vector

$$\mathbf{u}_\theta = -\sin\theta\,\mathbf{i} + \cos\theta\,\mathbf{j}$$

is the unit vector in the plane obtained by rotating $\mathbf{u}_r$ counterclockwise through an angle of measure $\pi/2$.

2. a) Use the equations

$$\mathbf{a} = a_r\mathbf{u}_r + a_\theta\mathbf{u}_\theta$$

and

$$\mathbf{a} = \frac{d^2x}{dt^2}\mathbf{i} + \frac{d^2y}{dt^2}\mathbf{j},$$

together with the values of $\mathbf{u}_r$ and $\mathbf{u}_\theta$, to show that

$$a_r = \frac{d^2x}{dt^2}\cos\theta + \frac{d^2y}{dt^2}\sin\theta,$$

$$a_\theta = -\frac{d^2x}{dt^2}\sin\theta + \frac{d^2y}{dt^2}\cos\theta.$$

b) Use the results in (a) to derive (5–77).

3. In Fig. 5–7 let $P$ be the point $(x, y)$ and $l$ the line $x = d$ ($d > 0$).

a) Show that the general equation of the conic with focus at the origin, directrix $l$, and eccentricity $e$ is )

$$(1 - e^2)x^2 + 2\,de^2x + y^2 = d^2e^2.$$

b) Show that when $0 < e < 1$ the equation in (a) can be rewritten as

$$\frac{(x + b^2/d)^2}{a^2} + \frac{y^2}{b^2} = 1$$

with $a^2 = d^2e^2/(1 - e^2)^2$ and $b^2 = a^2(1 - e^2)$, and hence that the locus is an ellipse.

c) Show that the equation in (a) describes a parabola when $e = 1$ and a hyperbola when $e > 1$.

4. *Kepler's Third Law*. Let

$$\frac{x^2}{a^2} + \frac{y^2}{b^2} = 1$$

be the equation in rectangular coordinates of the ellipse with semimajor axis $a$, semiminor axis $b$, foci at $(\pm c, 0)$, where $c^2 = a^2 - b^2$. Let

$$r = \frac{ed}{1 + e \cos \theta}$$

be the polar-coordinate equation of an ellipse with eccentricity $e$, and suppose that its semimajor axis and semiminor axis are also of lengths $a$ and $b$, respectively.
a) Show that $a = de/(1 - e^2)$. [*Hint*: Note that $a = \frac{1}{2}[r(0) + r(\pi)]$.]
b) Show that $c = de^2/(1 - e^2)$, and that $b = a\sqrt{1 - e^2}$. [*Hint*: $a = c + r(0)$.]
c) If $P$ denotes the time it takes a planet to complete one revolution about the sun, show that Kepler's second law implies that

$$P^2 = 4\pi^2 \left(\frac{de}{h^2}\right) a^3.$$

[*Hint*: The area of an ellipse with semimajor axis $a$ and semiminor axis $b$ is $\pi ab$.]
d) Establish Kepler's third law:

$$P^2 = \frac{\pi^2}{2\Gamma M} L^3,$$

where $L$ is the length of the major axis of the planet's orbit.

5. According to the law of conservation of energy, the total energy of a planet in orbit about the sun is a constant $E_0$. That is, the total energy does not depend on the position the planet occupies in its orbit at a particular time. On the other hand, $E_0$ is the sum of two energies that do vary with time and position, the *kinetic energy* of motion and the *potential energy* of position, where

   i) the kinetic energy is defined by

$$\text{K.E.} = \tfrac{1}{2}m \|\mathbf{v}\|^2,$$

   $\|\mathbf{v}\|^2$ being the square of the planet's speed (see Exercise 1), and

   ii) the potential energy is the negative of the work required to move the planet against the gravitational attraction from its orbit to infinity:

$$\text{P.E.} = -\int_r^\infty \frac{\Gamma M m}{r^2}\, dr = -\frac{\Gamma M m}{r}.$$

a) Use the fact that

$$\mathbf{v} = \frac{d\mathbf{r}}{dt} = r\frac{d\theta}{dt}\mathbf{u}_\theta + \frac{dr}{dt}\mathbf{u}_r$$

to show that

$$\text{K.E.} = \tfrac{1}{2}m\left[r^2\left(\frac{d\theta}{dt}\right)^2 + \left(\frac{dr}{dt}\right)^2\right].$$

b) Show that when $\theta = 0$ the equations

$$r = \frac{h^2/\Gamma M}{1 + e \cos \theta} \quad \text{and} \quad \text{K.E.} + \text{P.E.} = E_0$$

reduce to

$$r = \frac{h^2/\Gamma M}{1 + e} \quad \text{and} \quad \frac{mh^2}{2r^2} - \frac{\Gamma Mm}{r} = E_0,$$

respectively.

c) Eliminate $r$ between the equations in (b) to show that

$$e = \sqrt{1 + E_0(2h^2/\Gamma^2 M^2 m)},$$

and hence deduce that $E_0 < 0$ for all planets in the solar system.

d) Prove that the speed $\|\mathbf{v}\|$ of a planet at any point in its orbit is

$$\|\mathbf{v}\| = M\Gamma \left(\frac{2}{r} - \frac{1}{a}\right),$$

where $a$ is the length of the semimajor axis of the planet's orbit.

e) Show that the total energy $E_0$ of a planet in orbit about the sun is

$$E_0 = -\frac{\Gamma Mm}{2a}.$$

5:1-7 in quick review

Euler's Formula        $e^{ix} = \cos x + i \sin x$

$(D - \alpha)$
$\qquad$ if $\alpha_1$, $\alpha_2$

Real $\alpha_1 \neq \alpha_2$

Real $\alpha_1 = \alpha_2 = \alpha$

complex $\alpha_1 = a + bi$
$\qquad \alpha_2 = a - bi$

then gen soln.

$C_1 e^{\alpha_1 x} + C_2 e^{\alpha_2 x}$

$(c_1 + c_2 x) e^{\alpha x}$

$e^{ax}(c_1 \cos bx + c_2 \sin bx)$

To solve 2nd
order L.D.Q

$r > 0$ + an integ. $\alpha, a, b$ are real w/ $b \neq 0$

$(D - \alpha)^r (x^{r-1} e^{\alpha x}) = 0$

Annihilation
factors

$\begin{cases} (D^2 - 2aD + a^2 + b^2)^r (x^{r-1} e^{as} \sin bx) = 0 \\ (D^2 - 2aD + a^2 + b^2)^r (x^{r-1} e^{as} \cos bx) = 0 \end{cases}$

pg 159

variation
of parameters

$C_1'(x) = \dfrac{-h(x) \, y_2(x)}{w[y_1(x), y_2(x)]}$

$C_2'(x) = \dfrac{h(x) \, y_1(x)}{w[y_1(x) y_2(x)]}$

where $h(x)$ orig. rt-hand side

$y_p = C_1 y_1 + C_2 y_2$ (product)

$y = y_p + C_1 y_1 + C_2 y_2$

plug in for $y_1 + y_2$
leave $C_1, \& C_2$

# The Laplace
# Transform
# 6

## 6-1 PIECEWISE CONTINUOUS FUNCTIONS AND THE LAPLACE TRANSFORM

In the preceding chapter we saw that the solution of an initial-value problem for a normal $n$th-order linear differential equation

$$Ly = h(t)$$

can be found by applying the *integral operator* $G$ to $h$, where

$$G[h](t) = \int_{t_0}^{t} K(t, s)h(s)\, ds,$$

and the *kernel* $K(t, s)$ of $G$ is the Green's function for $L$ for initial-value problems.* The operator $G$ is only one of a host of integral operators that have been devised to solve problems in analysis. In this chapter we shall introduce another: the *Laplace transform*. But before doing so we shall enlarge the space from which $h = h(t)$ may be chosen to include functions with jump discontinuities like those exhibited by the **square-wave function** whose graph is shown in Fig. 6-1. Such functions arise in the study of electrical and mechanical systems that respond to *input* or *forcing functions* with such discontinuities. As we shall see, the Laplace transform is ideally suited to handling initial-value problems involving forcing functions of this type.

**Definition 6-1**  *A real-valued function $f$ is said to be **piecewise continuous** on $[a, b]$ if*

i)  *$f$ is continuous at all but a finite number of points of $[a, b]$, and*

ii)  *the limits*

$$f(t_0^+) = \lim_{h \to 0^+} f(t_0 + h), \qquad f(t_0^-) = \lim_{h \to 0^+} f(t_0 - h) \qquad (6\text{–}1)$$

---

* In the discussion of the Laplace transform it is traditional to use the letter $t$, for time, as the independent variable in all differential equations.

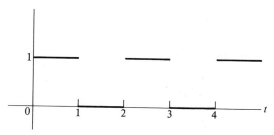

**Figure 6–1**

*exist at each point $t_0$ in $[a, b]$. (Only one of these limits is meaningful, of course, when $t_0$ is an endpoint of $[a, b]$.)*

If $f$ is piecewise continuous on every interval $[a, b]$, with $b > a$, we say that $f$ is **piecewise continuous on [a, $\infty$)**. The notation $h \to 0^+$ used in this definition means that $h$ approaches zero through positive values only, and the limits appearing in (6–1) are called, respectively, the *right-* and *left-hand limits* of $f$ at $t_0$. Furthermore, $f$ is continuous at $t_0$ if and only if $f(t_0)$ exists and

$$f(t_0^+) = f(t_0^-) = f(t_0).$$

The following facts about piecewise continuous functions are of particular importance and will be used repeatedly hereafter, usually without explicit mention.

1. If $f$ is piecewise continuous on $[a, b]$, then

$$\int_a^b f(t)\, dt$$

exists and is independent of whatever values, if any, $f$ assumes at its points of discontinuity. Indeed, if $f$ has discontinuities at $t_0 < t_1 < \cdots < t_n$, and possibly at $a$ and $b$ as well, then

$$\int_a^b f(t)\, dt = \lim_{h \to 0^+} \left[ \int_{a+h}^{t_0-h} f(t)\, dt + \int_{t_0+h}^{t_1-h} f(t)\, dt + \cdots + \int_{t_n+h}^{b-h} f(t)\, dt \right].$$

2. If $f$ and $g$ are piecewise continuous on $[a, b]$, then so is their product $fg$. In particular,

$$\int_a^b f(t)g(t)\, dt$$

exists whenever $f$ and $g$ are piecewise continuous on $[a, b]$.

3. The set of piecewise continuous functions on $[a, b]$ is a vector space under the usual definitions of addition and scalar multiplication in a function space:

$$(f + g)(t) = f(t) + g(t) \qquad \text{and} \qquad (\alpha f)(t) = \alpha f(t).$$

Notice that this space contains the space of continuous functions on $[a, b]$ as one of its subspaces.

This said, we come to the basic definition of the chapter.

**Definition 6–2**  *Let $f$ be piecewise continuous on $[0, \infty)$. Then the* **Laplace transform** *of $f$, denoted by $\mathscr{L}[f]$ or $\mathscr{L}[f](s)$, is the function of $s$ defined by*

$$\mathscr{L}[f](s) = \int_0^\infty e^{-st} f(t)\, dt. \tag{6–2}$$

We recall that an improper integral of this type is evaluated according to the rule

$$\int_0^\infty e^{-st} f(t)\, dt = \lim_{t_0 \to \infty} \int_0^{t_0} e^{-st} f(t)\, dt. \tag{6–3}$$

The integral is said to *converge* for a particular value of $s$ if and only if this limit exists. The following example shows how this definition is used.

**Example**  Find the Laplace transform of

$$f(t) = \cos at,$$

where $a$ is a constant.

*Solution.*  By (6–2) and (6–3) we have

$$\mathscr{L}[\cos at](s) = \int_0^\infty e^{-st} \cos at\, dt = \lim_{t_0 \to \infty} \int_0^{t_0} e^{-st} \cos at\, dt$$

$$= \lim_{t_0 \to \infty} \left[ \frac{e^{-st}}{s^2 + a^2} (a \sin at - s \cos at) \right]_0^{t_0}$$

$$= \lim_{t_0 \to \infty} \left[ \frac{e^{-st_0}}{s^2 + a^2} (a \sin at_0 - s \cos at_0) + \frac{s}{s^2 + a^2} \right]$$

$$= \frac{s}{s^2 + a^2} + \lim_{t_0 \to \infty} \left[ \frac{e^{-st_0}}{s^2 + a^2} (a \sin at_0 - s \cos at_0) \right].$$

To evaluate the limit appearing in this expression, we observe that

i) when $s > 0$, $e^{-st_0} \to 0$ as $t_0 \to \infty$,

ii) when $s = 0$, $e^{-st_0} = 1$ for all $t_0$, and

iii) when $s < 0$, $e^{-st_0} \to \infty$ as $t_0 \to \infty$.

Thus the limit exists only when $s > 0$, in which case its value is zero. It follows that the Laplace transform of $\cos at$ is the function $s/(s^2 + a^2)$ *restricted to the interval* $(0, \infty)$. In other words,

$$\mathscr{L}[\cos at] = \frac{s}{s^2 + a^2}, \qquad s > 0. \tag{6–4}$$

This argument is typical of how formulas for Laplace transforms are derived.

## EXERCISES

Which of the following functions are piecewise continuous on $[0, \infty)$ and which are not?

1. $f(t) = \dfrac{1}{t}$

2. $f(t) = [t]$, the greatest integer less than $t$

3. $f(t) = \ln(1 + t^2)$

4. $f(t) = \dfrac{t + 1}{t - 1}$

5. $f(t) = \sin\dfrac{1}{t}$

6. $f(t) = \begin{cases} 0, & t \text{ an integer} \\ 1, & \text{otherwise} \end{cases}$

7. $f(t) = \begin{cases} 0, & t = 1/n, \quad n = 1, 2, \ldots \\ 1, & \text{otherwise} \end{cases}$

8. $f(t) = \dfrac{\sin t}{t^n}$, $n$ a positive integer

9. $f(t) = e^{1/t}$

Evaluate $\int_0^2 f(t)\, dt$ for the functions in Exercises 10 through 13.

10. $f(t) = \begin{cases} t, & 0 \leqslant t \leqslant 1 \\ t - 1, & 1 < t < 2 \end{cases}$

11. $f(t) = |\cos \pi t|$

12. $f(t) = [2t]$ (see Exercise 2)

13. $f(t) = |t(1 - t)|$

Compute the Laplace transforms of the functions in Exercises 14 through 23.

14. $f(t) = 1$

15. $f(t) = t$

16. $f(t) = e^{at}$

17. $f(t) = \sin at$

18. $f(t) = t^2$

19. $f(t) = \sinh t$

20. $f(t) = \cosh t$

21. $f(t) = e^{bt} - e^{at}$

22. $f(t) = \begin{cases} 1, & 0 < t \leqslant 1 \\ 0, & t > 1 \end{cases}$

23. $f(t) = \begin{cases} 0, & 0 \leqslant t \leqslant 1 \\ t, & t > 1 \end{cases}$

24. Show that $f(t) = e^{t^2}$ does not have a Laplace transform.

## 6–2 FUNCTIONS OF EXPONENTIAL ORDER; PROPERTIES OF THE LAPLACE TRANSFORM

Before going on to derive the formulas that make the Laplace transform such a valuable tool for solving initial-value problems, we turn our attention to three general questions that require at least partial answers.

1. What conditions should be imposed on $f$ to guarantee that $\mathscr{L}[f]$ exists?

2. What, if anything, can be said about the function $\mathscr{L}[f]$ when it does exist?

3. What properties does the operator $\mathscr{L}$ enjoy? Is it linear? Is it one-to-one?

We shall discuss each of these questions briefly in turn.

**The Domain of** $\mathscr{L}$   Unfortunately, it is impossible to give a complete description of the domain of $\mathscr{L}$ in a book at this level. Instead, we shall content ourselves by describing a subset of that domain which includes most of the functions we shall encounter in the study of linear differential equations. (The exceptions will be treated on their own merits as they arise.) To this end we introduce the following definition.

> **Definition 6-3**   A function $f$ is said to be of **exponential order** on $[0, \infty)$ if there exist constants $C$ and $\alpha$ such that
>
> $$|f(t)| \leqslant Ce^{\alpha t} \tag{6-5}$$
>
> for all $t > 0$ where $f$ is defined.

**Example 1**   Show that for all real numbers $a$ and $b$, and all nonnegative integers $n$, the function

$$f(t) = t^n e^{at} \sin bt$$

is of exponential order on $[0, \infty)$.

*Solution.*   We divide the argument into cases depending on the value of $a$.

CASE 1.   $a > 0$. Here we observe that

$$\left| \frac{t^n e^{at} \sin bt}{e^{2at}} \right| \leqslant \frac{t^n e^{at}}{e^{2at}} = \frac{t^n}{e^{at}},$$

and that $\lim_{t \to \infty} (t^n/e^{at}) = 0$. (Why?) Hence there exists a $t_0 > 0$ such that for all $t > t_0$,

$$\left| \frac{t^n e^{at} \sin bt}{e^{2at}} \right| < 1$$

or, alternatively,

$$|t^n e^{at} \sin bt| < e^{2at}.$$

It now follows from the fact that $|t^n e^{at} \sin bt|$ is bounded on $(0, t_0]$ that we can find a constant $C$ such that

$$|t^n e^{at} \sin bt| \leqslant Ce^{2at}$$

for all $t > 0$, as required. (See Exercise 2.)

CASE 2.   $a \leqslant 0$. In this case

$$|t^n e^{at} \sin bt| \leqslant t^n,$$

and the inequality $t^n < e^t$ for all $t > t_0 > 0$ implies the existence of a constant $C$ such that

$$|t^n e^{at} \sin bt| \leqslant Ce^t$$

for all $t > 0$.

*Remark.*   Since the above argument also shows that $t^n e^{at} \cos bt$ is of exponential order on $[0, \infty)$, *all the functions*

$$t^n, \quad e^{at}, \quad \sin bt, \quad \cos bt, \quad t^n e^{at} \sin bt, \quad t^n e^{at} \cos bt,$$

*familiar from the study of constant-coefficient linear differential equations are of exponential order.* This fact will prove to be important as we continue.

**Example 2** The function

$$f(t) = e^{t^2}$$

is *not* of exponential order on $[0, \infty)$ because

$$\lim_{t \to \infty} \frac{e^{t^2}}{e^{\alpha t}} = \lim_{t \to \infty} e^{t(t-\alpha)} = \infty$$

for all $\alpha$. (See Exercise 3(b).)

Our next theorem shows the relationship of functions of exponential order to the Laplace transform.

**Theorem 6–1** *If $f$ is piecewise continuous and of exponential order on $[0, \infty)$, then $\mathscr{L}[f]$ exists. In fact, there exists a constant $\alpha$ (which may be $-\infty$) such that*

$$\int_0^\infty e^{-st} f(t) \, dt$$

*converges for all $s > \alpha$.*

*Proof.* This result is an immediate consequence of a theorem that is the integral analog of the comparison test for the convergence of infinite series:

*If $f$ and $g$ are integrable on every interval of the form $[a, b]$, where $a$ is fixed and $b > a$ is arbitrary, and if $|f(t)| \leqslant g(t)$ for all $t \geqslant a$, then $\int_a^\infty f(t) \, dt$ exists whenever $\int_a^\infty g(t) \, dt$ exists.*

To apply the test, we choose $C$ and $\alpha$ so that $|f(t)| \leqslant Ce^{\alpha t}$ for all $t > 0$. Then

$$\int_0^\infty e^{-st}(Ce^{\alpha t}) \, dt = C \int_0^\infty e^{-(s-\alpha)t} \, dt$$

$$= \lim_{t_0 \to \infty} \frac{C}{s - \alpha} \left[ 1 - e^{-(s-\alpha)t_0} \right]$$

$$= \frac{C}{s - \alpha} \quad \text{if } s > \alpha,$$

and the comparison theorem implies that $\int_0^\infty e^{-st} f(t) \, dt$ exists for all $s > \alpha$. ∎

On the strength of this result we can assert that the domain of $\mathscr{L}$ includes the vector space $\mathscr{E}$ of all piecewise continuous functions of exponential order on $[0, \infty)$. In fact, as we have already suggested, it includes more, but for most of our purposes $\mathscr{E}$ is large enough.

**The Function $\mathscr{L}[f]$** Theorem 6–1 asserts not only that $\mathscr{L}[f]$ exists, but that it is always defined on an interval of the form $(\alpha, \infty)$. Actually we can say more, for in the course of proving the theorem we found that there exist constants $C$ and $\alpha$

such that

$$|\mathscr{L}[f](s)| \leqslant \frac{C}{s - \alpha}$$

for all $s > \alpha$. Thus, if we pass to the limit as $s \to \infty$, we obtain

**Theorem 6–2**   *If f is piecewise continuous and of exponential order on* $[0, \infty)$, *then*

$$\lim_{s \to \infty} \mathscr{L}[f](s) = 0.$$

On the strength of this result we can assert that such functions as

$$1, \quad s, \quad \sin s, \quad \frac{s}{s + 1}, \quad e^s$$

cannot appear as Laplace transforms of piecewise continuous functions of exponential order, because none of them approaches zero as $s \to \infty$. There is nothing, however, to prevent one of these functions from being the Laplace transform of a function that is *not* of exponential order on $[0, \infty)$.

**The Operator** $\mathscr{L}$   We have seen that $\mathscr{L}[f]$ exists whenever $f$ is piecewise continuous and of exponential order on $[0, \infty)$. Thus $\mathscr{L}$ can be viewed as an operator, or transformation, defined on the vector space $\mathscr{E}$, and as such it is linear:

$$\mathscr{L}[f + g] = \mathscr{L}[f] + \mathscr{L}[g]$$

and

$$\mathscr{L}[\alpha f] = \alpha \mathscr{L}[f].$$

(See Exercise 7.) For reasons which by now require no explanation, we ask whether $\mathscr{L}$ is one-to-one. Does

$$\mathscr{L}[f] = \mathscr{L}[g]$$

imply that $f(t) = g(t)$ for all $t \geqslant 0$? It is easy to see that the answer to this question is no. For if $f$ and $g$ are two functions in $\mathscr{E}$ whose values differ *only* at their points of discontinuity, then the fact that $\mathscr{L}$ is an integral operator implies that $\mathscr{L}[f] = \mathscr{L}[g]$ even though $f \neq g$. But two such functions are very nearly identical since they agree everywhere, except possibly at a finite number of isolated points, on any finite interval of the nonnegative $t$-axis. Thus, if this is the worst that can happen, we would be justified in asserting that for all practical purposes $\mathscr{L}$ *is* one-to-one. The following theorem, whose proof we shall omit, guarantees this to be the case.

**Theorem 6–3. Lerch's theorem**   *If f and g are piecewise continuous and of exponential order on* $[0, \infty)$, *and if*

$$\mathscr{L}[f] = \mathscr{L}[g],$$

*then, with the possible exception of their points of discontinuity,* $f(t) = g(t)$ *for all* $t \geqslant 0$. *In particular,* $\mathscr{L}$ *is one-to-one on the space of **continuous** functions of exponential order on* $[0, \infty)$.

This theorem implies that whenever an operator equation of the form

$$\mathscr{L}[y] = \phi$$

can be solved for $y$, the solution is essentially unique. We call any solution of this equation the **inverse Laplace transform** of $\phi$, and denote it by $\mathscr{L}^{-1}[\phi]$. The inverse Laplace transform is characterized by the property that

$$\mathscr{L}^{-1}[\phi] = y \qquad \text{if and only if} \qquad \mathscr{L}[y] = \phi.$$

It, too, is linear provided we restrict attention to inverse transforms that are continuous. (See Theorem 3–4.)

## EXERCISES

1. Prove that (a) $f(t) = \ln(1 + t)$ and (b) $f(t) = t^{1/2}$ are of exponential order on $[0, \infty)$.

2. Let $f$ be piecewise continuous on $[0, \infty)$, and suppose that there exist constants $C$ and $\alpha$, and a $t_0 > 0$, such that

$$|f(t)| \leqslant Ce^{\alpha t} \qquad \text{whenever } t_0 < t.$$

   Prove that $f$ is of exponential order on $[0, \infty)$.

3. Let $f$ be piecewise continuous on $[0, \infty)$. Prove that:
   a) $f$ is of exponential order whenever there exists a constant $\alpha$ such that

$$\lim_{t \to \infty} \frac{f(t)}{e^{\alpha t}} = 0;$$

   b) $f$ is *not* of exponential order if

$$\lim_{t \to \infty} \frac{f(t)}{e^{\alpha t}} = \infty$$

   for *all* $\alpha > 0$.

4. Use the results of the preceding exercise to prove that $e^{t^{\alpha}}$ is of exponential order if $\alpha \leqslant 1$ and not if $\alpha > 1$.

5. It can be shown that if $f$ and $g$ are integrable on $[a, 1]$, $0 < a \leqslant 1$, and if $|f(t)| \leqslant g(t)$ for all $0 < t \leqslant 1$, then $\int_0^1 f(t)\,dt$ exists whenever $\int_0^1 g(t)\,dt$ exists.
   a) Use this result to prove that $\mathscr{L}[t^{-1/2}]$ exists. [*Hint*: $\int_0^\infty f(t)\,dt = \int_0^1 f(t)\,dt + \int_1^\infty f(t)\,dt$.]
   b) What conclusion can you draw about the domain of $\mathscr{L}$ from the result of part (a)?

6. Let $f$ be piecewise continuous and of exponential order on $[0, \infty)$.
   a) Prove that $\lim_{s \to \infty} s\mathscr{L}[f]$ is bounded.
   b) Use the result in (a) together with Theorem 6–2 to conclude that $s^\alpha$ is not the Laplace transform of a piecewise continuous function of exponential order for any $\alpha > -1$.

\* 7. a) Let $f(t) = \cos at$ and $g(t) = -\cos at$. Show that $\mathscr{L}[f] + \mathscr{L}[g] \neq \mathscr{L}[f + g]$. [*Hint*: Recall that two functions are equal if and only if they have the same domain and assume the same value at each point in this domain.]
   b) Let $\mathscr{F}$ be the set of all functions whose domain includes an interval of the form $(s_0, \infty)$, where $s_0 \geqslant -\infty$, let $f$ and $g$ be any two functions in $\mathscr{F}$, and define $f + g$ to be the function whose domain is the intersection of the domains of $f$ and $g$ and whose value

at any point $s$ in that intersection is $f(s) + g(s)$. Prove that with this definition of addition and the usual definition of scalar multiplication, $\mathscr{F}$ is a vector space.

c) Show that $\mathscr{L}$ is a linear transformation from $\mathscr{E}$ to $\mathscr{F}$.

## 6-3  ELEMENTARY FORMULAS

In Section 6-1 we saw that

$$\mathscr{L}[\cos at] = \frac{s}{s^2 + a^2}, \qquad s > 0. \tag{6-6}$$

This is only one of a long list of elementary formulas involving the Laplace transform, among which are

$$\mathscr{L}[1] = \frac{1}{s}, \qquad s > 0, \tag{6-7}$$

$$\mathscr{L}[e^{at}] = \frac{1}{s - a}, \qquad s > a, \tag{6-8}$$

$$\mathscr{L}[\sin at] = \frac{a}{s^2 + a^2}, \qquad s > 0, \tag{6-9}$$

$$\mathscr{L}[t^n] = \frac{n!}{s^{n+1}}, \qquad s > 0, \tag{6-10}$$

*memorize these*

where in the last formula $n$ is a nonnegative integer and $0! = 1$.

The derivation of each of these formulas is straightforward. For instance, (6-7) follows from (6-6) by setting $a = 0$, while (6-8) is derived as follows:

$$\mathscr{L}[e^{at}] = \int_0^\infty e^{-st} e^{at} \, dt = \lim_{t_0 \to \infty} \int_0^{t_0} e^{-(s-a)t} \, dt$$

$$= \lim_{t_0 \to \infty} \left[ \frac{1}{s - a} (1 - e^{-(s-a)t_0}) \right]$$

$$= \frac{1}{s - a}, \qquad s > a.$$

Formulas (6-9) and (6-10) will be established presently, and a more comprehensive list of such results appears in the Table of Laplace Transforms on pp. 262-264.

Although we will find these formulas valuable as we continue, they fail to show how the Laplace transform can be used to solve initial-value problems for linear differential equations. The following theorem provides the needed connection and will be of decisive importance in the work to come.

***Theorem 6-4*** *Let $f$ be a continuous function on $(0, \infty)$ which has a piecewise continuous derivative of exponential order on $[0, \infty)$. Then both $\mathscr{L}[f]$ and*

$\mathscr{L}[f']$ exist, and

$$\mathscr{L}[f'] = s\mathscr{L}[f] - f(0^+),  \qquad (6\text{–}11)$$

where $f(0^+) = \lim_{t \to 0^+} f(t)$.

More generally, if $f, f', \ldots, f^{(n-1)}$ are continuous and if $f^{(n)}$ is piecewise continuous and of exponential order on $[0, \infty)$, then

$$\mathscr{L}[f''] = s^2\mathscr{L}[f] - sf(0^+) - f'(0^+),  \qquad (6\text{–}12)$$

$$\mathscr{L}[f^{(n)}] = s^n\mathscr{L}[f] - s^{n-1}f(0^+) - \cdots - f^{(n-1)}(0^+). \qquad (6\text{–}13)$$

*Proof.* The proof of the fact that $\mathscr{L}[f]$ exists when $f'$ is piecewise continuous and of exponential order is left as Exercise 24. To establish (6–11), we integrate by parts, as follows:

$$\begin{aligned}
\mathscr{L}[f'] &= \int_0^\infty e^{-st}f'(t)\, dt \\
&= e^{-st}f(t) \Big|_0^\infty + s\int_0^\infty e^{-st}f(t)\, dt \\
&= s\mathscr{L}[f] + e^{-st}f(t) \Big|_0^\infty .
\end{aligned}$$

Since $f$ is of exponential order (Exercise 24), $\lim_{t \to \infty} e^{-st}f(t) = 0$ whenever $s$ is sufficiently large (Exercise 25). Thus

$$\begin{aligned}
e^{-st}f(t) \Big|_0^\infty &= \lim_{t \to \infty} e^{-st}f(t) - \lim_{t \to 0^+} e^{-st}f(t) \\
&= -\lim_{t \to 0^+} e^{-st}f(t) = -f(0^+),
\end{aligned}$$

where the last equality takes account of the fact that $f$ may have a jump discontinuity at the origin. Thus

$$\mathscr{L}[f'] = s\mathscr{L}[f] - f(0^+),$$

as asserted. Finally, (6–12) and (6–13) can be established by repeated use of this result.  ∎

If $f$ and $f'$ are *continuous* at $t = 0$, then (6–11) and (6–12) become, respectively,

$$\mathscr{L}[f'] = s\mathscr{L}[f] - f(0),  \qquad (6\text{–}14)$$

and

$$\mathscr{L}[f''] = s^2\mathscr{L}[f] - sf(0) - f'(0).  \qquad (6\text{–}15)$$

A generalization of (6–11) to functions with jump discontinuities is given in Exercise 26 at the end of the section.

**Example 1**   If $f(t) = -(\cos at)/a, a \neq 0$, then

$$f'(t) = \sin at$$

and (6–14) and (6–6) yield

$$\mathcal{L}[\sin at] = -\frac{s}{a}\mathcal{L}[\cos at] + \frac{1}{a}$$

$$= -\frac{s}{a}\left(\frac{s}{s^2 + a^2}\right) + \frac{1}{a}$$

$$= \frac{a}{s^2 + a^2}, \qquad s > 0.$$

This establishes (6–9).

**Example 2**   Since $D^n(t^n) = n!$, (6–7) implies that

$$\mathcal{L}[D^n(t^n)] = \mathcal{L}[n!] = n!\mathcal{L}[1] = \frac{n!}{s}, \qquad s > 0.$$

On the other hand, (6–13) yields

$$\mathcal{L}[D^n(t^n)] = s^n\mathcal{L}[t^n] - s^{n-1} \cdot 0 - \cdots - 0$$
$$= s^n\mathcal{L}[t^n].$$

Thus $s^n\mathcal{L}[t^n] = n!/s$ for every nonnegative integer $n$, and

$$\mathcal{L}[t^n] = \frac{n!}{s^{n+1}}, \qquad s > 0,$$

which proves (6–10).

The next example illustrates how the Laplace transform is used to solve an initial-value problem.

**Example 3**   Solve the initial-value problem

$$y'' - y = 1, \qquad y(0) = 0, \qquad y'(0) = 1.$$

*Solution.*   We begin by applying the operator $\mathcal{L}$ to both sides of the equation to obtain

$$\mathcal{L}[y''] - \mathcal{L}[y] = \mathcal{L}[1]$$

or, since $\mathcal{L}[1] = 1/s$,

$$\mathcal{L}[y''] - \mathcal{L}[y] = \frac{1}{s}.$$

But by (6–15),

$$\mathcal{L}[y''] = s^2\mathcal{L}[y] - sy(0) - y'(0),$$

and the given initial conditions yield

$$\mathcal{L}[y''] = s^2\mathcal{L}[y] - 1.$$

Hence

$$s^2\mathcal{L}[y] - 1 - \mathcal{L}[y] = \frac{1}{s},$$

and it follows that

$$\mathcal{L}[y] = \frac{1}{s(s-1)}. \tag{6–16}$$

It remains for us to find the inverse transform of $1/s(s-1)$. We use the method of partial fractions to rewrite (6–16) as

$$\mathscr{L}[y] = \frac{1}{s-1} - \frac{1}{s},$$

from which it follows that

$$y = e^t - 1.$$

Notice the way in which the differentiation formula automatically incorporated the initial conditions into the solution of the problem. This is just one of the many valuable features of the Laplace transform.

As one might expect, there are companion formulas to (6–11), (6–12), and (6–13) for integrals.

***Theorem 6–5***   *If $f$ is piecewise continuous and of exponential order on $[0, \infty)$, then so is $\int_0^t f(x)\,dx$, and*

$$\mathscr{L}\left[\int_0^t f(x)\,dx\right] = \frac{1}{s}\mathscr{L}[f]. \tag{6–17}$$

*More generally,*

$$\mathscr{L}\left[\underbrace{\int_0^t \cdots \int_0^t f(x)\,dx \cdots dx}_{n \text{ times}}\right] = \frac{1}{s^n}\mathscr{L}[f]. \tag{6–18}$$

*Proof.*   The proof of the fact that $\int_0^t f(x)\,dx$ is piecewise continuous and of exponential order is left as Exercise 27. To prove the basic formula, we use integration by parts again, this time with

$$u(t) = \int_0^t f(x)\,dx \qquad \text{and} \qquad dv = e^{-st}\,dt.$$

Then

$$\mathscr{L}\left[\int_0^t f(x)\,dx\right] = \int_0^\infty e^{-st}\int_0^t f(x)\,dx\,dt$$

$$= -\frac{1}{s}e^{-st}\int_0^t f(x)\,dx \Big|_0^\infty + \frac{1}{s}\int_0^\infty e^{-st}f(t)\,dt$$

$$= \frac{1}{s}\mathscr{L}[f] - \frac{1}{s}e^{-st}\int_0^t f(x)\,dx \Big|_0^\infty.$$

But as $t \to \infty$, $e^{-st}\int_0^t f(x)\,dx \to 0$ whenever $s$ is sufficiently large. Thus the last term in the preceding expression is zero, and

$$\mathscr{L}\left[\int_0^t f(x)\,dx\right] = \frac{1}{s}\mathscr{L}[f].$$

Finally, (6–18) is established by iterating this result.   ∎

**Example 4**   Since

$$\int_0^t \cos ax\, dx = \frac{1}{a}\sin at,$$

(6–17) yields

$$\mathcal{L}\left[\frac{1}{a}\sin at\right] = \frac{1}{s}\mathcal{L}[\cos at] = \frac{1}{s}\left(\frac{s}{s^2 + a^2}\right)$$

$$= \frac{1}{s^2 + a^2}, \qquad s > 0.$$

Thus

$$\mathcal{L}[\sin at] = \frac{a}{s^2 + a^2}, \qquad s > 0,$$

which again proves (6–9).

**Example 5**   Use the first integration formula to find $\mathcal{L}[te^t]$.

*Solution.*   Since

$$\int_0^t xe^x\, dx = te^t - e^t + 1,$$

(6–17) yields

$$\frac{1}{s}\mathcal{L}[te^t] = \mathcal{L}[te^t - e^t + 1].$$

But

$$\mathcal{L}[te^t - e^t + 1] = \mathcal{L}[te^t] - \mathcal{L}[e^t] + \mathcal{L}[1]$$

$$= \mathcal{L}[te^t] - \frac{1}{s-1} + \frac{1}{s},$$

and so we have

$$\frac{1}{s}\mathcal{L}[te^t] = \mathcal{L}[te^t] - \frac{1}{s-1} + \frac{1}{s}, \qquad s > 1.$$

Solving this equation for $\mathcal{L}[te^t]$, we obtain

$$\mathcal{L}[te^t] = \frac{1}{(s-1)^2}, \qquad s > 1.$$

Notice that this result can also be derived from the differentiation formula (6–11).

## EXERCISES

Find the Laplace transform of each of the following functions.

1. $f(t) = \sin^2 t$   Done in class as an example
2. $f(t) = \cos^2 t$
3. $f(t) = \sinh at$
4. $f(t) = \cosh at$
5. $f(t) = e^{at} - e^{bt}$
6. $f(t) = \sin(t + a)$   sum rule here
7. $f(t) = (t + a)^n$, $n$ a positive integer   binomial formula
8. $f(t) = t \sin at$   differentiation
9. $f(t) = t \cos at$
10. $f(t) = t^2 \sin at$

11. $f(t) = t^2 \cos at$

12. $f(t) = \cos^3 t$    *when 2 functs are multiplied tahedei.*

13. $f(t) = e^t \sin t$

14. $f(t) = t^2 e^{at}$

15. $f(t) = \begin{cases} 0, & t \leqslant a \\ 1, & t > a \end{cases}$

Use the Laplace transform to solve the initial-value problems in Exercises 16 through 23.

16. $y'' - 3y' + 2y = 0$;    $y(0) = 3,\ y'(0) = 4$

*use partial fractions to get from $\mathscr{L}[y]$ to $t$ terms*

17. $y'' + y = t$;    $y(0) = -1,\ y'(0) = 3$

18. $y'' + 6y' + 8y = 4(4t + 3)$;    $y(0) = 2,\ y'(0) = -6$

19. $4y'' + y = -2$;    $y(0) = 0,\ y'(0) = \frac{1}{2}$

20. $y'' - y' - 6y = 3t^2 + t - 1$;    $y(0) = -1,\ y'(0) = 6$

21. $y''' + y'' + 4y' + 4y = -2$;    $y(0) = 0,\ y'(0) = 1,\ y''(0) = -1$

22. $y''' - 2y'' - y' + 2y = t$;    $y(0) = y'(0) = y''(0) = 0$

23. $y''' - y'' + 9y' - 9y = 0$;    $y(0) = 0,\ y'(0) = 3,\ y''(0) = 0$

24. Suppose that $f$ is continuous on $(0, \infty)$, and that $f'$ is of exponential order. Integrate the inequality

$$-Ce^{\alpha t} \leqslant f'(t) \leqslant Ce^{\alpha t}$$

to deduce that $f$ is of exponential order on $[0, \infty)$.

25. Prove that if $f(t)$ is of exponential order on $[0, \infty)$, then $\lim_{t \to \infty} e^{-st} f(t) = 0$ whenever $s$ is sufficiently large. How large is "sufficiently large"?

26. Suppose that the function in Theorem 6-4 has a jump discontinuity at $t_0 > 0$. Prove that

$$\mathscr{L}[f'] = s\mathscr{L}[f] - f(0^+) - e^{-st_0}[f(t_0^+) - f(t_0^-)].$$

27. a) Prove that $\int_0^t f(x)\, dx$ is piecewise continuous and of exponential order on $[0, \infty)$ whenever the same is true of $f$.
    b) Deduce (6-13) from (6-11).
    c) Deduce (6-18) from (6-17).

28. Let $f$ be piecewise continuous and of exponential order on $[0, \infty)$, and suppose that $a$ is a nonnegative real number. Prove that

$$\mathscr{L}\left[\int_a^t f(x)\, dx\right] = \frac{1}{s}\mathscr{L}[f] - \frac{1}{s}\int_0^a f(x)\, dx.$$

## 6-4 FURTHER PROPERTIES OF THE LAPLACE TRANSFORM

The crucial step in using the Laplace transform to solve an initial-value problem consists of finding the inverse transform of a function $\phi$. In practice this is done by converting $\phi$ to a form in which its inverse can be recognized, usually with the aid of a table like the one on pp. 262–264 at the end of the chapter. In this section we shall derive a number of the more important formulas in that table and illustrate how they are used in computations.

We begin with a theorem that permits us to compute the Laplace transform of $e^{at}f(t)$ whenever the Laplace transform of $f$ is known.

*φ is any function*

**Theorem 6–6.  *The first shift theorem*  If**

$$\mathscr{L}[f] = \phi(s),$$

then

$$\mathscr{L}[e^{at}f(t)] = \phi(s - a). \tag{6-19}$$

*Proof.*  By definition,

$$\mathscr{L}[e^{at}f(t)] = \int_0^\infty e^{-st}e^{at}f(t)\, dt$$

$$= \int_0^\infty e^{-(s-a)t}f(t)\, dt$$

$$= \mathscr{L}[f(t)](s - a)$$

$$= \phi(s - a). \quad\blacksquare$$

When written in terms of inverse transforms this formula becomes

$$\mathscr{L}^{-1}[\phi(s - a)] = e^{at}\mathscr{L}^{-1}[\phi(s)] \tag{6-20}$$

or

$$\mathscr{L}^{-1}[\phi(s)] = e^{at}\mathscr{L}^{-1}[\phi(s + a)]. \tag{6-21}$$

**Example 1**   Since $\mathscr{L}[\cos 3t] = s/(s^2 + 9)$, (6–19) yields

$$\mathscr{L}[e^{2t}\cos 3t] = \frac{s - 2}{(s - 2)^2 + 9} = \frac{s - 2}{s^2 - 4s + 13}.$$

**Example 2**   Find

$$\mathscr{L}^{-1}\left[\frac{2s + 3}{s^2 - 4s + 20}\right].$$

*Solution.*  The $s^2 - 4s$ in the denominator suggests that we complete the square to rewrite the function in terms of the variable $s - 2$:

$$\frac{2s + 3}{s^2 - 4s + 20} = \frac{2(s - 2) + 7}{(s - 2)^2 + 16}$$

$$= 2\left[\frac{s - 2}{(s - 2)^2 + 16}\right] + \frac{7}{4}\left[\frac{4}{(s - 2)^2 + 16}\right].$$

Now we are in a position to use known results. Since

$$\mathscr{L}^{-1}\left[\frac{s - 2}{(s - 2)^2 + 16}\right] = e^{2t}\cos 4t$$

and

$$\mathscr{L}^{-1}\left[\frac{4}{(s - 2)^2 + 16}\right] = e^{2t}\sin 4t,$$

we see that

$$\mathscr{L}^{-1}\left[\frac{2s + 3}{s^2 - 4s + 20}\right] = 2e^{2t}\cos 4t + \tfrac{7}{4}e^{2t}\sin 4t.$$

In many situations, especially in physics, an input or forcing function begins to act at time $t = a > 0$ rather than at $t = 0$ as we have been assuming until now. Such a *right-shift* or *time delay* can be accomplished in two ways, as suggested by Figs. 6–2 through 6–4. Figure 6–2 shows the graph of $f$, which we assume to be defined on $[0, \infty)$. Figure 6–3 shows the same graph shifted $a$ units to the right with the value 0 assigned to the shifted function in the interval $[0, a]$. Figure 6–4 shows the graph that is obtained by annihilating $f$ to the left of $a$ but leaving $f$ unchanged for $t > a$. To express $f_1$ and $f_2$ in terms of $f$, we introduce the so-called

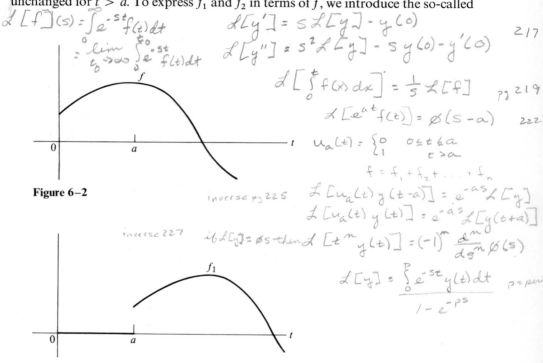

$$\mathcal{L}[f](s) = \int_0^\infty e^{-st} f(t)\,dt$$
$$= \lim_{t_0 \to \infty} \int_0^{t_0} e^{-st} f(t)\,dt$$

$$\mathcal{L}[y'] = s\mathcal{L}[y] - y(0)$$
$$\mathcal{L}[y''] = s^2 \mathcal{L}[y] - sy(0) - y'(0) \qquad 217$$

$$\mathcal{L}\left[\int_0^t f(x)\,dx\right] = \frac{1}{s}\mathcal{L}[f] \qquad pg\ 219$$

$$\mathcal{L}[e^{at} f(t)] = \phi(s - a) \qquad 222$$

$$u_a(t) = \begin{cases} 0 & 0 \le t \le a \\ 1 & t > a \end{cases}$$

$$f = f_1 + f_2 + \cdots + f_n$$

*inverse pg 225*

$$\mathcal{L}[u_a(t)\, y(t-a)] = e^{-as} \mathcal{L}[y]$$
$$\mathcal{L}[u_a(t)\, y(t)] = e^{-as} \mathcal{L}[y(t+a)]$$

*inverse 227*    if $\mathcal{L}[y] = \phi s$ then $\mathcal{L}[t^m y(t)] = (-1)^m \dfrac{d^m}{ds^m} \phi(s)$

$$\mathcal{L}[y] = \frac{\int_0^P e^{-st} y(t)\,dt}{1 - e^{-ps}} \qquad p = per$$

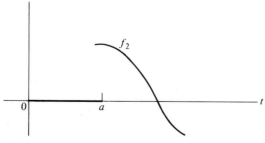

**Figure 6–2**

**Figure 6–3**

**Figure 6–4**

**unit step function** $u_a$ defined by the rule

$$u_a(t) = \begin{cases} 0, & 0 \leqslant t \leqslant a, \\ 1, & t > a. \end{cases} \qquad (6\text{-}22)$$

(See Fig. 6–5.) Then it is clear that

$$f_1(t) = u_a(t)f(t - a)$$

and

$$f_2(t) = u_a(t)f(t).$$

(Recall that the change of variable from $t$ to $t - a$ has the effect of shifting a graph $a$ units to the right if $a > 0$.) The second shift theorem enables us to express the Laplace transforms of $f_1$ and $f_2$ in terms of the Laplace transform of $f$.

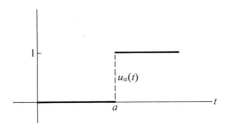

**Figure 6–5**

**Theorem 6–7.  The second shift theorem**  *If $u_a$ is the unit step function defined by (6–22), then*

$$\mathscr{L}[u_a(t)f(t - a)] = e^{-as}\mathscr{L}[f(t)] \qquad (6\text{-}23)$$

*and*

$$\mathscr{L}[u_a(t)f(t)] = e^{-as}\mathscr{L}[f(t + a)]. \qquad (6\text{-}24)$$

*Proof.*  First of all,

$$\mathscr{L}[u_a(t)f(t - a)] = \int_0^\infty e^{-st}u_a(t)f(t - a)\, dt$$

$$= \int_a^\infty e^{-st}f(t - a)\, dt.$$

Now we make the change of variable $x = t - a$ in the integral to obtain

$$\mathscr{L}[u_a(t)f(t - a)] = \int_0^\infty e^{-s(x + a)}f(x)\, dx$$

$$= e^{-as}\int_0^\infty e^{-sx}f(x)\, dx$$

$$= e^{-as}\mathscr{L}[f],$$

and (6–23) is proved. The second formula follows immediately from this result.  ∎

When (6–23) is rewritten in terms of inverse transforms, it becomes

$$\mathscr{L}^{-1}[e^{-as}\mathscr{L}[f(t)]] = u_a(t)f(t-a) \qquad (6\text{–}25)$$

or

$$\mathscr{L}^{-1}[e^{-as}\phi(s)] = u_a(t)\mathscr{L}^{-1}[\phi](t-a). \qquad (6\text{–}26)$$

Similar versions of (6–24) can also be given.

**Example 3**    If $f(t) = u_a(t)\sin t$, then

$$\begin{aligned}
\mathscr{L}[f] &= e^{-as}\mathscr{L}[\sin(t+a)] \\
&= e^{-as}\mathscr{L}[\cos a \sin t + \sin a \cos t] \\
&= e^{-as}(\cos a\,\mathscr{L}[\sin t] + \sin a\,\mathscr{L}[\cos t]) \\
&= \frac{e^{-as}(\cos a + s \sin a)}{s^2 + 1}.
\end{aligned}$$

**Example 4**    Find the Laplace transform of the function whose graph is shown in Fig. 6–6.

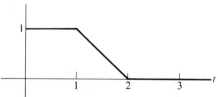

**Figure 6–6**

*Solution.*    We can view $f$ as the sum of the functions

$$f_1(t) = 1, \qquad t \geqslant 0,$$

$$f_2(t) = -u_1(t) \cdot (t-1) = \begin{cases} 0, & 0 \leqslant t \leqslant 1, \\ -(t-1), & t > 1, \end{cases}$$

$$f_3(t) = u_2(t) \cdot (t-2) = \begin{cases} 0, & 0 \leqslant t \leqslant 2, \\ t-2, & t \geqslant 2. \end{cases}$$

(See Fig. 6–7.) Hence

$$\begin{aligned}
\mathscr{L}[f] &= \mathscr{L}[1] - \mathscr{L}[u_1(t) \cdot (t-1)] + \mathscr{L}[u_2(t) \cdot (t-2)] \\
&= \mathscr{L}[1] - e^{-s}\mathscr{L}[t] + e^{2s}\mathscr{L}[t] \\
&= \frac{1}{s} - \frac{e^{-s}}{s^2} + \frac{e^{-2s}}{s^2} = \frac{s - e^{-s} + e^{-2s}}{s^2}.
\end{aligned}$$

**Example 5**    Find

$$\mathscr{L}^{-1}\left[\frac{e^{-3s}}{s^2 + 6s + 10}\right].$$

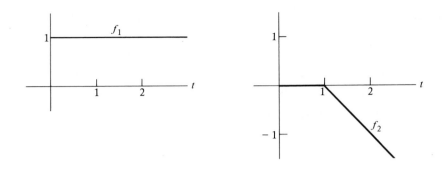

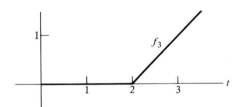

**Figure 6–7**

*Solution.* By (6–26) we have

$$\mathcal{L}^{-1}\left[\frac{e^{-3s}}{s^2 + 6s + 10}\right] = u_3(t)\mathcal{L}^{-1}\left[\frac{1}{s^2 + 6s + 10}\right](t - 3).$$

But by (6–20)

$$\mathcal{L}^{-1}\left[\frac{1}{s^2 + 6s + 10}\right] = \mathcal{L}^{-1}\left[\frac{1}{(s + 3)^2 + 1}\right]$$

$$= e^{-3t}\mathcal{L}^{-1}\left[\frac{1}{s^2 + 1}\right] = e^{-3t}\sin t.$$

Hence

$$\mathcal{L}^{-1}\left[\frac{e^{-3s}}{s^2 + 6s + 10}\right] = u_3(t)e^{-3(t - 3)}\sin (t - 3).$$

Our next result is somewhat akin to Theorem 6–6 in that it permits us to compute the Laplace transform of $t^n f(t)$ when $\mathcal{L}[f]$ is known.

**Theorem 6–8**  If $\mathcal{L}[f] = \phi(s)$, then

$$\mathcal{L}[t^n f(t)] = (-1)^n \frac{d^n}{ds^n} \phi(s). \tag{6–27}$$

*Proof.* This result is established by differentiating both sides of the equation

$$\phi(s) = \int_0^\infty e^{-st} f(t)\, dt$$

*n* times with respect to *s*:

$$\frac{d}{ds}\phi(s) = \frac{d}{ds}\int_0^\infty e^{-st} f(t)\, dt$$

$$= \int_0^\infty \frac{\partial}{\partial s}[e^{-st} f(t)]\, dt$$

$$= -\int_0^\infty e^{-st} t f(t)\, dt$$

$$= -\mathcal{L}[t f(t)],$$

and so forth. ∎

This time the companion formula is

$$\mathcal{L}^{-1}\left[\frac{d^n}{ds^n}\phi(s)\right] = (-1)^n t^n \mathcal{L}^{-1}[\phi(s)]. \qquad (6\text{–}28)$$

**Example 6**

$$\mathcal{L}[t \sin t] = -\frac{d}{ds}\mathcal{L}[\sin t]$$

$$= -\frac{d}{ds}\left(\frac{1}{s^2 + 1}\right)$$

$$= \frac{2s}{(s^2 + 1)^2}.$$

**Example 7**

$$\mathcal{L}[t^n] = (-1)^n \frac{d^n}{ds^n}\mathcal{L}[1] = (-1)^n \frac{d^n}{ds^n}\left(\frac{1}{s}\right)$$

$$= (-1)^n(-1)^n \frac{n!}{s^{n+1}}$$

$$= \frac{n!}{s^{n+1}},$$

which again proves (6–10).

**Example 8**   Find

$$\mathcal{L}^{-1}\left[\frac{1}{(s^2 + 1)^2}\right].$$

*Solution.* Since

$$\frac{1}{(s^2 + 1)^2} = -\frac{1}{2s}\frac{d}{ds}\left(\frac{1}{s^2 + 1}\right),$$

(6–17) yields

$$\mathcal{L}^{-1}\left[\frac{1}{(s^2 + 1)^2}\right] = -\frac{1}{2}\int_0^t \mathcal{L}^{-1}\left[\frac{d}{ds}\left(\frac{1}{s^2 + 1}\right)\right] dt.$$

But (6–28) with $n = 1$ yields

$$\mathcal{L}^{-1}\left[\frac{d}{ds}\left(\frac{1}{s^2 + 1}\right)\right] = -t\mathcal{L}^{-1}\left[\frac{1}{s^2 + 1}\right] = -t \sin t.$$

Hence

$$\mathcal{L}^{-1}\left[\frac{1}{(s^2 + 1)^2}\right] = -\frac{1}{2}\int_0^t (-t \sin t)\, dt = \frac{1}{2}\int_0^t t \sin t\, dt$$

$$= -\tfrac{1}{2}t \cos t + \tfrac{1}{2}\sin t.$$

Our final formula reduces the computation of the Laplace transform of a periodic function* to the evaluation of an integral over a finite interval. It reads as follows.

**Theorem 6–9**   *If f is of exponential order and is periodic with period p, then*

$$\mathcal{L}[f] = \frac{\int_0^p e^{-st}f(t)\, dt}{1 - e^{-ps}}. \tag{6–29}$$

*Proof.*   By definition

$$\mathcal{L}[f] = \int_0^\infty e^{-st}f(t)\, dt$$

$$= \int_0^p e^{-st}f(t)\, dt + \int_p^{2p} e^{-st}f(t)\, dt + \cdots$$

$$+ \int_{np}^{(n+1)p} e^{-st}f(t)\, dt + \cdots.$$

We now set $x + np = t$ in the $(n + 1)$st integral of this series to obtain

$$\int_{np}^{(n+1)p} e^{-st}f(t)\, dt = \int_0^p e^{-s(x+np)}f(x + np)\, dx$$

$$= e^{-nps}\int_0^p e^{-sx}f(x)\, dx,$$

the last step following from the periodicity of $f$. Hence

$$\mathcal{L}[f] = \int_0^p e^{-sx}f(x)\, dx + e^{-ps}\int_0^p e^{-sx}f(x)\, dx + \cdots$$

$$+ e^{-nps}\int_0^p e^{-sx}f(x)\, dx + \cdots$$

$$= [1 + e^{-ps} + e^{-2ps} + \cdots]\int_0^p e^{-sx}f(x)\, dx.$$

---

* A piecewise continuous function $f$ is said to be **periodic** with period $p > 0$ if $f(t + p) = f(t)$ for all values of $t$ where $f$ is defined.

But the sum of the geometric series

$$1 + e^{-ps} + \cdots + e^{-nps} + \cdots$$

is $1/(1 - e^{-ps})$, and it follows that

$$\mathscr{L}[f] = \frac{\int_0^p e^{-st}f(t)\,dt}{1 - e^{-ps}},$$

as asserted.  ∎

**Example 9**    Find the Laplace transform of the function whose graph is shown in Fig. 6–8.

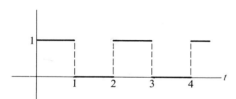

**Figure 6–8**

*Solution.*    Since $f$ is periodic with period 2,

$$\mathscr{L}[f] = \frac{\int_0^2 e^{-st}f(t)\,dt}{1 - e^{-2s}} = \frac{\int_0^1 e^{-st}\,dt}{1 - e^{-2s}} = \frac{1 - e^{-s}}{s(1 - e^{-2s})} = \frac{1}{s(1 + e^{-s})}.$$

## EXERCISES

Find the Laplace transform of each of the following functions.

1. $e^{2t} \sin 3t$  2. $3e^{-t} \cos 2t$  3. $t^3 \sin 3t$

4. $t^2 e^t \cos t$  5. $e^{-3t} \cos (2t + 4)$  6. $te^t(d/dt)(\sin 2t)$

7. $te^{2t}f'(t)$  8. $(D^3 + 1)f(t)$, where $D = d/dt$

9. $f(t) = \begin{cases} 0, & t < \frac{1}{2} \\ 1 + t, & t \geqslant \frac{1}{2} \end{cases}$  10. $f(t) = \begin{cases} t, & t < 2 \\ 2, & t > 2 \end{cases}$

11. $f(t) = \begin{cases} \sin t, & t < 2\pi \\ 0, & t > 2\pi \end{cases}$  12. $f(t) = \begin{cases} 0, & t < \pi/2 \\ \cos t, & \pi/2 \leqslant t \leqslant 3\pi/2 \\ 0, & t > 3\pi/2 \end{cases}$

13. $f(t) = \begin{cases} t, & t < 2 \\ 8 - 3t, & 2 \leqslant t \leqslant 3 \\ t - 4, & 3 < t \leqslant 4 \\ 0, & t > 4 \end{cases}$  14. $\sin t \cos t$

15. $2e^{2t} \sin t \cos t$  16. $\sin^2 t$ [*Hint*: What is the period of $\sin^2 t$?]

17. $|\sin t|$  18. $f(t)$, as shown in Fig. 6–9

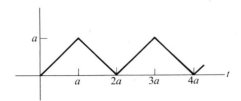

**Figure 6–9**

19. $\int_0^t te^{2t} \sin t\, dt$

20. $e^{-3t} \int_0^t t \cos 4t\, dt$

21. $t^2 \int_1^t t \sin t\, dt$

22. $\dfrac{d^2}{dt^2} \int_0^t e^{-t} \cos 3t\, dt$

23. $te^t \int_a^t t \dfrac{d}{dt}(e^{2t} \sin t)\, dt$

24. Find the Laplace transform of the *staircase function*

$$f(t) = n + 1, \qquad n < t < n + 1, \qquad n = 0, 1, 2, \ldots,$$

shown in Fig. 6–10.

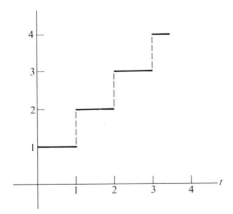

**Figure 6–10**

Find the inverse Laplace transform of each of the following functions.

25. $\dfrac{1}{s(s+1)}$

26. $\dfrac{3}{(s-1)^2}$

27. $\dfrac{1}{s(s+2)^2}$

28. $\dfrac{5}{s^2(s-5)^2}$

29. $\dfrac{1}{(s-a)^n}$,    $n \geqslant 1$

30. $\dfrac{1}{(s-a)(s-b)}$,    $a, b$ constants

31. $\dfrac{1}{s^2 + 4s + 29}$

32. $\dfrac{2s}{2s^2 + 1}$

33. $\dfrac{2s}{(s^2 + 1)^2}$ [Hint: See Example 8.]

34. $\dfrac{1}{(s^2 + 4)^3}$

35. $\dfrac{3s^2}{(s^2 + 1)^2}$ [Hint: First expand in partial fractions.]

36. $\dfrac{2s^3}{(s^2 + 1)^3}$

37. $\dfrac{3e^{-2s}}{3s^2 + 1}$

38. $\dfrac{e^{-s}}{s^4 + 1}$

39. $\dfrac{1}{s^4 + 1}$ [Hint: $s^4 + 1 = (s^4 + 2s^2 + 1) - 2s^2.$]

40. $\dfrac{3s}{(s+1)^4}$ [Hint: See Exercise 35.]

41. $\dfrac{1 + e^{-s}}{s}$

42. $\dfrac{e^{-s}}{(s-1)(s-2)}$

43. $\dfrac{1}{s^3 + a^3}$

44. $\dfrac{s^2}{s^3 + a^3}$

45. $\ln\dfrac{s+3}{s+2}$ $\left[ \text{Hint:} \dfrac{d}{ds} \ln\dfrac{s+3}{s+2} = -\dfrac{1}{(s+3)(s+2)}. \right]$

46. $\ln\dfrac{s^2 + 1}{s(s+3)}$

*47. Let $P(s)$ be a polynomial of degree less than $n$, and let

$$Q(s) = (s - a_1)(s - a_2)\cdots(s - a_n),$$

where the $a_i$ are distinct. Prove that

$$\mathscr{L}^{-1}\left[\frac{P(s)}{Q(s)}\right] = \sum_{i=1}^{n} \frac{P(a_i)}{Q'(a_i)} e^{a_i t}.$$

*48. Show that for any integer $n \geqslant 1$ and any $a \neq 0$,

$$\mathscr{L}^{-1}\left[\frac{1}{(s^2 + a^2)^{n+1}}\right] = \frac{1}{2n}\int_0^t t\mathscr{L}^{-1}\left[\frac{1}{(s^2 + a^2)^n}\right] dt.$$

*49. a) Use the result of Exercise 48 to show that

$$\mathscr{L}^{-1}\left[\frac{1}{(s^2 + a^2)^{n+1}}\right] = \frac{1}{2^n a n!} \underbrace{\int_0^t t \int_0^t t \cdots \int_0^t t}_{n \text{ times}} \sin at\, dt\, dt \cdots dt$$

b) Derive a similar formula for $\mathscr{L}^{-1}[s/(s^2 + a^2)^{n+1}]$.

## 6–5 THE LAPLACE TRANSFORM AND INITIAL-VALUE PROBLEMS

In Section 6–3 we introduced the technique of using the Laplace transform to solve initial-value problems. In this section we shall illustrate how the formulas we now have in hand can be used to solve more elaborate problems.

**Example 1**   Solve

$$y'' + 4y' + 13y = 2t + 3e^{-2t} \cos 3t,$$

$$y(0) = 0, \qquad y'(0) = -1.$$

(6–30)

*Solution.*   We take the Laplace transform of both sides of the differential equation and apply the initial conditions to obtain

$$s^2 \mathscr{L}[y] + 1 + 4s\mathscr{L}[y] + 13\mathscr{L}[y] = \frac{2}{s^2} + \frac{3(s + 2)}{(s + 2)^2 + 9}.$$

Hence

$$\mathscr{L}[y] = -\frac{1}{s^2 + 4s + 13} + \frac{2}{s^2(s^2 + 4s + 13)} + \frac{3(s + 2)}{(s^2 + 4s + 13)^2}.$$

We must now find the inverse transform of the terms on the right-hand side of this equation.

The first can be disposed of without difficulty, since

$$\frac{1}{s^2 + 4s + 13} = \frac{1}{(s + 2)^2 + 9} = \frac{1}{3}\left(\frac{3}{(s + 2)^2 + 9}\right),$$

so that

$$\mathscr{L}^{-1}\left[-\frac{1}{s^2 + 4s + 13}\right] = -\tfrac{1}{3}e^{-2t} \sin 3t.$$

To handle the second, we use the method of partial fractions, as follows:

$$\frac{2}{s^2(s^2 + 4s + 13)} = \frac{A}{s} + \frac{B}{s^2} + \frac{Cs + D}{s^2 + 4s + 13};$$

whence

$$As(s^2 + 4s + 13) + B(s^2 + 4s + 13) + (Cs + D)s^2 = 2$$

or

$$(A + C)s^3 + (4A + B + D)s^2 + (13A + 4B)s + 13B = 2.$$

In order that this equation hold identically in *s*, we must have

$$A + C = 0,$$

$$4A + B + D = 0,$$

$$13A + 4B = 0,$$

$$13B = 2.$$

It follows that $A = -\frac{8}{169}, B = \frac{2}{13}, C = \frac{8}{169}, D = \frac{6}{169}$. Thus

$$\frac{2}{s^2(s^2 + 4s + 13)} = -\frac{8}{169}\left(\frac{1}{s}\right) + \frac{2}{13}\left(\frac{1}{s^2}\right) + \frac{1}{169}\left(\frac{8s + 6}{s^2 + 4s + 13}\right)$$

$$= -\frac{8}{169}\left(\frac{1}{s}\right) + \frac{2}{13}\left(\frac{1}{s^2}\right) + \frac{8}{169}\left(\frac{s + 2}{(s + 2)^2 + 9}\right)$$

$$- \frac{10}{3(169)}\left(\frac{3}{(s + 2)^2 + 9}\right)$$

and

$$\mathscr{L}^{-1}\left[\frac{2}{s^2(s^2 + 4s + 13)}\right] = -\frac{8}{169} + \frac{2}{13}t + \frac{8}{169}e^{-2t}\cos 3t - \frac{10}{507}e^{-2t}\sin 3t.$$

Finally, since

$$\frac{3(s + 2)}{(s^2 + 4s + 13)^2} = -\frac{3}{2}\frac{d}{ds}\left(\frac{1}{s^2 + 4s + 13}\right) = -\frac{1}{2}\frac{d}{ds}\left(\frac{9}{(s + 2)^2 + 9}\right),$$

we can apply Formulas (6–20) and (6–28) to obtain

$$\mathscr{L}^{-1}\left[\frac{3(s + 2)}{(s^2 + 4s + 13)^2}\right] = \frac{1}{2}te^{-2t}\sin 3t.$$

Combining these results, we see that the solution of (6–30) is

$$y = -\frac{179}{507}e^{-2t}\sin 3t + \frac{8}{169}e^{-2t}\cos 3t + \frac{1}{2}te^{-2t}\sin 3t + \frac{2}{13}t - \frac{8}{169}.$$

Anyone who feels overwhelmed by the details of these computations should compare them with those involved in the method of variation of parameters or undetermined coefficients. The argument given above is simpler and much more straightforward than either of the other two.

**Example 2**   In this example we show how the method of Laplace transforms can be used to find the *general solution* of a differential equation. As an illustration, we again solve

$$(D - \alpha)^2 y = 0, \tag{6–31}$$

where $\alpha$ is an arbitrary real number.

*Solution.*   In the absence of specific initial conditions we choose them arbitrarily as $y(0) = c_1$, $y'(0) = c_2$. Taking the Laplace transform of both sides of (6–31) and applying the initial conditions, we have

$$s^2\mathscr{L}[y] - c_1s - c_2 - 2\alpha(s\mathscr{L}[y] - c_1) + \alpha^2\mathscr{L}[y] = 0.$$

Hence

$$\mathscr{L}[y] = \frac{c_1 s + c_2 - 2\alpha c_1}{s^2 - 2\alpha s + \alpha^2}$$

$$= \frac{c_1 s - c_1 \alpha}{(s - \alpha)^2} + \frac{c_2 - c_1 \alpha}{(s - \alpha)^2}$$

$$= \frac{c_1}{s - \alpha} + (c_2 - c_1 \alpha) \frac{1}{(s - \alpha)^2}.$$

But

$$\mathscr{L}^{-1}\left[\frac{1}{s - \alpha}\right] = e^{\alpha t}, \quad \text{and} \quad \mathscr{L}^{-1}\left[\frac{1}{(s - \alpha)^2}\right] = t e^{\alpha t}.$$

Hence

$$y = c_1 e^{\alpha t} + (c_2 - c_1 \alpha) t e^{\alpha t} = c_3 e^{\alpha t} + c_4 t e^{\alpha t}$$

is the general solution of (6–31).

As our final example in this section we use the Laplace transform to solve an initial-value problem with a discontinuous input function $h$.

**Example 3**    Solve

$$y'' + y = \begin{cases} t, & 0 \leqslant t \leqslant \pi, \\ \sin t, & t > \pi, \end{cases} \quad y(0) = A, \quad y'(0) = B.$$

*Solution.*    We illustrate two methods of solution.

METHOD 1.    On the interval $[0, \pi]$ where $h(t) = t$, we have

$$s^2 \mathscr{L}[y] - As - B + \mathscr{L}[y] = \mathscr{L}[t]$$

or

$$(s^2 + 1)\mathscr{L}[y] = As + B + \frac{1}{s^2}.$$

It follows that

$$\mathscr{L}[y] = \frac{As}{s^2 + 1} + \frac{B}{s^2 + 1} + \frac{1}{s^2(s^2 + 1)}.$$

Therefore, since

$$\frac{1}{s^2(s^2 + 1)} = \frac{1}{s^2} - \frac{1}{s^2 + 1},$$

we have

$$y(t) = A\mathscr{L}^{-1}\left[\frac{s}{s^2 + 1}\right] + B\mathscr{L}^{-1}\left[\frac{1}{s^2 + 1}\right] + \mathscr{L}^{-1}\left[\frac{1}{s^2}\right] - \mathscr{L}^{-1}\left[\frac{1}{s^2 + 1}\right]$$

$$= A \cos t + B \sin t + t - \sin t$$

$$= A \cos t + (B - 1) \sin t + t, \quad 0 \leqslant t \leqslant \pi. \tag{6–32}$$

This completes the first part of the solution.

To continue, we differentiate (6–32) to obtain

$$y'(t) = -A \sin t + (B - 1) \cos t + 1,$$

and then observe that

$$y(\pi) = \pi - A, \qquad y'(\pi) = 2 - B,$$

which are the initial conditions the solution must satisfy at $t = \pi$. Thus, if the initial-value problem

$$y'' + y = \sin(t + \pi);$$
$$y(0) = \pi - A, \qquad y'(0) = 2 - B, \qquad (6\text{–}33)$$

has the solution $y = y(t)$ for $t > 0$, then $y(t - \pi)$ will be the solution of the original problem for $t > \pi$.

Applying the Laplace transform to (6–33), and recalling that $\sin(t + \pi) = -\sin t$, we obtain

$$s^2 \mathscr{L}[y] - (\pi - A)s - (2 - B) + \mathscr{L}[y] = -\frac{1}{s^2 + 1}$$

or

$$\mathscr{L}[y] = \frac{(\pi - A)s}{s^2 + 1} + \frac{2 - B}{s^2 + 1} - \frac{1}{(s^2 + 1)^2}.$$

From this and the formulas developed earlier in the chapter we find that

$$y(t) = (\pi - A) \cos t + (2 - B) \sin t - \tfrac{1}{2} \sin t + \tfrac{1}{2} t \cos t.$$

Hence

$$y(t - \pi) = -(\pi - A) \cos t - (2 - B) \sin t + \tfrac{1}{2} \sin t - \tfrac{1}{2}(t - \pi) \cos t,$$

and the solution of the problem is

$$y(t) = \begin{cases} A \cos t + (B - 1) \sin t + t, & 0 \leqslant t \leqslant \pi, \\ \left( A - \dfrac{\pi}{2} - \dfrac{t}{2} \right) \cos t + \left( B - \dfrac{3}{2} \right) \sin t, & t > \pi. \end{cases}$$

METHOD 2.   We use the unit step function $u_\pi$ to rewrite the problem as

$$y'' + y = t + u_\pi(t)(\sin t - t)$$
$$y(0) = A, \qquad y'(0) = B.$$

Now we apply the Laplace transform and use Formula (6–24) to obtain

$$s^2 \mathscr{L}[y] - As - B + \mathscr{L}[y] = \frac{1}{s^2} + e^{-\pi s} \mathscr{L}[\sin(t + \pi) - (t + \pi)]$$

$$= \frac{1}{s^2} - e^{-\pi s} \left( \frac{1}{s^2 + 1} + \frac{1}{s^2} + \frac{\pi}{s} \right).$$

Hence

$$\mathcal{L}[y] = \frac{As}{s^2 + 1} + \frac{B}{s^2 + 1} + \frac{1}{s^2(s^2 + 1)}$$
$$- e^{-\pi s}\left[\frac{1}{(s^2 + 1)^2} + \frac{1}{s^2(s^2 + 1)} + \frac{\pi}{s(s^2 + 1)}\right],$$

and a straightforward computation using Formula (6–25), among others, yields

$$y(t) = A \cos t + (B - 1) \sin t + t$$
$$- u_\pi(t)\left[\left(\frac{t}{2} + \frac{\pi}{2}\right) \cos t + \tfrac{1}{2} \sin t + t\right],$$

which agrees with our earlier solution. (See Exercise 1.)

## EXERCISES

1. Carry out the computations leading to the second form of the solution of the initial-value problem in Example 3.

Use the Laplace transform to solve each of the following initial-value problems.

2. $y'' + 6y' + 5y = t$;     $y(0) = y'(0) = 0$

3. $y'' + 2y' + y = e^t$;     $y(0) = y'(0) = 0$

4. $y' + 3y = t \sin at$;     $y(0) = -1$

5. $y'' + 2y' + 3y = 3t$;     $y(0) = 0, y'(0) = 1$

6. $y'' - 4y' + 4y = 2e^{2t} + \cos t$;     $y(0) = \frac{3}{25}, y'(0) = -\frac{4}{25}$

7. $y' + ky = h(t)$, where $k$ is a constant and $h$ is the square-wave function of Fig. 6–1; $y(0) = 0$

8. $y'' - 2y' + y = te^t \sin t$;     $y(0) = y'(0) = 0$

9. $y''' - y'' + 4y' - 4y = -3e^t + 4e^{2t}$;     $y(0) = 0, y'(0) = 5, y''(0) = 3$

10. $y^{(iv)} + 3y''' + y'' - 3y' - 2y = t$;     $y(0) = y'(0) = y''(0) = y'''(0) = 0$

11. $y'' + 4y' + 4y = \begin{cases} 0, & 0 \leqslant t \leqslant 2, \\ e^{-(t-2)}, & t > 2; \end{cases}$     $y(0) = 1, y'(0) = -1$

12. $y'' + y = t^2 + 1$;     $y(\pi) = \pi^2, y'(\pi) = 2\pi$

13. $y'' - y = -10 \sin 2t$;     $y(\pi) = -1, y'(\pi) = 0$

14. $y^{(iv)} + y = \begin{cases} 0, & 0 \leqslant t \leqslant 1, \\ t - 1, & t > 1; \end{cases}$     $y(0) = y'(0) = 1, y''(0) = y'''(0) = 0$

15. $y'' + 2y' + y = \begin{cases} 1, & 0 \leqslant t \leqslant 1, \\ 0, & t > 1; \end{cases}$     $y(0) = 0, y'(0) = 1$

16. $y'' + y = \begin{cases} 1, & 0 \leqslant t \leqslant \pi/2, \\ 0, & t > \pi/2; \end{cases}$     $y(0) = 0, y'(0) = 1$

17. $y'' + y = u_\pi(t)$;     $y(0) = 1, y'(0) = 0$

18. Use the Laplace transform to solve

$$\frac{dy}{dt} + 2y + \int_0^t y(t)\, dt = \begin{cases} t, & 0 \leqslant t \leqslant 1, \\ 2 - t, & 1 \leqslant t \leqslant 2, \\ 0, & t > 2, \end{cases}$$

given that $y(0) = 1$.

19. a) Let $y$ be a function of exponential order, and suppose that $y$ is a solution of the Euler equation

$$t^2 y'' + aty' + by = 0,$$

where $a$ and $b$ are constants. Show that $\mathscr{L}[y]$ also satisfies an Euler equation.
b) Generalize the result obtained in part (a) to solutions of Euler equations of arbitrary order.

20. Show that if $f$ is continuous for all $t > 0$, and if $f$ and $f'$ are of exponential order, then

$$\lim_{s \to \infty} s\mathscr{L}[f] = f(0^+).$$

*21. The differential equation

$$ty'' + y' + ty = 0$$

is known as *Bessel's equation of order zero*. (See p. 295.) This equation has a solution $J_0$ that is defined for all $t \geqslant 0$ and satisfies the condition $J_0(0) = 1$.
a) Prove that

$$\mathscr{L}[J_0] = \frac{1}{\sqrt{1 + s^2}}.$$

[*Hint*: Show that $\mathscr{L}[J_0]$ is a solution of the differential equation

$$(1 + s^2)\phi' + s\phi = 0,$$

and apply the result of Exercise 20.]
b) Find a power series expansion for $J_0$. [*Hint*: Expand $s\mathscr{L}[J_0]$ as a binomial series.]

## 6–6 THE CONVOLUTION THEOREM

In this section we shall establish the most important single property of the Laplace transform, the so-called *convolution formula*. Far from being just a computational device, as were the formulas of Section 6–4, the convolution formula is important in theoretical work in analysis. Moreover, as we shall see, it is ideally suited to the task of constructing inverses for constant-coefficient linear differential operators. The formula in question reads as follows.

**Theorem 6–10** *Let $f$ and $g$ be piecewise continuous functions of exponential order, and suppose that*

$$\mathscr{L}[f] = \phi(s), \qquad \mathscr{L}[g] = \psi(s).$$

*Then*

$$\mathscr{L}\left[\int_0^t f(t - \xi)g(\xi)\, d\xi\right] = \phi(s)\psi(s). \tag{6–34}$$

*used backwards for*
*invers Laplace of product.*

When written in terms of inverse transforms, (6–34) becomes

$$\mathcal{L}^{-1}[\phi(s)\psi(s)] = \int_0^t f(t - \xi)g(\xi)\,d\xi. \tag{6–35}$$

In this form it asserts that if we know the inverse transforms, $f$ and $g$, of the functions $\phi$ and $\psi$, we can express the inverse transform of the product $\phi(s)\psi(s)$ as an integral involving $f$ and $g$. The integral in question is called the **convolution** of $f$ and $g$ and is denoted by $f * g$:

$$(f * g)(t) = \int_0^t f(t - \xi)g(\xi)\,d\xi. \tag{6–36}$$

With this notation, we can write (6–35) as

$$f * g = \mathcal{L}^{-1}[\phi(s)\psi(s)]$$

or, even more suggestively,

$$\mathcal{L}^{-1}[\phi(s)\psi(s)] = \mathcal{L}^{-1}[\phi(s)] * \mathcal{L}^{-1}[\psi(s)]. \tag{6–37}$$

*Proof*  Using the definition of $\mathcal{L}$, we have

$$\mathcal{L}\left[\int_0^t f(t - \xi)g(\xi)\,d\xi\right] = \int_0^\infty e^{-st}\int_0^t f(t - \xi)g(\xi)\,d\xi\,dt$$
$$= \int_0^\infty \int_0^t e^{-st}f(t - \xi)g(\xi)\,d\xi\,dt,$$

where the integration is performed over the region of the $t\xi$-plane described by the inequalities

$$0 \leqslant \xi \leqslant t, \qquad 0 \leqslant t < \infty$$

(see Fig. 6–11). But this region is also described by

$$\xi \leqslant t < \infty, \qquad 0 \leqslant \xi < \infty.$$

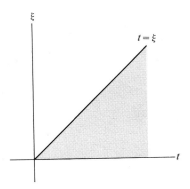

**Figure 6–11**

Hence the preceding iterated integral may be written as

$$\int_0^\infty \int_\xi^\infty e^{-st} f(t - \xi) g(\xi) \, dt \, d\xi \, *$$

or

$$\int_0^\infty g(\xi) \left( \int_\xi^\infty e^{-st} f(t - \xi) \, dt \right) d\xi.$$

We now make the change of variable $u = t - \xi$ in $\int_\xi^\infty e^{-st} f(t - \xi) \, dt$ to obtain

$$\int_\xi^\infty e^{-st} f(t - \xi) \, dt = \int_0^\infty e^{-s(u + \xi)} f(u) \, du.$$

Hence

$$
\begin{aligned}
\mathcal{L}\left[ \int_0^t f(t - \xi) g(\xi) \, d\xi \right] &= \int_0^\infty g(\xi) \left( \int_0^\infty e^{-s(u+\xi)} f(u) \, du \right) d\xi \\
&= \int_0^\infty e^{-s\xi} g(\xi) \left( \int_0^\infty e^{-su} f(u) \, du \right) d\xi \\
&= \int_0^\infty e^{-su} f(u) \, du \int_0^\infty e^{-s\xi} g(\xi) \, d\xi \\
&= \mathcal{L}[f] \mathcal{L}[g],
\end{aligned}
$$

and the proof is complete. ∎

The operation of convolution can be viewed as a multiplication on the vector space $\mathscr{E}$ of piecewise continuous functions of exponential order, with $f * g$ as the product of $f$ and $g$, and it is well worthwhile to investigate its properties. For instance, Formula (6–35) and the equality $\mathcal{L}[f]\mathcal{L}[g] = \mathcal{L}[g]\mathcal{L}[f]$ imply at once that

$$\int_0^t f(t - \xi) g(\xi) \, d\xi = \int_0^t g(t - \xi) f(\xi) \, d\xi.$$

Hence

$$f * g = g * f,$$

and convolution is commutative. It is also associative and distributive, since

$$f * (g * h) = (f * g) * h \quad \text{and} \quad f * (g + h) = f * g + f * h$$

(see Exercises 14 and 15). Much of the advanced work connected with the convolution integral is devoted to studying the behavior of $\mathscr{E}$ under this multiplication.

**Example 1** Find

$$\mathcal{L}^{-1}\left[ \frac{1}{s(s^2 + 1)} \right].$$

*Solution.* It is clear that this problem can be solved by separating $1/(s(s^2 + 1))$ into partial fractions as $(1/s) - (s/(s^2 + 1))$ and applying our earlier formulas.

---

\* It can be shown that the interchange of the order of integration is permissible at this point.

But it can be handled just as well with (6–36) and (6–37):

$$\mathscr{L}^{-1}\left[\frac{1}{s(s^2 + 1)}\right] = \mathscr{L}^{-1}\left[\frac{1}{s}\right] * \mathscr{L}^{-1}\left[\frac{1}{s^2 + 1}\right]$$

$$= 1 * \sin t$$

$$= \int_0^t \sin \xi \, d\xi$$

$$= 1 - \cos t.$$

**Example 2**   Solve the initial-value problem

$$(D^2 + D - 6)y = h(t), \qquad y(0) = y'(0) = 0, \tag{6–38}$$

given that $h$ is a function of exponential order.

*Solution.*   We apply the operator $\mathscr{L}$ to obtain

$$(s^2 + s - 6)\mathscr{L}[y] = \mathscr{L}[h]$$

or

$$\mathscr{L}[y] = \frac{\mathscr{L}[h]}{s^2 + s - 6},$$

from which it follows that

$$y = \mathscr{L}^{-1}\left[\frac{1}{(s - 2)(s + 3)}\right] * h(t).$$

Finally, since

$$\mathscr{L}^{-1}\left[\frac{1}{(s - 2)(s + 3)}\right] = \mathscr{L}^{-1}\left[\frac{1}{5(s - 2)} - \frac{1}{5(s + 3)}\right]$$

$$= \tfrac{1}{5}e^{2t} - \tfrac{1}{5}e^{-3t},$$

we have

$$y = \int_0^t \left[\tfrac{1}{5}e^{2(t - \xi)} - \tfrac{1}{5}e^{-3(t - \xi)}\right]h(\xi) \, d\xi, \tag{6–39}$$

and the problem is solved pending explicit knowledge of $h$.

If, for example, $h(t) = 1$, then

$$y = \int_0^t \left[\tfrac{1}{5}e^{2(t - \xi)} - \tfrac{1}{5}e^{-3(t - \xi)}\right] d\xi$$

$$= \tfrac{1}{5}e^{2t} \int_0^t e^{-2\xi} \, d\xi - \tfrac{1}{5}e^{-3t} \int_0^t e^{3\xi} \, d\xi$$

$$= \tfrac{1}{10}e^{2t} + \tfrac{1}{15}e^{-3t} - \tfrac{1}{6}.$$

You may want to compare the formula

$$G(t, \xi) = \mathscr{L}^{-1}\left[\frac{1}{s^2 + s + 6}\right](t - \xi) = \tfrac{1}{5}e^{2(t - \xi)} - \tfrac{1}{5}e^{-3(t - \xi)}$$

with the Green's function for the operator $D^2 + D - 6$ for initial-value problems presented in Section 5–5. We shall explore the connection between Green's functions and the Laplace transform in the next section.

### EXERCISES

Use the convolution formula to find the inverse Laplace transform of each of the following functions.

1. $\dfrac{\mathscr{L}[f]}{s^2 + 1}$

2. $\dfrac{e^{-3s}\mathscr{L}[f]}{s^3}$

3. $\dfrac{1}{s^2(s + 1)}$

4. $\dfrac{s}{(s^2 + 1)^2}$

5. $\dfrac{3s^2}{(s^2 + 1)^2}$

6. $\dfrac{1}{(s - a)(s - b)}, \quad a \neq b$

Evaluate each of the following.

7. $e^{at} * e^{bt}$

8. $t * \cos at$

9. $\sin at * \cos bt$

10. $t * e^{at}$

11. $f(t - 1) * e^{-t}g(t + 1)$

12. $f(-t) * (\sin t)g(t^2)$

13. Prove directly that $f * g = g * f$, that is,

$$\int_0^t f(t - \xi)g(\xi)\,d\xi = \int_0^t g(t - \xi)f(\xi)\,d\xi.$$

[*Hint*: Make the substitution $u = t - \xi$.]

14. Prove that $f * (g + h) = f * g + f * h$.

15. Prove that $f * (g * h) = (f * g) * h$.

16. Find $1 * 1$ and $1 * 1 * 1$.

17. Derive a formula for $1 * 1 * 1 * \cdots * 1$   ($n$ factors).

\* 18. Solve the initial-value problem $y'' + y = h(t); y(0) = 1, y'(0) = 0; h(t)$ periodic with period $2\pi$, and

$$h(t) = \begin{cases} 1, & 0 \leq t \leq \pi \\ 0, & \pi < t < 2\pi \end{cases}$$

19. Suppose that $f(t)$ is of exponential order and that $\lim_{t \to 0^+} f(t)/t$ exists. Assuming that the order of integration may be reversed in the computation, prove that

$$\mathscr{L}\left[\frac{f(t)}{t}\right] = \int_s^\infty \mathscr{L}[f]\,ds.$$

Use the result of Exercise 19 to compute the Laplace transform of each of the following functions.

20. $t\left(=\dfrac{t^2}{t}, \text{ with } \mathscr{L}[t^2] = \dfrac{2}{s^3}\right)$

21. $\dfrac{\sin at}{t}$

22. $\dfrac{e^t - 1}{t}$

23. $\dfrac{1 - \cos 3t}{t^2}$

24. a) It can be proved that whenever $\int_0^\infty [f(t)/t]\, dt$ exists, the formula in Exercise 19 remains valid with $s$ set equal to zero. Use this fact to show that

$$\int_0^\infty \frac{f(t)}{t}\, dt = \int_0^\infty \mathscr{L}[f]\, ds.$$

   b) Use the result obtained in (a) to prove that

$$\int_0^\infty \frac{\sin t}{t}\, dt = \frac{\pi}{2}.$$

25. Equations of the form

$$\phi(t) + \int_0^t K(t - \xi)\phi(\xi)\, d\xi = h(t),$$

where $h$ and $K$ are known, and $\phi$ is unknown, are called **Volterra integral equations**. They arise frequently in applied mathematics, and an extensive literature exists about them. Use the Laplace transform to solve the following Volterra equations.

   a) $\phi(t) + \displaystyle\int_0^t (t - \xi)\phi(\xi)\, d\xi = \sin 2t$

   b) $\phi(t) + \displaystyle\int_0^t (t - \xi)\phi(\xi)\, d\xi = 1$

   c) $\phi(t) + \displaystyle\int_0^t e^{-(t-\xi)}\phi(\xi)\, d\xi = \cos t$

   d) $\phi(t) + \displaystyle\int_0^t e^{-(t-\xi)}\phi(\xi)\, d\xi = \sin 2t$

   e) $\phi(t) + \displaystyle\int_0^t (t + 1 - \xi)\phi(\xi)\, d\xi = 1 + \sin t$

26. Show that every Volterra equation of the type defined in Exercise 25 can be viewed as a (linear) operator equation on a suitable vector space of functions.

## *6–7 GREEN'S FUNCTIONS AND CONVOLUTION

In Chapter 5 we saw that if $L$ is a normal $n$th-order linear differential operator acting on $\mathscr{C}^{(n)}(I)$, then the integral operator defined for each $h$ in $\mathscr{C}(I)$ by

$$G[h](t) = \int_{t_0}^t K(t, \xi)h(\xi)\, d\xi, \tag{6–40}$$

where $K$ is the Green's function for $L$ for initial-value problems, is a (right) inverse for $L$ for the initial-value problem

$$Ly = h; \qquad y(t_0) = y'(t_0) = \cdots = y^{(n-1)}(t_0) = 0.$$

The Green's function $K(t, \xi)$ appearing in (6–40) can be constructed by the method of variation of parameters from a basis for the solution space of the equation $Ly = 0$, and as we saw earlier, is uniquely determined by $L$ in the sense that it is independent of the particular basis and initial point used to compute it.

   In this section we shall use the convolution formula to derive an extremely efficient method of computing Green's functions for operators with *constant coefficients*. One of the principal advantages of this method is that it does not use a basis for the solution space of $Ly = 0$, and therefore eliminates the tedious computations involved in the method of variation of parameters. As such it is an excellent example of the efficiency that results from using the "coordinate-free" techniques of linear algebra.

The result we seek is an easy consequence of our earlier work with Green's functions. Indeed, let

$$L = D^n + a_{n-1}D^{n-1} + \cdots + a_0 \qquad (6\text{–}41)$$

be a normal $n$th-order constant-coefficient linear differential operator, and consider the initial-value problem

$$Ly = h; \qquad y(0) = y'(0) = \cdots = y^{(n-1)}(0) = 0, \qquad (6\text{–}42)$$

where $h$ belongs to the vector space $\mathscr{E}$. When we apply the Laplace transform to (6–42), we obtain

$$p(s)\mathscr{L}[y] = \mathscr{L}[h],$$

where

$$p(s) = s^n + a_{n-1}s^{n-1} + \cdots + a_0$$

is the characteristic polynomial of $L$. It follows that

$$\mathscr{L}[y] = \frac{1}{p(s)}\,\mathscr{L}[h];$$

and thus if

$$\mathscr{L}^{-1}\left[\frac{1}{p(s)}\right] = g(t),$$

the convolution formula yields

$$y(t) = \int_0^t g(t-\xi)h(\xi)\,d\xi. \qquad (6\text{–}43)$$

In view of this result let us now introduce the (linear) operator $G$ defined by the equation*

$$G[h] = \int_0^t g(t-\xi)h(\xi)\,d\xi. \qquad (6\text{–}44)$$

When combined with our earlier work on Green's functions, the fact that $G[h]$ is the solution of (6–42) implies that $g(t-\xi)$ is the Green's function for $L$ for initial-value problems. (See Exercises 12 and 13.) Thus we have

**Theorem 6–11** *Let $L$ be a constant-coefficient linear differential operator whose characteristic polynomial is*

$$p(s) = s^n + a_{n-1}s^{n-1} + \cdots + a_0.$$

---

\* The assumptions made in the discussion leading to this formula imply that $G$ is defined on the space of piecewise continuous functions of exponential order on $[0, \infty)$. In Exercise 14 at the end of this section we show how this restriction can be relaxed.

*Then if*

$$g(t) = \mathscr{L}^{-1} \left[ \frac{1}{p(s)} \right],$$

*the function $g(t - \xi)$ is the Green's function for L for initial-value problems on $(-\infty, \infty)$.*

**Example 1**    Find the Green's function for the operator

$$L = D^2 - 2aD + a^2 + b^2, \qquad b \neq 0.$$

*Solution.*    Since

$$\mathscr{L}^{-1} \left[ \frac{1}{s^2 - 2a + a^2 + b^2} \right] = \mathscr{L}^{-1} \left[ \frac{1}{(s - a)^2 + b^2} \right] = \frac{1}{b} e^{at} \sin bt,$$

the desired Green's function is

$$g(t - \xi) = \frac{1}{b} e^{a(t - \xi)} \sin b(t - \xi).$$

**Example 2**    Solve the initial-value problem

$$y'' - 2y' + 5y = e^t; \qquad y(0) = y'(0) = 0.$$

*Solution.*    The result of the preceding example implies that

$$g(t - \xi) = \tfrac{1}{2} e^{t - \xi} \sin 2(t - \xi)$$

is the Green's function for the problem. Hence

$$
\begin{aligned}
y &= \frac{1}{2} \int_0^t e^{t - \xi} \sin 2(t - \xi) e^\xi \, d\xi \\
&= \frac{e^t}{2} \int_0^t \sin 2(t - \xi) \, d\xi \\
&= \frac{e^t}{2} \left( \sin 2t \int_0^t \cos 2\xi \, d\xi - \cos 2t \int_0^t \sin 2\xi \, d\xi \right) \\
&= \frac{e^t}{4} \left[ \sin^2 2t + \cos 2t (\cos 2t - 1) \right] \\
&= \frac{e^t}{4} (1 - \cos 2t).
\end{aligned}
$$

The next example illustrates how to handle a problem in which the initial conditions are not imposed at $t = 0$ and the initial values are different from zero.

**Example 3**    Solve the initial-value problem

$$y'' - 2y' + 5y = e^t; \qquad y(\pi) = 2, \quad y'(\pi) = 3.$$

*Solution.*    We begin by observing that

$$y = y_h + G[e^\xi],$$

where $G$ is the inverse of $L = D^2 - 2D + 5$ for initial-value problems at $t_0 = \pi$, and $y_h$ is the solution of

$$y'' - 2y' + 5y = 0; \qquad y(\pi) = 2, \quad y'(\pi) = 3.$$

By the results of Example 2 the Green's function for $L$ is

$$g(t - \xi) = \tfrac{1}{2}e^{t-\xi} \sin 2(t - \xi).$$

Hence

$$G[e^{\xi}] = \frac{1}{2} \int_{\pi}^{t} e^{t-\xi} \sin 2(t - \xi)e^{\xi} \, d\xi$$

$$= \frac{e^t}{2} \int_{\pi}^{t} \sin 2(t - \xi) \, d\xi$$

$$= \frac{e^t}{2} \left[ \sin 2t \int_{\pi}^{t} \cos 2\xi \, d\xi - \cos 2t \int_{\pi}^{t} \sin 2\xi \, d\xi \right]$$

$$= \frac{e^t}{4} \left[ \sin^2 2t + \cos 2t(\cos 2t - 1) \right]$$

$$= \frac{e^t}{4} (1 - \cos 2t).$$

We still need to compute $y_h$ and we do this by observing that $y_h(t) = y(t - \pi)$, where $y$ is the solution of

$$y'' - 2y' + 5y = 0; \qquad y(0) = 2, \quad y'(0) = 3.$$

Using the Laplace transform, we obtain

$$(s^2 - 2s + 5)\mathscr{L}[y] = 2s + 1,$$

and it follows that

$$\mathscr{L}[y] = \frac{2s + 1}{s^2 - 2s + 5} = \frac{2(s - 1)}{(s - 1)^2 + 2^2} + \frac{3}{(s - 1)^2 + 2^2}.$$

Thus

$$y(t) = 2\mathscr{L}^{-1}\left[ \frac{s - 1}{(s - 1)^2 + 2^2} \right] + 3\mathscr{L}^{-1}\left[ \frac{1}{(s - 1)^2 + 2^2} \right]$$

$$= 2e^t \cos 2t + \tfrac{1}{2}e^t \sin 2t,$$

and

$$y_h(t) = y(t - \pi) = 2e^{t-\pi} \cos 2(t - \pi) + \tfrac{1}{2}e^{t-\pi} \sin 2(t - \pi).$$
$$= e^{t-\pi}(2 \cos 2t + \tfrac{1}{2} \sin 2t).$$

Combining these results, we have

$$y = e^{t-\pi}(2 \cos 2t + \tfrac{1}{2} \sin 2t) + \frac{e^t}{4} (1 - \cos 2t).$$

## EXERCISES

1. Let $L = D^n + a_{n-1}D^{n-1} + \cdots + a_0$. Verify that when the Laplace transform is applied to the initial-value problem

$$Ly = h; \qquad y(0) = y'(0) = \cdots = y^{(n-1)}(0) = 0,$$

   it yields the equation

$$(s^n + a_{n-1}s^{n-1} + \cdots + a_0)\mathscr{L}[y] = \mathscr{L}[h].$$

Use the technique developed in this section to find the Green's function for each of the following linear differential operators.

2. $D^2 + D$

3. $D^2 + 6D + 13$

4. $D^2 + \frac{1}{2}D - \frac{1}{2}$

5. $D^2 + \frac{1}{3}D + \frac{1}{36}$

6. $4D^3 - D$

7. $D^3 + 1$

8. $(D^2 + 1)^2$

9. $D^4 + 1$

10. $(D^2 - 4D + 20)^2$

11. Solve the initial-value problem

$$(D^2 - 1)y = \begin{cases} 0, & 0 \leqslant t \leqslant 1, \\ t - 1, & t > 1, \end{cases} \qquad y(0) = y'(0) = 0,$$

   by (a) using Laplace transforms directly; (b) determining a Green's function for $D^2 - 1$; and (c) solving the initial-value problem

$$(D^2 - 1)y = t - 1, \qquad y(1) = a, \ y'(1) = b,$$

   with an appropriate choice of constants $a, b$.

12. Prove that the function $g(t) = \mathscr{L}^{-1}[1/p(s)]$ defined in the text satisfies the initial-value problem

$$Ly = 0;$$
$$y(0) = \cdots = y^{(n-2)}(0) = 0, \quad y^{(n-1)}(0) = 1,$$

   on the interval $[0, \infty)$.

13. Show that the methods of Sections 6–3 and 6–4, when used to compute $\mathscr{L}^{-1}[1/p(s)]$, lead to the unique solution on the entire interval $(-\infty, \infty)$ of the initial-value problem of Exercise 12. (This justifies the statement that $g(t - \xi)$ is *the* Green's function for $L$ on $(-\infty, \infty)$, where $g(t) = \mathscr{L}^{-1}[1/p(s)]$.)

14. Use Laplace transforms to derive (6–44) for any function $h$ which is piecewise continuous on $[0, \infty)$. [*Hint*: First consider the solution of (6–42) with $h$ replaced by

$$H(t) = \begin{cases} h(t), & 0 \leqslant t \leqslant a, \\ 0, & t > a, \end{cases}$$

   where $a$ is a constant $> 0$.]

15. If $L$ is a constant coefficient operator, show that $y_p$ is a solution of the initial-value problem

$$Ly = h;$$
$$y(t_0) = c_0, \ y'(t_0) = c_1, \ \ldots, \ y^{(n-1)}(t_0) = c_{n-1},$$

if and only if $y_p(t) = Y(t - t_0)$, where $Y(t)$ satisfies

$$Ly = h(t + t_0);$$
$$y(0) = c_0, \quad y'(0) = c_1, \quad \ldots, \quad y^{(n-1)}(0) = c_{n-1}.$$

Use the method given at the end of this section to solve each of the following initial-value problems.

16. $(D^2 + 1)y = e^{t-1}$;    $y(1) = y'(1) = 0$
17. $(2D^2 + D - 1)y = \sin t$;    $y(\pi) = y'(\pi) = 0$
18. $(4D^2 + 16D + 17)y = t^2 - 1$;    $y(a) = y'(a) = 0$
19. $(D^2 - 3D - 4)y = e^{-t}$;    $y(2) = 3, y'(2) = 0$
20. $(2D^2 - 3D + 1)y = t$;    $y(1) = 0, y'(1) = -1$
21. $(4D^2 - 4D + 37)y = e^{t/2} \cos 3t$;    $y(a) = 7, y'(a) = -2, \quad a > 0$
22. $(D^3 + 1)y = te^t$;    $y(1) = y'(1) = 0, y''(1) = 1$
23. $D^2(D^2 + 1)^2 y = h(t)$;    $y(a) = y'(a) = y''(a) = y'''(a) = y^{(iv)}(a) = 0, y^{(v)}(a) = 1$

## 6–8 THE VIBRATING SPRING; IMPULSE FUNCTIONS

A simple yet surprisingly fruitful system for the study of motion is provided by an elastic spring that vibrates in the vertical direction with an object of mass $m$ attached. We shall study the motion of this system under the hypothesis (known as *Hooke's law*) that whenever the spring is stretched beyond its natural or unweighted length, a restoring force arises in the spring that is proportional to the amount by which the spring is stretched. Thus, if the spring with weight attached comes into equilibrium at the point $y = 0$ located $d$ units below the natural length of the spring (Fig. 6–12), the downward force of gravity must be balanced by the upward restoring force provided by the spring, and we have

$$mg = kd, \tag{6–45}$$

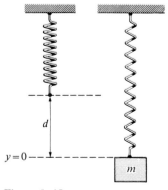

**Figure 6–12**

where $k$, the constant of proportionality in Hooke's law, is positive and measures the "stiffness" of the spring. It is traditional to refer to $k$ as the *spring constant*. We propose to study the response of this system when the weight is initially at rest in the equilibrium position and then is subjected to the influence of an external force $h = h(t)$ that varies with time.

**The General Case**   At time $t \geqslant 0$, with the weight at the point $y = y(t)$ and the positive $y$-direction measured upward, the forces acting on the weight are:

$-mg$, the downward force of gravity,
$k(d - y)$, the upward restoring force of the spring, and
$h(t)$.

(See Fig. 6–13, drawn with $h(t)$ negative.) Hence by Newton's second law,

$$m\frac{d^2 y}{dt^2} = -mg + k(d - y) + h(t),$$

and it follows from (6–45) and the fact that the system was initially at rest in the equilibrium position that the motion is described by the initial-value problem

$$m\frac{d^2 y}{dt^2} + ky = h(t); \qquad y(0) = y'(0) = 0. \tag{6–46}$$

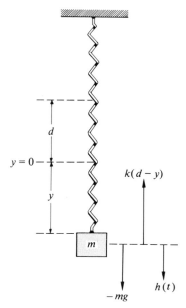

**Figure 6–13**

To solve this problem, we apply the Laplace transform, to obtain

$$\mathscr{L}[y] = \frac{\mathscr{L}[h]}{ms^2 + k}. \tag{6-47}$$

Hence

$$y = \mathscr{L}^{-1}\left[\frac{h}{ms^2 + k}\right] * h(t),$$

and since

$$\mathscr{L}^{-1}\left[\frac{1}{ms^2 + k}\right] = \mathscr{L}^{-1}\left[\frac{\sqrt{k/m}}{\sqrt{km}(s^2 + k/m)}\right] = \frac{1}{\sqrt{km}}\sin\sqrt{\frac{k}{m}}\,t,$$

we have

$$y = \frac{1}{\sqrt{km}}\int_0^t \sin\sqrt{\frac{k}{m}}(t - \xi)h(\xi)\,d\xi. \tag{6-48}$$

Let us now consider the response of the weighted spring to two distinct types of inputs.

**Sinusoidal impressed forces**  In many applications the impressed force is of the form

$$h(t) = A\sin\omega t,$$

where $A$ and $\omega$ are positive constants, in which case (6–48) yields the equation of motion

$$y = \frac{A}{\sqrt{km}}\int_0^t \sin\sqrt{\frac{k}{m}}(t - \xi)\sin\omega\xi\,d\xi.$$

Though this integral can be evaluated by elementary techniques, it is instructive to return to (6–47) and solve the problem directly. We then find that

$$\mathscr{L}[y] = \frac{1}{ms^2 + k}\mathscr{L}[A\sin\omega t] = \frac{1}{ms^2 + k}\cdot\frac{A\omega}{s^2 + \omega^2}.$$

There are two cases to consider, depending on the value of $\omega$.

CASE 1.   $\omega \neq \sqrt{k/m}$. Here $A\omega/[(ms^2 + k)(s^2 + \omega^2)]$ can be decomposed by the method of partial fractions as

$$\frac{A\omega}{k - m\omega^2}\left(\frac{1}{s^2 + \omega^2} - \frac{m}{ms^2 + k}\right),$$

and a routine computation yields

$$y = \frac{A\omega}{k - m\omega^2}\left[\frac{1}{\omega}\sin\omega t - \sqrt{\frac{m}{k}}\sin\sqrt{\frac{k}{m}}\,t\right].$$

In this case the motion is the superposition (sum) of the oscillations of two different frequencies*:

$$\frac{\omega}{2\pi} \quad \text{and} \quad \frac{1}{2\pi}\sqrt{\frac{k}{m}}.$$

The first of these is the frequency of the impressed force, while the second is the so-called *natural frequency* of the system. Finally, we observe that the amplitude of these vibrations is bounded, because it cannot exceed

$$\left|\frac{A\omega}{k - m\omega^2}\right|\left(\frac{1}{\omega} + \sqrt{\frac{m}{k}}\right).$$

(See Fig. 6–14, drawn for the case $m = k = 1$, $A = 3$, and $\omega = 2$.)

CASE 2.   $\omega = \sqrt{k/m}$. Here

$$\mathscr{L}[y] = \frac{A\sqrt{km}}{(ms^2 + k)^2} = \frac{A\sqrt{km}}{s} \cdot \frac{s}{(ms^2 + k)^2}.$$

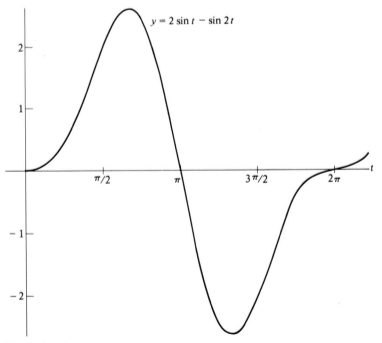

$y = 2\sin t - \sin 2t$

**Figure 6–14**

---

* The *frequency* $f$ of a periodic function is the number of cycles (periods) it completes in a unit time. Thus $fp = 1$, or $f = 1/p$, where $p$ is the period of the function.

Since

$$\mathscr{L}^{-1}\left[\frac{A\sqrt{km}}{s}\right] = A\sqrt{km}$$

and

$$\mathscr{L}^{-1}\left[\frac{s}{(ms^2 + k)^2}\right] = \mathscr{L}^{-1}\left[-\frac{1}{2m}\frac{d}{ds}\left(\frac{1}{ms^2 + k}\right)\right]$$

$$= \frac{t}{2m}\mathscr{L}^{-1}\left[\frac{1}{ms^2 + k}\right]$$

$$= \frac{t}{2m\sqrt{km}}\sin\sqrt{\frac{k}{m}}\,t,$$

the convolution formula yields

$$y = \frac{A}{2m}\int_0^t \xi \sin\sqrt{\frac{k}{m}}\,\xi\,d\xi$$

$$= \frac{A}{2m}\left(\frac{m}{k}\sin\sqrt{\frac{k}{m}}\,t - \sqrt{\frac{m}{k}}\,t\cos\sqrt{\frac{k}{m}}\,t\right).$$

It follows that

$$y = \frac{A}{2k}\sin\sqrt{\frac{k}{m}}\,t - \frac{A}{2\sqrt{km}}\,t\cos\sqrt{\frac{k}{m}}\,t.$$

This result shows that when the frequency of the impressed force is equal to the natural frequency of the system, the amplitude of the oscillations increases with time and the spring is eventually stretched beyond its elastic limit. (See Fig. 6–15, sketched for $A = k = m = 1$.) This is the phenomenon of resonance that we encountered earlier in our study of electrical circuits.

**Impulses**    A completely different situation arises when we describe the response of the weighted spring when the mass is struck a sharp blow from below at time $t = 0$ and is allowed to oscillate undisturbed thereafter. Before attempting to solve this problem we must decide on a mathematical description of the term "a sharp blow." What, precisely, does it mean? We take our cue from mechanics, where a constant force of magnitude $F$ that acts on an object of mass $m$ during a time interval $t$ is said to impart an **impulse**

$$I = Ft$$

to the object. Recalling that

$$F = \frac{d}{dt}(mv),$$

where $v$ is the magnitude of the velocity of the object, we see that $I$ is the total change in the momentum $mv$ of the object during the time interval $t$. We now agree that a sharp blow will consist of a constant force that acts for an arbitrarily short period of time on the object and imparts a predetermined impulse, or change of

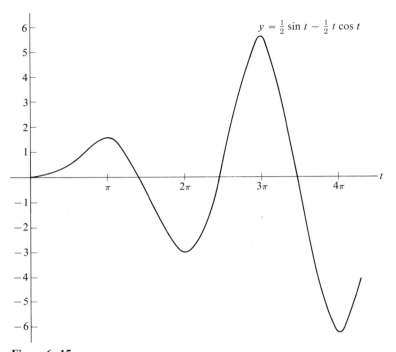

$$y = \tfrac{1}{2}\sin t - \tfrac{1}{2}t\cos t$$

**Figure 6–15**

momentum, to the object. Thus we introduce the forcing function (Fig. 6–16)

$$h_\varepsilon(t) = \begin{cases} 1/\varepsilon, & 0 \leqslant t \leqslant \varepsilon, \\ 0, & t > \varepsilon, \end{cases} \tag{6–49}$$

where $\varepsilon$ is an arbitrary positive constant, and study the response of the weighted spring to $h_\varepsilon$ as $\varepsilon \to 0$. For convenience we choose $h_\varepsilon$ so that it imparts a *unit* impulse to the system.

It follows from (6–48) that

$$y(t) = \begin{cases} \dfrac{1}{\varepsilon}\displaystyle\int_0^t \dfrac{1}{\sqrt{km}}\sin\sqrt{\dfrac{k}{m}}\,(t-\xi)\,d\xi, & 0 \leqslant t \leqslant \varepsilon, \\[2mm] \dfrac{1}{\varepsilon}\displaystyle\int_0^\varepsilon \dfrac{1}{\sqrt{km}}\sin\sqrt{\dfrac{k}{m}}\,(t-\xi)\,d\xi, & t > \varepsilon. \end{cases} \tag{6–50}$$

Hence, passing to the limit as $\varepsilon \to 0$, we obtain

$$y_0(t) = \begin{cases} 0, & t = 0, \\[2mm] \dfrac{1}{\sqrt{km}}\sin\sqrt{\dfrac{m}{k}}\,t, & t > 0. \end{cases} \tag{6–51}$$

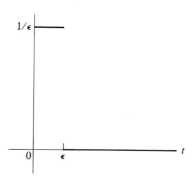

**Figure 6–16**

(See Exercise 2.) But all this is mere formalism, and though suggestive it does not establish $y_0$ as the solution of the problem we have been investigating. For as it was derived, $y_0$ is not the solution of *any* initial-value problem. Instead, it is the limit as $\varepsilon \to 0$ of the solutions $y_\varepsilon$ of

$$m\frac{d^2y}{dt^2} + ky = h_\varepsilon(t); \qquad y(0) = y'(0) = 0. \qquad (6\text{--}52)$$

Happily we do not have far to look to find an initial-value problem of which $y_0$ *is* the solution. We simply consider the problem that our discussion of impulse forces would lead us to pose in any event:

$$m\frac{d^2y}{dt^2} + ky = 0; \qquad y(0) = 0, \quad y'(0^+) = \frac{1}{m}. \qquad (6\text{--}53)$$

The task of verifying that $y_0$ is a solution of this problem is routine. The important point is that as such it can be interpreted as the response of a weighted spring that is:

1. in its equilibrium position at time zero: $y(0) = 0$;
2. given a unit momentum at that instant: $my'(0^+) = 1$;
3. left undisturbed thereafter: $h(t) = 0$.

And this, in mathematical terms, is the problem we set out to solve.

In many situations it is convenient to think of a unit impulse at $t = 0$ as arising from a *fictitious* function, $\delta(t)$, called the **Dirac delta function** and defined by the conditions

$$\delta(t) = 0 \text{ for all } t \neq 0 \qquad \text{and} \qquad \int_{-\infty}^{\infty} \delta(t)\, dt = 1. \qquad (6\text{--}54)$$

(The reason for calling $\delta$ a "fictitious" function is that no ordinary function can vanish everywhere except at $t = 0$ and still have a nonzero integral over the entire real line.) Nevertheless, if we accept the existence of such a function, the preceding discussion should imply that $y_0$ as defined by (6–51) is the solution of the initial-value problem

$$m \frac{d^2y}{dt^2} + ky = \delta(t), \qquad y(0) = y'(0) = 0, \qquad (6\text{–}55)$$

where the initial conditions state that the weight is at rest in its equilibrium position at $t = 0$.* Let us assume that it does, and that in addition $\delta$ has a Laplace transform. Then, applying $\mathscr{L}$ to (6–55), we obtain

$$ms^2 \mathscr{L}[y] + k\mathscr{L}[y] = \mathscr{L}[\delta(t)],$$

and

$$\mathscr{L}[y] = \frac{1}{ms^2 + k}\, \mathscr{L}[\delta(t)].$$

This, then, must be the Laplace transform of $y_0$. But since we know that

$$\mathscr{L}[y_0] = \frac{1}{ms^2 + k},$$

we are forced to conclude that

$$\mathscr{L}[\delta(t)] = 1. \qquad (6\text{–}56)$$

(For another approach that will yield the same result see Exercise 1 at the end of the section.) This apparent contradiction of Theorem 6–2 which asserts that $\mathscr{L}[f](s) \to 0$ as $s \to \infty$ is merely a reflection of the fact that $\delta$ is *not* a function in the usual sense of the term.

With this result in hand, together with the companion formula

$$\mathscr{L}[\delta(t - a)] = e^{-as} \qquad (6\text{–}57)$$

derived from the first shift theorem, we are in a position to solve initial-value problems involving impulse functions. The following example illustrates the technique.

**Example**  A spring in equilibrium with an object of mass $m$ is subjected at time $t = 0$ to the sinusoidal force $h(t) = A \sin \omega t$. At time $t = 9.3$ the mass is struck a sharp blow from below that instantaneously imparts 5 units of momentum to it. Describe the motion of the system.

*Solution.*  We must solve the initial-value problem

$$m \frac{d^2y}{dt^2} + ky = A \sin \omega t + 5\,\delta(t - 9.3);$$

$$y(0) = y'(0) = 0.$$

---

* There is a minor difficulty with the value $y'(0) = 0$, namely $y_0'(t)$ does not exist when $t = 0$. This difficulty can be circumvented by setting $y_0(t) = 0$ for $t \leqslant 0$ and agreeing that $y'(0) = \lim_{t \to 0^-} y_0'(t)$.

Applying the Laplace transform, we have

$$(ms^2 + k)\mathscr{L}[y] = \frac{A\omega}{s^2 + \omega^2} + 5e^{-9.3s}$$

or

$$\mathscr{L}[y] = \frac{A\omega}{(ms^2 + k)(s^2 + \omega^2)} + \frac{5e^{-9.3s}}{ms^2 + k}.$$

Thus

$$y = y_1 + y_2,$$

where

$$y_1 = \mathscr{L}^{-1}\left[\frac{A\omega}{(ms^2 + k)(s^2 + \omega^2)}\right], \qquad y_2 = 5\mathscr{L}^{-1}\left[\frac{e^{-9.3s}}{ms^2 + k}\right].$$

The first of these inverses was computed earlier in the section, while the second can be found by recalling that

$$\frac{1}{\sqrt{km}}\mathscr{L}\left[\sin\sqrt{\frac{m}{k}}\,t\right] = \frac{1}{ms^2 + k},$$

and applying the second shift theorem. It is

$$y_2 = \frac{5}{\sqrt{mk}}u_{9.3}(t)\sin\sqrt{\frac{k}{m}}\,(t - 9.3).$$

Hence, when $\omega \neq \sqrt{k/m}$,

$$y = \frac{A\omega}{k - m\omega^2}\left[\frac{1}{\omega}\sin\omega t - \sqrt{\frac{m}{k}}\sin\sqrt{\frac{k}{m}}\,t\right] + \frac{5}{\sqrt{mk}}u_{9.3}(t)\sin\sqrt{\frac{k}{m}}\,(t - 9.3).$$

Graphs of this solution and its derivative are sketched in Fig. 6–17. Note the discontinuity in the graph of $y'$ at the instant the impulse is applied, a fact predicted by our earlier analysis.

## EXERCISES

1. Show that the definition

$$\mathscr{L}[\delta(t)] = \lim_{\varepsilon \to 0}\mathscr{L}[h_\varepsilon(t)],$$

   where $h_\varepsilon$ is defined by (6–49), leads to the formula

$$\mathscr{L}[\delta(t)] = 1.$$

2. Verify that the limit given in (6–51) is correct.

3. Solve the initial-value problem

$$y''' + y = e^t + \delta(t - 1);$$
$$y(0) = 1, \qquad y'(0) = 1, \qquad y''(0) = 2.$$

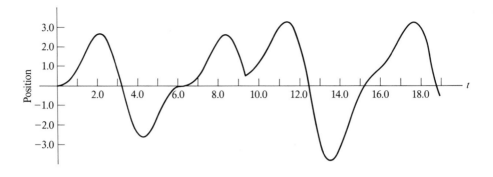

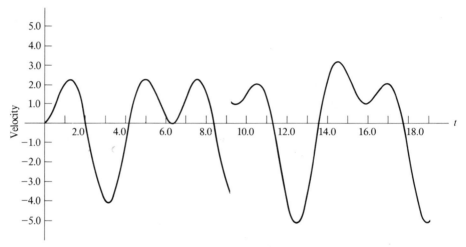

**Figure 6–17**

4.  The weighted spring discussed in this section is in equilibrium. At time $t = 0$ it is struck from above and instantaneously given two units of momentum. At time $t = a$ the mass is also subjected to the forcing function $\sin(t - a)$. Find the equation of motion of the system under the assumption that $m$ is different from the spring constant $k$.

5.  An object of mass 1 is attached to a spring whose spring constant is 4, and at time $t = 0$ is struck a blow from below that instantaneously imparts 1 unit of momentum to it. At time $t = \pi/2$ a force of magnitude $-\sin(t - (\pi/2))$ begins to act on the system. Find the equation of motion.

6.  A mass $m$ is suspended on a spring beneath a car which is moving with constant velocity $v$ along the track shown in Fig. 6–18. Suppose that at time $t = 0$ the mass is struck from below and instantaneously given one additional unit of momentum. Find the equation for the motion of $m$ in the vertical direction as a function of time.

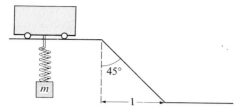

**Figure 6–18**

7. A unit mass is attached to a rigid spring whose spring constant is 3; it is then mounted in an elevator as shown in Fig. 6–19. At time $t = 0$ the elevator begins to descend with a constant velocity of 2 ft/sec, and at that moment the mass is struck a blow from above which instantaneously gives it one unit of momentum. Find the equation of motion of the mass as a function of time.

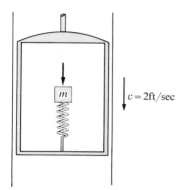

**Figure 6–19**

\* 8. Let $f$ be continuous at $t = a \geqslant 0$. Prove that

$$\int_0^\infty f(t)\, \delta(t - a)\, dt = f(a).$$

[*Hint*: Replace $\delta(t - a)$ by the function obtained by translating $h_\varepsilon$ as defined in the text $a$ units to the right, and take the limit as $\varepsilon \to 0$.]

*Remark.* This result generalizes the formula

$$\mathscr{L}[\delta(t - a)] = \int_0^\infty e^{-st}\, \delta(t - a)\, dt = e^{-as}.$$

9. Use the result of Exercise 8 to prove that $\delta(t)$ is the multiplicative identity for convolution:

$$\delta(t) * f(t) = f(t)$$

for all piecewise continuous $f$ of exponential order.

### 6–9  THE PRINCIPLE OF CONSERVATION OF ENERGY: THE TAUTOCHRONE

In this section we shall solve a famous classical problem by means of the convolution integral and the principle of **conservation of energy**:

> *The sum of the kinetic and potential energies of any dynamical system is a constant.*

We recall that the *kinetic energy* of an object of mass $m$ moving with velocity of magnitude $v$ is by definition

$$\tfrac{1}{2}mv^2, \tag{6–58}$$

and that near the surface of the earth the *potential energy* of an object of mass $m$ that is influenced only by the earth's gravitational field is

$$mgh, \tag{6–59}$$

where $h$ is the distance of the object above the position at which it would come to rest.

**The Tautochrone Problem**    A bead slides from rest under the influence of gravity down a smooth plane curve from the point $(x_0, y_0) \neq (0, 0)$ to the origin. Find the curve, if one exists, with the property that the time of descent of the bead is independent of its starting point. Such a curve is called a *tautochrone* from the Greek *tauto*, meaning "same," and *chronos*, meaning "time."

*Solution.*    Let $(x, y)$ be the position of the bead at some intermediate time before it reaches the origin and, as shown in Fig. 6–20, let $\sigma$ denote arc length along the curve measured from $(0, 0)$ to $(x, y)$. (We use the variable $\sigma$ for arc length instead of $s$ to avoid confusion with Laplace transforms.) We now apply the principle of conservation of energy in the form which asserts that the gain in kinetic energy of the bead must equal its loss of potential energy in falling from $(x_0, y_0)$ to $(x, y)$,

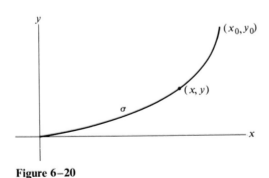

**Figure 6–20**

to obtain

$$\frac{1}{2} m \left(\frac{d\sigma}{dt}\right)^2 = mg(y_0 - y).$$

Hence

$$\frac{d\sigma}{dt} = -\sqrt{2g} \sqrt{y_0 - y}, \tag{6–60}$$

where the minus sign indicates that $\sigma$ is decreasing with increasing $t$. This equation can be rewritten as

$$\frac{d\sigma}{\sqrt{y_0 - y}} = -\sqrt{2g} \, dt, \tag{6–61}$$

and in this form it suggests that we view $\sigma$ as a function of $y$. With $\sigma = F(y)$, we rewrite (6–61) as

$$\frac{F'(y)}{\sqrt{y_0 - y}} \, dy = -\sqrt{2g} \, dt$$

and integrate over the (constant) duration $T_0$ of descent to obtain

$$\int_0^{y_0} \frac{F'(y)}{\sqrt{y_0 - y}} \, dy = -\sqrt{2g} \int_{T_0}^0 dt$$

or

$$T_0 = \frac{1}{\sqrt{2g}} \int_0^{y_0} (y_0 - y)^{-1/2} F'(y) \, dy.* \tag{6–62}$$

But (6–62) is the convolution of the functions $y^{-1/2}$ and $F'(y)$. Thus

$$\mathcal{L}[T_0] = \frac{1}{\sqrt{2g}} \mathcal{L}[y^{-1/2}] \mathcal{L}[F']$$

and

$$\mathcal{L}[F'] = \sqrt{2g} \, \frac{\mathcal{L}[T_0]}{\mathcal{L}[y^{-1/2}]}.$$

Next, we use the fact that

$$\mathcal{L}[T_0] = T_0/s,$$

and the formula

$$\mathcal{L}[y^{-1/2}] = \sqrt{\pi/s}, \tag{6–63}$$

proved in Exercise 2 at the end of the section, to obtain

$$\mathcal{L}[F'] = \sqrt{\frac{2g}{\pi}} \, T_0 s^{-1/2}.$$

---

* This equation is a special case of *Abel's integral equation*, which is encountered in solving a generalized version of the tautochrone problem. For a simple and elegant discussion of this problem, see *Differential Equations, With Applications and Historical Notes*, by G. F. Simmons, McGraw-Hill, New York, 1972.

Thus

$$F' = \sqrt{\frac{2g}{\pi}}\, T_0 \mathscr{L}^{-1}\left[\frac{1}{s^{1/2}}\right],$$

and (6–63) yields

$$F' = \frac{\sqrt{2g}}{\pi}\, T_0 y^{-1/2}. \tag{6–64}$$

From this the solution can be obtained with ease.

Indeed, since $\sigma = F(y)$ measures arc length,

$$F'(y) = \frac{d\sigma}{dy} = \sqrt{1 + \left(\frac{dx}{dy}\right)^2},$$

and (6–64) yields

$$1 + \left(\frac{dx}{dy}\right)^2 = \frac{2gT_0}{\pi^2} \cdot \frac{1}{y}$$

or

$$\frac{dx}{dy} = \sqrt{\frac{k - y}{y}}, \tag{6–65}$$

where $k = (2gT_0^2)/\pi^2$. To solve Eq. (6–65), we make the substitution

$$y = k \sin^2 \frac{\theta}{2}$$

to obtain

$$dx = k \cos^2 \frac{\theta}{2}\, d\theta.$$

Hence

$$x = k \int \cos^2 \frac{\theta}{2}\, d\theta = \frac{k}{2} \int (1 + \cos\theta)\, d\theta$$

$$= \frac{k}{2}(\theta + \sin\theta) + c.$$

Finally, since the curve must pass through the origin, and $y = 0$ when $\theta = 0$, we must have $c = 0$. Thus

$$x = \frac{k}{2}(\theta + \sin\theta), \qquad y = \frac{k}{2}(1 - \cos\theta). \tag{6–66}$$

You may recognize (6–66) as the parametric equations of the *cycloid* shown in Fig. 6–21. This curve is generated by a point on the circumference of a circle of radius $k/2$ as the circle rolls without slipping along the line $y = k$. Thus the tautochrone is an arch of a cycloid the radius of whose generating circle is determined by $T_0$, the constant time of descent.

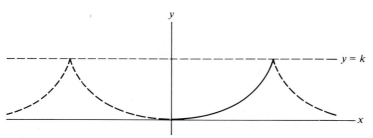

**Figure 6–21**

## EXERCISES

1. Verify that (6–66) are the parametric equations of the cycloid shown in Fig. 6–21 by deriving the equations of that curve.

2. Prove that

$$\mathscr{L}\left[\frac{1}{\sqrt{t}}\right] = \sqrt{\frac{\pi}{s}}.$$

[*Hint*: Make the substitution $t = x^2$ to deduce that

$$\mathscr{L}\left[\frac{1}{\sqrt{t}}\right] = 2\int_0^\infty e^{-sx^2}\,dx.$$

Then observe that

$$\left(\int_0^\infty e^{-sx^2}\,dx\right)^2 = \int_0^\infty e^{-sx^2}\,dx \int_0^\infty e^{-sy^2}\,dy = \int_0^\infty \int_0^\infty e^{-s(x^2+y^2)}\,dx\,dy,$$

and shift to polar coordinates.]

3. A certain factory requires $N(t)$ machines to be in operation at time $t$. It is known that in any time interval $[t_0, t_0 + t]$ only a fraction $F(t)$ of the machines that were new at time $t_0$ will still be in operation at time $t_0 + t$. We wish to find a replacement function $R$, where $R(t)$ is the number of machine replacements needed in the time interval $[0, t]$. Assume that the number of machines is so large that $N$, $F$, and $R$ may as well be continuous.

   a) Under the assumption that $R$ is differentiable show that the number of replacements required in the time interval $[v_k, v_k + \Delta_k v]$ is approximately

   $$R'(v_k)\,\Delta_k v,$$

   where $\Delta_k v = v_{k+1} - v_k$, and $0 < v_1 < \cdots < v_n = t$.

   b) How many of the replacements in (a) will still be in operation at time $t$?

   c) Given that all of the machines in the factory were new at time $t = 0$, use the result in (b) to show that

   $$N(t) \approx N(0)F(t) + \sum_{k=1}^{n} F(t - v_k)R'(v_k)\,\Delta_k v,$$

   and hence, in the limit,

   $$N(t) = N(0)F(t) + \int_0^t F(t - v)R'(v)\,dv.$$

d) Use the equation in (c) to find a formal expression for $\mathscr{L}[R]$ given that $R(0) = 0$.
e) Find $R$ when $N(t) = N_0$, a constant, and $F(t) = e^{-at}$, $a > 0$.
f) Find a formal expression for $R(t)$ when $N(t)$ is arbitrary and every machine has the same finite life span $L$.
g) Specialize the result in (f) to the case where $N(t) = N_0$, a constant.


## 6–10  TABLE OF LAPLACE TRANSFORMS

## A.  General Properties of the Transform

| Function | Transform |
|---|---|
| $f(t)$ | $\mathscr{L}[f] = \int_0^\infty e^{-st} f(t)\, dt$ |
| $\alpha f(t) + \beta g(t)$ | $\alpha \mathscr{L}[f] + \beta \mathscr{L}[g]$ |
| $f'(t)$ | $s\mathscr{L}[f] - f(0^+)$ |
| $f''(t)$ | $s^2 \mathscr{L}[f] - sf(0^+) - f'(0^+)$ |
| $f^{(n)}(t)$ | $s^n \mathscr{L}[f] - s^{n-1} f(0^+)$ $-s^{n-2} f'(0^+) - \cdots - f^{(n-1)}(0^+)$ |
| $\int_0^t f(t)\, dt$ | $\dfrac{1}{s} \mathscr{L}[f]$ |
| $\underbrace{\int_0^t \cdots \int_0^t f(t)\, dt \cdots dt}_{n \text{ times}}$ | $\dfrac{1}{s^n} \mathscr{L}[f]$ |
| $f(at)$ | $\dfrac{1}{a} \mathscr{L}[f]\left(\dfrac{s}{a}\right)$ |
| $e^{at} f(t)$ | $\mathscr{L}[f](s - a)$ |
| $u_a(t) f(t) = \begin{cases} 0, & 0 \leqslant t \leqslant a \\ f(t), & t > a \end{cases}$ | $e^{-as} \mathscr{L}[f(t + a)]$ |
| $u_a(t) f(t - a) = \begin{cases} 0, & 0 \leqslant t \leqslant a \\ f(t - a), & t > a \end{cases}$ | $e^{-as} \mathscr{L}[f]$ |
| $f(t)$ periodic with period $p$ | $\dfrac{\int_0^p e^{-st} f(t)\, dt}{1 - e^{-ps}}$ |
| $\int_0^t f(t - \xi) g(\xi)\, d\xi$ | $\mathscr{L}[f]\mathscr{L}[g]$ |
| $\dfrac{f(t)}{t}$  if $\lim\limits_{t \to 0^+} \dfrac{f(t)}{t}$ exists | $\int_s^\infty \mathscr{L}[f]\, ds$ |

## B. Table of Special Transforms

| Function | Transform |
|---|---|
| 1 | $\dfrac{1}{s}$ |
| $t^n$, $n$ a nonnegative integer | $\dfrac{n!}{s^{n+1}}$ |
| $t^\alpha$, $\alpha > -1$ | $\dfrac{\Gamma(\alpha + 1)}{s^{\alpha+1}}$<br><br>$\Gamma$ the generalized factorial or gamma function (p. 297) |
| $e^{at}$ | $\dfrac{1}{s - a}$ |
| $t^n e^{at}$, $n$ a nonnegative integer | $\dfrac{n!}{(s - a)^{n+1}}$ |
| $\sin at$ | $\dfrac{a}{s^2 + a^2}$ |
| $\cos at$ | $\dfrac{s}{s^2 + a^2}$ |
| $e^{at} \sin bt$ | $\dfrac{b}{(s - a)^2 + b^2}$ |
| $e^{at} \cos bt$ | $\dfrac{s - a}{(s - a)^2 + b^2}$ |
| $t \sin at$ | $\dfrac{2as}{(s^2 + a^2)^2}$ |
| $t \cos at$ | $\dfrac{s^2 - a^2}{(s^2 + a^2)^2}$ |
| $\sinh at$ | $\dfrac{a}{s^2 - a^2}$ |
| $\cosh at$ | $\dfrac{s}{s^2 - a^2}$ |
| $\dfrac{\sin at - at \cos at}{2a^3}$ | $\dfrac{1}{(s^2 + a^2)^2}$ |
| $\dfrac{\sin at + at \cos at}{2a}$ | $\dfrac{s^2}{(s^2 + a^2)^2}$ |
| $\dfrac{at \cosh at - \sinh at}{2a^3}$ | $\dfrac{1}{(s^2 - a^2)^2}$ |

| Function | Transform |
|---|---|
| $\dfrac{t \sinh at}{2a}$ | $\dfrac{s}{(s^2 - a^2)^2}$ |
| $\dfrac{\sinh at + at \cosh at}{2a}$ | $\dfrac{s^2}{(s^2 + a^2)^2}$ |
| $J_0(at)$<br>$J_0$ the Bessel function of order zero of the first kind (p. 297) | $\dfrac{1}{\sqrt{s^2 + a^2}}$ |
| $\delta(t)$ | $1$ |
| $\delta(t - a)$ | $e^{-as}$ |

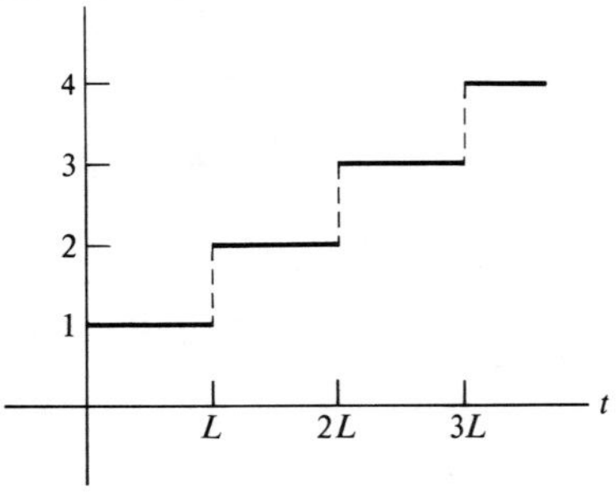

**Figure 6–22**

The staircase function

$$\sum_{n=0}^{\infty} u_{nL}(t - nL) \qquad\qquad \frac{1}{s(1 - e^{-Ls})}$$

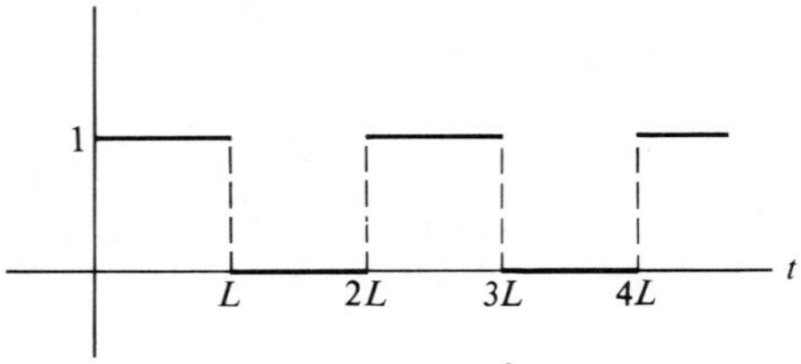

**Figure 6–23**

Assymetric square wave

$$\sum_{n=0}^{\infty} (-1)^n u_{nL}(t - nL) \qquad\qquad \frac{1}{s(1 + e^{-Ls})}$$

# Series
# Solutions

**7**

## 7–1 POWER SERIES

In Chapter 5 we reduced the task of solving a homogeneous linear differential equation with constant coefficients to a problem in algebra. Thus, in theory at least, we succeeded in solving all such equations. Except for a few very special cases, no reduction like this is possible for equations with variable coefficients. In fact, one of the distinctive features of equations with variable coefficients is that their solutions usually cannot be expressed in closed form in terms of elementary functions. As a result, it is impossible to "solve" such equations according to the interpretation which considers a solution of a differential equation to be a finite formula involving familiar functions. Nevertheless, with persistence and ingenuity it is possible to gain enough information about the solutions of any such equation to describe their behavior in detail, and to use them effectively in the analysis of other problems. The time has therefore come to drop the artificial restriction of seeking solutions within some preassigned collection of "known" functions, and to adopt instead the more liberal point of view which sees the solutions of a differential equation as functions *defined* by the equation itself and uses the equation to determine their properties.

One of the most powerful techniques available for studying functions defined by differential equations is to produce power series expansions of their solutions when such expansions exist. This is the technique we now propose to investigate, and we begin by recalling some basic facts about power series.

**I.** An expression of the form

$$\sum_{n=0}^{\infty} a_n x^n = a_0 + a_1 x + a_2 x^2 + \cdots, \qquad (7–1)$$

in which the $a_n$ are constants, is called a **power series in** $x$, while

$$\sum_{n=0}^{\infty} a_n (x - x_0)^n, \qquad (7–2)$$

in which $x_0$ is also a constant, is called a **power series in $x - x_0$**. Since (7–2) can always be transformed into (7–1) by the change of variable $u = x - x_0$, we shall restrict our attention for the most part to power series in $x$.

**II.** A power series in $x$ is said to **converge** when $x = x_1$ if

$$\sum_{n=0}^{\infty} a_n x_1^n \tag{7-3}$$

is a convergent series of real numbers, in which case (7–3) is called the **sum** of the series at $x_1$. Otherwise, the series is said to **diverge** at $x_1$.

Every power series in $x$ obviously converges when $x = 0$, and its sum at that point is $a_0$, the **constant term** of the series. More generally, every power series has an associated **radius of convergence** $R$, where $0 \leqslant R \leqslant \infty$, which is characterized by the property that the series converges when $|x| < R$ and diverges when $|x| > R$. In other words,

$$\sum_{n=0}^{\infty} a_n x^n$$

converges inside an interval of radius $R$ centered at 0 and diverges outside that interval. (Convergence or divergence at the endpoints of the interval must be determined on a case-by-case basis by examining the particular series in question.)

**III.** The radius of convergence of many power series can be found by means of the **ratio test**:

*If*

$$L = \lim_{n \to \infty} \left| \frac{a_{n+1} x_1^{n+1}}{a_n x_1^n} \right| = \lim_{n \to \infty} \left| \frac{a_{n+1}}{a_n} \right| |x_1|,$$

*then*

$$\sum_{n=0}^{\infty} a_n x_1^n$$

*converges when $L < 1$ and diverges when $L > 1$.*

This test immediately implies, for instance, that the series

$$\sum_{n=0}^{\infty} x^n = 1 + x + x^2 + \cdots,$$

converges for $|x| < 1$ and diverges for $|x| > 1$. Hence its radius of convergence is 1.

**IV.** A power series in $x$ with a positive radius of convergence $R$ defines a function $f$ in the interval $|x| < R$ by the rule

$$f(x) = a_0 + a_1 x + a_2 x^2 + \cdots = \sum_{n=0}^{\infty} a_n x^n. \tag{7-4}$$

This function is continuous and has derivatives of all orders everywhere in the interval. Moreover, these derivatives can be found by differentiating (7–4) term by term:

$$f'(x) = a_1 + 2a_2 x + 3a_3 x^2 + \cdots = \sum_{n=1}^{\infty} n a_n x^{n-1},$$

$$f''(x) = 2a_2 + 3 \cdot 2a_3 x + \cdots = \sum_{n=2}^{\infty} n(n-1) a_n x^{n-2},$$

(7–5)

and so on.

**V.** A function $f$ that can be represented by a convergent power series of the form

$$f(x) = \sum_{n=0}^{\infty} a_n (x - x_0)^n$$

for all $x$ in an open interval $I$ centered at $x_0$ is said to be **analytic at $x_0$**. In this case the coefficients of the series are *uniquely* determined by the formula

$$a_n = \frac{f^{(n)}(x_0)}{n!},$$

where $f^{(n)}(x_0)$ denotes the $n$th derivative of $f$ evaluated at $x_0$. In particular, if $f(x) = 0$ for all $x$ in $I$, then $a_n = 0$ for all $n$. Thus if

$$\sum_{n=0}^{\infty} a_n (x - x_0)^n = \sum_{n=0}^{\infty} b_n (x - x_0)^n$$

for all $x$ in $I$, then $a_n = b_n$ for all $n$ because

$$\sum_{n=0}^{\infty} a_n (x - x_0)^n - \sum_{n=0}^{\infty} b_n (x - x_0)^n = \sum_{n=0}^{\infty} (a_n - b_n)(x - x_0)^n = 0$$

for all $x$ in $I$. We shall use these facts repeatedly, though usually without explicit mention, in our work with differential equations.

It turns out that if $f$ is analytic at $x_0$, it is actually analytic at each point in its interval of convergence about $x_0$. Thus it is customary to speak of functions as being **analytic on an interval**, the phrase "analytic at $x_0$" being used only to direct attention to the point about which the series is expanded.

**VI.** Every polynomial in one variable is analytic on the entire real line, since

$$a_0 + a_1 x + \cdots + a_k x^k$$

can be viewed as a power series in $x$ with $a_n = 0$ for $n > k$. In fact, the notion of an analytic function can be seen as a generalization of the notion of a polynomial, and these two classes of functions have many properties in common. For instance, both are vector spaces in which addition and scalar multiplication are performed term by term, and in both there is a well-defined multiplication. Thus if $f$ and $g$ are analytic on an open interval $I$, then so are $f + g$, $\alpha f$ for any scalar $\alpha$, and $fg$. Moreover, if

$$f(x) = a_0 + a_1 x + \cdots = \sum_{n=0}^{\infty} a_n x^n$$

and

$$g(x) = b_0 + b_1 x + \cdots = \sum_{n=0}^{\infty} b_n x^n,$$

then $fg$ is computed according to the formula

$$(fg)(x) = a_0 b_0 + (a_0 b_1 + a_1 b_0)x + (a_0 b_2 + a_1 b_1 + a_2 b_0)x^2 + \cdots.$$

In other words,

$$(fg)(x) = \sum_{n=0}^{\infty} c_n x^n,$$

where

$$c_n = a_0 b_n + a_1 b_{n-1} + \cdots + a_n b_0.$$

**VII.** We have already observed that polynomials are analytic on the entire real line. So are the functions $e^x$, $\sin x$, and $\cos x$, and their power series expansions about $x = 0$ are

$$e^x = 1 + x + \frac{x^2}{2!} + \cdots = \sum_{n=0}^{\infty} \frac{x^n}{n!},$$

$$\sin x = x - \frac{x^3}{3!} + \frac{x^5}{5!} - \cdots = \sum_{n=0}^{\infty} (-1)^n \frac{x^{2n+1}}{(2n+1)!},$$

$$\cos x = 1 - \frac{x^2}{2!} + \frac{x^4}{4!} - \cdots = \sum_{n=0}^{\infty} (-1)^n \frac{x^{2n}}{(2n)!}.$$

These series are encountered so frequently that it is well worth the effort to remember them.

Rational functions (quotients of polynomials) are also analytic wherever they are defined. Thus

$$\frac{p(x)}{q(x)}$$

is analytic at $x = 0$ when $p$ and $q$ are polynomials and $q(0) \neq 0$. In more advanced work it is shown that the radius of convergence of the power series expansion of such a function about $x = 0$ is the distance *in the complex plane* from 0 to the nearest root of $q$. Thus, for instance, the power series expansion of

$$\frac{1}{x^2 + 1}$$

about $x = 0$ has radius of convergence 1 because $x^2 + 1$ has roots at $\pm i$.

**VIII.** Finally, we note that the **index of summation** $n$ in

$$\sum_{n=0}^{\infty} a_n x^n \tag{7–6}$$

is a "dummy variable," and can be changed whenever it is convenient to do so. For instance, if we replace $n$ by $n + 1$ in (7–6), we obtain

$$\sum_{n+1=0}^{\infty} a_{n+1} x^{n+1},$$

which can be rewritten as

$$\sum_{n=-1}^{\infty} a_{n+1} x^{n+1}.$$

This substitution has the effect of changing the index of summation in the original series by one. In our work with differential equations we shall use this maneuver to rewrite the formulas for the first and second derivatives of (7–6) as

$$f'(x) = \sum_{n=0}^{\infty} (n + 1)a_{n+1} x^n$$

and

$$f''(x) = \sum_{n=0}^{\infty} (n + 2)(n + 1)a_{n+2} x^n.$$

**EXERCISES**

1. a) Verify that the power series expansions for $e^x$, $\sin x$, and $\cos x$ given on p. 268 are correct.
   b) Prove that each of these series converges for all $x$.

2. a) Find the power series expansion of $\ln x$ about $x = 1$.
   b) What is the radius of convergence of the series in (a)?

3. Use the series

$$\frac{1}{1 + x} = 1 - x + x^2 - \cdots$$

   to deduce that

$$\frac{1}{(1 + x)^2} = 1 - 2x + 3x^2 - \cdots$$

   by (a) differentiation, and (b) squaring.

4. Assume that $\sin x$ and $\cos x$ are defined by means of their power series expansions.
   a) Prove that $D(\sin x) = \cos x$ and $D(\cos x) = -\sin x$.
   b) Use the results in (a) to show that $\sin x$ and $\cos x$ are solutions of $y'' + y = 0$.

5. Which of the following series are alternative versions of

$$\sum_{n=0}^{\infty} \frac{(-1)^n}{2^{2n}(n!)^2} x^{2n}?$$

a) $\displaystyle\sum_{n=1}^{\infty} \frac{(-1)^{n-1}}{4^{n-1}[(n-1)!]^2} x^{2(n-1)}$     b) $\displaystyle\sum_{n=2}^{\infty} \frac{(-1)^{n+2}}{4^{n+2}[(n+2)!]^2} x^{2(n+2)}$

c) $\displaystyle\sum_{n=0}^{\infty} \frac{(-1)^{2n}}{2^{4n}[(2n)!]^2} x^{4n}$     d) $16 \displaystyle\sum_{n=2}^{\infty} \frac{(-1)^n}{4^n[(n-2)!]^2} x^{2(n-2)}$

6. Rewrite each of the following series as directed.

a) $\displaystyle\sum_{n=2}^{\infty} \frac{(n+1)(n+4)}{n!} x^{n-2}$     as a series in $x^n$

b) $\displaystyle\sum_{n=0}^{\infty} \frac{2(n+1)}{n!} x^{n+1}$     as a series in $x^n$

c) $\displaystyle\sum_{n=3}^{\infty} \frac{2^n}{(n+1)!} x^{2(n+2)}$     as a series in $x^{2n}$

d) $\displaystyle\sum_{n=2}^{\infty} \frac{2n+1}{3} x^{2n-1}$     as a series in $x^{2n+1}$

e) $\displaystyle\sum_{n=1}^{\infty} \frac{(-1)^n}{2 \cdot 4 \cdots (2n)} x^{n-1}$     as a series in $x^{n+1}$

## 7–2 SERIES SOLUTIONS ABOUT AN ORDINARY POINT

We now resume the study of the equation

$$p(x)\frac{d^2y}{dx^2} + q(x)\frac{dy}{dx} + r(x)y = 0 \tag{7–7}$$

and impose the restriction that $p$, $q$, and $r$ be analytic on an open interval $I$ of the $x$-axis. As we shall see, the behavior of the solutions of (7–7) in a neighborhood of a point $x_0$ in $I$ depends in large measure on whether $p(x_0) = 0$ or not. In the former case $x_0$ is said to be a **singular point** for the equation; in the latter case it is said to be an **ordinary point**. We begin by considering solutions about ordinary points, the easier of the two cases.

When $p(x_0) \neq 0$, the continuity of $p$ implies the existence of an interval about $x_0$ in which $p(x) \neq 0$. Thus, if we restrict our attention to that interval, Eq. (7–7) can be rewritten as

$$\frac{d^2y}{dx^2} + P(x)\frac{dy}{dx} + Q(x)y = 0, \tag{7–8}$$

where $P(x) = q(x)/p(x)$ and $Q(x) = r(x)/p(x)$. In this form $x_0$ is an ordinary point for the equation if $P$ and $Q$ are analytic in an interval about $x_0$. The following theorem, which we state without proof, describes the solutions of (7–8) in this case.

**Theorem 7-1**   *If the coefficients P and Q of the equation*

$$\frac{d^2y}{dx^2} + P(x)\frac{dy}{dx} + Q(x)y = 0$$

*are analytic at $x_0$ and have power series expansions that converge in the interval $|x - x_0| < R$, then every solution of the equation is analytic at $x_0$, and its power series expansion also converges when $|x - x_0| < R$.* *

Although this theorem describes an interval about $x_0$ in which the power series expansion of a solution of (7–8) must converge, it makes no statement about the behavior of the series *outside* that interval. This omission, as we shall see, is inevitable because there are equations whose solutions have power series expansions that converge in a larger interval than the one described in the theorem.

Theorem 7–1 states that about $x_0$ the solutions of

$$\frac{d^2y}{dx^2} + P(x)\frac{dy}{dx} + Q(x)y = 0$$

can be written in the form

$$y = \sum_{n=0}^{\infty} a_n(x - x_0)^n,$$

where the $a_n$ are constants. The following examples show how the **method of undetermined coefficients** can be used to evaluate the $a_n$.

**Example 1**   Find the power series expansions about $x = 0$ of the general solution of

$$y'' + y = 0. \tag{7-9}$$

*Solution.*   By Theorem 7–1 the required solution has the form

$$y = \sum_{n=0}^{\infty} a_n x^n, \tag{7-10}$$

and converges for all $x$. Then

$$y'' = \sum_{n=2}^{\infty} n(n - 1)a_n x^{n-2},$$

and (7–9) implies that the $a_n$ must be chosen so that

$$\sum_{n=2}^{\infty} n(n - 1)a_n x^{n-2} + \sum_{n=0}^{\infty} a_n x^n = 0. \tag{7-11}$$

To facilitate collecting terms in this expression, we shift the index of summation in

---

\* A proof of this theorem can be found in the book by Coddington cited in the bibliography.

the first series and write the series in terms of $x^n$:

$$\sum_{n=2}^{\infty} n(n-1)a_n x^{n-2} = \sum_{n=0}^{\infty} (n+2)(n+1)a_{n+2} x^n.$$

Then (7–11) becomes

$$\sum_{n=0}^{\infty} (n+2)(n+1)a_{n+2} x^n + \sum_{n=0}^{\infty} a_n x^n = 0$$

or, by collecting like terms,

$$\sum_{n=0}^{\infty} [(n+2)(n+1)a_{n+2} + a_n]x^n = 0.$$

For this equation to be satisfied, *all* the coefficients of the series must be zero. Thus

$$(n+2)(n+1)a_{n+2} + a_n = 0$$

for all $n$, and it follows that

$$a_{n+2} = -\frac{a_n}{(n+2)(n+1)}, \qquad n = 0, 1, 2, \dots. \qquad (7-12)$$

This equation is an example of a **recurrence relation** or **finite difference equation**. It enables us to determine all the $a_n$ from $n = 2$ onward in terms of $a_0$ and $a_1$. They fall into two sets, depending on whether $n$ is even or odd, as follows:

| *n even* | *n odd* |
|---|---|
| $a_0$ | $a_1$ |
| $a_2 = -\dfrac{a_0}{2!}$ | $a_3 = -\dfrac{a_1}{3!}$ |
| $a_4 = -\dfrac{a_2}{4 \cdot 3} = \dfrac{a_0}{4!}$ | $a_5 = -\dfrac{a_3}{5 \cdot 4} = \dfrac{a_1}{5!}$ |
| $\vdots$ | $\vdots$ |
| $a_{2n} = -\dfrac{a_{2n-2}}{2n(2n-1)} = (-1)^n \dfrac{a_0}{(2n)!}$ | $a_{2n+1} = -\dfrac{a_{2n-1}}{(2n+1)(2n)} = (-1)^n \dfrac{a_1}{(2n+1)!}$ |

When we substitute these values in (7–10) and rearrange terms, we find that

$$y = a_0 \left(1 - \frac{x^2}{2!} + \frac{x^4}{4!} - \cdots\right) + a_1 \left(x - \frac{x^3}{3!} + \frac{x^5}{5!} - \cdots\right),$$

where $a_0$ and $a_1$ are arbitrary constants. This, of course, is just another way of writing

$$y = a_0 \cos x + a_1 \sin x.$$

**Example 2**  Find the general solution of

$$y'' + xy' + y = 0. \tag{7-13}$$

*Solution.*  Theorem 7–1 again guarantees that the solutions of this equation have power series expansions of the form

$$y = \sum_{n=0}^{\infty} a_n x^n$$

that converge for all values of $x$. We now substitute this series and its first two derivatives in the differential equation to obtain

$$\sum_{n=2}^{\infty} n(n-1)a_n x^{n-2} + \sum_{n=1}^{\infty} na_n x^n + \sum_{n=0}^{\infty} a_n x^n = 0.$$

Next, we shift the index of summation in the first series and collect like terms to obtain first

$$\sum_{n=0}^{\infty} (n+2)(n+1)a_{n+2} x^n + \sum_{n=1}^{\infty} na_n x^n + \sum_{n=0}^{\infty} a_n x^n = 0,$$

and then

$$a_0 + 2a_2 + \sum_{n=1}^{\infty} (n+1)[(n+2)a_{n+2} + a_n]x^n = 0.$$

Thus

$$a_0 + 2a_2 = 0,$$
$$(n+2)a_{n+2} + a_n = 0, \qquad n \geqslant 1,$$

and we are again in a position to determine all the $a_n$ in terms of $a_0$ and $a_1$. They are as follows:

| $n$ even | $n$ odd |
|---|---|
| $a_0$ | $a_1$ |
| $a_2 = -\dfrac{a_0}{2}$ | $a_3 = -\dfrac{a_1}{3}$ |
| $a_4 = -\dfrac{a_2}{4} = \dfrac{a_0}{2 \cdot 4}$ | $a_5 = -\dfrac{a_3}{5} = \dfrac{a_1}{3 \cdot 5}$ |
| . | . |
| . | . |
| . | . |
| $a_{2n} = (-1)^n \dfrac{a_0}{2 \cdot 4 \cdots (2n)}$ | $a_{2n+1} = (-1)^n \dfrac{a_1}{3 \cdot 5 \cdots (2n+1)}$ |

Thus

$$y = a_0 \left( 1 - \frac{x^2}{2} + \frac{x^4}{2 \cdot 4} - \cdots \right) + a_1 \left( x - \frac{x^3}{3} + \frac{x^5}{3 \cdot 5} - \cdots \right)$$

$$= a_0 \sum_{n=0}^{\infty} (-1)^n \frac{x^{2n}}{2^n n!} + a_1 \sum_{n=0}^{\infty} (-1)^n \frac{x^{2n+1}}{1 \cdot 3 \cdots (2n+1)},$$

where $a_0$ and $a_1$ are arbitrary constants, and the equation has been "solved."

At this point anyone not used to working with power series may wonder whether "solutions" of the type just produced have more than decorative value. They do, though the deeper reasons behind this fact depend on results from the theory of functions of a complex variable. It is worth mentioning, however, that a power series can be used to generate numerical approximations to the function it represents, and an estimate of the error incurred in making the approximation is usually easy to obtain by means of the remainder term in Taylor's formula. This, for instance, is how tables of values for functions are often produced.

In addition, the power series expansion of a solution of a differential equation often reveals properties of the solution that would be exceedingly difficult to discover by other means. For this reason an analysis of the functions defined by a differential equation usually begins with an attempt to produce a power series expansion of those functions. At the same time, however, we should point out that power series solutions are not always as easy to come by as the examples we have given might suggest. The difficulty arises in connection with the recurrence relation which can easily be too complicated to yield a formula for the coefficients of the series. Examples are given in the exercises that follow.

Finally, we note that the method of undetermined coefficients can be extended to include linear differential equations of any order, homogeneous or not, so long as every function in sight is analytic at $x_0$. And, of course, it can also be used to solve initial-value problems. Again examples are provided in the exercises.

**EXERCISES**

1. Use the method of undetermined coefficients to find the series expansion about $x = 0$ of the general solution of $y' - y = 0$.

2. Use the method of undetermined coefficients to find the series expansion about $x = 0$ of the general solution of $y' + y = x$.

3. a) Use the method of undetermined coefficients to find the series expansion about $x = 0$ of the solution of
$$(1 + x)y' - py = 0; \qquad y(0) = 1,$$
where $p$ is an arbitrary real number.

   b) What is the radius of convergence of the series in (a) when (i) $p$ is an integer? (ii) $p$ is not an integer?

c) The differential equation in (a) is a first-order linear equation. Solve it by standard methods, and use your solution to derive the *binomial formula* and the *binomial series*.

Find the power series expansion about $x = 0$ of the general solution of each equation in Exercises 4 through 10. What does Theorem 7–1 predict about the interval of convergence of each of the series?

4. $y'' - y = 0$

5. $(x^2 + 1)y'' - 4xy' + 6y = 0$

6. $(x^2 + 1)y'' - 6y = 0$

7. $y'' + x^2y' + 2xy = 0$

8. $(x^2 + 1)y'' + xy' - y = 0$

9. $y'' - 3xy = 0$

\* 10. $y''' - 3xy' - y = 0$

11. Find the first four nonzero terms in the power series expansion about $x = 0$ of the solution of
$$3y'' - y' + (x + 1)y = 1; \qquad y(0) = y'(0) = 0.$$

12. Find the first four nonzero terms in the power series expansion about $x = 0$ of the solution of
$$y'' + (\sin x)y = 0; \qquad y(0) = 1, \qquad y'(0) = 0.$$
[*Hint*: Expand $\sin x$ as a power series about $x = 0$.]

13. Find the first four nonzero terms in the power series expansion about $x = 0$ of the solution of
$$3y''' - xy' + x^2y = e^x;$$
$$y(0) = y'(0) = 0, \qquad y''(0) = \tfrac{1}{4}.$$
[*Hint*: Expand $e^x$ as a power series about $x = 0$.]

14. Find all the terms up to and including those of degree 4 in the power series expansion about $x = 1$ of the solution of
$$xy'' + y' + xy = 0; \qquad y(1) = 0, \qquad y'(1) = -1.$$

## \*7–3 LEGENDRE'S EQUATION AND THE LEGENDRE POLYNOMIALS

Over the past 200 years or so a small number of second-order linear differential equations with variable coefficients have been found to be of fundamental importance in applied mathematics. As a result, the solutions of these equations have been intensively studied, and out of this study has arisen one of the most fascinating branches of higher mathematics, the so-called *theory of special functions*. In a book at this level we can do no more than take a glimpse into this field. Yet, even that, inadequate as it will be to show the scope and variety of the subject, may prove worthwhile by illustrating how the techniques we have developed so far can be used in the sustained study of the solutions of an individual differential equation.

The equation we have chosen to examine in this section is

$$(1 - x^2)y'' - 2xy' + p(p + 1)y = 0, \qquad (7\text{–}14)$$

where $p$ is a nonnegative real number. Despite the fact that as a differential equation (7–14) is of order 2, it is known as **Legendre's equation of order $p$**. It has singular points at $x = \pm 1$ and ordinary points elsewhere. In particular, the origin is an ordinary point, and there the solutions of Legendre's equation have power series expansions that converge when $|x| < 1$. It is these expansions that we now seek. To this end, we set

$$y = \sum_{n=0}^{\infty} a_n x^n \tag{7-15}$$

and substitute the series and its first two derivatives in (7–14) to obtain

$$(1 - x^2) \sum_{n=2}^{\infty} n(n - 1)a_n x^{n-2} - 2x \sum_{n=1}^{\infty} na_n x^{n-1}$$

$$+ p(p + 1) \sum_{n=0}^{\infty} a_n x^n = 0,$$

or

$$\sum_{n=2}^{\infty} n(n - 1)a_n x^{n-2} - \sum_{n=2}^{\infty} n(n - 1)a_n x^n - \sum_{n=1}^{\infty} 2na_n x^n$$

$$+ \sum_{n=0}^{\infty} p(p + 1)a_n x^n = 0.$$

By shifting the index of summation in the first series in this expression we find that

$$\sum_{n=0}^{\infty} (n + 2)(n + 1)a_{n+2} x^n - \sum_{n=2}^{\infty} n(n - 1)a_n x^n - \sum_{n=1}^{\infty} 2na_n x^n$$

$$+ \sum_{n=0}^{\infty} p(p + 1)a_n x^n = 0.$$

Next, we collect like terms to obtain

$$2a_2 + p(p + 1)a_0 + [3 \cdot 2a_3 + (p + 2)(p - 1)a_1]x$$

$$+ \sum_{n=2}^{\infty} \{(n + 2)(n + 1)a_{n+2} + [p(p + 1) - n(n + 1)]a_n\}x^n = 0.$$

Finally, since

$$p(p + 1) - n(n + 1) = (p + n + 1)(p - n),$$

the preceding equation implies that

$$2a_2 + p(p + 1)a_0 = 0$$
$$3 \cdot 2a_3 + (p + 2)(p - 1)a_1 = 0$$
$$(n + 2)(n + 1)a_{n+2} + (p + n + 1)(p - n)a_n = 0, \qquad n \geqslant 2.$$

These relations allow us to determine all the $a_n$ from $n = 2$ onward in terms of $a_0$ and $a_1$, as follows:

| $n$ even | $n$ odd |
|---|---|

$$a_0 \qquad\qquad\qquad\qquad\qquad\qquad a_1$$

$$a_2 = -\frac{(p + 1)p}{2!}a_0 \qquad\qquad a_3 = -\frac{(p + 2)(p - 1)}{3!}a_1$$

$$a_4 = \frac{(p + 3)(p + 1)(p)(p - 2)}{4!}a_0 \qquad a_5 = \frac{(p + 4)(p + 2)(p - 1)(p - 3)}{5!}a_1$$

$$\vdots \qquad\qquad\qquad\qquad\qquad\qquad \vdots$$

Hence in the interval $|x| < 1$ the general solution of Legendre's equation of order $p$ is

$$y = a_0\left[1 - \frac{(p + 1)p}{2!}x^2 + \frac{(p + 3)(p + 1)(p)(p - 2)}{4!}x^4 - \cdots\right]$$
$$+ a_1\left[x - \frac{(p + 2)(p - 1)}{3!}x^3\right.$$
$$\left. + \frac{(p + 4)(p + 2)(p - 1)(p - 3)}{5!}x^5 - \cdots\right]. \quad (7\text{-}16)$$

Although the bracketed series in this solution are rather unwieldy, we can easily show that if $p$ is not an integer, then each of them converges when $|x| < 1$ and diverges when $|x| > 1$. (See Exercise 2. The series also diverge when $x = \pm 1$, though this is not so easy to prove.) When $p$ is an integer $n$, however, one of the series terminates and thus is a polynomial. In fact, it is a polynomial of degree $n$ involving only even powers of $x$ when $n$ is even and only odd powers when $n$ is odd. These polynomials provide examples of power series solutions of a linear differential equation whose radii of convergence exceed the prediction of Theorem 7–1.

**The Legendre Polynomials**    The polynomial solutions of Legendre's equation are of particular interest because they appear in the solutions of a surprising number of problems in mathematical physics, ranging from descriptions of temperature distributions in spherical regions to the structure of atoms. The first step in studying these polynomials is to find a reasonably efficient way of computing them, since the series expansions in (7–16) are anything but that. The following definition leads rather quickly to a formula that meets this need (see (7–19)).

**Definition 7–1**    *Let $P_0(x)$, $P_1(x)$, . . . be the sequence of polynomials defined by*

$$P_n(x) = \frac{1}{2^n n!}\frac{d^n}{dx^n}(x^2 - 1)^n. \qquad (7\text{-}17)$$

*Then $P_n(x)$ is called the **Legendre polynomial of degree n**, and (7–17) is known as **Rodrigues' formula for $P_n(x)$.***

(The coefficient $1/(2^n n!)$, which at this point merely seems to complicate (7–17), actually simplifies the task of working with the Legendre polynomials.) From (7–17) we find that

$$P_0(x) = 1, \qquad\qquad P_1(x) = x,$$
$$P_2(x) = \tfrac{3}{2}x^2 - \tfrac{1}{2}, \qquad\quad P_3(x) = \tfrac{5}{2}x^3 - \tfrac{3}{2}x,$$
$$P_4(x) = \tfrac{35}{8}x^4 - \tfrac{15}{4}x^2 + \tfrac{3}{8}, \qquad P_5(x) = \tfrac{63}{8}x^5 - \tfrac{35}{4}x^3 + \tfrac{15}{8}x.$$

The graphs of the first five of these polynomials are shown in Fig. 7–1.

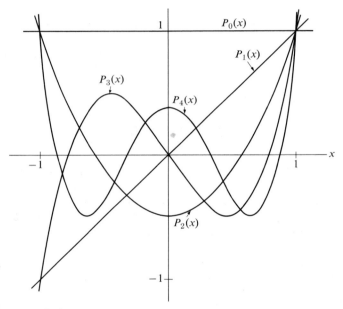

**Figure 7–1**

We begin our discussion of the Legendre polynomials by showing that they are in fact solutions of Legendre's equation.

Let

$$w = (x^2 - 1)^n,$$

and let $w^{(k)}$ denote the $k$th derivative of $w$. Then

$$w^{(1)} = 2nx(x^2 - 1)^{n-1},$$

and

$$(x^2 - 1)w^{(1)} - 2nxw = 0.$$

Repeated differentiation yields

$$(x^2 - 1)w^{(2)} - 2x(n - 1)w^{(1)} - 2nw = 0$$

. 
. 
. 

$$(x^2 - 1)w^{(k+2)} - 2x[n - (k + 1)]w^{(k+1)}$$
$$- 2[n + (n - 1) + \cdots + (n - k)]w^{(k)} = 0.$$

But since

$$n + (n - 1) + \cdots + (n - k) = \frac{(2n - k)(k + 1)}{2}$$

(Exercise 8), we have

$$(x^2 - 1)w^{(k+2)} - 2x[n - (k + 1)]w^{(k+1)} - (2n - k)(k + 1)w^{(k)} = 0.$$

We now set $k = n$ and multiply the resulting equation by $-1/(2^n n!)$. This yields

$$(1 - x^2)\frac{w^{(n+2)}}{2^n n!} - 2x\frac{w^{(n+1)}}{2^n n!} + n(n + 1)\frac{w^{(n)}}{2^n n!} = 0,$$

and since

$$P_n = \frac{w^{(n)}}{2^n n!},$$

we have

$$(1 - x^2)P_n'' - 2xP_n' + n(n + 1)P_n = 0.$$

This, of course, is just what we wished to show: $P_n$ is a solution of Legendre's equation of order $n$. Indeed, in view of the results established earlier in this section it is not difficult to prove that up to constant multiples $P_n$ is the *only* polynomial solution of Legendre's equation of order $n$.

Our next result, which may seem bizarre at first sight, is actually one of the most important properties of the Legendre polynomials.

***Theorem 7-2***    *If $m \neq n$, then*

$$\int_{-1}^{1} P_m(x)P_n(x) \, dx = 0. \tag{7-18}$$

*Proof.*    We start with the equations

$$(1 - x^2)P_m'' - 2xP_m' + m(m + 1)P_m = 0,$$
$$(1 - x^2)P_n'' - 2xP_n' + n(n + 1)P_n = 0,$$

multiply the first by $-P_n$, the second by $P_m$, and add to obtain

$$(1 - x^2)(P_m P_n'' - P_m'' P_n) - 2x(P_m P_n' - P_m' P_n)$$
$$= [m(m + 1) - n(n + 1)]P_m P_n.$$

But this equation can be written as

$$\frac{d}{dx}\left[(1 - x^2)(P_m P_n' - P_m' P_n)\right] = [m(m + 1) - n(n + 1)]P_m P_n,$$

and when we integrate both sides from $-1$ to 1 we see that

$$[m(m + 1) - n(n + 1)] \int_{-1}^{1} P_m(x)P_n(x)\, dx = 0.$$

Hence when $m \neq n$,

$$\int_{-1}^{1} P_m(x)P_n(x)\, dx = 0,$$

as asserted. ∎

One consequence of Theorem 7–2 is the following rather surprising fact.

**Corollary**  *Let Q be a polynomial of degree less than n, where $n \geq 1$. Then*

$$\int_{-1}^{1} P_n(x)Q(x)\, dx = 0.$$

*Proof.* The polynomials $P_0, P_1, \ldots, P_n$ are a basis for the vector space of polynomials of degree $\leq n$. (See Exercise 8 of Section 2–7.) Thus, if $Q$ is of degree $m$,

$$Q(x) = \alpha_0 P_0(x) + \cdots + \alpha_m P_m(x),$$

where $\alpha_0, \ldots, \alpha_m$ are constants, and it follows that

$$\int_{-1}^{1} P_n(x)Q(x)\, dx = \alpha_0 \int_{-1}^{1} P_n(x)P_0(x)\, dx + \cdots + \alpha_m \int_{-1}^{1} P_n(x)P_m(x)\, dx.$$

Since each of the integrals on the right-hand side of this equation vanishes when $n > m$,

$$\int_{-1}^{1} P_n(x)Q(x)\, dx = 0. \quad ∎$$

The following recurrence relation, whose proof we omit, is still another consequence of Theorem 7–2.

*For all integers $n > 0$,*

$$P_{n+1}(x) = \frac{2n + 1}{n + 1}\, xP_n(x) - \frac{n}{n + 1}\, P_{n-1}(x), \tag{7–19}$$

*where $P_0(x) = 1$ and $P_1(x) = x$.*

This identity plays a key role in the study of the Legendre polynomials because, as is indicated in the exercises, it is well suited to computation.

We conclude this discussion by establishing

**Theorem 7–3**  *The polynomial $P_n$ has n distinct zeros in the open interval $(-1, 1)$.*

*Proof.* Since $P_0(x) = 1$ the assertion is true for $n = 0$. Next, if $n \geqslant 1$, the corollary to Theorem 7–2 implies that

$$\int_{-1}^{1} P_n(x) \, dx = \int_{-1}^{1} P_n(x)P_0(x) \, dx = 0.$$

Hence $P_n$ must change sign *at least once* in the interval $(-1, 1)$.

Now let $x_1 < x_2 < \cdots < x_m$ be the distinct roots of $P_n$ in $(-1, 1)$, and consider the polynomial

$$Q(x) = (x - x_1) \cdots (x - x_m).$$

Observe that $m \leqslant n$ and that $P_n$ and $Q$ change sign at the same points of $(-1, 1)$, since $P_n$ has no repeated roots in that interval (Exercise 4). Hence the product $P_n(x)Q(x)$ is either positive or negative throughout the entire interval, and

$$\int_{-1}^{1} P_n(x)Q(x) \, dx \neq 0.$$

But by the Corollary to Theorem 7–2, this is impossible if $m < n$. Hence $m = n$, and the theorem is proved. ∎

## EXERCISES

### Series Solutions and Legendre's Equation

1. a) Prove that the function

$$\frac{x}{2} \ln\left(\frac{1 + x}{1 - x}\right) - 1$$

   is a solution of Legendre's equation of order one on $(-1, 1)$.

   b) Use the result in (a) to write the general solution of Legendre's equation of order one in closed form.

2. a) Prove that in the general solution of Legendre's equation of order $p$

$$a_{2n} = (-1)^n \frac{(p + 2n - 1)(p + 2n - 3) \cdots (p + 1)(p)(p - 2) \cdots (p - 2n + 2)}{(2n)!} a_0,$$

   and

$$a_{2n+1} = (-1)^{2n+1} \frac{(p + 2n)(p + 2n - 2) \cdots (p + 2)(p - 1) \cdots (p - 2n + 1)}{(2n + 1)!} a_1.$$

   b) Show that each of the bracketed series in (7–16), the general solution of Legendre's equation, converges for $|x| < 1$ and diverges for $|x| > 1$.

3. The second-order linear differential equation

$$y'' - 2xy' + 2py = 0,$$

   where $p$ is a nonnegative constant, is known as **Hermite's equation of order** $p$. It is encountered, among other places, in quantum mechanics.

a) Prove that $y$ is a solution of this equation if and only if $u = e^{-x^2/2}y$ is a solution of

$$u'' + (2p + 1 - x^2)u = 0.$$

b) Use the method of undetermined coefficients to find the series expansion of the general solution of Hermite's equation about $x = 0$.

c) Show that Hermite's equation of order $n$ has polynomial solutions for each integer $n \geqslant 0$.

4. The second-order linear differential equation

$$(1 - x^2)y'' - xy' + p^2y = 0,$$

where $p$ is a nonnegative constant, is known as **Chebyshev's equation of order $p$**.

a) Use the method of undetermined coefficients to find the series expansion of the general solution of Chebyshev's equation about $x = 0$.

b) Show that Chebyshev's equation of order $n$ has polynomial solutions for each integer $n \geqslant 0$.

5. Find a necessary and sufficient condition for a differential equation of the form

$$(x^2 + \alpha)y'' + \beta xy' + \gamma y = 0,$$

where $\alpha, \beta, \gamma$ are constants, to have a polynomial solution of degree $n$.

## The Legendre Polynomials

6. Use the recurrence relation for the Legendre polynomials to compute $P_n$ for $2 \leqslant n \leqslant 5$.

7. Use the recurrence relation for the Legendre polynomials to prove that for all integers $n \geqslant 0$:

a) $P_n(1) = 1$

b) $P_n(-1) = (-1)^n$

c) $P_n(0) = \begin{cases} 0, & n \text{ odd} \\ (-1)^{n/2} \dfrac{1 \cdot 3 \cdot 5 \cdots (n-1)}{2 \cdot 4 \cdot 6 \cdots n}, & n \text{ even.} \end{cases}$

8. Let $k$ and $n$ be nonnegative integers with $k \leqslant n$. Prove that

$$n + (n-1) + \cdots + (n - k) = \frac{(2n - k)(k + 1)}{2}.$$

[*Hint*: $1 + 2 + \cdots + n = n(n+1)/2$.]

9. At one point in the proof of Theorem 7–3 we asserted that $P_n$ has no repeated roots. Justify this assertion.

10. Let $w = (x^2 - 1)^n$, and let $w^{(n)}$ denote the $n$th derivative of $w$.

a) Use integration by parts to prove that

$$\int_{-1}^{1} w^{(n)}w^{(n)} \, dx = (2n!) \int_{-1}^{1} (1 - x)^n(1 + x)^n \, dx.$$

b) Prove that

$$\int_{-1}^{1} (1 - x)^n(1 + x)^n \, dx = \frac{(n!)^2}{(2n)!(2n + 1)} 2^{2n+1}.$$

c) Use the results obtained in parts (a) and (b) to deduce that

$$\int_{-1}^{1} [P_n(x)]^2 \, dx = \frac{2}{2n + 1}.$$

\*11. a) Let $x_0, \ldots, x_m$ be distinct numbers between $-1$ and 1. Find a polynomial $Q_i$ of degree $m$ that has zeros at $x_0, \ldots, x_{i-1}, x_{i+1}, \ldots, x_m$ and takes the value $a_i$ at $x_i$.
   b) Let $Q$ denote the sum of the polynomials $Q_i$ defined in part (a). Show that $Q(x_i) = a_i$ for $0 \leqslant i \leqslant m$, and prove that there is no other polynomial of degree $\leqslant m$ with this property.

12. Let $F(x)$ be *any* polynomial of degree $\leqslant 2m + 1$, and let $x_0, \ldots, x_m$ be the roots of $P_{m+1}$. Divide $F$ by $P_{m+1}$ to obtain

$$F(x) = P_{m+1}(x)P(x) + R(x),$$

where $R$ is either zero or of degree $\leqslant m$.
   a) Prove that

$$\int_{-1}^{1} F(x) \, dx = \int_{-1}^{1} R(x) \, dx.$$

   b) Prove that $R$ is the polynomial constructed in Exercise 11(b) if $a_i = F(x_i)$, $1 \leqslant i \leqslant m$.
   c) Prove that

$$\int_{-1}^{1} F(x) \, dx = \sum_{i=0}^{m} I_i F(x_i)$$

where $I_0, \ldots, I_m$ are constants that depend *only* on $x_0, \ldots, x_m$, respectively, and not on $F$.

## 7-4 SINGULAR POINTS

We recall that $x_0$ is a *singular point* for the equation

$$p(x)\frac{d^2 y}{dx^2} + q(x)\frac{dy}{dx} + r(x)y = 0 \qquad (7-20)$$

if $p$, $q$, and $r$ are analytic at $x_0$, and $p(x_0) = 0$. In general there is very little that can be said about the nature of the solutions of a differential equation about a singular point, as Exercises 6 and 7 at the end of this section suggest. There is, however, one important exception, namely the case where $x_0$ is a "regular" singular point in the sense of the following definition.

**Definition 7 – 2**   *A point $x_0$ is said to be a **regular singular point** for Eq. (7–20) if and only if the equation can be written in the form*

$$(x - x_0)^2 \frac{d^2 y}{dx^2} + (x - x_0)P(x)\frac{dy}{dx} + Q(x)y = 0, \qquad (7-21)$$

*where $P$ and $Q$ are analytic at $x_0$. A singular point that is not regular is said to be **irregular**.*

In the following discussion we shall limit ourselves to equations that have a regular singular point at the origin, in which case (7–21) becomes

$$x^2 y'' + xP(x)y' + Q(x)y = 0. \qquad (7\text{–}22)$$

As we observed earlier, this limitation involves no loss of generality since the change of variable $u = x - x_0$ will move a singularity from $x_0$ to 0.

**Example 1**    The simplest example of a second-order equation with a regular singular point at the origin is the Euler equation

$$x^2 y'' + axy' + by = 0,$$

where $a$ and $b$ are (real) constants. We have seen that the solution space of this equation is spanned by the pair of functions $y_1$ and $y_2$ constructed from the roots of the equation

$$v(v - 1) + av + b = 0$$

according to the following scheme. (See p. 187.)

| Roots $v_1$, $v_2$ | Solutions $y_1$, $y_2$ |
|---|---|
| Real, $v_1 \neq v_2$ | $y_1(x) = \lvert x \rvert^{v_1}$, $y_2(x) = \lvert x \rvert^{v_2} \cdot$ |
| Real, $v_1 = v_2 = v$ | $y_1(x) = \lvert x \rvert^{v}$, $y_2(x) = \lvert x \rvert^{v} \ln \lvert x \rvert$ |
| Complex conjugates: | $y_1(x) = \lvert x \rvert^{\alpha} \sin (\beta \ln \lvert x \rvert)$ |
| $v_1 = \alpha + \beta i,\ v_2 = \alpha - \beta i$ | $y_2(x) = \lvert x \rvert^{\alpha} \cos (\beta \ln \lvert x \rvert)$ |

Except for the fact that they can be expressed in closed form, the solutions of Euler's equation are typical of the solutions of second-order linear differential equations with a regular singular point at the origin. In the next two sections we shall see that about $x = 0$ the solution space of such an equation is always spanned by a pair of functions that depend on the roots of a polynomial of degree two. These functions usually involve powers of $\lvert x \rvert$ when $x < 0$ and, under certain circumstances, a logarithmic term as well. The following example illustrates the technique, known as the **method of Fröbenius**, that is used to find these solutions.

**Example 2**    Find the general solution of

$$x^2 y'' + x(x - \tfrac{1}{2})y' + \tfrac{1}{2}y = 0 \qquad (7\text{–}23)$$

on $(0, \infty)$ and $(-\infty, 0)$.

*Solution.*    We begin by considering the interval $x > 0$, where we seek a solution of the form

$$y = x^{v} \sum_{n=0}^{\infty} a_n x^n = \sum_{n=0}^{\infty} a_n x^{n+v}, \qquad (7\text{–}24)$$

where $a_0 \neq 0$ and $v$ is arbitrary. (This particular guess as to the form of a solution is motivated by our study of series solutions and the results obtained for the

Euler equation.) Then

$$y' = \sum_{n=0}^{\infty} (n + v)a_n x^{n+v-1} = x^v \sum_{n=0}^{\infty} (n + v)a_n x^{n-1},$$

$$y'' = \sum_{n=0}^{\infty} (n + v)(n + v - 1)a_n x^{n+v-2}$$

$$= x^v \sum_{n=0}^{\infty} (n + v)(n + v - 1)a_n x^{n-2},$$

and (7–23) implies that

$$\sum_{n=0}^{\infty} (n + v)(n + v - 1)a_n x^n + \sum_{n=0}^{\infty} (n + v)a_n x^{n+1}$$

$$- \frac{1}{2} \sum_{n=0}^{\infty} (n + v)a_n x^n + \frac{1}{2} \sum_{n=0}^{\infty} a_n x^n = 0.$$

But since

$$\sum_{n=0}^{\infty} (n + v)a_n x^{n+1} = \sum_{n=1}^{\infty} (n + v - 1)a_{n-1} x^n,$$

the preceding expression can be rewritten as

$$\left[ v(v - 1) - \tfrac{1}{2}v + \tfrac{1}{2} \right]a_0 + \sum_{n=1}^{\infty} \left[ (n + v)(n + v - 1) - \tfrac{1}{2}(n + v) + \tfrac{1}{2} \right]a_n x^n$$

$$+ \sum_{n=1}^{\infty} (n + v - 1)a_{n-1} x^n = 0.$$

Thus, if (7–24) is to be a solution of (7–23), we must have

$$v(v - 1) - \tfrac{1}{2}v + \tfrac{1}{2} = 0, \tag{7–25}$$

since by assumption $a_0 \neq 0$, and

$$\left[ (n + v)(n + v - 1) - \tfrac{1}{2}(n + v) + \tfrac{1}{2} \right]a_n + (n + v - 1)a_{n-1} = 0 \tag{7–26}$$

when $n \geq 1$.

Equation (7–25) determines the admissible values of $v$ for this problem as $\tfrac{1}{2}$ and 1, and is known as the **indicial equation** associated with (7–23). Substituting these values in (7–26), we find that when $v = \tfrac{1}{2}$,

$$a_n = -\frac{a_{n-1}}{n},$$

and when $v = 1$,

$$a_n = -\frac{2}{2n + 1} a_{n-1}.$$

From these we obtain the following two sets of values for the $a_n$, both expressed in terms of $a_0$:

| $v = \frac{1}{2}$ | | $v = 1$ |
|---|---|---|
| $a_1 = -a_0$ | | $a_1 = -\frac{2}{3} a_0$ |
| $a_2 = \dfrac{a_0}{2!}$ | | $a_2 = \dfrac{2^2}{3 \cdot 5} a_0$ |
| $a_3 = -\dfrac{a_0}{3!}$ | | $a_3 = -\dfrac{2^3}{3 \cdot 5 \cdot 7} a_0$ |
| . | | . |
| . | | . |
| . | | . |

Finally, setting $a_0 = 1$, we conclude that each of the following series *formally* satisfies Eq. (7–23):

$$y_1 = x^{1/2} \left( 1 - x + \frac{x^2}{2!} - \frac{x^3}{3!} + \cdots \right),$$

$$y_2 = x \left( 1 - \frac{2x}{3} + \frac{(2x)^2}{3 \cdot 5} - \frac{(2x)^3}{3 \cdot 5 \cdot 7} + \cdots \right).$$

At this point we have proved that *if*

$$x^2 y'' + x(x - \tfrac{1}{2})y' + \tfrac{1}{2}y = 0$$

has a solution of the form

$$y = x^v \sum_{n=0}^{\infty} a_n x^n,$$

with $a_0 = 1$, then this solution *must* be one of the two series found above. As yet, however, we have no guarantee that either of these series actually is a solution of the given equation, since it is conceivable that both of them might diverge for all $x > 0$. (This is what we meant a moment ago when we said that these series "formally" satisfy (7–23)). Fortunately, an easy computation by the ratio test disposes of this difficulty. Both series converge for all $x > 0$, and since $y_1$ and $y_2$ are also linearly independent in $\mathscr{C}(0, \infty)$, the general solution of (7–23) for $x > 0$ is

$$y = c_1 y_1 + c_2 y_2,$$

where $c_1$ and $c_2$ are arbitrary constants.

Finally, to remove the restriction on the interval, we observe that the preceding argument still holds if we replace $x^\nu$ by $|x|^\nu$ throughout; that is, by $(-x)^\nu$ for $x < 0$ (Exercise 9). Thus the general solution of $(7\text{–}23)$ on any interval not containing the origin is

$$y = c_1 |x|^{1/2} \sum_{n=0}^{\infty} (-1)^n \frac{x^n}{n!} + c_2 |x| \sum_{n=0}^{\infty} (-1)^n \frac{(2x)^n}{1 \cdot 3 \cdot 5 \cdots (2n+1)}.$$

## EXERCISES

Find and classify the singular points for the equations in Exercises 1 through 5.

1. $x^3(x^2 - 1)y'' - x(x + 1)y' - (x - 1)y = 0$
2. $(x^4 - 1)y'' + xy' = 0$
3. $x^3(x - 1)y'' + (x - 1)y' + 2xy = 0$
4. $x(1 - x)y'' + (1 - 5x)y' - 4y = 0$
5. Legendre's equation:
$$(1 - x^2)y'' - 2xy' + p(p + 1)y = 0.$$

6. a) Verify that $y = x$ is a solution of

$$y'' + \frac{1}{x^2} y' - \frac{1}{x^3} y = 0,$$

   and then use the technique of reduction of order to find the general solution of the equation.
   b) Discuss the behavior of the general solution in (a) as $x \to 0$.

7. Show that the equation

$$x^3 y'' + y = 0$$

   does not have a nontrivial series solution of the form

$$\sum_{n=0}^{\infty} a_n x^{n+\nu}$$

   for any real number $\nu$.

8. a) Verify that both of the series

$$y_1 = x^{1/2} \left( 1 - x + \frac{x^2}{2!} - \frac{x^3}{3!} + \cdots \right)$$

   and

$$y_2 = x \left( 1 - \frac{2x}{3} + \frac{(2x)^2}{3 \cdot 5} - \frac{(2x)^3}{3 \cdot 5 \cdot 7} + \cdots \right)$$

   converge for all $x > 0$.
   b) Show that the functions $y_1$ and $y_2$ defined in (a) are linearly independent in $\mathscr{C}(0, \infty)$.

9. Make the change of variable

$$x = -t, \qquad x < 0,$$

in (7–23), and deduce that the general solution of the equation on $(-\infty, 0)$ is

$$y = c_1|x|^{1/2} \sum_{n=0}^{\infty} (-1)^n \frac{x^n}{n!} + c_2|x| \sum_{n=0}^{\infty} (-1)^n \frac{(2x)^n}{1 \cdot 3 \cdots (2n + 1)}.$$

Use the method of Fröbenius to find two linearly independent solutions of each of the following equations for $x > 0$.

10. $2x^2y'' + xy' - y = 0$

11. $9x^2y'' + 3x(x + 3)y' - (4x + 1)y = 0$

12. $xy'' + \dfrac{x + 1}{2} y' - y = 0$

13. $8x^2y'' - 2x(x - 1)y' + (x + 1)y = 0$

14. $4x^2y'' + x(2x + 7)y' + 6y = 0$

## 7–5  SOLUTIONS ABOUT A REGULAR SINGULAR POINT: THE GENERAL CASE

In this section we shall indicate how the method of Fröbenius can always be used to obtain *at least one* solution of

$$x^2y'' + xP(x)y' + Q(x)y = 0 \qquad (7\text{–}27)$$

about $x = 0$ whenever $P$ and $Q$ are analytic at that point. Once again we begin by letting $x$ be positive, in which case we seek a solution of (7–27) of the form

$$y = x^v \sum_{n=0}^{\infty} a_n x^n \qquad (7\text{–}28)$$

with $a_0 \neq 0$. Then

$$y' = x^v \sum_{n=0}^{\infty} (n + v)a_n x^{n-1},$$

$$y'' = x^v \sum_{n=0}^{\infty} (n + v)(n + v - 1)a_n x^{n-2},$$

and (7–27) yields

$$\sum_{n=0}^{\infty} (n + v)(n + v - 1)a_n x^n + P(x) \sum_{n=0}^{\infty} (n + v)a_n x^n$$

$$+ Q(x) \sum_{n=0}^{\infty} a_n x^n = 0. \qquad (7\text{–}29)$$

But since $P$ and $Q$ are analytic at $x = 0$ it follows that

$$P(x) = \sum_{n=0}^{\infty} p_n x^n \qquad \text{and} \qquad Q(x) = \sum_{n=0}^{\infty} q_n x^n,$$

where both series converge in an interval $|x| < R$, $R > 0$. Thus (7–29) can be rewritten as

$$\sum_{n=0}^{\infty} (n + v)(n + v - 1)a_n x^n + \left( \sum_{n=0}^{\infty} (n + v)a_n x^n \right) \left( \sum_{n=0}^{\infty} p_n x^n \right)$$

$$+ \left( \sum_{n=0}^{\infty} a_n x^n \right) \left( \sum_{n=0}^{\infty} q_n x^n \right) = 0,$$

and if we now carry out the indicated multiplications according to the formula given on p. 268, we obtain

$$\sum_{n=0}^{\infty} \left[ (n + v)(n + v - 1)a_n + \sum_{j=0}^{n} (j + v)a_j p_{n-j} + \sum_{j=0}^{n} a_j q_{n-j} \right] x^n = 0.$$

Hence (7–28) will satisfy Eq. (7–27) *formally* in the interval $0 < x < R$ if

$$(n + v)(n + v - 1)a_n + \sum_{j=0}^{n} [(j + v)p_{n-j} + q_{n-j}]a_j = 0 \qquad (7\text{–}30)$$

for all $n \geqslant 0$.

When $n = 0$, (7–30) reduces to

$$v(v - 1) + p_0 v + q_0 = 0, \qquad (7\text{–}31)$$

and when $n \geqslant 1$ it can be written as

$$[(n + v)(n + v - 1) + p_0(n + v) + q_0]a_n$$

$$+ \sum_{j=0}^{n-1} [(j + v)p_{n-j} + q_{n-j}]a_j = 0. \quad (7\text{–}32)$$

The first of these relations is known as the **indicial equation** associated with (7–27); its roots determine the admissible values of $v$ in (7–28). Notice that since $p_0$ and $q_0$ are the constant terms in the series expansions of $P$ and $Q$, (7–31) may be re-written as

$$v(v - 1) + P(0)v + Q(0) = 0. \qquad (7\text{–}33)$$

Thus, when $P$ and $Q$ are explicitly given, the indicial equation associated with (7–27) can be written down at sight.

To continue, we set

$$I(v) = v(v - 1) + p_0 v + q_0,$$

and let $v_1$ and $v_2$ denote the roots of the equation $I(v) = 0$. Moreover, we suppose that $v_1$ and $v_2$ have been labeled so that $\text{Re}(v_1) \geqslant \text{Re}(v_2)$.* Then, when $v = v_1$,

---

\* By $\text{Re}(v)$ we mean the real part of the complex number $v$. Thus if $v = \alpha + \beta i$, $\text{Re}(v) = \alpha$. The real part of a real number is, of course, the number itself.

(7–32) becomes

$$I(n + v_1)a_n + \sum_{j=0}^{n-1} [(j + v_1)p_{n-j} + q_{n-j}]a_j = 0, \qquad n \geqslant 1, \qquad (7\text{–}34)$$

and since the choice of $v_1$ implies that $I(n + v_1) \neq 0$ for $n > 0$, we can solve (7–34) for $a_n$ to obtain

$$a_n = -\frac{1}{I(n + v_1)} \sum_{j=0}^{n-1} [(j + v_1)p_{n-j} + q_{n-j}]a_j, \qquad n \geqslant 1.$$

This relation determines all the $a_n$ from $n = 1$ onward in terms of $a_0$, and yields a *formal* solution of (7–27) in the interval $0 < x < R$. Moreover, if $x^v$ is replaced by $|x|^v$ throughout these computations, we obtain a formal solution in the interval $-R < x < 0$. Finally, the resulting series is known to converge when $0 < |x| < R$, and is therefore a solution of (7–27).*

To complete the discussion, we must now find a second solution of (7–27) that is linearly independent of the one just obtained. We attempt to repeat the preceding argument using the second root $v_2$ of the indicial equation. If $v_2 = v_1$, we get nothing new, of course. However, if $v_2 \neq v_1$, (7–34) becomes

$$I(n + v_2)a_n + \sum_{j=0}^{n-1} [(j + v_2)p_{n-j} + q_{n-j}]a_j = 0,$$

and can again be solved for $a_n$ *provided*

$$I(n + v_2) \neq 0$$

*for all $n > 0$. But when $n > 0$, $I(n + v_2) = 0$ if and only if $n + v_2 = v_1$; that is, if and only if $v_1 - v_2 = n$. Thus our technique will yield a second solution of (7–27) for $0 < |x| < R$ *whenever the roots of the indicial equation $I(v) = 0$ do not differ by an integer.* In this case it is easy to show that the (particular) solutions $y_1$ and $y_2$ obtained by setting $a_0 = 1$ in these series are linearly independent, and hence that the general solution of (7–27) about the origin is

$$y = c_1 y_1 + c_2 y_2,$$

where $c_1$ and $c_2$ are arbitrary constants. This, for instance, is what happened in the example in the preceding section. In the next section we shall discuss the so-called **exceptional cases** in which $v_2 - v_1$ is an integer.

**EXERCISES**

1. Find the indicial equation associated with the regular singular point at $x = 0$ for each of the following equations.
   a) $x^2y'' - 2x(x + 1)y' + (x - 1)y = 0$
   b) $x^2y'' - 2xy' + y = 0$

---

* A proof of this fact can also be found in the book by Coddington cited in the bibliography.

c) $xy'' + (1 - x)y' + 2y = 0$

d) $y'' + \left(1 + \dfrac{1}{4x^2}\right)y = 0$

e) $x(x - 2)y'' + 2(x - 1)y' + 2y = 0$

2. Prove that when $x < 0$ the change of variable $x = -t$ will transform

$$x^2 \frac{d^2y}{dx^2} + xP(x)\frac{dy}{dx} + Q(x)y = 0 \tag{A}$$

to

$$t^2 \frac{d^2y}{dt^2} + tP(-t)\frac{dy}{dt} + Q(-t)y = 0. \tag{B}$$

Find the indicial equation associated with (B), and deduce that if $y = y(x)$ is a solution of (A) on $(0, \infty)$, then $y = y(|x|)$ is a solution of (A) on $(-\infty, 0)$.

Find two linearly independent solutions about $x = 0$ for each of the following equations, and determine the values of $x$ for which these solutions are valid.

3. $x(x - 4)y'' + (x - 2)y' - 4y = 0$

4. $2x^2y'' + 5xy' - 2y = 0$

5. $8x(x + 4)y'' - 8y' + y = 0$

6. $2xy'' + 3y' - \dfrac{1}{x - 1}y = 0$

7. $3x^2y'' - \dfrac{2x}{x - 1}y' + \dfrac{2}{x - 1}y = 0$

8. $2x^2y'' + x(x - 1)y' + 2y = 0$

9. $4x^2(x + 1)y'' + x(3x - 1)y' + y = 0$

10. $x^2y'' + xy' + (x^2 - \frac{1}{4})y = 0$

11. $3x^2y'' + xy' - (1 + x)y = 0$

Compute the values of $a_1$, $a_2$, and $a_3$ in the series solutions of each of the following equations given that $a_0 = 1$.

12. $16x^2y'' - 4x(x^2 - 4)y' - y = 0$

13. $x^2y'' + x(x + 1)y' + y = 0$

14. $8x^2(x - 2)y'' + 2xy' - (\cos x)y = 0$

15. $x^2y'' + xe^xy' + y = 0$

16. $x^2(x^2 - 1)y'' - xy' - 2y = 0$

* 17. a) Show that $x = 1$ is a regular singular point for Legendre's equation

$$(1 - x^2)y'' - 2xy' + p(p + 1)y = 0,$$

and find a solution about this point in the form

$$y = |x - 1|^\nu \sum_{n=0}^{\infty} a_n(x - 1)^n.$$

b) Determine the values of $x$ for which this solution is valid.

### 7-6 SOLUTIONS ABOUT A REGULAR SINGULAR POINT: THE EXCEPTIONAL CASES

To complete the discussion of solutions about a regular singular point it remains to consider the case where $v_1$ and $v_2$, the roots of the indicial equation, differ by an integer. Our experience with the Euler equation suggests that a solution involving a logarithmic term should arise when $v_1 = v_2$, and, as we shall see, this can also happen when $v_1 \neq v_2$. The following theorem gives a complete description of the situation, both for the case already treated and for each of the exceptional cases.

**Theorem 7-4**  *Let*

$$x^2 y'' + xP(x)y' + Q(x)y = 0 \qquad (7\text{--}35)$$

*be a second-order homogeneous linear differential equation whose coefficients are analytic in the interval $|x| < R$, $R > 0$, and let $v_1$ and $v_2$ be the roots of the equation*

$$v(v-1) + P(0)v + Q(0) = 0,$$

*where $v_1$ and $v_2$ are labeled so that $\mathrm{Re}(v_1) \geqslant \mathrm{Re}(v_2)$. Then (7–35) has two linearly independent solutions $y_1$ and $y_2$, valid for $0 < |x| < R$, whose form depends on $v_1$ and $v_2$ in the following way.*

CASE 1.  $v_1 - v_2$ *not an integer.*

$$y_1 = |x|^{v_1} \sum_{n=0}^{\infty} a_n x^n, \qquad a_0 \neq 0,$$

$$y_2 = |x|^{v_2} \sum_{n=0}^{\infty} b_n x^n, \qquad b_0 \neq 0.$$

CASE 2.  $v_1 = v_2 = v$.

$$y_1 = |x|^{v} \sum_{n=0}^{\infty} a_n x^n, \qquad a_0 \neq 0,$$

$$y_2 = |x|^{v} \sum_{n=1}^{\infty} b_n x^n + y_1(x) \ln |x|.$$

CASE 3.  $v_1 - v_2$ *a positive integer.*

$$y_1 = |x|^{v_1} \sum_{n=0}^{\infty} a_n x^n, \qquad a_0 \neq 0,$$

$$y_2 = |x|^{v_2} \sum_{n=0}^{\infty} b_n x^n + cy_1(x) \ln |x|, \qquad b_0 \neq 0, \ c \ a \ (fixed) \ constant.^*$$

---

\* Note that when $c = 0$ the solution $y_2$ does not contain a logarithmic term.

*Finally, the values of the coefficients in each of these series are uniquely deter-
mined up to an arbitrary constant, and can be found directly from the differential
equation by the method of undetermined coefficients.*

We shall not attempt to prove this theorem, but shall instead present an argument that suggests why a solution involving a logarithmic term arises when
$v_1 = v_2$. The argument goes as follows.

As before, we begin by attempting to determine the $a_n$ in

$$x^v \sum_{n=0}^{\infty} a_n x^n$$

so that the resulting expression satisfies (7–35) for $0 < x < R$. This time, however,
we also regard $v$ as a variable and set

$$y(x, v) = x^v \sum_{n=0}^{\infty} a_n x^n. \tag{7–36}$$

Moreover, we assume from the outset that $a_0 = 1$. Then if $L$ denotes the linear
differential operator $x^2 D^2 + x P(x) D + Q(x)$, our earlier discussion implies that

$$Ly(x, v) = I(v)x^v + x^v \sum_{n=1}^{\infty} \left\{ I(n + v)a_n + \sum_{j=0}^{n-1} [(j + v)p_{n-j} + q_{n-j}]a_j \right\} x^n, \tag{7–37}$$

where the $p_{n-j}$ and $q_{n-j}$ are the coefficients of the power series expansions of $P$
and $Q$ about the origin. We now use the recurrence relation

$$I(n + v)a_n + \sum_{j=0}^{n-1} [(j + v)p_{n-j} + q_{n-j}]a_j = 0$$

to determine $a_1, a_2, \ldots$ *in terms of v* so that every term but the first on the right-
hand side of (7–37) vanishes. If the resulting expressions are denoted by $a_n(v)$ and
substituted in (7–36), we obtain a function

$$y_1(x, v) = x^v \left[ 1 + \sum_{n=1}^{\infty} a_n(v)x^n \right] \tag{7–38}$$

with the property that

$$Ly_1 = I(v)x^v. \tag{7–39}$$

But $v_1$ is a double root of the equation $I(v) = 0$. Therefore, $I(v) = (v - v_1)^2$ and

$$Ly_1 = (v - v_1)^2 x^v. \tag{7–40}$$

Thus $Ly_1 = 0$ when $v = v_1$, and the expression $y_1(x, v_1)$ formally satisfies the
equation $Ly = 0$. This, of course, agrees with our earlier results.

The idea behind obtaining a second solution in this case originates with the observation that when (7–40) is differentiated *with respect to* $v$ its right-hand side still vanishes when $v = v_1$. Indeed,

$$\frac{\partial}{\partial v}(v - v_1)^2 x^v = x^v(v - v_1)[2 + (v - v_1)\ln x].$$

But since

$$\frac{\partial}{\partial v}[Ly_1(x, v)] = L\left[\frac{\partial}{\partial v}y_1(x, v)\right],$$

(7–40) implies that

$$L\left[\frac{\partial}{\partial v}y_1(x, v)\right] = 0$$

when $v = v_1$. (See Exercise 13.) Thus, if we differentiate (7–38) term by term with respect to $v$ and then set $v = v_1$, the resulting expression will also formally satisfy the equation $Ly = 0$ when $0 < x < R$. Denoting this expression by $y_2(x, v_1)$, we have

$$y_2(x, v_1) = \frac{\partial}{\partial v}y_1(x, v)\Big|_{v=v_1}$$

$$= x^{v_1}\sum_{n=1}^{\infty} a_n'(v_1)x^n + y_1(x, v_1)\ln x.$$

This is precisely the form of the second solution given in the statement of Theorem 7–4 under Case 2.

### EXERCISES

Find two linearly independent solutions on the positive $x$-axis for each of the equations in Exercises 1 through 10.

1. $x^2y'' + x(x - 1)y' + (1 - x)y = 0$     2. $xy'' + (1 - x)y' - y = 0$

3. $x^2y'' + 3xy' + (x + 1)y = 0$     4. $x^2y'' + 2x^2y' - 2y = 0$

5. $xy'' - (x + 3)y' + 2y = 0$     6. $xy'' + (2x + 3)y' + 4y = 0$

7. $xy'' + (x^3 - 1)y' + x^2y = 0$     8. $x^2y'' - 2x^2y' + 2(2x - 1)y = 0$

9. $xy'' + (1 - x)y' + 3y = 0$     10. $x^2y'' + x^2y' + (3x - 2)y = 0$

11. Use Theorem 7–4 to determine the form of two linearly independent solutions about $x = 0$ for the equation
$$x^2y'' + xy' + (x^2 - p^2)y = 0,$$
where $p$ is a real number. Do not compute the solutions.

12. Prove that the solutions $y_1$ and $y_2$ given in Cases 2 and 3 of Theorem 7–4 are linearly independent in $\mathscr{C}(0, R)$ and $\mathscr{C}(-R, 0)$.

13. Verify that

$$\frac{\partial}{\partial v}[Ly(x, v)] = L\left[\frac{\partial}{\partial v} y(x, v)\right]$$

when $y(x, v)$ is defined by (7–36), and

$$L = x^2 D^2 + x P(x) D + Q(x).$$

## 7–7 BESSEL'S EQUATION

The differential equation

$$x^2 y'' + x y' + (x^2 - p^2) y = 0, \tag{7-41}$$

where $p$ is a nonnegative real number, is known as **Bessel's equation of order $p$**. It is one of the most important differential equations in applied mathematics, and as a consequence its solutions, which are called **Bessel functions**, have been intensively studied. In this section and the next we shall derive some of the more elementary properties of these functions, including their series expansions about the regular singular point at $x = 0$. For simplicity we shall confine our attention to the nonnegative $x$-axis.

The indicial equation associated with (7–41) is

$$v^2 - p^2 = 0,$$

which has the roots $\pm p$. Thus Bessel's equation of order $p$ has a solution of the form

$$y = x^p \sum_{k=0}^{\infty} a_k x^k,$$

with $a_0 \neq 0$.* To evaluate the $a_k$ in this series, we observe that

$$(x^2 - p^2) y = x^p \sum_{k=2}^{\infty} a_{k-2} x^k - x^p \sum_{k=0}^{\infty} p^2 a_k x^k,$$

$$x y' = x^p \sum_{k=0}^{\infty} (k + p) a_k x^k,$$

$$x^2 y'' = x^p \sum_{k=0}^{\infty} (k + p)(k + p - 1) a_k x^k.$$

When these expressions are substituted in Bessel's equation, we find that

$$\sum_{k=0}^{\infty} [(k + p)(k + p - 1) + (k + p) - p^2] a_k x^k + \sum_{k=2}^{\infty} a_{k-2} x^k = 0$$

---

\* Because it is traditional to use the letter $n$ rather than $p$ when Bessel's equation is of integral order, we shall use $k$ as the index of summation throughout this discussion.

or

$$(2p + 1)a_1 x + \sum_{k=2}^{\infty} [k(2p + k)a_k + a_{k-2}]x^k = 0.$$

From this it follows that

$$a_1 = 0 \quad \text{and} \quad a_k = -\frac{a_{k-2}}{k(2p + k)}, \quad k \geqslant 2.$$

Thus

$$a_1 = a_3 = a_5 = \cdots = 0,$$

while

$$a_2 = -\frac{a_0}{2(2p + 2)},$$

$$a_4 = -\frac{a_0}{2 \cdot 4 \cdot (2p + 2)(2p + 4)},$$

.
.
.

$$a_{2k} = (-1)^k \frac{a_0}{2 \cdot 4 \cdots (2k)(2p + 2)(2p + 4) \cdots (2p + 2k)}$$

$$= (-1)^k \frac{a_0}{2^{2k}k!(p + 1)(p + 2) \cdots (p + k)}.$$

Hence

$$y = a_0 \sum_{k=0}^{\infty} \frac{(-1)^k x^{2k+p}}{2^{2k}k!(p + 1)(p + 2) \cdots (p + k)}, \tag{7-42}$$

where $a_0 \neq 0$ is an arbitrary constant.

From this point onward the discussion divides into cases that depend on the value of $p$.

## A. Bessel Functions of the First Kind

*1. p = n, an integer* In this case (7–42) assumes a particularly simple form when we set

$$a_0 = \frac{1}{2^n n!}. \tag{7-43}$$

The corresponding solution of Bessel's equation is denoted by $J_n(x)$ and is called the **Bessel function of order $n$ of the first kind**:

$$J_n(x) = \sum_{k=0}^{\infty} \frac{(-1)^k}{k!(n + k)!} \left(\frac{x}{2}\right)^{2k+n}. \tag{7-44}$$

In particular, the series expansion of $J_0$, which, with $J_1$, is the most important of the Bessel functions, is

$$J_0(x) = \sum_{k=0}^{\infty} \frac{(-1)^k}{(k!)^2} \left(\frac{x}{2}\right)^{2k}$$

$$= 1 - \frac{x^2}{2^2} + \frac{x^4}{2^2 \cdot 4^2} - \frac{x^6}{2^2 \cdot 4^2 \cdot 6^2} + \cdots. \tag{7–45}$$

Graphs of $J_0$, $J_1$, and $J_2$ are shown in Fig. 7–2. Note the oscillatory behavior of these functions. This phenomenon will be discussed in some detail in the next chapter.

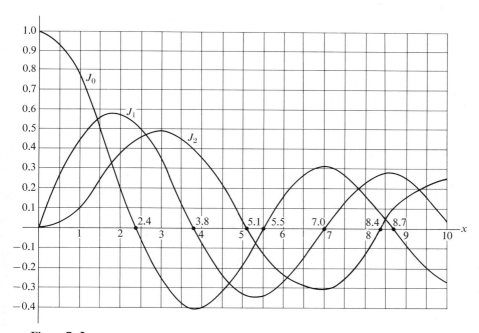

**Figure 7–2**

**2. *p* not an integer; the gamma function**    To obtain a formula for $J_p$ when $p$ is not an integer that is analogous to the formula for $J_n$, we must generalize the notion of the factorial function to include nonintegral values of its argument. Such a generalization was originally discovered by Euler, and is known as the **gamma function**. It is defined by the improper integral

$$\Gamma(p) = \int_0^{\infty} t^{p-1} e^{-t}\, dt, \qquad p > 0, \tag{7–46}$$

which converges for all $p > 0$.

The fact that $\Gamma(p)$ generalizes the factorial function is a consequence of the *functional equation*

$$\Gamma(p + 1) = p\Gamma(p). \tag{7-47}$$

To prove (7–47), we use integration by parts to evaluate $\Gamma(p + 1)$, as follows:

$$\Gamma(p + 1) = \int_0^\infty t^p e^{-t}\, dt$$

$$= -t^p e^{-t}\Big|_0^\infty + p\int_0^\infty t^{p-1} e^{-t}\, dt$$

$$= p\int_0^\infty t^{p-1} e^{-t}\, dt$$

$$= p\Gamma(p).$$

Thus since

$$\Gamma(1) = \int_0^\infty e^{-t}\, dt = 1,$$

we have

$$\Gamma(2) = 1\Gamma(1) = 1,$$
$$\Gamma(3) = 2\Gamma(2) = 2 \cdot 1,$$
$$\Gamma(4) = 3\Gamma(3) = 3 \cdot 2 \cdot 1,$$

and in general

$$\Gamma(n + 1) = n!, \qquad n \text{ a nonnegative integer.}$$

This, of course, is what we wished to show.

For future reference we also note that

$$\Gamma(p + 1)[(p + 1)(p + 2) \cdots (p + k)] = \Gamma(p + 2)[(p + 2) \cdots (p + k)]$$

$$\vdots$$

$$= \Gamma(p + k + 1). \tag{7-48}$$

Next, we observe that when (7–47) is rewritten as

$$\Gamma(p) = \frac{\Gamma(p + 1)}{p}, \tag{7-49}$$

it can be used to assign a value to $\Gamma(p)$ for *nonintegral negative* values of $p$. Indeed, (7–49) can be used to define $\Gamma(p)$ for $-1 < p < 0$, since $\Gamma(p + 1)$ is already defined in that interval. Then, continuing in the same fashion, $\Gamma(p)$ can be successively defined for $-2 < p < -1$, $-3 < p < -2$, and so forth.

Finally, it is easy to show that $\Gamma(p)$ is unbounded in the neighborhood of 0 and in the neighborhood of every negative integer. More precisely,

$$\lim_{p \to 0^+} \Gamma(p) = +\infty, \qquad \lim_{p \to 0^-} \Gamma(p) = -\infty,$$

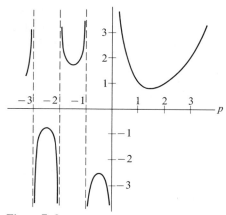

**Figure 7–3**

with similar results about $-1, -2, \ldots$. A graph of the gamma function appears in Fig. 7–3.

We now return to the discussion of Bessel's equation, and to (7–42) when $p$ is not an integer. This time we set

$$a_0 = \frac{1}{2^p \Gamma(p + 1)},$$

and use (7–48) to obtain

$$J_p(x) = \sum_{k=0}^{\infty} \frac{(-1)^k}{\Gamma(k + 1)\Gamma(k + p + 1)} \left(\frac{x}{2}\right)^{2k+p}, \tag{7–50}$$

which is known as the **Bessel function of order $p$ of the first kind**. It is defined for all $x$, and reduces to our earlier formula for $J_n$ when $p$ is a nonnegative integer $n$.

To continue, we seek a second solution of Bessel's equation of order $p$ that is linearly independent of $J_p$. The case where $p$ is not an integer is easy. We simply replace $p$ by $-p$ in the series for $J_p$ to obtain

$$J_{-p}(x) = \sum_{k=0}^{\infty} \frac{(-1)^k}{\Gamma(k + 1)\Gamma(k - p + 1)} \left(\frac{x}{2}\right)^{2k-p}. \tag{7–51}$$

It is not difficult to verify that $J_{-p}$ is a solution of Bessel's equation of order $p$, and that it is linearly independent of $J_p$. (See Exercise 7.) Thus when $p$ is not an integer the general solution of Bessel's equation can be written in the form

$$y = c_1 J_p + c_2 J_{-p}, \tag{7–52}$$

where $c_1$ and $c_2$ are arbitrary constants. Note, however, that because of the term involving $x^{-p}$ in (7–51), $J_{-p}$ is unbounded near the origin and is undefined at $x = 0$. Thus (7–52) is a solution of Bessel's equation only when $x > 0$.

## B. Bessel Functions of the Second Kind

It is tempting to try to modify the definition of $J_{-p}$ so that when $p$ is an integer $n$ it will yield a solution of Bessel's equation that is linearly independent of $J_n$. The most obvious way to try to do so is to agree that

$$J_{-n}(x) = \lim_{p \to n} J_{-p}(x). \tag{7–53}$$

It is easy to show that this limit exists, but unfortunately it turns out that

$$J_{-n}(x) = (-1)^n J_n(x). \tag{7–54}$$

(See Exercise 5.) Thus $J_{-n}$ is linearly dependent on $J_n$ and our attempt has failed.

At this point we could, of course, return to first principles and obtain the required solution by using the method of undetermined coefficients as described under the exceptional cases of Theorem 7–4. But the corresponding computations are rather complicated, and what is worse, they result in functions that are not particularly well suited for computation. As a result we need to find a different approach to this problem. Surprisingly enough, a simple variant of the maneuver that just failed with $J_{-p}$ will produce the solution we seek. In outline, the argument goes like this. If $p$ is not an integer, then the function

$$Y_p(x) = \frac{J_p(x) \cos p\pi - J_{-p}(x)}{\sin p\pi} \tag{7–55}$$

is defined for all $x > 0$ and, being a linear combination of $J_p$ and $J_{-p}$, is a solution of Bessel's equation of order $p$. The important facts about $Y_p$ are that it has a limit as $p \to n$, and that in the limit it is still a solution of Bessel's equation. Thus the function

$$Y_n(x) = \lim_{p \to n} Y_p(x), \qquad n = 0, 1, 2, \ldots, \tag{7–56}$$

is a solution of Bessel's equation of order $n$. What is more, it is linearly independent of $J_n$. The function $Y_n$ is known as the **Bessel function of the second kind of order** $n$.

The task of producing a series expansion for $Y_n$ is complicated and involves results that would take us too far afield to develop here. The computation begins with the observation that (7–55) assumes the indeterminant form $\frac{0}{0}$ when $p = n$, and hence that the limit in (7–56) can be evaluated by l'Hôpital's rule. Thus

$$Y_n(x) = \frac{1}{\pi} \left[ \frac{\partial}{\partial p} J_p(x) - (-1)^n \frac{\partial}{\partial p} J_{-p}(x) \right]_{p=n}. \tag{7–57}$$

If the series for $J_p$ and $J_{-p}$ are substituted in this expression, differentiated, and then reindexed with $n$ in place of $p$, it can be shown that

$$Y_n(x) = \frac{2}{\pi} J_n(x) \left( \ln \frac{x}{2} + \gamma \right) - \frac{1}{\pi} \sum_{k=0}^{n-1} \frac{(n-k-1)!}{k!} \left( \frac{x}{2} \right)^{2k-n}$$

$$- \frac{1}{\pi(n!)} \left( 1 + \frac{1}{2} + \frac{1}{3} + \cdots + \frac{1}{n} \right) \left( \frac{x}{2} \right)^n$$

$$- \frac{1}{\pi} \sum_{k=1}^{\infty} \frac{(-1)^k}{k!(n+k)!} \left( \sum_{j=1}^{k} \frac{1}{j} + \sum_{j=1}^{k+n} \frac{1}{k} \right) \left( \frac{x}{2} \right)^{2k+n},$$

where $\gamma = 0.57721566\ldots = \lim_{n \to \infty} (1 + \frac{1}{2} + \cdots + 1/n - \ln n)$ is *Euler's constant*. In particular, when $n = 0$

$$Y_0(x) = \frac{2}{\pi} J_0(x) \left( \ln \frac{x}{2} + \gamma \right) - \frac{2}{\pi} \sum_{k=1}^{\infty} \frac{(-1)^k}{(k!)^2} \left( 1 + \frac{1}{2} + \cdots + \frac{1}{k} \right) \left( \frac{x}{2} \right)^{2k}.$$

Graphs of $Y_0$ and $Y_1$ are shown in Fig. 7-4.

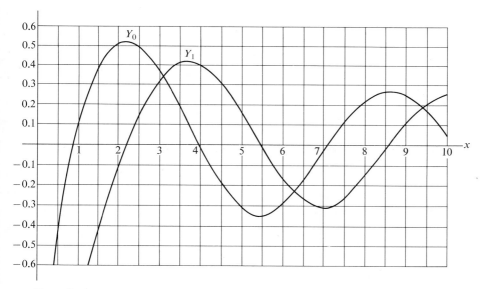

**Figure 7-4**

**EXERCISES**

1. a) Make a change of variable in the integral that defines the gamma function and show
   that

$$\Gamma(\tfrac{1}{2}) = 2 \int_0^{\infty} e^{-x^2} \, dx.$$

b) Use the result in (a) to deduce that

$$[\Gamma(\tfrac{1}{2})]^2 = 4 \int_0^\infty \int_0^\infty e^{-(x^2+y^2)} \, dx \, dy.$$

c) Evaluate the integral in (b) by changing to polar coordinates, and deduce that

$$\Gamma(\tfrac{1}{2}) = \sqrt{\pi}.$$

2.  a) Use the result of Exercise 1(c) to find the values of $\Gamma(\tfrac{3}{2})$ and $\Gamma(-\tfrac{1}{2})$.
    b) Show that if $n$ is a positive integer, then

$$\Gamma(n + \tfrac{1}{2}) = \frac{(2n)!}{2^{2n}n!} \sqrt{\pi}.$$

3. Show that $\lim_{p \to 0^+} \Gamma(p) = +\infty$, and $\lim_{p \to 0^-} \Gamma(p) = -\infty$.

4. Verify the formula for $J_n$ given on p. 296.

5. Show that if

$$J_{-n}(x) = \lim_{p \to n} J_{-p}(x), \qquad n = 1, 2, \ldots,$$

then

$$J_{-n}(x) = (-1)^n J_n(x).$$

6. Show that if

$$Y_n(x) = \lim_{p \to n} Y_p(x), \qquad n = 0, 1, 2, \ldots,$$

where $Y_p$ is defined by (7–55), then

$$Y_n(x) = \frac{1}{\pi} \left[ \frac{\partial}{\partial p} J_p(x) - (-1)^n J_{-p}(x) \right]_{p=n}.$$

7.  a) Prove that $J_p$ and $J_{-p}$ are linearly independent in $\mathscr{C}(0, \infty)$ when $p$ is not an integer.
    b) Show that $J_{-p}$ as defined by (7–51) is a solution of Bessel's equation of order $p$.

8. Show that the general solution of Bessel's equation of order $p$ can be written in the form

$$y = c_1 J_p(x) + c_2 J_p(x) \int \frac{dx}{x[J_p(x)]^2},$$

where $c_1$ and $c_2$ are arbitrary constants.

* 9.  a) Prove that the Wronskian of $J_p$ and $J_{-p}$ satisfies the differential equation

$$\frac{d}{dx} [xW(J_p, J_{-p})] = 0,$$

and hence deduce that

$$W(J_p, J_{-p}) = \frac{c}{x}, \qquad c \text{ a constant}.$$

   b) Use the series expansions for $J_p$, $J_{-p}$, and their derivatives to conclude that whenever $p$ is not an integer

$$c = -\frac{2}{\Gamma(1 - p)\Gamma(p)},$$

where $c$ is the constant in (a).

10. The differential equation

$$x^2 y'' + xy' - (x^2 + p^2)y = 0, \qquad p \geqslant 0,$$

is called the **modified Bessel equation of order** $p$, and its solutions are known as **modified Bessel functions**.

a) Show that

$$I_0(x) = 1 + \frac{x^2}{2^2} + \frac{x^4}{2^2 \cdot 4^2} + \frac{x^6}{2^2 \cdot 4^2 \cdot 6^2} + \cdots$$

is a solution of the modified Bessel equation of order 0.

b) Show that

$$I_1(x) = \frac{x}{2} + \frac{x^3}{2^2 \cdot 4} + \frac{x^5}{2^2 \cdot 4^2 \cdot 6} + \cdots$$

is a solution of the modified Bessel equation of order 1.

* 11. Show that when $p$ is not an integer the general solution of the modified Bessel equation of order $p$ is

$$y = c_1 I_p + c_2 I_{-p},$$

where $c_1$ and $c_2$ are arbitrary constants, and

$$I_p(x) = \sum_{k=0}^{\infty} \frac{1}{\Gamma(k+1)\Gamma(k+p+1)} \left(\frac{x}{2}\right)^{2k+p},$$

$$I_{-p}(x) = \sum_{k=0}^{\infty} \frac{1}{\Gamma(k+1)\Gamma(k-p+1)} \left(\frac{x}{2}\right)^{2k-p}.$$

* 12. Theorem 7–4 asserts that Bessel's equation of order 0 has a solution of the form

$$y_2(x) = \sum_{k=1}^{\infty} b_k x^k + J_0(x) \ln x.$$

Use the method of undetermined coefficients to show that

$$y_2(x) = \sum_{k=1}^{\infty} \frac{(-1)^{k+1}}{(k!)^2} \left(1 + \frac{1}{2} + \cdots + \frac{1}{k}\right) \left(\frac{x}{2}\right)^{2k} + J_0(x) \ln x.$$

13. a) Prove that when $x > 0$ the general solution of

$$y'' + \frac{2\alpha - 1}{x} y' + \beta^2 y = 0$$

is

$$y = x^\alpha Z_\alpha(\beta x),$$

where $Z_\alpha$ denotes the general solution of Bessel's equation of order $\alpha$, and $\beta \neq 0$ is a (real) constant.

b) Use the result in (a) to find the general solution of

$$y'' + \frac{a}{x} y' + by = 0, \qquad x > 0,$$

where $a$ and $b > 0$ are constants.

c) Use the result in (b) to deduce that

$$J_{1/2}(x) = \sqrt{\frac{2}{\pi x}}\, \sin x.$$

14. a) Show that the general solution of the equation

$$u'' + x^2 u = 0$$

is

$$u = \sqrt{x}\, Z_{1/4}\!\left(\frac{x^2}{2}\right),$$

where $Z_{1/4}$ is the general solution of Bessel's equation of order $1/4$.

b) Make the change of variable $y = -u'/u$, and use the result in (a) to solve the Riccati equation

$$y' = x^2 + y^2.$$

15. Show that the general solution of **Airy's equation**

$$y'' + xy = 0$$

is

$$y = x^{1/2}[c_1 J_{1/3}(\tfrac{2}{3}x^{3/2}) + c_2 J_{-1/3}(\tfrac{2}{3}x^{3/2})].$$

## *7–8 PROPERTIES OF BESSEL FUNCTIONS

As with the Legendre polynomials, there is an almost endless list of formulas and identities that involve Bessel functions. In this section we shall establish a few of them.

**Recurrence Relations**   The various recurrence relations that involve Bessel functions follow from the differentiation formulas

$$\frac{d}{dx}\left[x^p J_p(x)\right] = x^p J_{p-1}(x) \tag{7–58}$$

and

$$\frac{d}{dx}\left[x^{-p} J_p(x)\right] = -x^{-p} J_{p+1}(x). \tag{7–59}$$

To prove (7–58), we multiply the series expansion of $J_p$ by $x^p$ and differentiate:

$$\frac{d}{dx}\left[x^p J_p(x)\right] = \frac{d}{dx} \sum_{k=0}^{\infty} \frac{(-1)^k}{\Gamma(k+1)\Gamma(k+p+1)}\left(\frac{x}{2}\right)^{2k+2p}$$

$$= \sum_{k=0}^{\infty} \frac{(-1)^k (k+p)}{\Gamma(k+1)\Gamma(k+p+1)}\left(\frac{x}{2}\right)^{2k+2p-1}$$

$$= x^p \sum_{k=0}^{\infty} \frac{(-1)^k}{\Gamma(k+1)\Gamma(k+p)}\left(\frac{x}{2}\right)^{2k+p-1}$$

$$= x^p J_{p-1}(x).$$

The proof of (7–59) is similar.

When the derivatives appearing in (7–58) and (7–59) are expanded and the results are simplified, we obtain

$$xJ'_p + pJ_p = xJ_{p-1} \tag{7–60}$$

and

$$xJ'_p - pJ_p = -xJ_{p+1}. \tag{7–61}$$

The basic recurrence relations for the Bessel functions follow at once from these results. By subtracting (7–61) from (7–60) we obtain

$$2pJ_p = xJ_{p-1} + xJ_{p+1};$$

by adding (7–60) to (7–61) we obtain

$$2J'_p = J_{p-1} - J_{p+1}.$$

Hence

$$J_{p+1} = \frac{2p}{x} J_p - J_{p-1} \tag{7–62}$$

and

$$J'_p = \tfrac{1}{2}(J_{p-1} - J_{p+1}). \tag{7–63}$$

*Remark.* Formulas (7–58) through (7–63) also hold for the functions $Y_p$.

**Bessel Functions of Half-Integral Order**    When $p = \tfrac{1}{2}$, (7–50) becomes

$$J_{1/2}(x) = \sum_{k=0}^{\infty} \frac{(-1)^k}{\Gamma(k+1)\Gamma(\tfrac{3}{2}+k)} \left(\frac{x}{2}\right)^{2k+1/2}$$

$$= \sqrt{\frac{x}{2}} \sum_{k=0}^{\infty} \frac{(-1)^k}{2^{2k}\Gamma(k)\Gamma(\tfrac{3}{2}+k)} x^{2k}.$$

But

$$\Gamma(\tfrac{3}{2}+k) = \Gamma\left(\frac{3}{2}\right)\left[\frac{3}{2} \cdot \frac{5}{2} \cdots \frac{2k+1}{2}\right]$$

$$= \Gamma\left(\frac{3}{2}\right)\left[\frac{3 \cdot 5 \cdots (2k+1)}{2^k}\right]$$

$$= \frac{\sqrt{\pi}}{2}\left[\frac{3 \cdot 5 \cdots (2k+1)}{2^k}\right],$$

where the last step follows from the result in Exercise 2 of the preceding section. Hence

$$J_{1/2}(x) = \sqrt{\frac{x}{2}} \cdot \frac{2}{\sqrt{\pi}} \sum_{k=0}^{\infty} \frac{(-1)^k}{2^k k! \, 3 \cdot 5 \cdots (2k+1)} x^{2k}$$

$$= \sqrt{\frac{2x}{\pi}} \sum_{k=0}^{\infty} (-1)^k \frac{x^{2k}}{(2k+1)!} = \sqrt{\frac{2}{\pi x}} \sum_{k=0}^{\infty} (-1)^k \frac{x^{2k+1}}{(2k+1)!},$$

and we have proved that

$$J_{1/2}(x) = \sqrt{\frac{2}{\pi x}} \sin x. \tag{7–64}$$

A similar argument reveals that

$$J_{-1/2}(x) = \sqrt{\frac{2}{\pi x}} \cos x, \tag{7–65}$$

and it now follows from (7–62) that every Bessel function of the first kind of order $n + \frac{1}{2}$, $n$ an integer, can be written as a finite sum of terms involving powers of $x$ and the sine or cosine. For instance,

$$J_{3/2}(x) = \frac{1}{x} J_{1/2}(x) - J_{-1/2}(x) = \sqrt{\frac{2}{\pi x}} \left( \frac{\sin x}{x} - \cos x \right)$$

and

$$J_{5/2}(x) = \frac{3}{x} J_{3/2}(x) - J_{1/2}(x) = \sqrt{\frac{2}{\pi x}} \left( \frac{3 \sin x}{x^2} - \frac{3 \cos x}{x} - \sin x \right).$$

It can be shown that Bessel functions of half-integral order are the only Bessel functions that can be expressed in closed form in terms of elementary functions.

**EXERCISES**

1. Prove that $d/dx[x^{-p}J_p(x)] = -x^{-p}J_{p+1}(x)$.

2. Verify that

$$J_{-1/2}(x) = \sqrt{\frac{2}{\pi x}} \cos x.$$

3. Express $J_3$ and $J_4$ in terms of $J_0$ and $J_1$.

4. Express $J_{-3/2}$ and $J_{-5/2}$ in closed form.

5. a) Use the recurrence relation for the Bessel functions to show that

$$\frac{J_{p-1}(x)}{J_p(x)} = \frac{2p}{x} - \cfrac{1}{\cfrac{2p+2}{x} - \cfrac{1}{\cfrac{2p+4}{x} - \cdots}}$$

   b) Use the result in (a) to deduce that

$$\tan x = \cfrac{1}{\cfrac{1}{x} - \cfrac{1}{\cfrac{3}{x} - \cfrac{1}{\cfrac{5}{x} - \cdots}}}$$

6. Show that

$$\int J_1(x)\,dx = -J_0(x) + c \quad \text{and} \quad \int xJ_0(x)\,dx = xJ_1(x) + c,$$

where $c$ is an arbitrary constant.

7. Show that

$$\int x^2 J_0(x)\,dx = x^2 J_1(x) + xJ_0(x) - \int J_0(x)\,dx.$$

8. Show that:
   a) $J_n'' = \frac{1}{4}(J_{n-2} - 2J_n + J_{n+2})$,
   b) $J_n''' = \frac{1}{8}(J_{n-3} - 3J_{n-1} + 3J_{n+1} - J_{n+3})$.

9. Show that

$$\int J_0(x) \sin x\,dx = xJ_0(x) \sin x - xJ_1(x) \cos x + c$$

and

$$\int J_0(x) \cos x\,dx = xJ_0(x) \cos x + xJ_1(x) \sin x + c,$$

where $c$ is an arbitrary constant.

* 10. By multiplying the power series for $e^{xt/2}$ and $e^{-x/2t}$, show that

$$e^{x(t-1/t)/2} = \sum_{n=-\infty}^{\infty} J_n(x)t^n.$$

The function $e^{x(t-1/t)/2}$ is known as the **generating function** for the $J_n$.

11. Use the identity

$$e^{x(t-1/t)/2}e^{y(t-1/t)/2} = e^{(x+y)(t+1/t)/2}$$

and the result of Exercise 10 to prove the **addition formula**

$$J_n(x + y) = \sum_{k=-\infty}^{\infty} J_k(x) J_{n-k}(y), \quad n = 0, \pm 1, \pm 2, \ldots .$$

* 12. a) Prove that the Laplace transform of the function $x^p J_p(\lambda x)$ is given by the formula

$$\mathscr{L}[x^p J_p(\lambda x)] = \frac{\lambda^p \Gamma(2p + 1)}{2^p \Gamma(p + 1)(s^2 + \lambda^2)^{(2p+1)/2}}$$

for all nonnegative values of $x$ and $\lambda$ and all nonnegative integers $p$. [*Hint*: Show that $y = x^p J_p(\lambda x)$ is a solution of the equation

$$xy'' + (1 - 2p)y' + \lambda^2 xy = 0,$$

and then deduce that

$$\frac{d\mathscr{L}[y]}{\mathscr{L}[y]} + (1 + 2p)\frac{s\,ds}{s^2 + \lambda^2} = 0.$$

Solve this equation for $\mathscr{L}[y]$ and complete the proof by evaluating the constant of integration.]

b) Use the result of part (a) to show that

$$\mathcal{L}[J_n(\lambda x)] = \frac{\lambda^n}{(s^2 + \lambda^2)^{1/2}[s + (s^2 + \lambda^2)^{1/2}]^n}.$$

13. a) Use the convolution theorem and Exercise 12(a) to show that

$$\int_0^x J_0(\lambda) J_0(t - \lambda) \, d\lambda = \sin x.$$

b) Solve the initial-value problem

$$y'' + y = J_0(x); \qquad y(0) = y'(0) = 0,$$

by Laplace transform methods, and then use the result to deduce that

$$\int_0^x \sin (t - \lambda) J_0(\lambda) \, d\lambda = x J_1(x).$$

The following exercises refer to the modified Bessel functions $I_p$ and $I_{-p}$ defined in Exercises 10 and 11 of the preceding section.

14. Show that

$$I_{1/2}(x) = \sqrt{\frac{2}{\pi x}} \sinh x \qquad \text{and} \qquad I_{-1/2}(x) = \sqrt{\frac{2}{\pi x}} \cosh x.$$

15. Show that

$$\frac{d}{dx}[x^p I_p(x)] = x^p I_{p-1}(x) \qquad \text{and} \qquad \frac{d}{dx}[x^{-p} I_p(x)] = x^{-p} I_{p+1}(x).$$

16. a) Derive a recurrence relation that expresses $I_{p+1}$ in terms of $I_p$ and $I_{p-1}$.
    b) Derive a recurrence relation that expresses $I_p'$ in terms of $I_{p-1}$ and $I_{p+1}$.

# Oscillation
# Theory
**8**

## 8–1 THE SEPARATION THEOREM

In this chapter we continue our study of the second-order linear equation

$$y'' + P(x)y' + Q(x)y = 0 \qquad (8-1)$$

and establish a number of remarkable facts concerning the zeros of solutions of such equations.

**Theorem 8–1** *If $y_1$ and $y_2$ are linearly independent solutions of (8–1), then $y_1$ and $y_2$ cannot both vanish at the same point.*

*Proof.* This theorem is an immediate consequence of the fact that the Wronskian of $y_1$ and $y_2$ cannot vanish. For if $y_1(a) = y_2(a) = 0$, then

$$W[y_1(a), y_2(a)] = y_1(a)y_2'(a) - y_2(a)y_1'(a) = 0,$$

a contradiction. ∎

Theorem 8–1 asserts that the zeros of two linearly independent solutions of (8–1) are distinct. The next theorem gives far more precise information about the zeros of such solutions.

**Theorem 8–2. The Sturm separation theorem** *If $y_1$ and $y_2$ are linearly independent solutions of*

$$y'' + P(x)y' + Q(x)y = 0$$

*on an interval I, then their zeros alternate on I in the sense that $y_1$ vanishes exactly once between any two consecutive zeros of $y_2$, and conversely.*

*Proof.* Let $a < b$ be points of $I$ such that

$$y_2(a) = y_2(b) = 0,$$

and suppose that $y_2$ does not vanish between $a$ and $b$. We must show that there

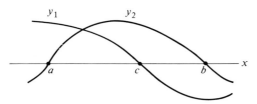

**Figure 8–1**

exists *exactly one* point $c$ with $a < c < b$ and $y_1(c) = 0$. (See Fig. 8–1.) Again the argument is based on an analysis of the Wronskian.

The linear independence of $y_1$ and $y_2$ implies that

$$W[y_1(x), y_2(x)] = y_1(x)y_2'(x) - y_2(x)y_1'(x)$$

does not vanish anywhere on $I$. Thus, since the values of the Wronskian at $a$ and $b$ are

$$y_1(a)y_2'(a) \quad \text{and} \quad y_1(b)y_2'(b),$$

the numbers $y_1(a)$, $y_1(b)$, $y_2'(a)$, and $y_2'(b)$ are all different from zero.

Next, since $W$ is continuous as well as nonzero on $I$, it must have the same sign throughout $I$.* In particular, it must have the same sign at $a$ as it does at $b$. On the other hand, the fact that $a$ and $b$ are *consecutive* zeros of $y_2$ implies that $y_2'$ must have opposite signs at $a$ and $b$ (i.e., the graph of $y_2$ must be increasing at $a$ and decreasing at $b$ or *vice versa*.) Hence to keep the Wronskian from changing sign, $y_1(a)$ and $y_1(b)$ must also have opposite signs, and the continuity of $y_1$ implies that $y_1$ must vanish *at least once* between $a$ and $b$.

Finally, by reversing the roles of $y_1$ and $y_2$ in the preceding argument, we conclude that $y_2$ must have at least one zero between every pair of consecutive zeros of $y_1$. Hence $y_1$ cannot vanish more than once between $a$ and $b$, and we are done. ∎

**Example 1**   Since the sine and cosine functions are solutions of

$$y'' + y = 0 \tag{8-2}$$

that are linearly independent on $(-\infty, \infty)$, the separation theorem implies that the zeros of these functions alternate on $(-\infty, \infty)$. More generally, the same conclusion holds for the zeros of

$$y_1 = a_1 \sin x + a_2 \cos x \quad \text{and} \quad y_2 = b_1 \sin x + b_2 \cos x$$

whenever $a_1 b_2 \neq a_2 b_1$, for then $y_1$ and $y_2$ are also linearly independent solutions of (8–2). (See Exercise 1.)

---

   *  This assertion follows from the fact that a continuous function assumes all values between its maximum and minimum on a closed interval $[a, b]$.

**Example 2**   The functions

$$y_1 = e^x \qquad \text{and} \qquad y_2 = e^{-x}$$

are solutions of $y'' - y = 0$ that are linearly independent on $(-\infty, \infty)$, and hence their zeros alternate on $(-\infty, \infty)$. This assertion is certainly true, though vacuously so, since neither $y_1$ nor $y_2$ vanishes for any value of $x$.

Example 2 reveals one of the major limitations of the separation theorem, namely that it says nothing at all about the *number* of zeros of solutions of $y'' + P(x)y' + Q(x)y = 0$. We shall consider this question in Section 8–3.

## EXERCISES

1. Show that the functions

$$y_1 = a_1 \sin x + a_2 \cos x \qquad \text{and} \qquad y_2 = b_1 \sin x + b_2 \cos x$$

are linearly independent in $\mathscr{C}(-\infty, \infty)$ whenever $a_1 b_2 \neq b_2 a_1$.

* 2. Prove that if $q(x) < 0$ for all $x$ in $I$, then every nontrivial solution of

$$y'' + q(x)y = 0$$

has at most one zero in $I$. [*Hint*: Consider the second derivative of a solution $y = y(x)$ that vanishes at a point $x_0$ in $I$.]

3. a) Find a function $v = v(x)$ with the property that the change of variable $y = uv$ transforms

$$y'' + P(x)y' + Q(x)y = 0$$

    into

$$u'' + q(x)u = 0.$$

  b) Find $q(x)$ in terms of $P(x)$ and $Q(x)$.

4. Use the substitution in Exercise 3 to show that Bessel's equation of order $p$ can be written in the form

$$u'' + \left(1 + \frac{1 - 4p^2}{4x^2}\right)u = 0.$$

## 8–2  STANDARD AND SELF-ADJOINT FORM

In this section we shall derive two variants of the equation

$$y'' + P(x)y' + Q(x)y = 0 \tag{8–3}$$

that will prove to be essential to our later discussion of the zeros of the solutions of such equations.

**Standard Form**   It turns out that the solutions of distinct second-order linear differential equations can best be compared when the equations are in the form

$$u'' + q(x)u = 0, \tag{8–4}$$

with no term involving a first derivative. Every normal second-order equation can be converted to this form by making the change of variable

$$y(x) = u(x)v(x),$$

with a suitably chosen $v$. For then

$$y' = uv' + u'v, \qquad y'' = uv'' + 2u'v' + u''v,$$

and (8–3) becomes

$$vu'' + (2v' + Pv)u' + (v'' + Pv' + Qv)u = 0.$$

Thus, if we now set $2v' + Pv = 0$, which implies that

$$v = e^{-\frac{1}{2}\int P(x)\,dx},$$

then (8–3) reduces to

$$u'' + q(x)u = 0,$$

where

$$q(x) = Q(x) - \left(\frac{P(x)}{2}\right)^2 - \frac{P'(x)}{2}.$$

(See Exercise 3 of the preceding section.)

Equation (8–4) is known as the **standard form** of (8–3).* For future reference we observe that since $v$ never vanishes, the change of variable $y = uv$ has no effect on the zeros of the solutions of (8–3) in the sense that if $y$ is a solution of (8–3) and $u$ is the corresponding solution of (8–4) then $y(x_0) = 0$ if and only if $u(x_0) = 0$.

**Self-Adjoint Form**   The second variant of

$$y'' + P(x)y' + Q(x)y = 0$$

is known as the **self-adjoint** form of this equation and is especially important in advanced work. To derive it we set

$$p(x) = e^{\int P(x)\,dx}.$$

Then

$$\frac{d}{dx}\left(p(x)\frac{dy}{dx}\right) = p(x)\frac{d^2y}{dx^2} + P(x)p(x)\frac{dy}{dx}$$

$$= p(x)\left[\frac{d^2y}{dx^2} + P(x)\frac{dy}{dx}\right],$$

and since $p(x) \neq 0$ for all $x$, we see that $y'' + P(x)y' + Q(x)y = 0$ can be written as

$$\frac{d}{dx}\left(p(x)\frac{dy}{dx}\right) + p(x)Q(x)y = 0$$

---

\* Some authors use the term "normal form" in referring to (8–4), and say that (8–3) is in "standard form."

or, more simply,

$$\frac{d}{dx}\left(p(x)\frac{dy}{dx}\right) + q(x)y = 0, \tag{8–5}$$

where $q(x) = p(x)Q(x)$.

**Example**   Find the standard and self-adjoint forms of Bessel's equation

$$x^2y'' + xy' + (x^2 - p^2)y = 0. \tag{8–6}$$

*Solution.*   We first rewrite (8–6) as

$$y'' + \frac{1}{x}y' + \left(\frac{x^2 - p^2}{x^2}\right)y = 0 \tag{8–7}$$

and then make the change of variable $y = uv$ to eliminate the term involving the first derivative. It follows that $v = x^{-1/2}$, so that

$$y = \frac{u}{\sqrt{x}}.$$

Then

$$y' = -\frac{u}{2x^{3/2}} + \frac{u'}{x^{1/2}}, \qquad y'' = \frac{3u}{4x^{5/2}} - \frac{u'}{x^{3/2}} + \frac{u''}{x^{1/2}},$$

and (8–7) becomes

$$u'' + \left(1 + \frac{1 - 4p^2}{4x^2}\right)u = 0. \tag{8–8}$$

A similar computation yields

$$\frac{d}{dx}\left(x\frac{dy}{dx}\right) + \left(\frac{x^2 - p^2}{x}\right)y = 0 \tag{8–9}$$

as the self-adjoint form of Bessel's equation.

**EXERCISES**

Write each of the following equations in (a) standard form and (b) self-adjoint form.

1. $(1 - x^2)y'' - 2xy' + 6y = 0$        2. $x^2y'' - 2x^3y' - (4 - x^2)y = 0$
3. $(x^3 - 2)y'' - x^2y' - 3y = 0$        4. $2x(\ln x)y'' + 3y' - (\sin x)y = 0$
5. $(x + 1)y'' - y' + 2xy = 0$
6. Verify that

$$\frac{d}{dx}\left(x\frac{dy}{dx}\right) + \left(\frac{x^2 - p^2}{x}\right)y = 0$$

    is the self-adjoint form of Bessel's equation of order $p$.

7. Prove that if $y_1$ and $y_2$ are solutions of

$$\frac{d}{dx}\left(p(x)\frac{dy}{dx}\right) + q(x)y = 0,$$

then

$$p[y_1 y_2' - y_1' y_2] = c,$$

where $c$ is a constant. This result is known as **Abel's identity**.

* 8. a) Show that with the change of variable

$$t = \int \frac{dx}{p(x)}$$

the equation

$$\frac{d}{dx}\left(p(x)\frac{dy}{dx}\right) + q(x)y = 0$$

assumes the form

$$\frac{d^2 y}{dt^2} + Q(t)y = 0.$$

b) Find $Q(t)$ in the preceding equation.


### 8-3  THE COMPARISON THEOREM

We now resume the study of the zeros of solutions of second-order linear differential equations. We begin by proving

**Theorem 8-3**  *Every nontrivial solution $y = y(x)$ of a second-order linear differential equation has at most a finite number of zeros in any closed interval.*

*Proof.*  This theorem is a consequence of the theorem from advanced calculus which asserts that every sequence of points in a *closed* interval has at least one limit point in that interval. For if $y$ had infinitely many zeros in $[a, b]$, then the result just cited would imply the existence of a point $x_0$ and a sequence $\{x_n\}$ of zeros of $y$, all in $[a, b]$, with $x_n \neq x_0$ and $\lim_{n \to \infty} x_n = x_0$. Then since $y$ is continuous,

$$y(x_0) = \lim_{n \to \infty} y(x_n) = 0,$$

and since $y$ is differentiable at $x_0$,

$$y'(x_0) = \lim_{x_n \to x_0} \frac{y(x_n) - y(x_0)}{x_n - x_0} = 0.$$

By the corollary of Theorem 4-5 these results imply that $y(x) = 0$ for all $x$ in $[a, b]$, contradicting the fact that $y$ is nontrivial.  ∎

In particular, this theorem allows us to talk about *consecutive* zeros of an arbitrary nontrivial solution of a second-order linear differential equation, and we need to do so in order to prove the following important result.

**Theorem 8-4.  The Sturm comparison theorem**  Let $y_1$ and $y_2$ be, respectively, nontrivial solutions of

$$y'' + q_1(x)y = 0 \qquad and \qquad y'' + q_2(x)y = 0$$

*on an interval I, and suppose that $q_1(x) > q_2(x)$ everywhere on I. Then there is **at least one** zero of $y_1$ between every two zeros of $y_2$.*

*Proof.* Let $a < b$ be consecutive zeros of $y_2$, and suppose that $y_1$ does *not* vanish in the interval $(a, b)$. Since the zeros of a function $y$ are the same as the zeros of $-y$, we may assume that $y_1$ and $y_2$ are both positive everywhere in $(a, b)$. Then, arguing as in the proof of the Sturm separation theorem, we have

$$W(a) = y_1(a)y_2'(a) \qquad \text{and} \qquad W(b) = y_1(b)y_2'(b), \qquad (8\text{–}10)$$

where $W(a)$ and $W(b)$ denote the values of the Wronskian of $y_1$ and $y_2$ at $a$ and $b$. But

$$\frac{d}{dx} W[y_1, y_2] = \frac{d}{dx}(y_1 y_2' - y_2 y_1')$$

$$= y_1 y_2'' - y_2 y_1''$$

$$= -y_1 q_2 y_2 + y_2 q_1 y_1$$

$$= y_1 y_2(q_1 - q_2) > 0,$$

because $q_1(x) > q_2(x)$ everywhere on $I$. This inequality implies that the Wronskian of $y_1$ and $y_2$ is an *increasing* function on $(a, b)$. On the other hand, because $y_2$ is positive on $(a, b)$ with zeros at $a$ and $b$, $y_2'(a) \geq 0$ and $y_2'(b) \leq 0$. Thus we may infer from (8–10) that $W(a) \geq 0$ and $W(b) \leq 0$, so that $W$ cannot be an increasing function on $(a, b)$. Therefore, the assumption that $y_1$ does not vanish in $(a, b)$ must be false, and the proof is complete. ∎

If we agree to say that a nontrivial solution of a differential equation **oscillates** on an interval $I$ if it has *at least two* zeros in $I$, then the comparison theorem can be paraphrased by saying that the larger the coefficient of $y$ in $y'' + q(x)y = 0$ the more rapidly the solution oscillates, if it oscillates at all, of course.

**Example 1** The functions

$$y_1 = \sin x \qquad \text{and} \qquad y_2 = \sin 2x$$

are (nontrivial) solutions

$$y'' + y = 0 \qquad \text{and} \qquad y'' + 4y = 0,$$

respectively. Since every zero of $\sin x$ is also a zero of $\sin 2x$, the comparison theorem allows us to conclude that $\sin 2x$ oscillates at least twice as rapidly as $\sin x$, in the sense that it has at least one zero between every two zeros of $\sin x$. In this case, of course, there is exactly one zero of $\sin 2x$ between two consecutive zeros of $\sin x$. The same conclusions hold for any pair of nontrivial solutions

$$y_1 = a_1 \sin x + a_2 \cos x \qquad \text{and} \qquad y_2 = b_1 \sin 2x + b_2 \cos 2x$$

of these equations.

**Example 2** Show that every nontrivial solution of

$$y'' + q(x)y = 0 \qquad\qquad (8\text{–}11)$$

has at most one zero on any interval in which $q(x) < 0$.

*Solution.* When the comparison theorem is applied to $(8-11)$ and the equation

$$y'' = 0, \qquad\qquad (8-12)$$

it implies that on an interval where $q(x) < 0$ *every* solution of $(8-11)$ must vanish at least once between successive zeros of a solution of $(8-12)$. The assertion now follows from the fact that $y'' = 0$ has solutions (namely $y = c$) that do not vanish on *any* interval.

It is instructive to compare the argument given here with the one suggested in Exercise 2 of Section $8-1$.

### EXERCISES

1.  a)  Show that every nontrivial solution of $y'' + (\sinh x)y = 0$ has at most one zero in $(-\infty, 0)$, and infinitely many zeros in $(0, \infty)$.
    b)  Show that the distance between successive zeros of the solutions in (a) approaches zero as $x \to \infty$.

2.  Prove that every nontrivial solution of Hermite's equation

$$y'' - 2xy' + 2py = 0,$$

    where $p$ is a nonnegative constant, has at most finitely many zeros.

3.  a)  Let $f$ be continuous on $(0, \infty)$, and suppose that $f(x) \geqslant \varepsilon > 0$ for all $x > 0$. Prove that every solution of

$$y'' + f(x)y = 0$$

    has infinitely many zeros in $(0, \infty)$.
    b)  Does the conclusion in (a) remain true if $f(x) \geqslant 0$ in $(0, \infty)$? Why?

4.  Show that the comparison theorem does not necessarily hold when $q_1(x) \geqslant q_2(x)$, by considering the functions

$$q_1(x) = \begin{cases} 1, & |x| \leqslant \pi/2, \\ 1 - \pi/2 + |x|, & |x| > \pi/2, \end{cases}$$

    and $q_2(x) = 1$.

### 8–4 THE SONIN-POLYA THEOREM

In this section we shall prove a rather surprising result about the magnitude of the oscillations of any (nontrivial) solution of a certain class of self-adjoint equations.

**Theorem 8–5.  The Sonin-Polya Theorem** *Let $p$ and $q$ be functions that are continuously differentiable on $I$, and suppose that*

  i)  $p(x) > 0$,

 ii)  $q(x) \neq 0$, *and*

iii)  $p(x)q(x)$ *is nonincreasing (nondecreasing) for all $x$ in $I$.*

*Then the absolute values of the relative maxima and minima of every nontrivial solution of the equation*

$$\frac{d}{dx}\left(p(x)\frac{dy}{dx}\right) + q(x)y = 0 \qquad (8-13)$$

*are nondecreasing (nonincreasing) as x increases.*

*Proof.* Let $y$ be a nontrivial solution of (8–13), and consider the function

$$F(x) = [y(x)]^2 + \frac{1}{p(x)q(x)}[p(x)y'(x)]^2.$$

Then

$$F' = 2yy' + \frac{2py'}{pq}(py')' - \frac{(pq)'}{(pq)^2}(py')^2,$$

and since by assumption $(py')' = -qy,$

$$F'(x) = -\left(\frac{y'}{q}\right)^2 \frac{d}{dx}(pq).$$

Now suppose that $pq$ is nonincreasing on $I$. Then $(d/dx)(pq) \leqslant 0$ and $F$ is non-decreasing on $I$ (i.e., $F' \geqslant 0$ on $I$). Hence the same is true of any sequence of values of $F$ computed at points $x_1 < x_2 < \cdots$. If, in particular, $x_1, x_2, \ldots$ are the points at which $y$ has a relative maximum or minimum, then $y'(x_i) = 0$, $F(x_i) = y(x_i)^2$, and we have

$$y(x_1)^2 \leqslant y(x_2)^2 \leqslant \cdots.$$

Thus

$$|y(x_1)| \leqslant |y(x_2)| \leqslant \cdots,$$

as asserted. A similar argument applies when $pq$ is nondecreasing, and the proof is complete.  ∎

**Example 1**  As with the separation theorem, the conclusion of the Sonin-Polya theorem may be only vacuously true even though the hypotheses of the theorem are satisfied. This is the case, for instance, when $p(x) = 1$ and $q(x) = -1$. For then (8–13) becomes

$$y'' - y = 0,$$

and none of its solutions oscillate.

**Example 2**  A more interesting example is provided by the nontrivial solutions of Bessel's equation of order $p$, all of which, as we shall see in the next section, oscillate on $(0, \infty)$. Since the self-adjoint form of Bessel's equation is

$$\frac{d}{dx}\left(x\frac{dy}{dx}\right) + \left(\frac{x^2 - p^2}{x}\right)y = 0, \qquad (8-14)$$

and since $p(x)q(x) = x^2 - p^2$ is increasing on $(p, \infty)$, the Sonin-Polya theorem implies that the magnitude of oscillations of every (nontrivial) solution of (8–14) is nonincreasing on this interval. In fact, it can be shown that the magnitude of these oscillations actually decreases with increasing $x$. (See Fig. 7–2, Fig. 7–4, and Exercise 1.)

## EXERCISES

1. Prove that the sequence of absolute values described in the Sonin-Polya theorem is decreasing (increasing) when $pq$ is increasing (decreasing).

2. Prove that no nontrivial solution of

$$\frac{d}{dx}\left(p(x)\frac{dy}{dx}\right) + q(x)y = 0, \qquad p(x) > 0,$$

can oscillate in an interval on which $q(x) < 0$.

## 8–5 THE ZEROS OF BESSEL FUNCTIONS

Bessel functions arise so often in mathematical physics and elsewhere that the number and location of their zeros have been a subject of considerable interest. The following theorem guarantees that they exist and describes how they are situated on the positive $x$-axis.

**Theorem 8–6**  *Every nontrivial solution $Z_p$ of Bessel's equation of order $p$ has infinitely many zeros on the positive $x$-axis. Moreover, the distance between successive zeros of $Z_p$ is*

  i) *less than $\pi$ if $0 \leqslant p \leqslant \frac{1}{2}$,*

  ii) *equal to $\pi$ if $p = \frac{1}{2}$,*

  iii) *greater than $\pi$ if $p > \frac{1}{2}$.*

*Finally, the distance between successive zeros approaches $\pi$ as $x \to \infty$.*

*Proof.*  We work with Bessel's equation in standard form:

$$u'' + \left(1 + \frac{1 - 4p^2}{4x^2}\right)u = 0. \tag{8–15}$$

CASE 1.  $0 \leqslant p < \frac{1}{2}$. In this case

$$1 + \frac{1 - 4p^2}{4x^2} > 1,$$

and comparison with the solutions of $u'' + u = 0$ implies that $Z_p$ has at least one zero in every interval of the positive $x$-axis of length $\pi$.

CASE 2.  $p = \frac{1}{2}$. Here (8–15) reduces to $u'' + u = 0$, and the assertion of the theorem is immediate.

CASE 3.  $p > \frac{1}{2}$. In this case

$$1 + \frac{1 - 4p^2}{4x^2} < 1,$$

and we argue as follows. Let $p$ be fixed, and observe that $(1 - 4p^2)/4x^2 \to 0$ as

$x \to \infty$. Hence, given $0 < \varepsilon < 1$, there exists an $x_0 > 0$ such that

$$1 + \frac{1 - 4p^2}{4x^2} > 1 - \varepsilon$$

for all $x > x_0$. Now we apply the comparison theorem to (8–15) and $u'' + (1 - \varepsilon)u = 0$ in the interval $(x_0, \infty)$ to conclude that $Z_p$ has at least one zero between successive zeros of every function of the form

$$u = c_1 \sin \sqrt{1 - \varepsilon}\, x + c_2 \cos \sqrt{1 - \varepsilon}\, x,$$

where $c_1$ and $c_2$ are not both zero.

This proves the first statement in the theorem. We leave the proof of the assertions concerning the distance between successive zeros of $Z_p$ and the limit of that distance as $x \to \infty$ as exercises. ∎

When read in conjunction with Example 2 of the preceding section, this theorem asserts that every nontrivial solution of Bessel's equation is of a *damped oscillatory* nature. Actually this description can be made much more precise, for it can be shown that

$$Z_p = \frac{A_p}{\sqrt{x}} \sin (x + \omega_p) + \frac{r_p(x)}{\sqrt{x^3}},$$

where $A_p$ and $\omega_p$ are constants whose values depend on $p$, and $r_p$ is a function, again dependent on $p$, which is *bounded* as $x \to \infty$. Thus for large values of $x$ the function $Z_p$ differs very little from the damped sinusoidal function

$$\frac{A_p}{\sqrt{x}} \sin (x + \omega_p).$$

In particular, $Z_p(x) \to 0$ as $x \to \infty$.

We conclude our discussion of this sequence of ideas by proving the following result.

**Theorem 8–7**  *The zeros of $J_p$ and $J_{p+1}$ are distinct and alternate on the positive x-axis.*

*Proof.*  By (7–61)

$$xJ_p'(x) = pJ_p(x) - xJ_{p+1}(x).$$

Hence if both $J_p$ and $J_{p+1}$ were to vanish at some $x_0 > 0$, the derivative $J_p'$ would also vanish at $x_0$, and the uniqueness theorem for solutions of initial-value problems would imply that $J_p = 0$ for all $x$. Since this is nonsense, the zeros of $J_p$ and $J_{p+1}$ are distinct.

To complete the proof, let $x_0 < x_1$ be consecutive zeros of $J_p$. Then by the preceding identity,

$$J_{p+1}(x_0) = -J_p'(x_0) \qquad \text{and} \qquad J_{p+1}(x_1) = -J_p'(x_1).$$

Since by assumption, $J'_p(x_0)$ and $J'_p(x_1)$ have opposite signs, the equalities above imply that $J_{p+1}$ must vanish at least once between $x_0$ and $x_1$. A similar argument using (7–60) shows that $J_p$ must vanish between consecutive zeros of $J_{p+1}$, and the proof is complete.  ∎

### EXERCISES

1. Prove that for any real number $\alpha$,

$$u'' + u = 0$$

   has a solution whose zeros are at $\alpha, \alpha \pm \pi, \alpha \pm 2\pi, \ldots$.

2. Use the result of Exercise 1 to prove that the distance between successive positive zeros of a solution of Bessel's equation of order $p$ is
   a) less than $\pi$ when $0 \leqslant p < \frac{1}{2}$, and
   b) greater than $\pi$ when $p > \frac{1}{2}$.

3. Prove that the distance between successive zeros of any nontrivial solution of Bessel's equation approaches $\pi$ as $x \to \infty$.

4. Discuss the behavior of the positive zeros of $J''_p$.

5. Discuss the oscillatory behavior of the nontrivial solutions of :
   a) $2x^2 y'' + 6xy' + (1 - 1/x^2)y = 0$,
   b) $y'' - y' + e^x(1 - x)y = 0$,
   c) $x^2 y'' + xy' + (x^3 - 1)y = 0$.

6. Prove that the function
$$F(x) = \alpha J_p(x) + \beta J'_p(x)$$

   has infinitely many zeros on the positive $x$-axis for all $p \geqslant 0$ and all constants $\alpha$ and $\beta$.

7. a) Discuss the oscillatory behavior of the nontrivial solutions of Airy's equation

$$y'' + xy = 0.$$

   b) Show that the general solution of Airy's equation is

$$y = \sqrt{x}\, Z_{1/3}(\tfrac{2}{3}x^{3/2}),$$

   where $Z_{1/3}$ is the general solution of Bessel's equation of order $\frac{1}{3}$.

# Matrices and Systems of Linear Equations　　9

## 9–1 LINEAR TRANSFORMATIONS AND BASES

Differential equations often appear in systems that consist of several equations that must be solved simultaneously. When all the equations in such a system are linear the techniques for solving the system are strongly algebraic and make extensive use of matrices. We therefore devote this chapter to the fundamentals of matrix algebra and systems of ordinary equations. Systems of linear differential equations will be treated in the following chapter.

Let $\mathscr{V}_1$ be a finite-dimensional vector space, and $A$ a linear transformation from $\mathscr{V}_1$ to an arbitrary vector space $\mathscr{V}_2$. Then if $\mathbf{e}_1, \ldots, \mathbf{e}_n$ is a basis for $\mathscr{V}_1$, and $\mathbf{x} = x_1\mathbf{e}_1 + \cdots + x_n\mathbf{e}_n$ is a vector in $\mathscr{V}_1$, the linearity of $A$ implies that

$$A(\mathbf{x}) = x_1 A(\mathbf{e}_1) + \cdots + x_n A(\mathbf{e}_n). \tag{9–1}$$

Among other things, this equation tells us that the value of $A$ on $\mathbf{x}$ is *completely determined* by the values of $A$ on the basis vectors $\mathbf{e}_1, \ldots, \mathbf{e}_n$ in the sense that as soon as $A(\mathbf{e}_1), \ldots, A(\mathbf{e}_n)$ are known, $A(\mathbf{x})$ is also known for every $\mathbf{x}$ in $\mathscr{V}_1$. In addition, (9–1) tells us that $\mathscr{I}(A)$, the image of $A$ in $\mathscr{V}_2$, is spanned by the vectors $A(\mathbf{e}_1), \ldots, A(\mathbf{e}_n)$ and hence is a finite-dimensional subspace of $\mathscr{V}_2$ whose dimension does not exceed the dimension of $\mathscr{V}_1$.

The dimension of the image of $A$ is related to the dimension of $\mathscr{V}_1$ and the dimension of $\mathscr{N}(A)$, the null space of $A$, in the following way.

**Theorem 9–1**　*Let $A: \mathscr{V}_1 \to \mathscr{V}_2$ be a linear transformation defined on a finite-dimensional vector space $\mathscr{V}_1$. Then both the image of $A$ and the null space of $A$ are finite dimensional, and*

$$\dim \mathscr{I}(A) + \dim \mathscr{N}(A) = \dim \mathscr{V}_1. \tag{9–2}$$

*Proof.*　Since $\mathscr{V}_1$ is finite dimensional, so is $\mathscr{N}(A)$. Hence by Theorem 2–8 there exists a basis $\mathbf{e}_1, \ldots, \mathbf{e}_m, \mathbf{e}_{m+1}, \ldots, \mathbf{e}_n$ for $\mathscr{V}_1$ whose first $m$ vectors are a basis for $\mathscr{N}(A)$. (If $\mathscr{N}(A)$ is the trivial subspace of $\mathscr{V}_1$, the vectors $\mathbf{e}_1, \ldots, \mathbf{e}_n$ can be any basis. for $\mathscr{V}_1$.) $\mathscr{I}(A)$ is then spanned by the vectors $A(\mathbf{e}_{m+1}), \ldots, A(\mathbf{e}_n)$ alone, and we will be done once we have shown that these vectors are linearly independent in $\mathscr{V}_2$.

To accomplish this, suppose that

$$\alpha_{m+1} A(\mathbf{e}_{m+1}) + \cdots + \alpha_n A(\mathbf{e}_n) = \mathbf{0}.$$

Then

$$A(\alpha_{m+1}\mathbf{e}_{m+1} + \cdots + \alpha_n\mathbf{e}_n) = \mathbf{0},$$

so that the sum $\alpha_{m+1}\mathbf{e}_{m+1} + \cdots + \alpha_n\mathbf{e}_n$ also belongs to $\mathscr{N}(A)$. Hence $\alpha_{m+1}\mathbf{e}_{m+1} + \cdots + \alpha_n\mathbf{e}_n$ is a linear combination of $\mathbf{e}_1, \ldots, \mathbf{e}_n$, and the linear independence of $\mathbf{e}_1, \ldots, \mathbf{e}_n$ now implies that $\alpha_{m+1} = \cdots = \alpha_n = 0$, as required. ∎

The dimension of the image of a linear transformation is known as its **rank**, and the dimension of the null space of the transformation is known as its **nullity**. In these terms (9–2) is often stated as follows:

*Rank plus nullity equals dimension of domain.*

We have seen that when $\mathscr{V}_1$ is finite dimensional and $A: \mathscr{V}_1 \to \mathscr{V}_2$ is linear, the value of $A(\mathbf{x})$ for any $\mathbf{x}$ in $\mathscr{V}_1$ is completely determined by the values of $A$ on a basis for $\mathscr{V}_1$. In fact, it is not difficult to show that we can create a linear transformation from $\mathscr{V}_1$ to $\mathscr{V}_2$ merely by assigning these values in advance. For if $\mathbf{y}_1, \ldots, \mathbf{y}_n$ are any vectors in $\mathscr{V}_2$, however chosen, then the mapping $A: \mathscr{V}_1 \to \mathscr{V}_2$ defined for each $\mathbf{x} = x_1\mathbf{e}_1 + \cdots + x_n\mathbf{e}_n$ in $\mathscr{V}_1$ by

$$A(\mathbf{x}) = x_1\mathbf{y}_1 + \cdots + x_n\mathbf{y}_n \tag{9–3}$$

is linear. (See Exercise 6.) Since these facts will be of paramount importance as we continue, we summarize them here as a theorem.

**Theorem 9–2** *Every linear transformation from a finite-dimensional vector space $\mathscr{V}_1$ to a vector space $\mathscr{V}_2$ is completely determined by its values on a basis for $\mathscr{V}_1$. Moreover, these values can be chosen arbitrarily in $\mathscr{V}_2$, with different choices yielding different transformations.*

**Example**  Let $A: \mathscr{R}^2 \to \mathscr{R}^1$ be the linear transformation defined by

$$A(\mathbf{e}_1) = 2, \qquad A(\mathbf{e}_2) = -1,$$

where $\mathbf{e}_1$ and $\mathbf{e}_2$ are the standard basis vectors in $\mathscr{R}^2$. Then if $\mathbf{x} = x_1\mathbf{e}_1 + x_2\mathbf{e}_2$, we have

$$A(\mathbf{x}) = x_1 A(\mathbf{e}_1) + x_2 A(\mathbf{e}_2) = 2x_1 - x_2,$$

and $A$ is completely determined by the ordered pair $(2, -1)$.

Similarly, if

$$\mathbf{e}_1' = \mathbf{e}_1 + \mathbf{e}_2, \qquad \mathbf{e}_2' = -\mathbf{e}_1,$$

then $\{\mathbf{e}_1', \mathbf{e}_2'\}$ is also a basis for $\mathscr{R}^2$, and

$$A(\mathbf{e}_1') = A(\mathbf{e}_1) + A(\mathbf{e}_2) = 1,$$
$$A(\mathbf{e}_2') = -A(\mathbf{e}_1) = -2.$$

Thus the ordered pair which describes $A$ with respect to the basis $\mathbf{e}'_1$, $\mathbf{e}'_2$ is $(1, -2)$, and we see that *the description of a linear transformation by means of its values on a basis changes with a change of basis.* This fact will be important in our later work.

## EXERCISES

1. Let $A: \mathscr{R}^2 \to \mathscr{R}^2$ be the linear transformation defined by

$$A(\mathbf{e}_1) = 2\mathbf{e}_1, \qquad A(\mathbf{e}_2) = \mathbf{e}_1 + \mathbf{e}_2,$$

where $\mathbf{e}_1$ and $\mathbf{e}_2$ are the standard basis vectors. Find:
   a) $A(2, -1)$                            b) $A(x_1, x_2)$

2. Let $A: \mathscr{R}^2 \to \mathscr{R}^2$ be the linear transformation defined by

$$A(\mathbf{e}_1) = 2\mathbf{e}_1 + \mathbf{e}_2, \qquad A(\mathbf{e}_2) = 3\mathbf{e}_2.$$

   Find a vector $\mathbf{x}$ in $\mathscr{R}^2$ such that $A(\mathbf{x}) = 2\mathbf{x}$.

3. Find the rank and nullity of each of the following linear transformations.
   a) $D: \mathscr{P}_n \to \mathscr{P}_n$                     b) $A: \mathscr{P}_n \to \mathscr{R}^1$, $A(p) = p(0)$

4. Let $A: \mathscr{P}_3 \to \mathscr{R}^1$ be defined by $A(p) = \int_0^1 p(x)\, dx$.
   a) Find the rank and nullity of $A$.
   b) Find a basis for $\mathscr{N}(A)$.

5. Repeat Exercise 4 when $A: \mathscr{R}^3 \to \mathscr{R}^3$ is defined by

$$A(\mathbf{e}_1) = (1, 2, 0), \qquad A(\mathbf{e}_2) = (0, 1, 1), \qquad A(\mathbf{e}_3) = (-1, -2, 0).$$

6. Prove that (9–3) defines a linear transformation from $\mathscr{V}_1$ to $\mathscr{V}_2$.

7. Let $A: \mathscr{V} \to \mathscr{V}$ be a linear transformation on a finite-dimensional vector space $\mathscr{V}$. Prove that there exists a linear transformation $B: \mathscr{V} \to \mathscr{V}$ such that $BA = 0$ and rank $A$ + rank $B = \dim \mathscr{V}$.

8. Let $A$ and $B$ be linear transformations from $\mathscr{V}_1$ to $\mathscr{V}_2$, and assume that $\mathscr{V}_1$ is finite dimensional.
   a) Prove that rank $(A + B) \leqslant$ rank $A$ + rank $B$.
   b) Is nullity $(A + B) \leqslant$ nullity $A$ + nullity $B$? Why?

9. Let $A: \mathscr{V} \to \mathscr{V}$, where $A$ is linear and $\mathscr{V}$ is finite dimensional, and suppose that $\mathscr{N}(A) = \mathscr{I}(A)$.
   a) What can you say about the dimension of $\mathscr{V}$?
   b) Give an example of such a transformation.

## 9–2 THE MATRIX OF A LINEAR TRANSFORMATION

We continue our study of a linear transformation $A: \mathscr{V}_1 \to \mathscr{V}_2$ under the additional assumption that $\mathscr{V}_2$ is also finite dimensional. Our objective is to give a complete description of $A$ in terms of a pair of fixed bases, one in each space. To do so, however, we must distinguish between bases in a vector space that differ from one another only in the order in which their vectors are displayed, a distinction that is made by fixing the order of the vectors in a basis and referring to the result as an **ordered basis**.

Thus let $\mathscr{B}_1 = \{\mathbf{e}_1, \ldots, \mathbf{e}_n\}$ and $\mathscr{B}_2 = \{\mathbf{f}_1, \ldots, \mathbf{f}_m\}$ be ordered bases for $\mathscr{V}_1$ and $\mathscr{V}_2$, respectively. Then, as we have seen, $A$ is completely determined as soon as we know the values $A(\mathbf{e}_j)$, $1 \leqslant j \leqslant n$, in $\mathscr{V}_2$. But each of these vectors can be written uniquely as a linear combination of $\mathbf{f}_1, \ldots, \mathbf{f}_m$. Thus there exist (unique) scalars $\alpha_{ij}$ such that

$$
\begin{aligned}
A(\mathbf{e}_1) &= \alpha_{11}\mathbf{f}_1 + \alpha_{21}\mathbf{f}_2 + \cdots + \alpha_{m1}\mathbf{f}_m \\
A(\mathbf{e}_2) &= \alpha_{12}\mathbf{f}_1 + \alpha_{22}\mathbf{f}_2 + \cdots + \alpha_{m2}\mathbf{f}_m
\end{aligned}
\tag{9-4}
$$

$$
A(\mathbf{e}_n) = \alpha_{1n}\mathbf{f}_1 + \alpha_{2n}\mathbf{f}_2 + \cdots + \alpha_{mn}\mathbf{f}_m.
$$

For computational purposes it is convenient to display these scalars in the $m \times n$ ("$m$ by $n$") rectangular array

$$
\begin{bmatrix}
\alpha_{11} & \alpha_{12} & \cdots & \alpha_{1n} \\
\alpha_{21} & \alpha_{22} & \cdots & \alpha_{2n} \\
\cdot & \cdot & & \cdot \\
\cdot & \cdot & & \cdot \\
\cdot & \cdot & & \cdot \\
\alpha_{m1} & \alpha_{m2} & \cdots & \alpha_{mn}
\end{bmatrix}
\tag{9-5}
$$

whose *columns* are the coefficients of the individual equations in (9-4). The scalars themselves are called the **entries** of the array. Note that the first subscript on each entry in (9-5) indicates the row in which that entry appears, while the second indicates the column. With this convention in force, the entire array can be abbreviated as $(\alpha_{ij})$, it being understood that $i$ and $j$ range independently over the integers $1, \ldots, m$ and $1, \ldots, n$, respectively. We call (9-5) the **matrix of $A$ with respect to the ordered bases** $\mathscr{B}_1, \mathscr{B}_2$, and denote it by

$$[A : \mathscr{B}_1, \mathscr{B}_2],$$

or simply by $[A]$ when no confusion about the bases is possible. (When $A$ maps $\mathscr{V}$ into itself, and $\mathscr{B}_1 = \mathscr{B}_2 = \mathscr{B}$, the notation $[A : \mathscr{B}]$ is also used.) Furthermore, we agree that two $m \times n$ matrices are **equal** if and only if their corresponding entries are equal. In other words, $(\alpha_{ij}) = (\beta_{ij})$ if and only if both matrices are of the same size and $\alpha_{11} = \beta_{11}$, $\alpha_{12} = \beta_{12}, \ldots, \alpha_{mn} = \beta_{mn}$.

We have seen that every linear transformation from $\mathscr{V}_1$ to $\mathscr{V}_2$ determines a unique $m \times n$ matrix with respect to the ordered bases $\mathscr{B}_1$ and $\mathscr{B}_2$. Conversely, since the $\alpha_{ij}$ in (9-5) determine the $A(\mathbf{e}_j)$ completely, Theorem 9-2 implies that every $m \times n$ matrix determines a *unique* linear transformation from $\mathscr{V}_1$ to $\mathscr{V}_2$ relative to $\mathscr{B}_1$ and $\mathscr{B}_2$. These facts are fundamental in the theory of matrices. We therefore state them formally as a theorem.

**Theorem 9-3**  *Let $\mathscr{V}_1$ and $\mathscr{V}_2$ be finite-dimensional vector spaces of dimensions $n$ and $m$, respectively, and let $\mathscr{B}_1$ be an ordered basis for $\mathscr{V}_1$ and $\mathscr{B}_2$ an* *ordered basis for $\mathcal{V}_2$. Then relative to $\mathcal{B}_1$ and $\mathcal{B}_2$, every linear transformation from $\mathcal{V}_1$ to $\mathcal{V}_2$ determines a unique $m \times n$ matrix, and conversely, every $m \times n$ matrix determines a unique linear transformation from $\mathcal{V}_1$ to $\mathcal{V}_2$.*

It is important to realize that Theorem 9–3 does *not* assert that linear transformations have unique matrices. Indeed, such an assertion is false, since the matrix of a linear transformation will in general change with a change of basis. This explains the references to bases in the statement of the theorem.

**Example 1**   Let $\{e_1, e_2\}$ be the standard ordered basis for $\mathcal{R}^2$, and let $A$ reflect $\mathcal{R}^2$ across the $e_1$-axis. Then

$$A(e_1) = e_1 = 1 \cdot e_1 + 0 \cdot e_2, \qquad A(e_2) = -e_2 = 0 \cdot e_1 + (-1)e_2,$$

and the matrix of $A$ with respect to $\{e_1, e_2\}$ is

$$\begin{bmatrix} 1 & 0 \\ 0 & -1 \end{bmatrix}.$$

**Example 2**   Let $D: \mathcal{P}_n \to \mathcal{P}_n$ be the differentiation operator, and let

$$\mathcal{B} = \{1, x, x^2, \ldots, x^{n-1}\}$$

be the "standard" ordered basis for $\mathcal{P}_n$. Then

$$\begin{aligned}
D(1) &= 0 = 0 \cdot 1 + 0 \cdot x + \cdots + 0 \cdot x^{n-2} + 0 \cdot x^{n-1}, \\
D(x) &= 1 = 1 \cdot 1 + 0 \cdot x + \cdots + 0 \cdot x^{n-2} + 0 \cdot x^{n-1}, \\
D(x^2) &= 2x = 0 \cdot 1 + 2 \cdot x + \cdots + 0 \cdot x^{n-2} + 0 \cdot x^{n-1}, \\
&\;\; \cdot \\
&\;\; \cdot \\
&\;\; \cdot \\
D(x^{n-1}) &= (n-1)x^{n-2} = 0 \cdot 1 + 0 \cdot x + \cdots + (n-1)x^{n-2} + 0 \cdot x^{n-1},
\end{aligned}$$

and it follows that

$$[D: \mathcal{B}] = \begin{bmatrix} 0 & 1 & 0 & \cdots & & 0 \\ 0 & 0 & 2 & \cdots & & 0 \\ \cdot & \cdot & \cdot & & & \cdot \\ \cdot & \cdot & \cdot & & & \cdot \\ \cdot & \cdot & \cdot & & & \cdot \\ 0 & 0 & 0 & \cdots & n-1 \\ 0 & 0 & 0 & \cdots & & 0 \end{bmatrix}.$$

**Example 3**   If $I$ is the identity map on an $n$-dimensional vector space, then regardless of the basis used, the matrix for $I$ is the $n \times n$ **identity matrix**

$$\begin{bmatrix} 1 & & & & 0 \\ & 1 & & & \\ & & \cdot & & \\ & & & \cdot & \\ 0 & & & & 1 \end{bmatrix},$$

with ones along its **main diagonal** and zeros elsewhere. Similarly, the matrix of the zero transformation from $\mathscr{V}_1$ to $\mathscr{V}_2$ is always the $m \times n$ **zero matrix**, all of whose entries are zeros.

## EXERCISES

1. Let $A: \mathscr{V}_1 \to \mathscr{V}_2$ be the linear transformation defined by an $m \times n$ matrix $(\alpha_{ij})$ with respect to a pair of ordered bases, $\mathscr{B}_1$ for $\mathscr{V}_1$ and $\mathscr{B}_2$ for $\mathscr{V}_2$. Compute $A(\mathbf{x})$ when $\mathbf{x} = x_1\mathbf{e}_1 + \cdots + x_n\mathbf{e}_n$.

2. Let $A: \mathscr{R}^3 \to \mathscr{R}^3$ be defined by

$$A(x_1, x_2, x_3) = (x_1 + x_2, x_1 + x_3, 0).$$

Find $[A: \mathscr{B}_1, \mathscr{B}_2]$ when
   a) $\mathscr{B}_1$ and $\mathscr{B}_2$ are the standard (ordered) bases,
   b) $\mathscr{B}_1$ is the standard basis, $\mathscr{B}_2 = \{(1, 1, 0), (1, 0, 1), (0, 0, 1)\}$,
   c) $\mathscr{B}_1 = \{(1, 1, 0), (1, 0, 1), (0, 1, 1)\}$, $\mathscr{B}_2 = \{(1, 0, 0), (1, 1, 0), (1, 1, 1)\}$.

3. Let $A: \mathscr{P}_3 \to \mathscr{R}^1$ be defined by

$$A(p) = \int_0^1 p(x)\,dx.$$

Find $[A: \mathscr{B}_1, \mathscr{B}_2]$ when
   a) $\mathscr{B}_1 = \{1, x, x^2\}$, $\mathscr{B}_2 = \{1\}$,
   b) $\mathscr{B}_1 = \{1, x - 1, x(x - 1)\}$, $\mathscr{B}_2 = \{1\}$,
   c) $\mathscr{B}_1 = \{1, x - 1, x(x - 1)\}$, $\mathscr{B}_2 = \{2\}$.

4. Let $A: \mathscr{P}_3 \to \mathscr{P}_4$ be defined by

$$A(p) = \int_0^x p(t)\,dt.$$

Find $[A: \mathscr{B}_1, \mathscr{B}_2]$ when
   a) $\mathscr{B}_1 = \{1, x, x^2\}$, $\mathscr{B}_2 = \{1, x, x^2, x^3\}$,
   b) $\mathscr{B}_1 = \{1, x, x^2\}$, $\mathscr{B}_2 = \{1, x, x^2/2, x^3/3\}$,
   c) $\mathscr{B}_1 = \{2x - 1, 2x + 1, x(x + 1)\}$, $\mathscr{B}_2 = \{x - 1, x + 1, x^2, x^3/3\}$.

5. Let $A: \mathscr{P}_3 \to \mathscr{P}_4$ be defined by

$$A(p) = (x^2 - 1)p'(x).$$

Find $[A: \mathscr{B}_1, \mathscr{B}_2]$ when
   a) $\mathscr{B}_1 = \{1, x, x^2\}$, $\mathscr{B}_2 = \{1, x, x^2, x^3\}$,
   b) $\mathscr{B}_1 = \{1, x, x^2\}$, $\mathscr{B}_2 = \{1, x - 1, (x - 1)^2, (x - 1)^3\}$,
   c) $\mathscr{B}_1 = \{1, x - 1, (x - 1)^2\}$, $\mathscr{B}_2 = \{1, x - 2, (x - 2)^2, (x - 1)(x - 2)^2\}$.

6. Let $A: \mathscr{R}^3 \to \mathscr{R}^2$ be defined by

$$A(x_1, x_2, x_3) = (x_1 - x_2, 2x_2 - 3x_3).$$

Find $[A: \mathscr{B}_1, \mathscr{B}_2]$ when
   a) $\mathscr{B}_1$ and $\mathscr{B}_2$ are the standard (ordered) bases,
   b) $\mathscr{B}_1 = \{(1, 1, 0), (1, 0, 1), (0, 1, 1)\}$, $\mathscr{B}_2$ is the standard basis,
   c) $\mathscr{B}_1 = \{(1, 1, \frac{2}{3}), (2, -1, -1), (3, 2, \frac{1}{2})\}$, $\mathscr{B}_2 = \{(3, 1), (1, \frac{5}{3})\}$.

7. Let $\mathscr{B}_1$ and $\mathscr{B}_2$ be ordered bases for $\mathscr{V}_1$ and $\mathscr{V}_2$, respectively, and let $A: \mathscr{V}_1 \to \mathscr{V}_2$ be linear. What will happen to $[A: \mathscr{B}_1, \mathscr{B}_2]$ when the order of the vectors in $\mathscr{B}_1$ is changed? when the order of the vectors in $\mathscr{B}_2$ is changed?

## 9–3  ADDITION AND SCALAR MULTIPLICATION OF MATRICES

Let $\mathscr{L}(\mathscr{V}_1, \mathscr{V}_2)$ denote the vector space of all linear transformations from $\mathscr{V}_1$ to $\mathscr{V}_2$ (see Theorem 3–1), and let $\mathscr{M}_{mn}$ denote the set of all $m \times n$ matrices. Then if dim $\mathscr{V}_1 = n$ and dim $\mathscr{V}_2 = m$, Theorem 9–3 asserts that the function which associates each $A$ in $\mathscr{L}(\mathscr{V}_1, \mathscr{V}_2)$ with its matrix $[A : \mathscr{B}_1, \mathscr{B}_2]$ is a *one-to-one* mapping of $\mathscr{L}(\mathscr{V}_1, \mathscr{V}_2)$ *onto* $\mathscr{M}_{mn}$. This fact allows us to translate every statement about linear transformations into a statement about matrices. In particular, it allows us to convert $\mathscr{M}_{mn}$ into a vector space by using the addition and scalar multiplication in $\mathscr{L}(\mathscr{V}_1, \mathscr{V}_2)$ to define an addition and scalar multiplication for matrices. The conversion is carried out as follows.

Let $(\alpha_{ij})$ and $(\beta_{ij})$ be $m \times n$ matrices, and let $\sigma$ be a real number. Then relative to $\mathscr{B}_1$ and $\mathscr{B}_2$, there exist unique linear transformations $A$ and $B$ in $\mathscr{L}(\mathscr{V}_1, \mathscr{V}_2)$ such that

$$[A : \mathscr{B}_1, \mathscr{B}_2] = (\alpha_{ij}) \qquad \text{and} \qquad [B : \mathscr{B}_1, \mathscr{B}_2] = (\beta_{ij}).$$

It follows that

$$(A + B)(\mathbf{e}_j) = A(\mathbf{e}_j) + B(\mathbf{e}_j)$$

$$= \sum_{i=1}^{m} \alpha_{ij}\mathbf{f}_i + \sum_{i=1}^{m} \beta_{ij}\mathbf{f}_i$$

$$= \sum_{i=1}^{m} (\alpha_{ij} + \beta_{ij})\mathbf{f}_i,$$

and

$$(\sigma A)(\mathbf{e}_j) = \sigma A(\mathbf{e}_j)$$

$$= \sigma \sum_{i=1}^{m} \alpha_{ij}\mathbf{f}_i$$

$$= \sum_{i=1}^{m} (\sigma\alpha_{ij})\mathbf{f}_i.$$

Hence the matrices of $A + B$ and $\sigma A$ with respect to $\mathscr{B}_1$ and $\mathscr{B}_2$ are

$$(\alpha_{ij} + \beta_{ij}) \qquad \text{and} \qquad (\sigma\alpha_{ij}).$$

This suggests strongly that we give the following definition of addition and scalar multiplication in $\mathscr{M}_{mn}$.

**Definition 9–1**  *If $(\alpha_{ij})$ and $(\beta_{ij})$ are $m \times n$ matrices, and $\sigma$ is a real number, then*

$$(\alpha_{ij}) + (\beta_{ij}) = (\alpha_{ij} + \beta_{ij})$$

*and*

$$\sigma(\alpha_{ij}) = (\sigma\alpha_{ij}).$$

In other words, matrix addition and scalar multiplication are to be performed entry by entry, or termwise. Thus, for instance,

$$\begin{bmatrix} 1 & 3 & 0 \\ 2 & 1 & -4 \end{bmatrix} + \begin{bmatrix} -5 & -2 & 2 \\ 1 & 3 & 1 \end{bmatrix} = \begin{bmatrix} -4 & 1 & 2 \\ 3 & 4 & -3 \end{bmatrix}$$

and

$$\frac{1}{2}\begin{bmatrix} 1 & -2 & -3 \\ 0 & 6 & 2 \\ 5 & -2 & -4 \end{bmatrix} = \begin{bmatrix} \frac{1}{2} & -1 & -\frac{3}{2} \\ 0 & 3 & 1 \\ \frac{5}{2} & -1 & -2 \end{bmatrix}$$

Moreover, we now have

> **Theorem 9–4**   *The set $\mathcal{M}_{mn}$ of all $m \times n$ matrices is a vector space under the addition and scalar multiplication of Definition 9–1.*

Together, this result, Theorem 9–3, and the intervening argument imply that the vector spaces $\mathcal{L}(\mathcal{V}_1, \mathcal{V}_2)$ and $\mathcal{M}_{mn}$ are *algebraically identical* in the sense that any algebraic statement which is true for one is true for the other as well. As an example of how this fact can be used to give easy proofs of results that are not otherwise obvious, we now show that $\mathcal{L}(\mathcal{V}_1, \mathcal{V}_2)$ is finite dimensional and compute its dimension. To do so, we introduce the special matrices $\mathbf{E}_{ij}$, $1 \leq i \leq m$, $1 \leq j \leq n$, each of which has a 1 at the intersection of the $i$th row and $j$th column and zeros elsewhere:

$$\mathbf{E}_{ij} = \begin{bmatrix} 0 & \overset{j}{\vdots} & 0 \\ \cdot & \cdot\ 1\ \cdot & \cdot \\ 0 & \vdots & 0 \end{bmatrix} i \tag{9–6}$$

For instance, when $m = n = 2$ the $\mathbf{E}_{ij}$ are four in number:

$$\mathbf{E}_{11} = \begin{bmatrix} 1 & 0 \\ 0 & 0 \end{bmatrix}, \qquad \mathbf{E}_{12} = \begin{bmatrix} 0 & 1 \\ 0 & 0 \end{bmatrix},$$

$$\mathbf{E}_{21} = \begin{bmatrix} 0 & 0 \\ 1 & 0 \end{bmatrix}, \qquad \mathbf{E}_{22} = \begin{bmatrix} 0 & 0 \\ 0 & 1 \end{bmatrix}.$$

Then for each $(\alpha_{ij})$ in $\mathcal{M}_{mn}$ we have

$$(\alpha_{ij}) = \alpha_{11}\mathbf{E}_{11} + \alpha_{12}\mathbf{E}_{12} + \cdots + \alpha_{1n}\mathbf{E}_{1n} + \cdots$$
$$\vdots$$
$$+ \alpha_{m1}\mathbf{E}_{m1} + \alpha_{m2}\mathbf{E}_{m2} + \cdots + \alpha_{mn}\mathbf{E}_{mn},$$

and it follows that the $\mathbf{E}_{ij}$ span $\mathcal{M}_{mn}$. Moreover, since this expression is clearly the only way of writing $(\alpha_{ij})$ as a linear combination of the $\mathbf{E}_{ij}$, these matrices are a basis for $\mathcal{M}_{mn}$, and we have proved

**Theorem 9–5A**  $\mathcal{M}_{mn}$ *is a finite-dimensional vector space with dimension mn and ordered basis* $\{\mathbf{E}_{11}, \mathbf{E}_{12}, \ldots, \mathbf{E}_{mn}\}$.

Phrased in terms of linear transformations and the space $\mathcal{L}(\mathcal{V}_1, \mathcal{V}_2)$, the theorem reads as follows.

**Theorem 9–5B**  *If* $\mathcal{V}_1$ *and* $\mathcal{V}_2$ *are finite-dimensional vector spaces, then so is* $\mathcal{L}(\mathcal{V}_1, \mathcal{V}_2)$, *and*

$$\dim \mathcal{L}(\mathcal{V}_1, \mathcal{V}_2) = (\dim \mathcal{V}_1)(\dim \mathcal{V}_2).$$

We leave the task of finding the analogs in $\mathcal{L}(\mathcal{V}_1, \mathcal{V}_2)$ of the matrices $\mathbf{E}_{ij}$ as an exercise (Exercise 7). For future reference we will call the matrices $\mathbf{E}_{11}, \mathbf{E}_{12}, \ldots, \mathbf{E}_{mn}$, listed in this order, the **standard ordered basis** of $\mathcal{M}_{mn}$.

### EXERCISES

1. Let $A$ be the reflection of $\mathcal{R}^2$ about the line $y = x$, and let $B$ be the reflection of $\mathcal{R}^2$ in the origin. Find the matrix of $A + B$
   a)  with respect to the standard ordered basis in $\mathcal{R}^2$;
   b)  with respect to the basis $\{\mathbf{e}'_1 = (1, 1), \mathbf{e}'_2 = (-1, 1)\}$.

2. Let $A$ and $B$ be the linear transformations from $\mathcal{P}_3$ to $\mathcal{P}_4$ defined by

   $$A: p(x) \to xp(x) - p(1), \qquad B: p(x) \to (x - 1)p(x).$$

   Find the matrix of $2A - B$
   a)  with respect to the bases $\{1, x, x^2\}$ and $\{1, x, x^2, x^3\}$;
   b)  with respect to the bases $\{1, x - 1, (x - 1)^2\}$ and $\{1, x - 1, (x - 1)^2, (x - 1)^3\}$.

3. Let $A: \mathcal{M}_{22} \to \mathcal{M}_{23}$ be the linear transformation defined by

   $$A: \begin{bmatrix} \alpha_{11} & \alpha_{12} \\ \alpha_{21} & \alpha_{22} \end{bmatrix} \to \begin{bmatrix} \alpha_{11} & \alpha_{12} & 0 \\ \alpha_{21} & \alpha_{22} & 0 \end{bmatrix}.$$

   Find the matrix of $A$ with respect to the ordered bases

   $$\{\mathbf{E}_{11}, \mathbf{E}_{12}, \mathbf{E}_{21}, \mathbf{E}_{22}\} \qquad \text{and} \qquad \{\mathbf{E}_{11}, \mathbf{E}_{12}, \mathbf{E}_{13}, \mathbf{E}_{21}, \mathbf{E}_{22}, \mathbf{E}_{23}\}.$$

4. Repeat Exercise 3 for the linear transformation $A: \mathcal{M}_{22} \to \mathcal{M}_{22}$ defined by

   $$A: \begin{bmatrix} \alpha_{11} & \alpha_{12} \\ \alpha_{21} & \alpha_{22} \end{bmatrix} \to \begin{bmatrix} \alpha_{11} & \alpha_{12} \\ 0 & \alpha_{22} \end{bmatrix}$$

   with respect to the basis $\{\mathbf{E}_{11}, \mathbf{E}_{12}, \mathbf{E}_{21}, \mathbf{E}_{22}\}$.

5. What is the dimension of
   a)  $\mathcal{L}(\mathcal{M}_{22}, \mathcal{M}_{23})$? $= 2 \cdot 2 \cdot 2 \cdot 3$     b)  $\mathcal{L}(\mathcal{M}_{m1}, \mathcal{M}_{n1})$? $= mn$
   c)  $\mathcal{L}(\mathcal{M}_{mn}, \mathcal{M}_{mn})$? $= m^2, n^2$

6. Show that the matrices

   $$\begin{bmatrix} 2 & 1 \\ 0 & -1 \end{bmatrix}, \qquad \begin{bmatrix} -1 & -8 \\ 0 & -1 \end{bmatrix}, \qquad \begin{bmatrix} 1 & -2 \\ 0 & -1 \end{bmatrix}$$

   are linearly dependent in $\mathcal{M}_{22}$.

* 7. Let $\{e_1, \ldots, e_m\}$ and $\{f_1, \ldots, f_n\}$ be ordered bases for $\mathscr{V}_1$ and $\mathscr{V}_2$, respectively. Find the linear transformations in $\mathscr{L}(\mathscr{V}_1, \mathscr{V}_2)$ whose matrices with respect to these bases are the $E_{ij}$ defined in (9–6).

## 9–4 MATRIX MULTIPLICATION

Continuing in the spirit of the preceding section, we now propose to translate the multiplication (i.e., composition) of linear transformations into matrix terms. To do so, let

$$B: \mathscr{V}_1 \to \mathscr{V}_2 \qquad \text{and} \qquad A: \mathscr{V}_2 \to \mathscr{V}_3$$

be linear, let

$$\mathscr{B}_1 = \{e_1, \ldots, e_r\}, \qquad \mathscr{B}_2 = \{f_1, \ldots, f_n\}, \qquad \mathscr{B}_3 = \{g_1, \ldots, g_m\}$$

be ordered bases for $\mathscr{V}_1$, $\mathscr{V}_2$, and $\mathscr{V}_3$, respectively, and suppose that

$$[A: \mathscr{B}_2, \mathscr{B}_3] = \begin{bmatrix} \alpha_{11} & \cdots & \alpha_{1n} \\ \cdot & & \cdot \\ \cdot & & \cdot \\ \alpha_{m1} & \cdots & \alpha_{mn} \end{bmatrix} \qquad \text{and} \qquad [B: \mathscr{B}_1, \mathscr{B}_2] = \begin{bmatrix} \beta_{11} & \cdots & \beta_{1r} \\ \cdot & & \cdot \\ \cdot & & \cdot \\ \beta_{n1} & \cdots & \beta_{nr} \end{bmatrix}.$$

Then

$$A(f_k) = \sum_{i=1}^{m} \alpha_{ik} g_i \qquad \text{and} \qquad B(e_j) = \sum_{k=1}^{n} \beta_{kj} f_k,$$

and we have

$$
\begin{aligned}
AB(e_j) &= A(B(e_j)) \\
&= A\left(\sum_{k=1}^{n} \beta_{kj} f_k\right) \\
&= \sum_{k=1}^{n} \beta_{kj} A(f_k) \\
&= \sum_{k=1}^{n} \beta_{kj} \left(\sum_{i=1}^{m} \alpha_{ik} g_i\right) \\
&= \sum_{i=1}^{m} \left(\sum_{k=1}^{n} \alpha_{ik} \beta_{kj}\right) g_i.
\end{aligned}
$$

Thus the matrix of $AB$ with respect to $\mathscr{B}_1$ and $\mathscr{B}_3$ is the $m \times r$ matrix whose $ij$th entry is formed from the $i$th row of $[A: \mathscr{B}_2, \mathscr{B}_3]$ and the $j$th column of $[B: \mathscr{B}_1, \mathscr{B}_2]$ according to the formula

$$\sum_{k=1}^{n} \alpha_{ik} \beta_{kj} = \alpha_{i1} \beta_{1j} + \alpha_{i2} \beta_{2j} + \cdots + \alpha_{in} \beta_{nj}.$$

In view of these results we define multiplication as follows.

**Definition 9-2** The **product** of an $m \times n$ matrix $(\alpha_{ik})$ and an $n \times r$ matrix $(\beta_{kj})$ is the $m \times r$ matrix $(\sum_{k=1}^{n} \alpha_{ik}\beta_{kj})$:

$$
\begin{bmatrix} \alpha_{11} \cdots \alpha_{1n} \\ \cdot \quad\quad \cdot \\ \cdot \quad\quad \cdot \\ \cdot \quad\quad \cdot \\ \alpha_{m1} \cdots \alpha_{mn} \end{bmatrix} \cdot \begin{bmatrix} \beta_{11} \cdots \beta_{1r} \\ \cdot \quad\quad \cdot \\ \cdot \quad\quad \cdot \\ \cdot \quad\quad \cdot \\ \beta_{n1} \cdots \beta_{nr} \end{bmatrix}
$$

$$
= \begin{bmatrix} \alpha_{11}\beta_{11} + \cdots + \alpha_{1n}\beta_{n1} & \cdots & \alpha_{11}\beta_{1r} + \cdots + \alpha_{1n}\beta_{nr} \\ & \cdot & \\ & \cdot & \\ & \cdot & \\ \alpha_{m1}\beta_{11} + \cdots + \alpha_{mn}\beta_{n1} & \cdots & \alpha_{m1}\beta_{1r} + \cdots + \alpha_{mn}\beta_{nr} \end{bmatrix}.
$$

Note that the product of two matrices is defined *only* when the number of columns in the first is equal to the number of rows in the second. This restriction is the matrix analog of the fact that the product of two linear transformations is defined only when the image of the first is contained in the domain of the second.

**Example 1** Let

$$
[A] = \begin{bmatrix} 2 & -1 & 0 \\ 1 & 2 & -3 \end{bmatrix} \quad \text{and} \quad [B] = \begin{bmatrix} 1 & 3 \\ -2 & 1 \\ 0 & 4 \end{bmatrix}.
$$

Then $[A] \cdot [B]$ and $[B] \cdot [A]$ are both defined, and we have

$$
[A] \cdot [B] = \begin{bmatrix} 2 & -1 & 0 \\ 1 & 2 & -3 \end{bmatrix} \begin{bmatrix} 1 & 3 \\ -2 & 1 \\ 0 & 4 \end{bmatrix} = \begin{bmatrix} 2+2+0 & 6-1+0 \\ 1-4+0 & 3+2-12 \end{bmatrix} = \begin{bmatrix} 4 & 5 \\ -3 & -7 \end{bmatrix},
$$

$$
[B] \cdot [A] = \begin{bmatrix} 1 & 3 \\ -2 & 1 \\ 0 & 4 \end{bmatrix} \begin{bmatrix} 2 & -1 & 0 \\ 1 & 2 & -3 \end{bmatrix} = \begin{bmatrix} 2+3 & -1+6 & 0-9 \\ -4+1 & 2+2 & 0-3 \\ 0+4 & 0+8 & 0-12 \end{bmatrix} = \begin{bmatrix} 5 & 5 & -9 \\ -3 & 4 & -3 \\ 4 & 8 & -12 \end{bmatrix}.
$$

Thus $[A] \cdot [B] \neq [B] \cdot [A]$, and we see that *matrix multiplication is not commutative.*

Since matrix multiplication is simply a coordinatized version of the multiplication of linear transformations, all the results established earlier about products of linear transformations on finite-dimensional vector spaces remain true when phrased in matrix terms. For instance, we can assert without further proof that *matrix multiplication is associative and that it is distributive over addition.* We can also assert that an $n \times n$ matrix $[A]$ has an inverse $[A]^{-1}$ ("$A$ inverse") if and only if the linear transformation it defines in $\mathscr{R}^n$ with respect to the standard basis is one-to-one and onto. Such matrices are characterized by the property

$$
[A]^{-1}[A] = [A]^{-1}[A] = [I]
$$

and are of sufficient importance to merit a special name.

**Definition 9–3**   *An $n \times n$ matrix is said to be **nonsingular** if it has an inverse. Otherwise, it is said to be **singular**.*

In Section 2–10 we observed that $n$ vectors in $\mathscr{R}^n$ are linearly independent if and only if the $n \times n$ determinant formed from the components of the vectors is different from zero (Theorem 2–9). Since the columns of an $n \times n$ matrix can be viewed as the images of the standard basis vectors in $\mathscr{R}^n$ under the linear transformation on $\mathscr{R}^n$ defined by the matrix, and since this transformation is one-to-one and onto if and only if these images are linearly independent, we have the following test for nonsingularity.

**Theorem 9–6**   *An $n \times n$ matrix is nonsingular if and only if the determinant formed from the matrix is different from zero.*

**Example 2**   Singular or nonsingular?

$$
\text{a)} \begin{bmatrix} 1 & -2 \\ 3 & 4 \end{bmatrix} \qquad \text{b)} \begin{bmatrix} 2 & 1 & 0 \\ -1 & 0 & -1 \\ 1 & -1 & 3 \end{bmatrix}
$$

*Solution.*   Evaluate the determinant of each matrix:

a) $\begin{vmatrix} 1 & -2 \\ 3 & 4 \end{vmatrix} = 4 - (-6) = 10.$     The matrix is nonsingular.

b) $\begin{vmatrix} 2 & 1 & 0 \\ -1 & 0 & -1 \\ 1 & -1 & 3 \end{vmatrix} = -1 - 2 - (-3) = 0.$     The matrix is singular.

As we continue to work with matrices it will often prove convenient to display the components of a vector with respect to a given basis in column form rather than in row form, writing

$$
\mathbf{x} = \begin{bmatrix} x_1 \\ \cdot \\ \cdot \\ \cdot \\ x_n \end{bmatrix}
$$

instead of $\mathbf{x} = (x_1, \ldots, x_n)$. The advantage of writing the components of $\mathbf{x}$ in column form is that we can then write the image of $\mathbf{x}$ under a linear transformation as a matrix product. For instance, if $A : \mathscr{V}_1 \to \mathscr{V}_2$ is a linear transformation with matrix $[A : \mathscr{B}_1, \mathscr{B}_2] = (\alpha_{ij})$ and if the components of $\mathbf{x}$ are given relative to $\mathscr{B}_1$, then

$$
A(\mathbf{x}) = \begin{bmatrix} \alpha_{11} & \cdots & \alpha_{1n} \\ \cdot & & \cdot \\ \cdot & & \cdot \\ \cdot & & \cdot \\ \alpha_{m1} & \cdots & \alpha_{mn} \end{bmatrix} \begin{bmatrix} x_1 \\ \cdot \\ \cdot \\ \cdot \\ x_n \end{bmatrix}. \tag{9–7}
$$

(See Exercise 12.) In other words, to find the value of $A(\mathbf{x})$ when a matrix for $A$ is known, we simply compute the product of the matrix for $A$ and the corresponding column matrix for $\mathbf{x}$.

**Example 3**  We saw earlier that the matrix of the differentiation operator $D: \mathcal{P}_4 \to \mathcal{P}_4$ with respect to the ordered basis $\{1, x, x^2, x^3\}$ is

$$\begin{bmatrix} 0 & 1 & 0 & 0 \\ 0 & 0 & 2 & 0 \\ 0 & 0 & 0 & 3 \\ 0 & 0 & 0 & 0 \end{bmatrix}.$$

Thus since the polynomial $3 - 2x - 4x^2 + 2x^3$ can be represented relative to this basis by the column vector

$$\begin{bmatrix} 3 \\ -2 \\ -4 \\ 2 \end{bmatrix}$$

and since

$$\begin{bmatrix} 0 & 1 & 0 & 0 \\ 0 & 0 & 2 & 0 \\ 0 & 0 & 0 & 3 \\ 0 & 0 & 0 & 0 \end{bmatrix} \begin{bmatrix} 3 \\ -2 \\ -4 \\ 2 \end{bmatrix} = \begin{bmatrix} -2 \\ -8 \\ 6 \\ 0 \end{bmatrix},$$

we have

$$D(3 - 2x - 4x^2 + 2x^3) = -2 - 8x + 6x^2.$$

## EXERCISES

Evaluate the products in Exercises 1 through 4.

1. $\begin{bmatrix} 1 & -3 & 2 \\ 4 & 1 & -1 \\ 2 & -5 & 3 \end{bmatrix} \begin{bmatrix} 6 & 1 \\ -2 & 3 \\ 3 & -4 \end{bmatrix}$

2. $\begin{bmatrix} \frac{1}{3} & \frac{1}{6} & 0 \\ \frac{1}{6} & \frac{1}{2} & -\frac{1}{6} \\ 0 & \frac{1}{6} & \frac{1}{3} \end{bmatrix} \begin{bmatrix} \frac{7}{2} & -1 & -\frac{1}{2} \\ -1 & 2 & 1 \\ \frac{1}{2} & -1 & \frac{5}{2} \end{bmatrix}$

3. $[2 \quad \frac{1}{2} \quad -1 \quad 3] \begin{bmatrix} 4 \\ 0 \\ 2 \\ -1 \end{bmatrix}$

4. $\begin{bmatrix} 4 \\ 0 \\ 2 \\ -1 \end{bmatrix} [2 \quad \frac{1}{2} \quad -1 \quad 3]$

5. Find all $2 \times 2$ matrices that commute with

$$\begin{bmatrix} 1 & -1 \\ 0 & 2 \end{bmatrix}.$$

6. Find all $2 \times 2$ matrices whose squares are the $2 \times 2$ identity matrix.

7. Let $A: \mathcal{M}_{22} \to \mathcal{M}_{22}$ be defined by

$$A: \begin{bmatrix} \alpha_{11} & \alpha_{12} \\ \alpha_{21} & \alpha_{22} \end{bmatrix} \to \begin{bmatrix} 2 & 1 \\ 0 & -1 \end{bmatrix} \begin{bmatrix} \alpha_{11} & \alpha_{12} \\ \alpha_{21} & \alpha_{22} \end{bmatrix}.$$

a) Prove that $A$ is linear.

b) Find the matrix of $A$ with respect to the standard ordered basis $\{\mathbf{E}_{11}, \mathbf{E}_{12}, \mathbf{E}_{21}, \mathbf{E}_{22}\}$ of $\mathscr{M}_{22}$.

c) Use the matrix of (b) to find the image under $A$ of the matrix

$$\begin{bmatrix} 1 & 1 \\ -2 & 2 \end{bmatrix}.$$

8. Find the matrix relative to the standard ordered bases of $\mathscr{M}_{22}$ and $\mathscr{M}_{23}$ of the linear transformation $A: \mathscr{M}_{22} \to \mathscr{M}_{23}$ defined by

$$A: \begin{bmatrix} \alpha_{11} & \alpha_{12} \\ \alpha_{21} & \alpha_{22} \end{bmatrix} \to \begin{bmatrix} \alpha_{11} & \alpha_{12} \\ \alpha_{21} & \alpha_{22} \end{bmatrix} \begin{bmatrix} 1 & 0 & -1 \\ 2 & 1 & 1 \end{bmatrix}.$$

9. Find the inverse of

$$\begin{bmatrix} a & b \\ c & d \end{bmatrix}$$

given that $ad - bc \neq 0$.

10. Verify that the matrix

$$\begin{bmatrix} 2 & -1 & 0 \\ 0 & -1 & 1 \\ 1 & 0 & -1 \end{bmatrix}$$

is nonsingular and find its inverse.

11. Find all values of $\alpha$ and $\beta$ for which

$$\begin{bmatrix} 1 & 0 & \alpha \\ 0 & -1 & 0 \\ \alpha & \beta & 0 \end{bmatrix}$$

is nonsingular, and find the inverse.

12. Verify Eq. (9–7).

* 13. Prove the associativity of matrix multiplication directly from Definition 9–2.

## 9–5 CHANGE OF BASIS

As we have seen, the matrix of a linear transformation between two finite-dimensional vector spaces generally changes with a change of basis. In this section we show how to use this fact to simplify the matrix of a linear transformation by making appropriate basis changes.

Let $\mathscr{B}_1 = \{\mathbf{e}_1, \ldots, \mathbf{e}_n\}$ and $\mathscr{B}_2 = \{\mathbf{f}_1, \ldots, \mathbf{f}_m\}$ be ordered bases for $\mathscr{V}_1$ and $\mathscr{V}_2$, respectively, and let $A: \mathscr{V}_1 \to \mathscr{V}_2$ be a linear transformation with $[A: \mathscr{B}_1, \mathscr{B}_2] = (\alpha_{ij})$; that is,

$$A(\mathbf{e}_j) = \alpha_{1j}\mathbf{f}_1 + \cdots + \alpha_{mj}\mathbf{f}_m, \qquad 1 \leqslant j \leqslant n. \tag{9–8}$$

Using this notation, we investigate the way in which the matrix of $A$ changes under the following three types of basis change.

**Type 1.  Interchange Two Basis Vectors**   Let $\mathscr{B}'_2 = \{\mathbf{f}'_1, \ldots, \mathbf{f}'_m\}$ be the ordered basis for $\mathscr{V}_2$ obtained from $\mathscr{B}_2$ by interchanging the vectors $\mathbf{f}_i$ and $\mathbf{f}_k$, $i < k$. Then as an *ordered* basis,

$$\mathscr{B}'_2 = \{\mathbf{f}_1, \ldots, \mathbf{f}_k, \ldots, \mathbf{f}_i, \ldots, \mathbf{f}_m\},$$

and we have

$$\begin{aligned}
A(\mathbf{e}_j) &= \alpha_{1j}\mathbf{f}_1 + \cdots + \alpha_{ij}\mathbf{f}_i + \cdots + \alpha_{kj}\mathbf{f}_k + \cdots + \alpha_{mj}\mathbf{f}_m \\
&= \alpha_{1j}\mathbf{f}_1 + \cdots + \alpha_{kj}\mathbf{f}_k + \cdots + \alpha_{ij}\mathbf{f}_i + \cdots + \alpha_{mj}\mathbf{f}_m \\
&= \alpha_{1j}\mathbf{f}'_1 + \cdots + \alpha_{kj}\mathbf{f}'_i + \cdots + \alpha_{ij}\mathbf{f}'_k + \cdots + \alpha_{mj}\mathbf{f}'_m.
\end{aligned}$$

By comparing this expression with (9–8), we see that the matrix of $A$ with respect to $\mathscr{B}_1$ and $\mathscr{B}'_2$ can be obtained from the matrix $[A: \mathscr{B}_1, \mathscr{B}_2]$ by interchanging the $i$th and $k$th *rows*. (Recall that the components of $A(\mathbf{e}_j)$ appear as the $j$th *column* of the matrix for $A$.) Hence:

*Two $m \times n$ matrices which differ only in the order of their rows represent the same linear transformation from $\mathscr{V}_1$ to $\mathscr{V}_2$ with respect to different ordered bases in $\mathscr{V}_2$.*

An analogous result holds if $\mathscr{B}_1$ and $\mathscr{B}'_1$ are ordered bases for $\mathscr{V}_1$ which differ only in the order of two of their vectors. This time, however, *columns* in the matrices are interchanged.

**Type 2.  Multiply a Basis Vector by a Nonzero Constant**   Let $\mathscr{B}'_2$ be obtained from $\mathscr{B}_2$ by multiplying the $i$th basis vector of $\mathscr{B}_2$ by a nonzero constant $c$ and leaving the remaining vectors unchanged. Then $\mathscr{B}'_2$ is a basis for $\mathscr{V}_2$, and if we set $\mathscr{B}'_2 = \{\mathbf{f}'_1, \ldots, \mathbf{f}'_m\}$, we have $\mathbf{f}'_i = c\mathbf{f}_i$, $c \neq 0$, and $\mathbf{f}'_k = \mathbf{f}_k$ for all $k \neq i$. In this case

$$\begin{aligned}
A(\mathbf{e}_j) &= \alpha_{1j}\mathbf{f}_1 + \cdots + \alpha_{ij}\mathbf{f}_i + \cdots + \alpha_{mj}\mathbf{f}_m \\
&= \alpha_{1j}\mathbf{f}'_1 + \cdots + (\alpha_{ij}/c)\mathbf{f}'_i + \cdots + \alpha_{mj}\mathbf{f}'_m,
\end{aligned}$$

and it follows that the matrix $[A: \mathscr{B}_1, \mathscr{B}'_2]$ can be obtained from $[A: \mathscr{B}_1, \mathscr{B}_2]$ by multiplying each of the entries in the $i$th *row* of the latter matrix by $1/c$. Thus, since $c \neq 0$ was arbitrary, we have the following result:

*Two $m \times n$ matrices which differ from each other in that a row of one can be obtained from the corresponding row of the other by multiplying each of the entries in that row by a fixed nonzero constant represent the same linear transformation from $\mathscr{V}_1$ to $\mathscr{V}_2$ with respect to different ordered bases in $\mathscr{V}_2$.*

Again an analogous result holds for the columns of two matrices when a similar basis change is made in $\mathscr{V}_1$.

**Type 3.  Multiply One Basis Vector by a Constant and Add to Another**   The last of the basis changes we shall consider is effected by replacing the $i$th basis vector in $\mathscr{B}_2$ by $\mathbf{f}_i + c\mathbf{f}_k$, where $c$ is an arbitrary constant and $k \neq i$. Again it can be shown that the resulting set $\{\mathbf{f}_1, \ldots, \mathbf{f}_i + c\mathbf{f}_k, \ldots, \mathbf{f}_m\}$ is a basis for $\mathscr{V}_2$ (see Exercise 10).

Denoting this basis by $\mathscr{B}_2' = \{\mathbf{f}_1', \ldots, \mathbf{f}_m'\}$, we have

$$\mathbf{f}_i' = \mathbf{f}_i + c\mathbf{f}_k, \qquad k \neq i,$$
$$\mathbf{f}_l' = \mathbf{f}_l, \qquad\qquad l \neq i,$$

and

$$A(\mathbf{e}_j) = \alpha_{1j}\mathbf{f}_1 + \cdots + \alpha_{ij}\mathbf{f}_i + \cdots + \alpha_{kj}\mathbf{f}_k + \cdots + \alpha_{mj}\mathbf{f}_m$$
$$= \alpha_{1j}\mathbf{f}_1' + \cdots + \alpha_{ij}(\mathbf{f}_i' - c\mathbf{f}_k') + \cdots + \alpha_{kj}\mathbf{f}_k' + \cdots + \alpha_{mj}\mathbf{f}_m'$$
$$= \alpha_{1j}\mathbf{f}_1' + \cdots + \alpha_{ij}\mathbf{f}_i' + \cdots + (\alpha_{kj} - c\alpha_{ij})\mathbf{f}_k' + \cdots + \alpha_{mj}\mathbf{f}_m'.$$

Hence the matrix $[A : \mathscr{B}_1, \mathscr{B}_2']$ can be obtained from $[A : \mathscr{B}_1, \mathscr{B}_2]$ by multiplying each entry in the $i$th *row* of $[A : \mathscr{B}_1, \mathscr{B}_2]$ by $-c$ and adding the results to the corresponding entries in the $k$th row. Thus:

> *Two $m \times n$ matrices which differ from one another in that a row of one can be obtained from the corresponding row of the other by adding to the entries in that row a scalar multiple of the entries of some other row represent the same linear transformation from $\mathscr{V}_1$ to $\mathscr{V}_2$ with respect to different bases in $\mathscr{V}_2$.*

Here too a similar result holds for the columns of a pair of $m \times n$ matrices.

*Remark.* In addition to telling us how the matrix of a linear transformation changes under certain types of basis changes, the results above also tell us how the coordinates of vectors in $\mathscr{V}_2$ change under these basis changes. To see this, we merely have to replace the vector $A(\mathbf{e}_j)$ in the foregoing argument by an arbitrary column vector

$$\mathbf{y} = \begin{bmatrix} y_1 \\ \cdot \\ \cdot \\ \cdot \\ y_m \end{bmatrix}$$

in $\mathscr{V}_2$ and proceed as before.

The three types of basis changes just introduced, namely

  i) interchange the order of two vectors,
 ii) multiply a vector by a nonzero constant,
iii) multiply one basis vector by a constant and add to another,

are called **elementary basis changes**. The corresponding change which each induces in a matrix $[A : \mathscr{B}_1, \mathscr{B}_2]$ is said to be obtained by an **elementary row** or **elementary column operation**. Note that an elementary row operation on a matrix corresponds to an elementary basis change in the range space of the associated linear transformation, while an elementary column operation corresponds to an elementary basis change in the domain space. Two matrices that can be obtained from each other by a finite sequence of elementary row operations are said to be **row**

**equivalent**, and two matrices that can be obtained from each other by a finite sequence of elementary column operations are said to be **column equivalent**. If both types of operations are used, the matrices are simply said to be **equivalent**.

In view of the amount of attention we have devoted to these particular basis changes, the following theorem, whose proof is outlined in the exercises at the end of the section, will hardly come as a surprise.

**Theorem 9–7** *If $\mathscr{B}$ and $\mathscr{B}'$ are any two ordered bases in a finite-dimensional vector space, it is possible to pass from one to the other by a finite number of elementary basis changes.*

This theorem tells us that we can attack and solve the problem of producing simplified matrix representations of linear transformations by applying elementary row or column operations to matrices. But since our objective is to solve operator equations of the form $A(\mathbf{x}) = \mathbf{y}$, where $\mathbf{x}$ is unknown and $\mathbf{y}$ is known, we must restrict our attention to basis changes in the range space of $A$. Thus in simplifying matrices we shall allow ourselves to use only elementary row operations.

With this in mind, we continue the discussion of simplifying the matrix of a linear transformation with

**Definition 9–3** *If every entry below the main diagonal in an $m \times n$ matrix is zero, the matrix is said to be in **lower triangular form**.*

(The **main diagonal** of an $m \times n$ matrix consists of the entries $\alpha_{11}, \alpha_{22}, \dots$ ) In these terms our first theorem on simplified matrix representations of linear transformations is

**Theorem 9–8** *Every $m \times n$ matrix is row equivalent to an $m \times n$ matrix in lower triangular form in which the only nonzero entries on the main diagonal are ones.*

This theorem should be understood as asserting that if $A: \mathscr{V}_1 \to \mathscr{V}_2$ is a linear transformation between finite-dimensional vector spaces and if $\mathscr{B}_1$ is an ordered basis for $\mathscr{V}_1$, then it is always possible to choose an ordered basis $\mathscr{B}_2$ in $\mathscr{V}_2$ so that the matrix of $A$ with respect to $\mathscr{B}_1$ and $\mathscr{B}_2$ is in lower triangular form with only 1's or 0's on the main diagonal. Again the proof is left as an exercise. The following example shows how these reductions are performed, however, and also illustrates a special case of a technique that can be used to prove the theorem.

**Example 1** Use elementary row operations to reduce the matrix

$$\begin{bmatrix} 1 & -1 & 3 & 4 \\ 2 & 0 & 5 & 2 \\ -3 & 1 & 4 & 2 \end{bmatrix}$$

to the lower triangular form described in Theorem 9–8.

*Solution.* We display the reduction schematically as follows.

$$
\begin{bmatrix} 1 & -1 & 3 & 4 \\ 2 & 0 & 5 & 2 \\ -3 & 1 & 4 & 2 \end{bmatrix} \rightarrow \begin{bmatrix} 1 & -1 & 3 & 4 \\ 0 & 2 & -1 & -6 \\ -3 & 1 & 4 & 2 \end{bmatrix}
$$
Multiply row 1 by $-2$ and add to row 2.

$$
\rightarrow \begin{bmatrix} 1 & -1 & 3 & 4 \\ 0 & 2 & -1 & -6 \\ 0 & -2 & 13 & 14 \end{bmatrix}
$$
Multiply row 1 by 3 and add to row 3.

At this point every entry below the main diagonal in the first column is zero, while the entry on the diagonal is one. This completes the first stage of the reduction, and we now pass to the second column:

$$
\rightarrow \begin{bmatrix} 1 & -1 & 3 & 4 \\ 0 & 1 & -\frac{1}{2} & -3 \\ 0 & -2 & 13 & 14 \end{bmatrix}
$$
Multiply row 2 by $\frac{1}{2}$.

$$
\rightarrow \begin{bmatrix} 1 & -1 & 3 & 4 \\ 0 & 1 & -\frac{1}{2} & -3 \\ 0 & 0 & 12 & 8 \end{bmatrix}
$$
Multiply row 2 by 2 and add to row 3.

$$
\rightarrow \begin{bmatrix} 1 & -1 & 3 & 4 \\ 0 & 1 & -\frac{1}{2} & -3 \\ 0 & 0 & 1 & \frac{2}{3} \end{bmatrix}
$$
Multiply row 3 by $\frac{1}{12}$.

And with this we are done.

Actually we can use elementary row operations to reduce an $m \times n$ matrix to a much simpler form than the one described above. This form is known as *row-reduced echelon form* and can be described as follows.

**Definition 6–4**  *An $m \times n$ matrix is said to be in **row-reduced echelon form** if and only if the following conditions are satisfied:*

i)   *The first nonzero entry in any row which has nonzero entries is 1.*

ii)  *Every column which contains the first nonzero entry of some row has all of its other entries zero.*

iii) *Every row with all entries zero follows every row which has nonzero entries.*

iv)  *If the first nonzero entry in row i occurs in column j, then the first nonzero entry in row i + 1, if it exists, occurs in column k with k > j (whence the term "echelon").*

Despite this rather formidable description, matrices in row-reduced echelon form are really rather simple. Here are some examples:

$$\begin{bmatrix} 1 & 0 & 0 \\ 0 & 1 & 0 \\ 0 & 0 & 1 \end{bmatrix}, \quad \begin{bmatrix} 0 & 1 & 0 & -1 \\ 0 & 0 & 1 & 2 \\ 0 & 0 & 0 & 0 \end{bmatrix}, \quad \begin{bmatrix} 1 & 0 & 0 & 0 & 4 \\ 0 & 0 & 1 & 0 & -2 \\ 0 & 0 & 0 & 1 & 3 \end{bmatrix}.$$

By contrast, none of the following matrices are in row-reduced echelon form, because they respectively violate conditions (ii), (iii), and (iv) of the definition:

$$\begin{bmatrix} 1 & 0 & 1 & 3 \\ 0 & 1 & 0 & -2 \\ 0 & 0 & 1 & 0 \end{bmatrix}, \quad \begin{bmatrix} 0 & 1 & 0 & 2 \\ 0 & 0 & 0 & 0 \\ 0 & 0 & 1 & -1 \end{bmatrix}, \quad \begin{bmatrix} 0 & 1 & -1 & 2 \\ 1 & 0 & 0 & 3 \\ 0 & 0 & 0 & 0 \end{bmatrix}.$$

Now that we have described row-reduced echelon matrices, we can state

**Theorem 9–9** *Every $m \times n$ matrix is row equivalent to a matrix in row-reduced echelon form. Moreover, this matrix is unique.*

As the next example will suggest, the first assertion of this theorem can be proved simply by extending the argument used to prove Theorem 9–8.

**Example 2** Convert the matrix

$$\begin{bmatrix} 1 & -1 & 3 & 4 \\ 2 & 0 & 5 & 2 \\ -3 & 1 & 4 & 2 \end{bmatrix}$$

to row-reduced echelon form.

*Solution.* In Example 1 we obtained the equivalence

$$\begin{bmatrix} 1 & -1 & 3 & 4 \\ 2 & 0 & 5 & 2 \\ -3 & 1 & 4 & 2 \end{bmatrix} \rightarrow \begin{bmatrix} 1 & -1 & 3 & 4 \\ 0 & 1 & -\frac{1}{2} & -3 \\ 0 & 0 & 1 & \frac{2}{3} \end{bmatrix}.$$

We now continue as follows:

$$\rightarrow \begin{bmatrix} 1 & 0 & \frac{5}{2} & 1 \\ 0 & 1 & -\frac{1}{2} & -3 \\ 0 & 0 & 1 & \frac{2}{3} \end{bmatrix} \quad \begin{array}{l} \text{Add row 2} \\ \text{to row 1.} \end{array}$$

$$\rightarrow \begin{bmatrix} 1 & 0 & 0 & -\frac{2}{3} \\ 0 & 1 & -\frac{1}{2} & -3 \\ 0 & 0 & 1 & \frac{2}{3} \end{bmatrix} \quad \begin{array}{l} \text{Multiply row 3} \\ \text{by } -\frac{5}{2} \text{ and add} \\ \text{to row 1.} \end{array}$$

$$\rightarrow \begin{bmatrix} 1 & 0 & 0 & -\frac{2}{3} \\ 0 & 1 & 0 & -\frac{8}{3} \\ 0 & 0 & 1 & \frac{2}{3} \end{bmatrix} \quad \begin{array}{l} \text{Multiply row 3} \\ \text{by } \frac{1}{2} \text{ and add} \\ \text{to row 2.} \end{array}$$

This is the required form.

Notice that the last matrix in Example 2 has exactly three linearly independent columns:

$$\begin{bmatrix} 1 \\ 0 \\ 0 \end{bmatrix}, \quad \begin{bmatrix} 0 \\ 1 \\ 0 \end{bmatrix}, \quad \begin{bmatrix} 0 \\ 0 \\ 1 \end{bmatrix}.$$

Thus if $A: \mathcal{R}^4 \to \mathcal{R}^3$ is the linear transformation whose matrix with respect to the standard ordered bases (or any other pair of ordered bases for that matter) is row equivalent to

$$\begin{bmatrix} 1 & 0 & 0 & -\frac{2}{3} \\ 0 & 1 & 0 & -\frac{8}{3} \\ 0 & 0 & 1 & \frac{2}{3} \end{bmatrix},$$

then $A$ has rank 3 and therefore maps $\mathcal{R}^4$ onto $\mathcal{R}^3$. This is a special case of the following result, whose proof is obvious.

*Corollary 9–1* *The rank of a linear transformation from one finite-dimensional vector space to another is equal to the number of nonzero **rows** in the row-reduced echelon matrix of the transformation.*

We conclude this section with an example that illustrates another application of elementary row operations on matrices.

**Example 3** Show that the vectors

$$(2, 2, -1, 3), \quad (-1, -2, 1, -2), \quad (1, -2, 1, 0)$$

are linearly dependent in $\mathcal{R}^4$.

*Solution.* We display the vectors as the columns of a $4 \times 3$ matrix which we then convert to row-reduced echelon form:

$$\begin{bmatrix} 2 & -1 & 1 \\ 2 & -2 & -2 \\ -1 & 1 & 1 \\ 3 & -2 & 0 \end{bmatrix} \to \begin{bmatrix} 1 & 0 & 2 \\ 0 & 1 & 3 \\ 0 & 0 & 0 \\ 0 & 0 & 0 \end{bmatrix}.$$

But elementary row operations do not affect the linear independence or dependence of the columns of a matrix. Thus, since the third column of the second of these matrices is linearly dependent on the first two columns, the same is true of the third column of the first matrix. In fact, using the second matrix, we conclude that

$$2\begin{bmatrix} 2 \\ 2 \\ -1 \\ 3 \end{bmatrix} + 3\begin{bmatrix} -1 \\ -2 \\ 1 \\ -2 \end{bmatrix} = \begin{bmatrix} 1 \\ -2 \\ 1 \\ 0 \end{bmatrix}.$$

## EXERCISES

In Exercises 1 through 4 find the matrix in row-reduced echelon form that is row equivalent to the given matrix.

1. $\begin{bmatrix} 3 & -1 & 4 \\ 2 & 5 & 1 \\ 1 & 0 & -2 \end{bmatrix}$

2. $\begin{bmatrix} 2 & 1 & -3 & 4 \\ -1 & 5 & 2 & 1 \\ 3 & 8 & -4 & 9 \end{bmatrix}$

3. $\begin{bmatrix} -1 & 0 & 3 & -2 \\ 4 & 2 & -1 & -1 \\ 3 & 1 & 0 & 5 \\ 0 & 1 & 2 & -8 \end{bmatrix}$

4. $\begin{bmatrix} 2 & -1 & 4 \\ 3 & 2 & -2 \\ 5 & 1 & 0 \\ 2 & -1 & 1 \end{bmatrix}$

5. Find all $2 \times 3$ matrices in row-reduced echelon form.

6. For what values of $\alpha$ is

$$\begin{bmatrix} 1 & 0 & -2 \\ \alpha & 1 & -1 \\ 0 & \alpha & 1 \end{bmatrix}$$

row equivalent to the $3 \times 3$ identity matrix?

7. Find the rank and nullity of the linear transformations defined by the following matrices.

a) $\begin{bmatrix} 2 & 2 & -2 \\ 4 & -6 & 1 \\ -1 & 1 & -1 \end{bmatrix}$

b) $\begin{bmatrix} -2 & 1 & 3 \\ -6 & 3 & 9 \\ 4 & 2 & -6 \end{bmatrix}$

8. Repeat Exercise 7 for these matrices:

a) $\begin{bmatrix} -3 & 4 & 5 & 2 \\ 5 & 3 & 5 & -1 \\ 2 & 2 & -1 & 1 \\ -1 & -2 & 0 & -1 \end{bmatrix}$

b) $\begin{bmatrix} -4 & 6 & 0 & 2 & -1 \\ 3 & -1 & 2 & 1 & -3 \\ 6 & -2 & 0 & 2 & 1 \\ -1 & 4 & 2 & -1 & 0 \end{bmatrix}$

9. Which of the following sets of vectors are linearly independent in $\mathscr{R}^4$?
   a) $\{(2, 1, 1, 0), (-3, 4, 2, 1), (1, 3, 1, -1)\}$
   b) $\{(1, -1, 2, 3), (-2, 3, 2, 1), (1, 0, 8, 10)\}$
   c) $\{(1, 3, 1, 2), (1, -1, 2, 1), (3, -5, -4, 3), (0, 5, 4, 1)\}$
   d) $\{(1, 1, 0, 1), (2, -1, 1, 3), (3, -3, 2, 1), (-1, 4, 1, 2)\}$

10. Let $\mathscr{B} = \{\mathbf{f}_1, \ldots, \mathbf{f}_m\}$ be an ordered basis for a vector space $\mathscr{V}$, let $\mathbf{f}_i$ and $\mathbf{f}_k$ be two different vectors in $\mathscr{B}$, let $c$ be an arbitrary constant, and let $\mathscr{B}'$ be the set of vectors obtained from $\mathscr{B}$ by replacing $\mathbf{f}_i$ by the sum $\mathbf{f}_i + c\mathbf{f}_k$. That is, $\mathscr{B}' = \{\mathbf{f}'_1, \ldots, \mathbf{f}'_m\}$, where

$$\mathbf{f}'_j = \begin{cases} \mathbf{f}_j, & j \neq i, \\ \mathbf{f}_i + c\mathbf{f}_k, & j = i. \end{cases}$$

Prove that $\mathscr{B}'$ is also a basis for $\mathscr{V}$.

11. Let $\mathscr{V}_1$ and $\mathscr{V}_2$ be finite-dimensional vector spaces with dim $\mathscr{V}_1 = n$ and dim $\mathscr{V}_2 = m$, and suppose that $A: \mathscr{V}_1 \to \mathscr{V}_2$ is linear. Show that there exist bases $\mathscr{B}_1$ in $\mathscr{V}_1$ and $\mathscr{B}_2$

in $\mathscr{V}_2$ such that the only nonzero entries in the matrix for $A$, if any, are 1's on the main diagonal.

12. Prove Theorem 9–8. [*Hint*: Argue by induction on $m$, the number of rows in the matrix, and consider two cases: (i) every entry in the first column of the matrix is zero; (ii) the first column of the matrix contains at least one nonzero entry.]

\* 13. Prove Theorem 9–7. [*Hint*: Use induction and reason as follows. Let $\mathscr{B} = \{\mathbf{e}_1, \ldots, \mathbf{e}_n\}$ and $\mathscr{B}' = \{\mathbf{e}'_1, \ldots, \mathbf{e}'_n\}$, write $\mathbf{e}'_j = \sum \alpha_{ij}\mathbf{e}_i$, and let $\mathscr{W}$ be the subspace of $\mathscr{V}$ spanned by $\mathbf{e}_2, \ldots, \mathbf{e}_n$.
   a) Show that $\alpha_{1j} \neq 0$ for some $j$ and then reorder $\mathscr{B}'$ to obtain a basis $\{\mathbf{f}'_1, \ldots, \mathbf{f}'_n\}$ for $\mathscr{V}$ for which $\alpha'_{11} \neq 0$ in $\mathbf{f}'_1 = \sum \alpha'_{i1}\mathbf{e}_i$.
   b) Replace $\mathbf{f}'_j, j > 1$, by $\mathbf{f}^*_j = \mathbf{f}'_j + \beta_j\mathbf{f}'_1$, where $\beta_j$ is chosen so that $\mathbf{f}^*_j$ belongs to $\mathscr{W}$, to obtain a basis $\{\mathbf{f}^*_2, \ldots, \mathbf{f}^*_n\}$ for $\mathscr{W}$.
   c) Apply the induction assumption to obtain the basis $\{\mathbf{f}'_1, \mathbf{e}_2, \ldots, \mathbf{e}_n\}$ for $\mathscr{V}$ from $\{\mathbf{f}'_1, \mathbf{f}^*_2, \ldots, \mathbf{f}^*_n\}$ by a finite number of elementary basis changes.
   d) Show that a finite number of elementary basis changes can be used to pass from $\{\mathbf{f}'_1, \mathbf{e}_2, \ldots, \mathbf{e}_n\}$ to $\{\alpha\mathbf{e}_1, \mathbf{e}_2, \ldots, \mathbf{e}_n\}$, and then to $\{\mathbf{e}_1, \ldots, \mathbf{e}_n\}$.]

## 9–6 SYSTEMS OF LINEAR EQUATIONS

Let $A: \mathscr{R}^n \to \mathscr{R}^m$ be the linear transformation defined by the matrix

$$[A] = \begin{bmatrix} \alpha_{11} & \cdots & \alpha_{1n} \\ \cdot & & \cdot \\ \cdot & & \cdot \\ \cdot & & \cdot \\ \alpha_{m1} & \cdots & \alpha_{mn} \end{bmatrix}$$

and let

$$\mathbf{X} = \begin{bmatrix} x_1 \\ \cdot \\ \cdot \\ \cdot \\ x_n \end{bmatrix} \quad \text{and} \quad \mathbf{B} = \begin{bmatrix} \beta_1 \\ \cdot \\ \cdot \\ \cdot \\ \beta_m \end{bmatrix}$$

be column vectors in $\mathscr{R}^n$ and $\mathscr{R}^m$, respectively. Then the standard operator equation involving $A$ can be written in matrix form as

$$\mathbf{AX} = \mathbf{B}, \tag{9–9}$$

where for simplicity we have written $\mathbf{A}$ in place of $[A]$. When $\mathbf{X}$ is unknown and $\mathbf{B}$ is known, (9–9) is simply a compact version of the system of $m$ linear equations in $n$ unknowns

$$\alpha_{11}x_1 + \cdots + \alpha_{1n}x_n = \beta_1$$

$$\tag{9–10}$$

$$\alpha_{m1}x_1 + \cdots + \alpha_{mn}x_n = \beta_m.$$

Thus the task of solving such a system can be viewed as one of finding all vectors in $\mathscr{R}^n$ that are mapped by $A$ onto the vector $\mathbf{B}$ in $\mathscr{R}^m$. In particular, (9–10) will have a solution if and only if $\mathbf{B}$ is in the image of $A$ or, alternatively, if and only if $\mathbf{B}$ is a linear combination of the columns of the matrix for $A$. A system that has no solutions is said to be **incompatible** or **inconsistent**.

When $\mathbf{B} = \mathbf{0}$, the system (9–10) is homogeneous and always has $x_1 = \cdots = x_n = 0$ as a solution. This solution, which is known as the **trivial solution** of the system, will be unique if and only if $A$ is one-to-one or, alternatively, if and only if the column vectors formed from the matrix for $A$ are linearly independent. (See Theorem 9–1.) A related theorem is

> **Theorem 9–10**　If $A: \mathscr{R}^n \to \mathscr{R}^n$ is a linear transformation whose matrix $\mathbf{A}$ is nonsingular, then the system $\mathbf{AX} = \mathbf{B}$ has a unique solution.

(See Exercise 8.) In other words, if the number of unknowns in the system (9–10) equals the number of equations, then the system has a unique solution for any $\mathbf{B}$ if and only if $\mathbf{A}$ is nonsingular.

Suppose, on the other hand, that the number of unknowns in (9–10) is greater than the number of equations. Then since there are $n$ columns in the matrix for $A$, and since $n$ vectors in $\mathscr{R}^m$ *must* be linearly dependent when $n > m$, we have the following useful result.

> **Theorem 9–11**　A homogeneous system of linear equations has nontrivial solutions whenever the number of unknowns exceeds the number of equations.

Finally, the technique of using elementary row operations to simplify matrices can be used in a systematic and efficient way to solve systems of linear equations or to decide whether a system is incompatible. To do so, we form the so-called **augmented matrix**

$$\begin{bmatrix} \alpha_{11} & \cdots & \alpha_{1n} & \beta_1 \\ \cdot & & \cdot & \cdot \\ \cdot & & \cdot & \cdot \\ \cdot & & \cdot & \cdot \\ \alpha_{m1} & \cdots & \alpha_{mn} & \beta_m \end{bmatrix} \tag{9–11}$$

consisting of the coefficients and the right-hand side of the system, and proceed to bring this matrix to row-reduced echelon form, at which point the solutions or the incompatibility of the associated system of equations can be determined by inspection. The following examples illustrate this technique.

**Example 1**　Solve the system

$$\begin{aligned} x_1 - x_2 + 3x_3 &= 4 \\ 2x_1 \qquad\; + 5x_3 &= 2 \\ -3x_1 + x_2 + 4x_3 &= 2. \end{aligned} \tag{9–12}$$

*Solution.* The augmented matrix associated with this system is

$$\begin{bmatrix} 1 & -1 & 3 & 4 \\ 2 & 0 & 5 & 2 \\ -3 & 1 & 4 & 2 \end{bmatrix}, \tag{9-13}$$

which, by Example 2 of the preceding section, is row equivalent to

$$\begin{bmatrix} 1 & 0 & 0 & -\frac{2}{3} \\ 0 & 1 & 0 & -\frac{8}{3} \\ 0 & 0 & 1 & \frac{2}{3} \end{bmatrix}.$$

Thus (9–12) is equivalent (in the sense of having the same solutions) to the system

$$
\begin{aligned}
x_1 &= -\tfrac{2}{3} \\
x_2 &= -\tfrac{8}{3} \\
x_3 &= \tfrac{2}{3}.
\end{aligned}
$$

Actually we can solve a system such as this without going all the way to row-reduced echelon form. Lower triangular form will do. For instance, in the first example of the last section, we saw that (9–13) is row equivalent to

$$\begin{bmatrix} 1 & -1 & 3 & 4 \\ 0 & 1 & -\frac{1}{2} & -3 \\ 0 & 0 & 1 & \frac{2}{3} \end{bmatrix}.$$

Hence (9–12) is equivalent to

$$
\begin{aligned}
x_1 - x_2 + 3x_3 &= 4 \\
x_2 - \tfrac{1}{2}x_3 &= -3 \\
x_3 &= \tfrac{2}{3}.
\end{aligned}
$$

Working our way from bottom to top through this system, we find, as before, that $x_3 = \frac{2}{3}$, $x_2 = -\frac{8}{3}$, $x_1 = -\frac{2}{3}$.

**Example 2**   Solve the system of linear equations

$$
\begin{aligned}
x_1 + 2x_2 - x_3 + x_4 &= 1 \\
2x_1 + 3x_2 - x_3 - x_4 &= 0 \\
x_1 + 2x_2 \quad\quad + 3x_4 &= 2.
\end{aligned}
$$

Note that the number of unknowns in this system exceeds the number of equations; hence Theorem 9–11 implies that this system has infinitely many solutions if it has any at all.

*Solution.*   To solve the system, we bring the augmented matrix

$$\begin{bmatrix} 1 & 2 & -1 & 1 & 1 \\ 2 & 3 & -1 & -1 & 0 \\ 1 & 2 & 0 & 3 & 2 \end{bmatrix}$$

into lower triangular form by subtracting twice the first row from the second and subtracting the first row from the third. We then have

$$\begin{bmatrix} 1 & 2 & -1 & 1 & 1 \\ 0 & -1 & 1 & -3 & -2 \\ 0 & 0 & 1 & 2 & 1 \end{bmatrix}.$$

Hence the original system of equations is equivalent to

$$x_1 + 2x_2 - x_3 + x_4 = 1$$
$$x_2 - x_3 + 3x_4 = 2$$
$$x_3 + 2x_4 = 1,$$

or, alternatively, to

$$x_1 + 2x_2 - x_3 = 1 - x_4$$
$$x_2 - x_3 = 2 - 3x_4$$
$$x_3 = 1 - 2x_4.$$

From this we see that $x_4$ can be assigned values arbitrarily, after which $x_1$, $x_2$, and $x_3$ are uniquely determined as

$$x_1 = -4 + 7x_4$$
$$x_2 = 3 - 5x_4$$
$$x_3 = 1 - 2x_4.$$

Hence the solution set of the given system consists of all vectors in $\mathscr{R}^4$ of the form $(-4 + 7\alpha, 3 - 5\alpha, 1 - 2\alpha, \alpha)$, where $\alpha$ is an arbitrary real number. Geometrically these vectors lie on the line in 4-space through the point $(-4, 3, 1, 0)$ which is parallel to the line through the origin determined by the vector $(7, -5, -2, 1)$.

## EXERCISES

Solve, if possible, using matrix methods.

1.  $3x_1 - 2x_2 + x_3 = 4$
    $-2x_1 + x_2 - 3x_3 = 0$
    $4x_1 - 5x_2 + 8x_3 = -3$

2.  $5x_1 + 3x_2 - 2x_3 = 0$
    $-2x_1 - 2x_2 + x_3 = 4$
    $-8x_1 - 8x_2 + 3x_3 = -4$

3.  $2x_1 - x_2 - 4x_3 = -3$
    $x_1 + 2x_2 + 3x_3 = 1$
    $-3x_1 + x_2 + 5x_3 = -10$

4.  $3x_1 + x_2 + 2x_3 = 4$
    $x_2 + 2x_3 = 1$
    $x_1 + x_2 + 2x_3 = 2$

5.  $x_1 - 2x_2 + x_3 = 1$
    $2x_1 - 3x_2 + x_3 = 0$
    $-3x_1 + 2x_2 + 3x_3 = 1$
    $x_1 + x_2 - 2x_3 = 0$

6.  $2x_1 + x_2 + x_3 = 4$
    $x_2 + 3x_3 = 2$
    $2x_1 + x_2 + 2x_3 = 5$
    $x_1 + x_3 = 3$

*some answers are inconsistent*

7. What are the possible dimensions of the solution space of a homogeneous system of
   a) 2 linear equations in 4 unknowns?
   b) 3 linear equations in 4 unknowns?
   c) 4 linear equations in 4 unknowns?
8. Prove Theorem 9–10.


## 9–7  EIGENVALUES AND EIGENVECTORS

In the preceding section we saw how a system of linear equations can be solved by
the method of reduction to triangular form. In many situations this method is
tedious at best and ill suited for computations. For this reason, among others,
we now turn our attention to an entirely different technique which, though lacking
the generality of the method of reduction to triangular form, has the merit of being
efficient and easy to use.

The method we have in mind originates in the attempt to solve operator equations of the form

$$A\mathbf{x} = \lambda\mathbf{x}, \tag{9–14}$$

where $A$ is a linear transformation from $\mathscr{S}$ to $\mathscr{V}$, $\mathscr{S}$ being a subspace of $\mathscr{V}$, and
$\lambda$ is an unknown parameter which may assume real or complex values. Technically
the problem of solving (9–14) is known as the **eigenvalue problem** for the operator $A$,
and requires that we find all $\lambda$ for which (9–14) has a *nontrivial* solution and all
solutions corresponding to these $\lambda$. The motive for considering this equation
becomes clear if one imagines that $\mathbf{x}$ belongs to a basis for $\mathscr{S}$ which in turn is part
of a basis for $\mathscr{V}$. For then the matrix of $A$ with respect to these bases will have only
one nonzero entry in the column associated with $\mathbf{x}$. This entry will, of course, be $\lambda$,
and it will appear on the main diagonal of the matrix. We shall return to these
considerations in a moment, but first let us introduce some terminology.

> **Definition 9–5**   *The values of $\lambda$ for which Eq. (9–14) has nonzero solutions are
> called the **eigenvalues** (or **characteristic values**) of $A$, and for each eigenvalue $\lambda_0$,
> the nonzero vectors in $\mathscr{S}$ which satisfy the equation $A\mathbf{x} = \lambda_0\mathbf{x}$ are called the
> **eigenvectors** (or **characteristic vectors**) of $A$ **belonging to** $\lambda_0$.*

We call attention to the fact that by definition the zero vector is *never* an eigenvector
for $A$. Furthermore, zero is an eigenvalue for $A$ if and only if the equation $A\mathbf{x} = \mathbf{0}$
has nonzero solutions, that is, if and only if $A$ is *not* one-to-one. Failure to appreciate these distinctions is a frequent source of confusion.

If $\lambda_0$ is an eigenvalue for $A$ and $\mathbf{x}_0$ is an eigenvector belonging to $\lambda_0$, then

$$A(\alpha\mathbf{x}_0) = \alpha A(\mathbf{x}_0) = \alpha(\lambda_0\mathbf{x}_0) = \lambda_0(\alpha\mathbf{x}_0)$$

for all real numbers $\alpha$. Thus $\alpha\mathbf{x}_0$ is also an eigenvector belonging to $\lambda_0$ whenever $\alpha$
is different from zero. This fact, combined with the obvious fact that the sum of
two eigenvectors belonging to $\lambda_0$ is again an eigenvector belonging to $\lambda_0$, yields

**Lemma 9–1**    *The solution set of the equation* $A\mathbf{x} = \lambda_0\mathbf{x}$ *is a nontrivial subspace of* $\mathscr{S}$ *for each eigenvalue* $\lambda_0$ *of* $A$.

In other words, the zero vector together with the eigenvectors for $A$ which belong to $\lambda_0$ constitute a subspace of $\mathscr{S}$ (and hence, by implication, of $\mathscr{V}$ as well). We shall denote this subspace by $\mathscr{S}_{\lambda_0}$ and observe in passing that dim $\mathscr{S}_{\lambda_0} \geq 1$ for all $\lambda_0$. Geometrically, $A$ acts on $\mathscr{S}_{\lambda_0}$ by "stretching" each of its vectors by the scalar factor $\lambda_0$, as indicated in Fig. 9–1.

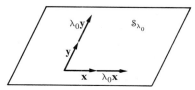

**Figure 9–1**

The fact that $A\mathbf{x}_0$ belongs to $\mathscr{S}_{\lambda_0}$ for all $\mathbf{x}_0$ in $\mathscr{S}_{\lambda_0}$ is sometimes expressed by saying that $\mathscr{S}_{\lambda_0}$ is "invariant" under $A$ in accordance with the following definition.

**Definition 9–6**    *Let* $A\colon \mathscr{S} \to \mathscr{V}$ *be a linear transformation, and suppose that* $\mathscr{S}$ *is a subspace of* $\mathscr{V}$. *Then a subspace* $\mathscr{W}$ *of* $\mathscr{S}$ *is said to be **invariant** under* $A$ *if and only if* $A\mathbf{w}$ *belongs to* $\mathscr{W}$ *for all* $\mathbf{w}$ *in* $\mathscr{W}$.

We hasten to point out that there is nothing in this definition which implies that the nonzero vectors in an invariant subspace for $A$ need be eigenvectors for $A$. Indeed, we shall see in a moment that such a conclusion is false. Instead, the implication goes the other way: the $\mathscr{S}_{\lambda_0}$ are invariant subspaces consisting of vectors with the special property that $A\mathbf{x} = \lambda_0\mathbf{x}$.

Having introduced the notion of invariant subspace, we can now rephrase the definition of an eigenvector for a linear transformation $A\colon \mathscr{S} \to \mathscr{V}$ to read: *A nonzero vector* $\mathbf{x}$ *in* $\mathscr{S}$ *is an eigenvector for* $A$ *if and only if the one-dimensional subspace of* $\mathscr{S}$ *spanned by* $\mathbf{x}$ *is invariant under* $A$. This observation is often useful in the search for eigenvectors.

**Example 1**    Let $\mathbf{e}_1, \mathbf{e}_2$ be the standard basis vectors in $\mathscr{R}^2$, and let $A\colon \mathscr{R}^2 \to \mathscr{R}^2$ be a reflection across the $\mathbf{e}_1$-axis; that is,

$$A\mathbf{e}_1 = \mathbf{e}_1, \qquad A\mathbf{e}_2 = -\mathbf{e}_2.$$

Then from geometric considerations alone it is clear that the only subspaces of $\mathscr{R}^2$ which are invariant under $A$ are (i) the trivial subspace, (ii) $\mathscr{R}^2$ itself, and (iii) the two one-dimensional subspaces spanned by $\mathbf{e}_1$ and $\mathbf{e}_2$. By the observation made a moment ago, the last two subspaces must consist of eigenvectors for $A$ and are the only such subspaces. Further, it is obvious from the definition of $A$ that these subspaces are associated with the eigenvalues 1 and $-1$, respectively.

**Example 2**    If $A: \mathcal{R}^2 \to \mathcal{R}^2$ is reflection across the origin, then *every* subspace of $\mathcal{R}^2$ is invariant under $A$. In this case $A\mathbf{x} = -\mathbf{x}$ for all $\mathbf{x}$, and $-1$ is the only eigenvalue for $A$, with $\mathcal{S}_{-1} = \mathcal{R}^2$. Note that the invariant subspace associated with the eigenvalue here is *two*-dimensional.

**Example 3**    Let $A$ be a rotation of $\mathcal{R}^2$ about the origin through an angle $\theta$. Then, if $\theta$ is not an integral multiple of $\pi$, there are no one-dimensional invariant subspaces, and $A$ has no eigenvectors at all.

We now prove a theorem which justifies introducing these notions in the first place.

‖  **Theorem 9–12**    *Any set of eigenvectors belonging to distinct eigenvalues of a linear transformation $A: \mathcal{S} \to \mathcal{V}$ is linearly independent in $\mathcal{S}$.*

(Note that this result would fail if $\mathbf{0}$ were allowed to be an eigenvector; thus the prejudice against the zero vector found in Definition 9–5.)

*Proof.*    The theorem is obviously true when applied to a single eigenvector. Beyond this we reason by induction.

Assume that the theorem has been proved for every set of $n - 1$ eigenvectors for $A$, where $n > 1$. Let $\mathbf{x}_1, \ldots, \mathbf{x}_n$ be $n$ eigenvectors belonging, respectively, to distinct eigenvalues $\lambda_1, \ldots, \lambda_n$, and suppose that

$$\alpha_1 \mathbf{x}_1 + \cdots + \alpha_{n-1} \mathbf{x}_{n-1} + \alpha_n \mathbf{x}_n = \mathbf{0}. \qquad (9\text{–}15)$$

When we apply $A$ to both sides of this equation, we find that

$$\alpha_1 A\mathbf{x}_1 + \cdots + \alpha_{n-1} A\mathbf{x}_{n-1} + \alpha_n A\mathbf{x}_n = \mathbf{0}$$

or

$$\alpha_1(\lambda_1 \mathbf{x}_1) + \cdots + \alpha_{n-1}(\lambda_{n-1}\mathbf{x}_{n-1}) + \alpha_n(\lambda_n \mathbf{x}_n) = \mathbf{0}. \qquad (9\text{–}16)$$

We can then multiply (9–15) by $\lambda_n$ and subtract the resulting equation from (9–16) to obtain

$$\alpha_1(\lambda_1 - \lambda_n)\mathbf{x}_1 + \cdots + \alpha_{n-1}(\lambda_{n-1} - \lambda_n)\mathbf{x}_{n-1} = \mathbf{0}.$$

But by assumption the vectors $\mathbf{x}_1, \ldots, \mathbf{x}_{n-1}$ are linearly independent. Hence each of the coefficients in this expression vanishes, and since $\lambda_i - \lambda_n \neq 0$ whenever $i \neq n$, we conclude that $\alpha_i = 0$ for $i = 1, \ldots, n - 1$. This, together with (9–15), implies that $\alpha_n$ is also zero, and we are done.  ∎

**EXERCISES**

1.  Every vector space $\mathcal{V}$ has at least two subspaces which are invariant under a linear transformation $A: \mathcal{V} \to \mathcal{V}$. What are they?

2.  Let $A$ be a linear transformation on a finite-dimensional vector space $\mathcal{V}$ (i.e., $A: \mathcal{V} \to \mathcal{V}$), and let $\mathcal{S}_1$ be invariant under $A$.
    a)  Prove that there exists a subspace $\mathcal{S}_2$ of $\mathcal{V}$ such that every vector $\mathbf{x}$ in $\mathcal{V}$ can be written in one and only one way as $\mathbf{x} = \mathbf{x}_1 + \mathbf{x}_2$ with $\mathbf{x}_1$ in $\mathcal{S}_1$, $\mathbf{x}_2$ in $\mathcal{S}_2$. [*Hint:* Choose an appropriate basis for $\mathcal{V}$.]

b)  Is the subspace $\mathscr{S}_2$ found in (a) necessarily unique? Why?

c)  Prove by example that it may be impossible to choose the subspace $\mathscr{S}_2$ of part (a) so that $\mathscr{S}_2$ is invariant under $A$. [*Hint*: Let $\mathscr{V} = \mathscr{P}_n$, the space of polynomials of degree $< n$, and let $A$ be the differentiation operator.]

3.  Let $\mathscr{S}_1$ and $\mathscr{S}_2$ be invariant under a linear transformation $A\colon \mathscr{S} \to \mathscr{V}$, $\mathscr{S}$ being a subspace of $\mathscr{V}$. Prove that the subspace of $\mathscr{S}$ spanned by $\mathscr{S}_1$ and $\mathscr{S}_2$ is also invariant under $A$.

4.  a)  Show that the null space of $A\colon \mathscr{V} \to \mathscr{V}$ is invariant under $A$.

b)  Let $A\colon \mathscr{V} \to \mathscr{V}$ be a linear transformation with the property that $A^2 = A$. Prove that the image of $A$ is an invariant subspace of $\mathscr{V}$ under $A$.

## 9-8  EIGENVECTORS IN FINITE-DIMENSIONAL SPACES

We have seen that eigenvectors belonging to distinct eigenvalues for a linear transformation are linearly independent. In the finite-dimensional case, this fact immediately yields the following theorem.

**Theorem 9–13**   *A linear transformation $A$ mapping an $n$-dimensional vector space $\mathscr{V}$ into itself has at most $n$ distinct eigenvalues. Moreover, when the number of distinct eigenvalues is equal to $n$, any complete set of eigenvectors, one for each eigenvalue, is a basis for $\mathscr{V}$, and the matrix of $A$ with respect to such a basis is*

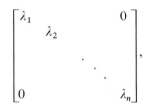

$$\begin{bmatrix} \lambda_1 & & & 0 \\ & \lambda_2 & & \\ & & \ddots & \\ 0 & & & \lambda_n \end{bmatrix},$$

*with the eigenvalues on the main diagonal and zeros elsewhere.*

As Examples 2 and 3 of the preceding section showed, such bases need not exist for a particular $A\colon \mathscr{V} \to \mathscr{V}$. But when they do, a number of pleasant things happen. For one thing, we can then solve operator equations involving $A$, and quite efficiently too. The following example illustrates the technique.

**Example 1**   Let $A$ be a linear transformation mapping $\mathscr{R}^3$ into itself and suppose that $A$ has distinct eigenvalues $\lambda_1, \lambda_2, \lambda_3$. Let $\mathbf{e}_1, \mathbf{e}_2, \mathbf{e}_3$ be eigenvectors belonging to these eigenvalues and consider the equation

$$A\mathbf{x} = \mathbf{y}, \tag{9–17}$$

where $\mathbf{y}$ is known, and $\mathbf{x}$ unknown. Then since $\{\mathbf{e}_1, \mathbf{e}_2, \mathbf{e}_3\}$ is a basis for $\mathscr{R}^3$, we have

$$\mathbf{x} = x_1\mathbf{e}_1 + x_2\mathbf{e}_2 + x_3\mathbf{e}_3,$$
$$\mathbf{y} = y_1\mathbf{e}_1 + y_2\mathbf{e}_2 + y_3\mathbf{e}_3,$$

and (9–17) can be written as

$$A(x_1\mathbf{e}_1 + x_2\mathbf{e}_2 + x_3\mathbf{e}_3) = y_1\mathbf{e}_1 + y_2\mathbf{e}_2 + y_3\mathbf{e}_3.$$

Hence

$$(x_1\lambda_1)\mathbf{e}_1 + (x_2\lambda_2)\mathbf{e}_2 + (x_3\lambda_3)\mathbf{e}_3 = y_1\mathbf{e}_1 + y_2\mathbf{e}_2 + y_3\mathbf{e}_3,$$

and it follows that $x_1, x_2, x_3$ must be chosen so that

$$x_1\lambda_1 = y_1, \qquad x_2\lambda_2 = y_2, \qquad x_3\lambda_3 = y_3.$$

In particular, we see that (9–17) has the *unique* solution

$$\mathbf{x} = \frac{y_1}{\lambda_1}\,\mathbf{e}_1 + \frac{y_2}{\lambda_2}\,\mathbf{e}_2 + \frac{y_3}{\lambda_3}\,\mathbf{e}_3,$$

whenever the $\lambda_i$ are different from zero. If, on the other hand, one of the $\lambda_i$, say $\lambda_1$, is zero, then (9–17) has no solutions at all *unless* $y_1 = 0$. In the latter case the equation $x_1\lambda_1 = y_1$ is satisfied for *all* values of $x_1$, and the solution set of (9–17) then consists of all vectors of the form

$$\mathbf{x} = x_1\mathbf{e}_1 + \frac{y_2}{\lambda_2}\,\mathbf{e}_2 + \frac{y_3}{\lambda_3}\,\mathbf{e}_3,$$

with $x_1$ arbitrary.

The generalization of these results to $n$-dimensional space is immediate and is left to the reader.

The technique introduced in Example 1 is known as the **eigenvalue method** for solving an operator equation. Its success depends on the existence of enough eigenvectors for $A$ to span $\mathscr{V}$, and on our ability to find them. Since both these questions can be settled by computing the eigenvalues of $A$, we now address ourselves to this problem.

In Section 9–6 we saw that an equation of the form $A\mathbf{x} = \mathbf{y}$, involving a linear transformation mapping an $n$-dimensional vector space $\mathscr{V}$ into itself, can be written in terms of a basis for $\mathscr{V}$ as a system of linear equations

$$\alpha_{11}x_1 + \alpha_{12}x_2 + \cdots + \alpha_{1n}x_n = y_1$$
$$\vdots$$
$$\alpha_{n1}x_1 + \alpha_{n2}x_2 + \cdots + \alpha_{nn}x_n = y_n,$$

where

$$\begin{bmatrix} \alpha_{11} & \alpha_{12} & \cdots & \alpha_{1n} \\ \cdot & \cdot & & \cdot \\ \cdot & \cdot & & \cdot \\ \cdot & \cdot & & \cdot \\ \alpha_{n1} & \alpha_{n2} & \cdots & \alpha_{nn} \end{bmatrix}$$

is the matrix of $A$, and $x_1, \ldots, x_n$ and $y_1, \ldots, y_n$ are the components of $\mathbf{x}$ and $\mathbf{y}$, all with respect to the chosen basis. In particular, this is true of the equation

$$A\mathbf{x} = \lambda\mathbf{x},$$

which can be rewritten as

$$(A - \lambda I)\mathbf{x} = \mathbf{0},$$

where $I$ is the identity transformation on $\mathscr{V}$. Noting that $A - \lambda I$ can be represented by the matrix

$$\begin{bmatrix} \alpha_{11} - \lambda & \alpha_{12} & \cdots & \alpha_{1n} \\ \alpha_{21} & \alpha_{22} - \lambda & \cdots & \alpha_{2n} \\ \cdot & & & \cdot \\ \cdot & & & \cdot \\ \cdot & & & \cdot \\ \alpha_{n1} & \alpha_{n2} & \cdots & \alpha_{nn} - \lambda \end{bmatrix},$$

we conclude that *the eigenvalues for $A$ are the values of $\lambda$ for which the system of homogeneous equations*

$$\begin{aligned}
(\alpha_{11} - \lambda)x_1 + \alpha_{12}x_2 + \cdots + \alpha_{1n}x_n &= 0 \\
\alpha_{21}x_1 + (\alpha_{22} - \lambda)x_2 + \cdots + \alpha_{2n}x_n &= 0 \\
&\phantom{=}\ \cdot \\
&\phantom{=}\ \cdot \\
&\phantom{=}\ \cdot \\
\alpha_{n1}x_1 + \alpha_{n2}x_2 + \cdots + (\alpha_{nn} - \lambda)x_n &= 0
\end{aligned} \tag{9–18}$$

*has **nontrivial** solutions*. But this will occur if and only if the determinant of the coefficients of (9–18) vanishes, that is, if and only if

$$\begin{vmatrix} \alpha_{11} - \lambda & \alpha_{12} & \cdots & \alpha_{1n} \\ \alpha_{21} & \alpha_{22} - \lambda & \cdots & \alpha_{2n} \\ \cdot & & & \cdot \\ \cdot & & & \cdot \\ \cdot & & & \cdot \\ \alpha_{n1} & \alpha_{n2} & \cdots & \alpha_{nn} - \lambda \end{vmatrix} = 0. \tag{9–19}$$

Thus the eigenvalues for $A$ can be computed by solving (9–19) for $\lambda$, and since the left-hand side of this equation is an $n$th-degree polynomial in $\lambda$, the computation can be done by the methods of elementary algebra (at least for small values of $n$). The polynomial appearing in (9–19) is known as the **characteristic polynomial** of the linear transformation $A$, and the equation itself is called the **characteristic equation** of $A$. As its name suggests, the characteristic polynomial is independent of the particular basis used to compute it—a fact which is proved in most texts on linear algebra.

    **Example 2**   Find the eigenvalues and eigenvectors for the linear transformation $A: \mathscr{R}^2 \to \mathscr{R}^2$, given that the matrix of $A$ with respect to the standard basis $\mathbf{e}_1, \mathbf{e}_2$ is

$$\begin{bmatrix} 1 & 0 \\ 2 & 1 \end{bmatrix}.$$

*Solution.*   In this case the characteristic equation of $A$ is

$$\begin{vmatrix} 1 - \lambda & 0 \\ 2 & 1 - \lambda \end{vmatrix} = 0$$

or

$$(1 - \lambda)^2 = 0.$$

Therefore, $\lambda = 1$ is the only eigenvalue for $A$. Hence a nonzero vector $\mathbf{x} = x_1\mathbf{e}_1 + x_2\mathbf{e}_2$ is an eigenvector for $A$ if and only if $A\mathbf{x} = \mathbf{x}$. When we rewrite this equation in matrix form as

$$\begin{bmatrix} 1 & 0 \\ 2 & 1 \end{bmatrix}\begin{bmatrix} x_1 \\ x_2 \end{bmatrix} = \begin{bmatrix} x_1 \\ x_2 \end{bmatrix},$$

we find that $x_1$ and $x_2$ must satisfy the equations

$$x_1 = x_1$$
$$2x_1 + x_2 = x_2.$$

Thus $x_1$ must be zero, while $x_2$ is arbitrary, and the eigenvectors for $A$ are of the form $x_2\mathbf{e}_2$, $x_2 \neq 0$. Finally, the associated invariant subspace $\mathscr{S}_1$ is the one-dimensional subspace of $\mathscr{R}^2$ spanned by $\mathbf{e}_2$.

**Example 3**    Find the eigenvalues and eigenvectors of the linear transformation $A$ on $\mathscr{R}^3$ whose matrix with respect to the standard basis $\mathbf{e}_1, \mathbf{e}_2, \mathbf{e}_3$ is

$$\begin{bmatrix} 0 & 0 & 1 \\ 0 & -1 & 0 \\ 2 & 2 & 1 \end{bmatrix}.$$

*Solution.*    Since the characteristic polynomial of $A$ is

$$\begin{vmatrix} -\lambda & 0 & 1 \\ 0 & -(1 + \lambda) & 0 \\ 2 & 2 & 1 - \lambda \end{vmatrix} = -(\lambda - 2)(\lambda + 1)^2,$$

the eigenvalues for $A$ are $\lambda = 2$ and $\lambda = -1$. To find the associated eigenvectors, we set $\mathbf{x} = x_1\mathbf{e}_1 + x_2\mathbf{e}_2 + x_3\mathbf{e}_3$, and solve the equation

$$A\mathbf{x} = \lambda\mathbf{x} \tag{9--20}$$

for $\lambda = 2$ and $\lambda = -1$.

In the first case, (9–20) becomes

$$\begin{bmatrix} 0 & 0 & 1 \\ 0 & -1 & 0 \\ 2 & 2 & 1 \end{bmatrix}\begin{bmatrix} x_1 \\ x_2 \\ x_3 \end{bmatrix} = \begin{bmatrix} 2x_1 \\ 2x_2 \\ 2x_3 \end{bmatrix},$$

and we have

$$x_3 = 2x_1$$
$$-x_2 = 2x_2$$
$$2x_1 + 2x_2 + x_3 = 2x_3.$$

Thus $x_2 = 0$, $2x_1 = x_3$, and the relevant eigenvectors are $x_1\mathbf{e}_1 + 2x_1\mathbf{e}_3$, where

$x_1$ is an arbitrary nonzero constant. Here the associated invariant subspace $\mathscr{S}_2$ is the one-dimensional subspace of $\mathscr{R}^3$ spanned by $\mathbf{e}_1 + 2\mathbf{e}_3$.

Finally, a similar computation reveals that the eigenvectors belonging to the eigenvalue $-1$ are of the form $x_1\mathbf{e}_1 - x_1\mathbf{e}_3$, with $x_1$ an arbitrary nonzero constant. The associated invariant subspace is the one-dimensional subspace of $\mathscr{R}^3$ spanned by the vector $\mathbf{e}_1 - \mathbf{e}_3$.

**Example 4**    In the preceding section we said that a rotation of $\mathscr{R}^2$ about the origin through an angle $\theta \neq n\pi$ has no (real) eigenvalues or eigenvectors. We can now establish this fact algebraically.

The matrix of a rotation $A$ of $\mathscr{R}^2$ with respect to the standard basis is

$$\begin{bmatrix} \cos\theta & -\sin\theta \\ \sin\theta & \cos\theta \end{bmatrix},$$

where $\theta$ is the angle of rotation. Hence the characteristic equation of $A$ is

$$\begin{vmatrix} \cos\theta - \lambda & -\sin\theta \\ \sin\theta & \cos\theta - \lambda \end{vmatrix} = 0,$$

and we have

$$\lambda^2 - 2(\cos\theta)\lambda + 1 = 0.$$

Thus

$$\lambda = \cos\theta \pm i\sin\theta,$$

and it follows that $\lambda$ is real if and only if $\theta = n\pi$. Moreover, when this is the case $\lambda$ assumes one of the values $\pm 1$ and has all of $\mathscr{R}^2$ as its invariant subspace. Otherwise, $\lambda$ is complex and $A$ has no eigenvectors.

**EXERCISES**

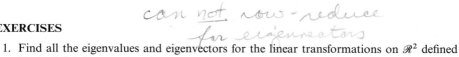

*can not row-reduce for eigenvectors*

1. Find all the eigenvalues and eigenvectors for the linear transformations on $\mathscr{R}^2$ defined by the following matrices.

   a) $\begin{bmatrix} 1 & 1 \\ 0 & 2 \end{bmatrix}$      b) $\begin{bmatrix} 1 & 0 \\ 0 & 0 \end{bmatrix}$      c) $\begin{bmatrix} 1 & 1 \\ 1 & 1 \end{bmatrix}$      d) $\begin{bmatrix} 1 & 4 \\ 1 & 1 \end{bmatrix}$

2. Find all the eigenvalues for the linear transformations on $\mathscr{R}^3$ defined by the following matrices, and in each case find the eigenvectors belonging to the real eigenvalues.

   a) $\begin{bmatrix} 2 & 0 & 1 \\ -1 & 2 & 3 \\ 1 & 0 & 2 \end{bmatrix}$          b) $\begin{bmatrix} 1 & 2 & 0 \\ 0 & 1 & 1 \\ 0 & 1 & 1 \end{bmatrix}$

   c) $\begin{bmatrix} -1 & 1 & 0 \\ 1 & 2 & 2 \\ 0 & -1 & 0 \end{bmatrix}$          d) $\begin{bmatrix} 2 & 0 & 0 \\ 0 & -1 & 0 \\ 1 & 0 & 1 \end{bmatrix}$

3. Repeat Exercise 2 for the following matrices.

a) $\begin{bmatrix} 1 & 0 & 1 \\ 0 & 1 & 0 \\ 1 & 0 & 1 \end{bmatrix}$
b) $\begin{bmatrix} 1 & 0 & 1 \\ 0 & 2 & 0 \\ 1 & 0 & -1 \end{bmatrix}$ .

c) $\begin{bmatrix} 5 & -6 & -6 \\ -1 & 4 & 2 \\ 3 & -6 & -4 \end{bmatrix}$
d) $\begin{bmatrix} 1 & 2 & 1 \\ 1 & 2 & 1 \\ 0 & 1 & 2 \end{bmatrix}$

4. Find the eigenvalues and eigenvectors for the linear transformation on $\mathscr{R}^4$ defined by the matrices:

a) $\begin{bmatrix} 1 & 0 & -1 & 0 \\ 0 & 1 & 1 & 0 \\ -1 & 1 & 2 & 1 \\ 0 & 0 & 1 & -1 \end{bmatrix}$
b) $\begin{bmatrix} -1 & 2 & 1 & 3 \\ 0 & 0 & -2 & 1 \\ 0 & 0 & 2 & -3 \\ 0 & 0 & 0 & 4 \end{bmatrix}$ .

5. Let $A: \mathscr{R}^3 \to \mathscr{R}^3$ be the linear transformation whose matrix with respect to the standard basis $e_1, e_2, e_3$ is

$$\begin{bmatrix} 1 & 0 & 2 \\ 0 & 1 & 0 \\ 2 & 0 & 1 \end{bmatrix}.$$

Use the eigenvalue method to solve the equation $A\mathbf{x} = \mathbf{y}$ for $\mathbf{x}$, given that:

a) $\mathbf{y} = 2e_1 + e_2$ 
b) $\mathbf{y} = e_1 + e_3$
c) $\mathbf{y} = 4e_1 - 2e_2 - 2e_3$

6. Repeat Exercise 5 when the matrix of $A$ is

$$\begin{bmatrix} 0 & 2 & 1 \\ 0 & 1 & 3 \\ 0 & 0 & 2 \end{bmatrix}$$

and

a) $\mathbf{y} = -2e_1 + e_2$ 
b) $\mathbf{y} = 4e_1 + 4e_2 + 2e_3$
c) $\mathbf{y} = e_1 + 9e_2 + 2e_3$.

7. Prove that every linear transformation on an *odd*-dimensional vector space has at least one real eigenvalue. Give a geometric interpretation of this result for linear transformations on $\mathscr{R}^3$.

8. Prove that each of the diagonal entries in an $n \times n$ matrix in triangular form is an eigenvalue for the linear transformation on $\mathscr{R}^n$ defined by the matrix.

* 9. a) Let $A$ be a linear transformation on a finite-dimensional vector space $\mathscr{V}$, and let $\lambda_0$ be a real eigenvalue for $A$ of multiplicity $m$, by which we mean that $(\lambda - \lambda_0)^m$ is a factor of the characteristic polynomial for $A$. Prove that the dimension of the invariant subspace associated with $\lambda_0$ is *at most* $m$. [*Hint*: Consider the characteristic polynomial of the linear transformation obtained by restricting $A$ to the subspace $\mathscr{S}_{\lambda_0}$.]

b) Give an example to show that this dimension can, in certain cases, be less than $m$. [*Hint*: Consider the operator $D$ on the space of polynomials $\mathscr{P}_n$.]

# Systems of Linear Differential Equations

**10**

## 10–1 INTRODUCTION

If each of the coefficients in a system of $m$ linear equations in $n$ unknowns is a linear differential operator defined on an interval $I$, and if each of the quantities on the right-hand side of the equations is a continuous function on $I$, we then have what is known as a system of $m$ linear differential equations in $n$ unknowns. Such a system has the form

$$L_{11}x_1 + \cdots + L_{1n}x_n = h_1(t)$$
$$\vdots$$
$$\tag{10–1}$$
$$\vdots$$
$$L_{m1}x_1 + \cdots + L_{mn}x_n = h_m(t),$$

where the $L_{ij}$ are linear differential operators defined on $I$, and $x_1, \ldots, x_n$ are unknown functions of $t$. As usual we say that a system like this is **homogeneous** if all of the $h_i$ are identically zero and **nonhomogeneous** otherwise. A **solution** of this system is an $n$-tuple of functions which satisfies each of the equations identically on $I$. Such systems are particularly important in applications and, as we shall see, arise in such diverse fields as biology, economics, and physics.

The formal similarity between (10–1) and an ordinary system of linear equations suggests that we rewrite (10–1) in matrix form as

$$\mathbf{LX} = \mathbf{H}(t), \tag{10–2}$$

where $\mathbf{L}$ is the $m \times n$ operator matrix

$$\begin{bmatrix} L_{11} & \cdots & L_{1n} \\ \cdot & & \cdot \\ \cdot & & \cdot \\ \cdot & & \cdot \\ L_{m1} & \cdots & L_{mn} \end{bmatrix},$$

355

and $\mathbf{X}$ and $\mathbf{H}(t)$ are the column vectors

$$\mathbf{X} = \begin{bmatrix} x_1 \\ \cdot \\ \cdot \\ \cdot \\ x_n \end{bmatrix} \qquad \text{and} \qquad \mathbf{H}(t) = \begin{bmatrix} h_1(t) \\ \cdot \\ \cdot \\ \cdot \\ h_m(t) \end{bmatrix}.$$

For then

$$\begin{bmatrix} L_{11} \cdots L_{1n} \\ \cdot \quad \cdot \\ \cdot \quad \cdot \\ \cdot \quad \cdot \\ L_{m1} \cdots L_{mn} \end{bmatrix} \begin{bmatrix} x_1 \\ \cdot \\ \cdot \\ \cdot \\ x_n \end{bmatrix} = \begin{bmatrix} L_{11}x_1 + \cdots + L_{1n}x_n \\ \cdot \qquad\qquad \cdot \\ \cdot \qquad\qquad \cdot \\ \cdot \qquad\qquad \cdot \\ L_{m1}x_1 + \cdots + L_{mn}x_n \end{bmatrix}, \qquad (10\text{-}3)$$

and (10-2) is an alternative version of (10-1). Of course, this kind of manipulation does not mean much until we have introduced suitable vector spaces and an appropriate linear transformation between them. But this is easily done. We just let $\mathscr{V}_m$ and $\mathscr{V}_n$ be the spaces of column vectors of the form

$$\mathbf{H}(t) = \begin{bmatrix} h_1(t) \\ \cdot \\ \cdot \\ \cdot \\ h_m(t) \end{bmatrix}, \qquad \mathbf{X}(t) = \begin{bmatrix} x_1(t) \\ \cdot \\ \cdot \\ \cdot \\ x_n(t) \end{bmatrix},$$

respectively, where the $h_i(t)$ are continuous on $I$ and the $x_i(t)$ sufficiently differentiable on $I$ so that the $L_{ij}$ can be applied to them, and where addition and scalar multiplication are defined componentwise, as usual. This done, we let $L: \mathscr{V}_n \to \mathscr{V}_m$ be the linear transformation defined by (10-3); that is,

$$L(\mathbf{X}) = \begin{bmatrix} L_{11}x_1 + \cdots + L_{1n}x_n \\ \cdot \qquad\qquad \cdot \\ \cdot \qquad\qquad \cdot \\ \cdot \qquad\qquad \cdot \\ L_{m1}x_1 + \cdots + L_{mn}x_n \end{bmatrix}.$$

Then the original system can indeed be written in vector form as $\mathbf{LX} = \mathbf{H}(t)$.

The virtue of this approach, besides an obvious economy in notation, is that it provides a conceptual setting for the study of systems of linear differential equations in which we can apply the known results for operator equations. But as it stands, (10-1) is far too general to permit a systematic analysis leading to specific techniques of solution. Hence in the sections which follow we shall devote our efforts to the study of more specialized systems for which detailed information can be obtained.

**EXERCISES**

1. Show that the $\mathcal{V}_m$ and $\mathcal{V}_n$ just defined are real vector spaces.
2. Show that the mapping $L: \mathcal{V}_n \to \mathcal{V}_m$ defined by Eq. (10–3) is linear.

## 10–2 RESULTS FROM THE GENERAL THEORY OF FIRST-ORDER SYSTEMS

In this section we shall consider systems of first-order equations of the form

$$x_1' = a_{11}(t)x_1 + \cdots + a_{1n}(t)x_n + b_1(t)$$

$$\vdots \tag{10–4}$$

$$x_n' = a_{n1}(t)x_1 + \cdots + a_{nn}(t)x_n + b_n(t),$$

in which the $a_{ij}(t)$ and $b_i(t)$ are continuous on an interval $I$. The matrix version of (10–4) is

$$\mathbf{X}' = \mathbf{A}(t)\mathbf{X} + \mathbf{B}(t), \tag{10–5}$$

where

$$\mathbf{A}(t) = \begin{bmatrix} a_{11}(t) \cdots a_{1n}(t) \\ \cdot \qquad \cdot \\ \cdot \qquad \cdot \\ \cdot \qquad \cdot \\ a_{n1}(t) \cdots a_{nn}(t) \end{bmatrix}$$

and

$$\mathbf{X} = \begin{bmatrix} x_1 \\ \cdot \\ \cdot \\ \cdot \\ x_n \end{bmatrix}, \qquad \mathbf{X}' = \begin{bmatrix} x_1' \\ \cdot \\ \cdot \\ \cdot \\ x_n' \end{bmatrix}, \qquad \mathbf{B}(t) = \begin{bmatrix} b_1(t) \\ \cdot \\ \cdot \\ \cdot \\ b_n(t) \end{bmatrix}.$$

As usual, an **initial-value problem** for such a system requires that we find a solution $\mathbf{X} = \mathbf{X}(t)$ of the system which satisfies an **initial condition** $\mathbf{X}(t_0) = \mathbf{X}_0$, where $t_0$ is a point in $I$ and

$$\mathbf{X}_0 = \begin{bmatrix} c_1 \\ \cdot \\ \cdot \\ \cdot \\ c_n \end{bmatrix}$$

is a point in $\mathcal{R}^n$.

Systems of this type are especially important in the theory of linear differential equations because, among other things, every normal $n$th-order linear differential equation can be transformed into a normal first-order system. We now prove this assertion as a theorem.

**Theorem 10–1**   *Every normal nth-order linear differential equation is equivalent to an $n \times n$ system of normal first-order linear differential equations.*

*Proof.*   Starting with

$$x^{(n)} + a_{n-1}(t)x^{(n-1)} + \cdots + a_0(t)x = h(t), \qquad (10\text{–}6)$$

let $x_1, \ldots, x_n$ be new variables defined by

$$x_1(t) = x(t)$$
$$x_2(t) = x'(t)$$
$$\cdot$$
$$\cdot$$
$$\cdot$$
$$x_n(t) = x^{(n-1)}(t).$$

Then (10–6) can be rewritten as

$$\begin{aligned}
x_1' &= x_2 \\
x_2' &= x_3 \\
&\ \cdot \\
&\ \cdot \\
&\ \cdot \\
x_{n-1}' &= x_n \\
x_n' &= -a_0(t)x_1 - a_1(t)x_2 - \cdots - a_{n-1}(t)x_n + h(t),
\end{aligned} \qquad (10\text{–}7)$$

which is a system of the required kind. Finally, from the very way in which (10–7) was derived it is clear that (10–7) and (10–6) have the same solutions.  ∎

This theorem tells us that the theory of first-order linear systems includes the theory of $n$th-order linear equations as a special case. The converse, however, is not true because there exist first-order systems that cannot be converted into a single $n$th-order equation.

As in the case of a single $n$th-order equation, the theory of first-order linear systems is based on an existence and uniqueness theorem.

**Theorem 10–2**   *Every initial-value problem*

$$\mathbf{X}' = \mathbf{A}(t)\mathbf{X} + \mathbf{B}(t), \qquad \mathbf{X}(t_0) = \mathbf{X}_0,$$

*involving a normal $n \times n$ system of first-order linear differential equations whose coefficients and right-hand sides are continuous on an interval $I$ has a unique solution on $I$.*

Assuming the truth of this result for the present (we will prove it in Chapter 12), we turn our attention to the homogeneous system

$$\mathbf{X}' = \mathbf{A}(t)\mathbf{X}, \qquad (10\text{–}8)$$

whose solution set is a subspace $\mathscr{W}$ of the vector space $\mathscr{V}_n$ defined in the preceding section. When combined with our earlier results on the dimension of the solution space of a normal homogeneous $n$th-order linear differential equation, Theorem 10–1 suggests that $\mathscr{W}$, too, is $n$-dimensional. Indeed, it is, and the proof follows easily from the next lemma.

**Lemma 10–1**    *Let $X_1, \ldots, X_k$ be solutions of $X' = A(t)X$ on $I$, and let $t_0$ be any point in $I$. Then $X_1, \ldots, X_k$ are linearly dependent in $\mathscr{V}_n$ if and only if the vectors $X_1(t_0), \ldots, X_k(t_0)$ are linearly dependent in $\mathscr{R}^n$.*

*Proof.*    Suppose that there exist constants $c_1, \ldots, c_k$, not all zero, such that

$$\sum_{j=1}^{k} c_j X_j(t) = 0$$

for all $t$ in $I$. Then for $t = t_0$ we have

$$\sum_{j=1}^{k} c_j X_j(t_0) = 0,$$

which means that the vectors $X_1(t_0), \ldots, X_k(t_0)$ are linearly dependent in $\mathscr{R}^n$. Conversely, suppose that

$$\sum_{j=1}^{k} c_j X_j(t_0) = 0,$$

where again the $c_j$ are constants and not all zero. Then since the $X_j(t)$ are solutions of $X' = A(t)X$, the vector

$$X(t) = \sum_{j=1}^{k} c_j X_j(t)$$

is a solution of the initial-value problem

$$X' = A(t)X, \qquad X(t_0) = 0.$$

But by Theorem 10–2 the only solution of this problem is the trivial solution $X(t) = 0$ for all $t$ in $I$. Hence

$$\sum_{j=1}^{k} c_j X_j(t) = 0,$$

and $X_1, \ldots, X_k$ are linearly dependent in $\mathscr{V}_n$.    ∎

To continue the discussion that preceded the lemma, we now prove

**Theorem 10–3**    *If $\mathscr{W}$ is the solution space of the $n \times n$ system*

$$X' = A(t)X,$$

*then* $\dim \mathscr{W} = n$.

*Proof.* In the first place the dimension of $\mathscr{W}$ cannot exceed $n$. For if it did we could find $n + 1$ linearly independent vectors $\mathbf{X}_1, \ldots, \mathbf{X}_{n+1}$ in $\mathscr{W}$, and the lemma would imply that the vectors $\mathbf{X}_1(t_0), \ldots, \mathbf{X}_{n+1}(t_0)$ are linearly independent in $\mathscr{R}^n$, which is impossible.

To complete the proof let $\mathbf{E}_i$ denote the $i$th standard basis vector in $\mathscr{R}^n$, and let $\mathbf{X}_i$ be the solution of the initial-value problem

$$\mathbf{X}' = \mathbf{A}(t)\mathbf{X}, \qquad \mathbf{X}(t_0) = \mathbf{E}_i.$$

Then since $\mathbf{E}_1, \ldots, \mathbf{E}_n$ are linearly independent in $\mathscr{R}^n$, Lemma 10–1 implies that $\mathbf{X}_1, \ldots, \mathbf{X}_n$ are linearly independent in $\mathscr{W}$. Thus the dimension of $\mathscr{W}$ must be at least $n$, and we are done. ∎

From here the theory of first-order systems of $n$ equations in $n$ unknowns develops in much the same way as the theory of a single $n$th-order equation. In particular, given $n$ solutions

$$\mathbf{X}_i(t) = \begin{bmatrix} x_{1i}(t) \\ \cdot \\ \cdot \\ \cdot \\ x_{ni}(t) \end{bmatrix}, \qquad 1 \leqslant i \leqslant n,$$

of $\mathbf{X}' = \mathbf{A}(t)\mathbf{X}$ on an interval $I$, the vectors $\mathbf{X}_1, \ldots, \mathbf{X}_n$ are a basis for the solution space of the equation if and only if their **Wronskian**

$$W[\mathbf{X}_1, \ldots, \mathbf{X}_n] = \begin{vmatrix} x_{11}(t) \cdots x_{1n}(t) \\ \cdot \qquad \cdot \\ \cdot \qquad \cdot \\ \cdot \qquad \cdot \\ x_{n1}(t) \cdots x_{nn}(t) \end{vmatrix} \tag{10–9}$$

never vanishes on $I$. The proof of this assertion depends on the fact that $W[\mathbf{X}_1, \ldots, \mathbf{X}_n]$ is identically zero on $I$ if and only if it vanishes at a single point of $I$. The details have been left as an exercise.

When $\mathbf{X}' = \mathbf{A}(t)\mathbf{X}$ is the first-order system derived from the normal $n$th-order equation

$$x^{(n)} + a_{n-1}(t)x^{(n-1)} + \cdots + a_0(t)x = 0, \tag{10–10}$$

$W[\mathbf{X}_1, \ldots, \mathbf{X}_n]$ is none other than the Wronskian of the $n$ solutions of (10–10) as defined in Chapter 4 (see Exercise 3). Hence the preceding results generalize known facts.

We conclude this section with a few remarks about solving nonhomogeneous first-order systems, which take the form

$$\mathbf{X}' = \mathbf{A}(t)\mathbf{X} + \mathbf{B}(t) \tag{10–11}$$

when expressed in matrix notation. As we know, every solution of such a system can be written in the form $\mathbf{X}_p + \mathbf{X}_h$ where $\mathbf{X}_p$ is a particular solution of (10–11) and $\mathbf{X}_h$ is a solution of the associated homogeneous system $\mathbf{X}' = \mathbf{A}(t)\mathbf{X}$. Thus if $\mathbf{X}_1, \ldots, \mathbf{X}_n$ is a basis for the solution space of

$$\mathbf{X}' = \mathbf{A}(t)\mathbf{X},$$

the general solution of (10–11) is

$$\mathbf{X}(t) = \mathbf{X}_p(t) + c_1\mathbf{X}_1(t) + \cdots + c_n\mathbf{X}_n(t),$$

the $c_i$ being arbitrary constants. Moreover, just as in the case of a single $n$th-order equation, a particular solution $\mathbf{X}_p$ of (10–11) can be obtained from a basis $\mathbf{X}_1, \ldots, \mathbf{X}_n$ of the associated homogeneous system by the method of variation of parameters. The procedure goes like this. First, form the vector

$$\mathbf{X}(t) = \sum_{i=1}^{n} c_i(t)\mathbf{X}_i(t) \tag{10–12}$$

and determine the $c_i(t)$ so that $\mathbf{X}(t)$ is a solution of (10–11). Then substitute (10–12) in (10–11) to obtain

$$\sum_{i=1}^{n} c_i(t)\mathbf{X}_i'(t) + \sum_{i=1}^{n} c_i'(t)\mathbf{X}_i(t) = \sum_{i=1}^{n} c_i(t)\mathbf{A}(t)\mathbf{X}_i(t) + \mathbf{B}(t).$$

Since $\mathbf{X}_i'(t) = \mathbf{A}(t)\mathbf{X}_i(t)$ for $i = 1, \ldots, n$, the equation above reduces to

$$\sum_{i=1}^{n} c_i'(t)\mathbf{X}_i(t) = \mathbf{B}(t),$$

which, in expanded form, reads

$$c_1'(t)x_{11}(t) + \cdots + c_n'(t)x_{1n}(t) = b_1(t)$$
$$\vdots$$
$$c_1'(t)x_{n1}(t) + \cdots + c_n'(t)x_{nn}(t) = b_n(t).$$

This system can be solved for $c_1'(t), \ldots, c_n'(t)$, since the determinant

$$W[\mathbf{X}_1, \ldots, \mathbf{X}_n] = \begin{vmatrix} x_{11}(t) \cdots x_{1n}(t) \\ \cdot \qquad \cdot \\ \cdot \qquad \cdot \\ \cdot \qquad \cdot \\ x_{n1}(t) \cdots x_{nn}(t) \end{vmatrix}$$

does not vanish anywhere on $I$. From these solutions we find $c_i(t)$ by integration, thereby determining a particular solution of (10–11).

As in Chapter 5, this method can be viewed as the construction of a (right) inverse for the linear transformation which sends every continuously differentiable vector $\mathbf{X}$ in $\mathscr{V}_n$ to the vector $\mathbf{X}' - \mathbf{A}(t)\mathbf{X}$. If we wished, we could push the analogy even further by introducing Green's functions for initial-value problems involving normal first-order systems and proving generalizations of all of our earlier results. But there is nothing new to be learned from such a discussion, and it is better to turn instead to the problem of devising specific techniques for solving systems of first-order linear differential equations.

### EXERCISES

1. Write the $n$th-order linear differential equation that corresponds to:

   a) $\begin{bmatrix} x_1' \\ x_2' \\ x_3' \end{bmatrix} = \begin{bmatrix} 0 & 1 & 0 \\ 0 & 0 & 1 \\ 2 & 1 & -3 \end{bmatrix} \begin{bmatrix} x_1 \\ x_2 \\ x_3 \end{bmatrix} + \begin{bmatrix} 0 \\ 0 \\ e^t \end{bmatrix}$

   b) $\begin{bmatrix} x_1' \\ x_2' \end{bmatrix} = \begin{bmatrix} 0 & 1 \\ -q(t) & -p(t) \end{bmatrix} \begin{bmatrix} x_1 \\ x_2 \end{bmatrix} + \begin{bmatrix} 0 \\ h(t) \end{bmatrix}$

2. Convert each of the following equations into a first-order linear system.

   a) $x'' + \dfrac{1}{t}x' + \left(\dfrac{t^2 - p^2}{t^2}\right)x = 0$

   b) $x''' + (\cos t)x'' + e^t x = t^2 + 1$

   c) $x^{(iv)} + e^t x'' = \cos t$

3. a) Write (10–7) in matrix form.

   b) Show that the initial condition

   $$\mathbf{X}(t_0) = \begin{bmatrix} c_1 \\ \cdot \\ \cdot \\ \cdot \\ c_n \end{bmatrix}$$

   for (10–7) is equivalent to the initial condition

   $$x(t_0) = c_1, \quad x'(t_0) = c_2, \quad \ldots, \quad x^{(n-1)}(t_0) = c_n$$

   for the corresponding $n$th-order linear differential equation.

   c) Let $\mathbf{X}' = \mathbf{A}(t)\mathbf{X}$ be the $n \times n$ first-order system derived from the equation

   $$x^{(n)} + a_{n-1}(t)x^{(n-1)} + \cdots + a_0(t)x = 0.$$

   Show that the Wronskian $W[\mathbf{X}_1, \ldots, \mathbf{X}_n]$ of any set of $n$ solutions of this system coincides with the Wronskian of the corresponding solutions of the $n$th-order equation as defined in Chapter 4.

4. Let

   $$\mathbf{X}_j(t) = \begin{bmatrix} x_{1j}(t) \\ \cdot \\ \cdot \\ \cdot \\ x_{nj}(t) \end{bmatrix}, \qquad 1 \leqslant j \leqslant n,$$

   be $n$ solutions of $\mathbf{X}' = \mathbf{A}(t)\mathbf{X}$ on an interval $I$.

a) Prove that these solutions are linearly independent in $\mathscr{V}_n$ if and only if their Wronskian, $W[\mathbf{X}_1, \ldots, \mathbf{X}_n]$, as defined by (10–9), does not vanish anywhere on $I$.

b) Prove that if $W$ vanishes at some point in $I$, it is identically zero on $I$.

5. a) Solve

$$\begin{bmatrix} x_1' \\ x_2' \end{bmatrix} = \begin{bmatrix} 0 & a \\ -a & 0 \end{bmatrix} \begin{bmatrix} x_1 \\ x_2 \end{bmatrix}, \qquad a \neq 0.$$

   b) Solve

$$\begin{bmatrix} x_1' \\ x_2' \end{bmatrix} = \begin{bmatrix} 0 & a \\ -a & 0 \end{bmatrix} \begin{bmatrix} x_1 \\ x_2 \end{bmatrix} + \begin{bmatrix} \cos t \\ 2 \sin t \end{bmatrix}, \qquad a \neq 0, \pm 1.$$

   c) Let $\mathbf{X}(\mathbf{X}_0, t)$ be the solution of the system of part (a) that satisfies the initial condition $\mathbf{X}(\mathbf{X}_0, 0) = \mathbf{X}_0$, and for each fixed $t_0 \geqslant 0$ let $F_{t_0}: \mathscr{R}^2 \to \mathscr{R}^2$ be defined by

$$F_{t_0}: \mathbf{X}_0 \to \mathbf{X}(\mathbf{X}_0, t_0).$$

   Prove that $F_t \circ F_s = F_{s+t}$, where $F_t \circ F_s$ denotes the composite of $F_s$ followed by $F_t$. [*Hint*: Use uniqueness.]

   d) Find $\mathbf{X}(\mathbf{X}_0, t)$ for the system in (a) when

$$\mathbf{X}_0 = \begin{bmatrix} \alpha \\ \beta \end{bmatrix},$$

   and show that $\mathbf{X}(\mathbf{X}_0, t) = \mathbf{M}(t)\mathbf{X}_0$, where $\mathbf{M}(t)$ is a $2 \times 2$ matrix whose entries are functions of $t$.

   e) Show that $\mathbf{M}(s + t) = \mathbf{M}(t)\mathbf{M}(s)$ for the matrix $\mathbf{M}$ in (d).

6. Assume that for each $\mathbf{X}_0$ in $\mathscr{R}^n$ the (not necessarily linear) system

$$x_1' = G_1(x_1, \ldots, x_n)$$

$$\cdot$$
$$\cdot$$
$$\cdot$$

$$x_n' = G_n(x_1, \ldots, x_n)$$

has a unique solution $\mathbf{X}(\mathbf{X}_0, t)$ such that $\mathbf{X}(\mathbf{X}_0, 0) = \mathbf{X}_0$. Let $F_{t_0}: \mathscr{R}^n \to \mathscr{R}^n$ be defined by

$$F_{t_0}: \mathbf{X}_0 \to \mathbf{X}(\mathbf{X}_0, t_0)$$

for each $t_0 \geqslant 0$. ($F_t$ is called a one-parameter family of transformations on $\mathscr{R}^n$.) Prove that

$$F_t \circ F_s = F_{s+t}.$$

7. Let $A = [a_{ij}(t)]$ and $B = [b_{ij}(t)]$. Show that if differentiation and integration of matrices are defined by

$$A' = \frac{dA}{dt} = [a_{ij}'(t)],$$

$$\int_{t_0}^t A \, dt = \left[ \int_{t_0}^t a_{ij}(s) \, ds \right],$$

then

$$(A + B)' = A' + B',$$

$$\int_{t_0}^t (A + B) \, dt = \int_{t_0}^t A \, dt + \int_{t_0}^t B \, dt$$

and

$$(AB)' = AB' + A'B.$$

(Note that we are not assuming $AB = BA$.)

* 8. Let $A(t) = [a_{ij}(t)]$ and $B(t) = \int_{t_0}^{t} A(t)\, dt$ (see Exercise 7). Assume that $AB = BA$ for all $t$ in some interval.

a) Prove that $\dfrac{d}{dt}(B^m) = mAB^{m-1}$.

b) Let $\mathbf{X}_0$ be a constant vector and define

$$\mathbf{X}_1(t) = \mathbf{X}_0 + \int_{t_0}^{t} A\mathbf{X}_0\, dt$$

$$\mathbf{X}_2(t) = \mathbf{X}_0 + \int_{t_0}^{t} A\mathbf{X}_1\, dt$$

$$\cdot$$
$$\cdot$$
$$\cdot$$

$$\mathbf{X}_m(t) = \mathbf{X}_0 + \int_{t_0}^{t} A\mathbf{X}_{m-1}\, dt,$$

Prove that

$$\mathbf{X}_m(t) = \left( I + B + \frac{1}{2}B^2 + \cdots + \frac{1}{m!}B^m \right)\mathbf{X}_0.$$

### 10–3 FIRST-ORDER SYSTEMS WITH CONSTANT COEFFICIENTS

The eigenvalue method introduced at the end of Chapter 9 is especially well suited to solving $n \times n$ systems of first-order linear differential equations with constant coefficients. Our discussion of these systems will divide into cases that depend on the nature of the eigenvalues of the coefficient matrix (by which we mean the eigenvalues of the linear transformation defined by the matrix), just as the discussion of $n$th-order linear differential equations divided into cases that depend on the nature of the roots of the associated characteristic equation. In fact, as Theorem 10–1 implies, the results we are about to derive include our earlier work as a special case.

We begin by considering the real eigenvalues of the coefficient matrix and, as usual, give most of our attention to the homogeneous case. Thus let

$$\mathbf{X}' = \mathbf{A}\mathbf{X} \tag{10–13}$$

be an $n \times n$ first-order linear system with coefficient matrix

$$\mathbf{A} = \begin{bmatrix} \alpha_{11} & \cdots & \alpha_{1n} \\ \cdot & & \cdot \\ \cdot & & \cdot \\ \cdot & & \cdot \\ \alpha_{n1} & \cdots & \alpha_{nn} \end{bmatrix}.$$

***Lemma 10–2***   *For each real eigenvalue $\lambda$ of $\mathbf{A}$ and each eigenvector*

$$
\mathbf{E}_\lambda = \begin{bmatrix} e_1 \\ \cdot \\ \cdot \\ \cdot \\ e_n \end{bmatrix}
$$

*belonging to $\lambda$, the function $\mathbf{X}_\lambda = \mathbf{E}_\lambda e^{\lambda t}$ is a solution of (10–13). Moreover, solutions formed in this way from distinct eigenvalues are linearly independent in $\mathscr{C}(-\infty, \infty)$.*

*Proof.*   First of all, $\mathbf{A}\mathbf{E}_\lambda = \lambda\mathbf{E}_\lambda$ because $\mathbf{E}_\lambda$ is an eigenvector belonging to $\lambda$. Thus

$$
\mathbf{A}\mathbf{X}_\lambda = \mathbf{A}\mathbf{E}_\lambda e^{\lambda t} = \lambda\mathbf{E}_\lambda e^{\lambda t} = \lambda\mathbf{X}_\lambda.
$$

On the other hand,

$$
\mathbf{X}'_\lambda = \lambda\mathbf{E}_\lambda e^{\lambda t} = \lambda\mathbf{X}_\lambda,
$$

and $\mathbf{X}'_\lambda = \mathbf{A}\mathbf{X}_\lambda$, as required.

Now let $\lambda_1, \ldots, \lambda_k$ be distinct (real) eigenvalues of $\mathbf{A}$ with associated eigenvectors $\mathbf{E}_{\lambda_1}, \ldots, \mathbf{E}_{\lambda_k}$, and suppose that

$$
c_1\mathbf{E}_{\lambda_1} e^{\lambda_1 t} + \cdots + c_k\mathbf{E}_{\lambda_k} e^{\lambda_k t} = \mathbf{0}.
$$

Then by setting $t = 0$, we have

$$
c_1\mathbf{E}_{\lambda_1} + \cdots + c_k\mathbf{E}_{\lambda_k} = \mathbf{0},
$$

and the linear independence of the $\mathbf{E}_{\lambda_i}$ in $\mathscr{R}^n$ implies that $c_1 = \cdots = c_k = 0$. Hence the $\mathbf{E}_{\lambda_i} e^{\lambda_i t}$ are linearly independent in $\mathscr{C}(-\infty, \infty)$.   ∎

This result, together with Theorem 10–3, immediately yields

***Theorem 10–4***   *If $\mathbf{A}$ has $n$ distinct real eigenvalues $\lambda_1, \ldots, \lambda_n$ and if $\mathbf{E}_{\lambda_1}, \ldots, \mathbf{E}_{\lambda_n}$ are eigenvectors belonging to these eigenvalues, then the general solution of the normal first-order system $\mathbf{X}' = \mathbf{A}\mathbf{X}$ is*

$$
\mathbf{X}(t) = c_1\mathbf{E}_{\lambda_1} e^{\lambda_1 t} + \cdots + c_n\mathbf{E}_{\lambda_n} e^{\lambda_n t},
$$

*where $c_1, \ldots, c_n$ are arbitrary constants.*

**Example 1**   Solve the first-order system

$$
\begin{aligned}
x'_1 &= x_1 + 3x_2 \\
x'_2 &= x_1 - x_2.
\end{aligned}
\tag{10–14}
$$

*Solution.*   Since this system can be written in matrix form as

$$
\begin{bmatrix} x'_1 \\ x'_2 \end{bmatrix} = \begin{bmatrix} 1 & 3 \\ 1 & -1 \end{bmatrix} \begin{bmatrix} x_1 \\ x_2 \end{bmatrix},
$$

its characteristic equation is

$$\begin{vmatrix} 1 - \lambda & 3 \\ 1 & -1 - \lambda \end{vmatrix} = 0.$$

Hence

$$\lambda^2 - 4 = 0,$$

and the eigenvalues for the system are $\lambda = \pm 2$. To find associated eigenvectors, we solve the matrix equations

$$\begin{bmatrix} 1 & 3 \\ 1 & -1 \end{bmatrix} \begin{bmatrix} e_1 \\ e_2 \end{bmatrix} = \begin{bmatrix} 2e_1 \\ 2e_2 \end{bmatrix} \quad \text{and} \quad \begin{bmatrix} 1 & 3 \\ 1 & -1 \end{bmatrix} \begin{bmatrix} e_1 \\ e_2 \end{bmatrix} = \begin{bmatrix} -2e_1 \\ -2e_2 \end{bmatrix}$$

for $e_1$ and $e_2$. An easy calculation reveals that

$$\mathbf{E}_2 = \begin{bmatrix} 3 \\ 1 \end{bmatrix} \quad \text{and} \quad \mathbf{E}_{-2} = \begin{bmatrix} -1 \\ 1 \end{bmatrix}$$

satisfy the equations, so the general solution of (10–14) is

$$\mathbf{X}(t) = c_1 \begin{bmatrix} 3 \\ 1 \end{bmatrix} e^{2t} + c_2 \begin{bmatrix} -1 \\ 1 \end{bmatrix} e^{-2t} = \begin{bmatrix} 3c_1 e^{2t} - c_2{}^{-2t} \\ c_1 e^{2t} + c_2 e^{-2t} \end{bmatrix},$$

where $c_1$ and $c_2$ are arbitrary constants.

Note that the solution of this problem was *entirely* algebraic; not a single derivative had to be taken nor a single integral evaluated. Such is the power of the eigenvalue method.

**Example 2**   Solve the system

$$\begin{aligned} x_1' &= x_1 + 3x_2 + \sin t \\ x_2' &= x_1 - x_2 - \cos t. \end{aligned} \tag{10–15}$$

*Solution.* A particular solution of this system can be obtained from the general solution of (10–14) by the method of variation of parameters described at the end of Section 10–2. In this case, however, it is easier to use undetermined coefficients.

We seek a solution of (10–15) of the form

$$\begin{aligned} x_1 &= A \sin t + B \cos t \\ x_2 &= C \sin t + D \cos t. \end{aligned}$$

When we substitute these expressions in the given equation and collect like terms we find that

$$(A + B + 3C + 1) \sin t + (-A + B + 3D) \cos t = 0$$
$$(A - C + D) \sin t + (B - C - D - 1) \cos t = 0.$$

Since these equations must hold for all values of $t$, the coefficients $A$, $B$, $C$, and $D$ must be chosen to make

$$
\begin{aligned}
A + B + 3C \quad\quad &= -1 \\
-A + B \quad\quad + 3D &= 0 \\
A \quad\quad - C + D &= 0 \\
B - C - D &= 1.
\end{aligned}
$$

The augmented matrix of this system of linear equations is

$$
\begin{bmatrix}
1 & 1 & 3 & 0 & -1 \\
-1 & 1 & 0 & 3 & 0 \\
1 & 0 & -1 & 1 & 0 \\
0 & 1 & -1 & -1 & 1
\end{bmatrix},
$$

and the row-reduced echelon form of the matrix is

$$
\begin{bmatrix}
1 & 0 & 0 & 0 & -\frac{1}{5} \\
0 & 1 & 0 & 0 & \frac{2}{5} \\
0 & 0 & 1 & 0 & -\frac{2}{5} \\
0 & 0 & 0 & 1 & -\frac{1}{5}
\end{bmatrix}.
$$

From this we read the values of $A$, $B$, $C$, and $D$ and conclude that

$$
\mathbf{X}_p(t) = \begin{bmatrix} -\frac{1}{5}\sin t + \frac{2}{5}\cos t \\ -\frac{2}{5}\sin t - \frac{1}{5}\cos t \end{bmatrix}
$$

is a particular solution of (10–15). The general solution of (10–15) therefore is

$$
\mathbf{X}(t) = \begin{bmatrix} 3c_1 e^{2t} - c_2 e^{-2t} - \frac{1}{5}\sin t + \frac{2}{5}\cos t \\ c_1 e^{2t} + c_2 e^{-2t} - \frac{2}{5}\sin t - \frac{1}{5}\cos t \end{bmatrix}.
$$

**EXERCISES**

Find the general solution of the systems in Exercises 1 through 8.

1. $x_1' = 2x_1 - x_2$
   $x_2' = 3x_1 - 2x_2$

2. $x_1' = 2x_1 + 3x_2$
   $x_2' = x_1 + 4x_2$

3. $x_1' = 6x_1 - 7x_2$
   $x_2' = x_1 - 2x_2$

4. $x_1' = -3x_1 + 8x_2$
   $x_2' = 2x_1 + 3x_2$

5. $x_1' = x_1 - x_2 + 4x_3$
   $x_2' = 3x_1 + 2x_2 - x_3$
   $x_3' = 2x_1 + x_2 - x_3$

6. $x_1' = -x_1 + x_3$
   $x_2' = -x_2 + x_3$
   $x_3' = x_1 + x_2$

7. $x_1' = 2x_1 - x_2 + t$
   $x_2' = 3x_1 - 2x_2 + 2t$

8. $x_1' = 2x_1 + x_2 + e^t$
   $x_2' = 4x_1 - x_2 - e^t$

9. Find the general solution of

$$x_1' = ax_1 + bx_2$$
$$x_2' = bx_1 + ax_2$$

given that $b \neq 0$.

10. Use the method of variation of parameters to find a particular solution of the system of Example 2.

11. Show that the system

$$tx_1' = \alpha_{11}x_1 + \alpha_{12}x_2,$$
$$tx_2' = \alpha_{21}x_1 + \alpha_{22}x_2$$

has a nontrivial solution of the form $\mathbf{X}(t) = \mathbf{X}_0 t^{\lambda}$, $t > 0$, whenever $\lambda$ is a root of the equation

$$\begin{vmatrix} \alpha_{11} - \lambda & \alpha_{12} \\ \alpha_{21} & \alpha_{22} - \lambda \end{vmatrix} = 0.$$

Use the result of Exercise 11 to find the general solution for $t > 0$ of each of the following systems.

12. $tx_1' = 2x_1 + x_2$
    $tx_2' = -2x_1 - 2x_2$

13. $tx_1' = 5x_1 - x_2$
    $tx_2' = 3x_1 + x_2$

## 10–4 COMPLEX AND REPEATED EIGENVALUES

It remains for us to consider those cases in which the characteristic equation of the coefficient matrix for

$$\mathbf{X}' = \mathbf{A}\mathbf{X}$$

has complex or repeated roots (or both). We begin with the complex case.

**Complex Eigenvalues**    Since the entries of $\mathbf{A}$ are real by assumption, the characteristic equation of $\mathbf{A}$ is a polynomial equation with real coefficients. Hence if $\lambda = \alpha + \beta i$, $\beta \neq 0$, is an eigenvalue for $\mathbf{A}$, the complex conjugate $\overline{\lambda} = \alpha - \beta i$ is also an eigenvalue of $\mathbf{A}$. To find eigenvectors for these eigenvalues, we extend the domain of the linear transformation $A$ associated with $\mathbf{A}$ to include $n$-tuples of *complex* numbers, as follows.

Let $\mathscr{C}^n$ denote the set of all column vectors

$$\mathbf{Z} = \begin{bmatrix} z_1 \\ \cdot \\ \cdot \\ \cdot \\ z_n \end{bmatrix},$$

where the $z_i$ are complex numbers, and let addition and scalar multiplication in

$\mathbf{C}^n$ be defined componentwise:

$$\begin{bmatrix} y_1 \\ \cdot \\ \cdot \\ \cdot \\ y_n \end{bmatrix} + \begin{bmatrix} z_1 \\ \cdot \\ \cdot \\ \cdot \\ z_n \end{bmatrix} = \begin{bmatrix} y_1 + z_1 \\ \cdot \\ \cdot \\ \cdot \\ y_n + z_n \end{bmatrix}, \qquad z \begin{bmatrix} z_1 \\ \cdot \\ \cdot \\ \cdot \\ z_n \end{bmatrix} = \begin{bmatrix} zz_1 \\ \cdot \\ \cdot \\ \cdot \\ zz_n \end{bmatrix},$$

where in the second equation $z$ is an arbitrary complex number.*

It is easy to verify that these definitions convert $\mathscr{C}^n$ into a vector space provided complex numbers are used in place of real numbers as the scalars of the original definition. In fact, once this change has been made, *all* the results previously established for $\mathscr{R}^n$ also hold for $\mathscr{C}^n$. In particular, if

$$\mathbf{A} = \begin{bmatrix} \alpha_{11} & \cdots & \alpha_{1n} \\ \cdot & & \cdot \\ \cdot & & \cdot \\ \cdot & & \cdot \\ \alpha_{n1} & \cdots & \alpha_{nn} \end{bmatrix}$$

is an $n \times n$ matrix with real or complex entries, then the mapping

$$\mathbf{A}: \begin{bmatrix} z_1 \\ \cdot \\ \cdot \\ \cdot \\ z_n \end{bmatrix} \rightarrow \begin{bmatrix} \alpha_{11}z_1 + \cdots + \alpha_{1n}z_n \\ \cdot \\ \cdot \\ \cdot \\ \alpha_{n1}z_1 + \cdots + \alpha_{nn}z_n \end{bmatrix}$$

is a linear transformation that maps $\mathscr{C}^n$ to itself. Again a number $\lambda$ (this time complex) is called an eigenvalue for $\mathbf{A}$, and $\mathbf{Z} \neq \mathbf{0}$ is called an eigenvector belonging to $\lambda$, if $\mathbf{AZ} = \lambda\mathbf{Z}$.

> **Lemma 10–3**   *Let $\mathbf{A}$ be an $n \times n$ matrix with real entries and suppose that $\lambda = \alpha + \beta i$, $\beta \neq 0$, is an eigenvalue for $\mathbf{A}$. Then if $\mathbf{Z}$ is an eigenvector belonging to $\lambda$, its complex conjugate $\bar{\mathbf{Z}}$ is an eigenvector belonging to the eigenvalue $\bar{\lambda} = \alpha - \beta i$.*

*Proof.* Since $\mathbf{AZ} = \lambda\mathbf{Z}$, we must have

$$\overline{\mathbf{AZ}} = \bar{\lambda}\bar{\mathbf{Z}}.$$

But the entries of $\mathbf{A}$ are real, so that $\bar{\mathbf{A}} = \mathbf{A}$ and the preceding equation reads

$$\mathbf{A}\bar{\mathbf{Z}} = \bar{\lambda}\bar{\mathbf{Z}}. \quad \blacksquare$$

---

\* Recall that complex numbers are multiplied by the rule

$$(a + bi)(c + di) = (ac - bd) + (ad + bc)i.$$

In short, multiply as usual and use the fact that $i^2 = -1$.

**Example 1**   Find the eigenvalues, and eigenvectors belonging to these eigenvalues, for the matrix

$$\mathbf{A} = \begin{bmatrix} 0 & -1 \\ 1 & 0 \end{bmatrix}.$$

*Solution.*   From the characteristic equation

$$\begin{vmatrix} -\lambda & -1 \\ 1 & -\lambda \end{vmatrix} = \lambda^2 + 1 = 0$$

we find that the eigenvalues for **A** are $\pm i$. To find an eigenvector belonging to the eigenvalue $i$, we solve the matrix equation

$$\begin{bmatrix} 0 & -1 \\ 1 & 0 \end{bmatrix} \begin{bmatrix} z_1 \\ z_2 \end{bmatrix} = \begin{bmatrix} iz_1 \\ iz_2 \end{bmatrix}.$$

In scalar form this equation reads

$$-z_2 = iz_1$$
$$z_1 = iz_2,$$

so that the eigenvectors belonging to the eigenvalue $i$ are

$$\begin{bmatrix} iz_2 \\ z_2 \end{bmatrix}, \quad z_2 \neq 0.$$

In particular,

$$\begin{bmatrix} i \\ 1 \end{bmatrix}$$

is such a vector, and Lemma 10–3 implies that

$$\begin{bmatrix} -i \\ 1 \end{bmatrix}$$

is an eigenvector belonging to the eigenvalue $-i$. The equation

$$\begin{bmatrix} 0 & -1 \\ 1 & 0 \end{bmatrix} \begin{bmatrix} -i \\ 1 \end{bmatrix} = \begin{bmatrix} -1 \\ -i \end{bmatrix} = -i \begin{bmatrix} -i \\ 1 \end{bmatrix}$$

provides a check that this is indeed the case.

Now that we know how to find eigenvectors of complex eigenvalues, we can prove the complex analog of Lemma 10–2. Specifically, we have

**Lemma 10–4**   *Let **A** be a real $n \times n$ matrix, and suppose that $\mathbf{E}_\lambda$ is an eigenvector in $\mathscr{C}^n$ of the complex eigenvalue $\lambda = \alpha + \beta i$ of **A**. Then*

$$\mathbf{E}_\lambda e^{\lambda t} \quad and \quad \overline{\mathbf{E}}_\lambda e^{\overline{\lambda} t}$$

*are solutions of the equation* $\mathbf{X}' = \mathbf{A}\mathbf{X}$.

*Proof.* The proof is the one we used before:

$$\frac{d}{dt}\,\mathbf{E}_\lambda e^{\lambda t} = \mathbf{E}_\lambda\frac{d}{dt}\,e^{(\alpha+\beta i)t} = \mathbf{E}_\lambda(\alpha+\beta i)e^{(\alpha+\beta i)t} = \lambda\mathbf{E}_\lambda e^{\lambda t},$$

while $\mathbf{AE}_\lambda e^{\lambda t} = \lambda\mathbf{E}e^{\lambda t}$. Thus

$$\frac{d}{dt}\,\mathbf{E}_\lambda e^{\lambda t} = \mathbf{AE}_\lambda e^{\lambda t}.$$

A similar argument can be given for the vector $\bar{\mathbf{E}}_\lambda e^{\bar\lambda t}$.*  ∎

Having come this far, we are now in a position to remove all references to complex numbers and complex-valued functions by invoking Euler's formula

$$e^{(\alpha+\beta i)} = e^\alpha(\cos\beta + i\sin\beta).$$

For then

$$\mathbf{E}_\lambda e^{\lambda t} = \mathbf{E}_\lambda e^{\alpha t}(\cos\beta t + i\sin\beta t),$$
$$\bar{\mathbf{E}}_\lambda e^{\bar\lambda t} = \bar{\mathbf{E}}_\lambda e^{\alpha t}(\cos\beta t - i\sin\beta t),$$

and since both of these functions are solutions of $\mathbf{X}' = \mathbf{AX}$, so are their linear combinations

$$\frac{1}{2}(\mathbf{E}_\lambda e^{\lambda t} + \bar{\mathbf{E}}_\lambda e^{\bar\lambda t}) = e^{\alpha t}\left[\frac{\mathbf{E}_\lambda + \bar{\mathbf{E}}_\lambda}{2}\cos\beta t + \frac{i(\mathbf{E}_\lambda - \bar{\mathbf{E}}_\lambda)}{2}\sin\beta t\right]$$

and

$$\frac{i}{2}(\mathbf{E}_\lambda e^{\lambda t} - \bar{\mathbf{E}}_\lambda e^{\bar\lambda t}) = e^{\alpha t}\left[\frac{i(\mathbf{E}_\lambda - \bar{\mathbf{E}}_\lambda)}{2}\cos\beta t - \frac{\mathbf{E}_\lambda + \bar{\mathbf{E}}_\lambda}{2}\sin\beta t\right].$$

Moreover, the coefficients that appear in these solutions are real since

$$\frac{z + \bar z}{2} = a \qquad\text{and}\qquad \frac{i(z - \bar z)}{2} = -b$$

whenever $z = a + bi$ and $\bar z = a - bi$ are complex conjugates. Thus we have proved

***Theorem 10–5*** *Let $\lambda = \alpha + \beta i$ be a complex eigenvalue for the $n \times n$ real matrix $\mathbf{A}$, and let $\mathbf{E}_\lambda$ be an eigenvector in $\mathscr{C}^n$ belonging to $\lambda$. Then the functions*

$$\mathbf{X}_1(t) = e^{\alpha t}(\mathbf{G}_\lambda\cos\beta t + \mathbf{H}_\lambda\sin\beta t),$$
$$\mathbf{X}_2(t) = e^{\alpha t}(\mathbf{H}_\lambda\cos\beta t - \mathbf{G}_\lambda\sin\beta t),$$

*where*

$$\mathbf{G}_\lambda = \frac{\mathbf{E}_\lambda + \bar{\mathbf{E}}_\lambda}{2} \qquad\text{and}\qquad \mathbf{H}_\lambda = \frac{i(\mathbf{E}_\lambda - \bar{\mathbf{E}}_\lambda)}{2},$$

*are linearly independent solutions of $\mathbf{X}' = \mathbf{AX}$.*

---

* Strictly speaking, we are allowing $\mathbf{X}' = \mathbf{AX}$ to have solutions which are functions of a complex variable. As we remarked earlier, we can legitimately do so as soon as we have defined the notion of differentiability for such functions.

**Example 2**   Solve the first-order system

$$x_1' = -x_2$$
$$x_2' = x_1. \tag{10–16}$$

*Solution.*   In matrix form this system is

$$\begin{bmatrix} x_1' \\ x_2' \end{bmatrix} = \begin{bmatrix} 0 & -1 \\ 1 & 0 \end{bmatrix} \begin{bmatrix} x_1 \\ x_2 \end{bmatrix},$$

and as in Example 1, it has $\lambda = i$ as an eigenvalue. Since

$$\mathbf{E}_i = \begin{bmatrix} i \\ 1 \end{bmatrix}$$

is an eigenvector belonging to this eigenvalue,

$$\frac{\mathbf{E}_i + \bar{\mathbf{E}}_i}{2} = \begin{bmatrix} 0 \\ 1 \end{bmatrix}, \qquad \frac{i(\mathbf{E}_i - \bar{\mathbf{E}}_i)}{2} = \begin{bmatrix} -1 \\ 0 \end{bmatrix}.$$

Thus by Theorem 10–5

$$\mathbf{X}_1(t) = \begin{bmatrix} 0 \\ 1 \end{bmatrix} \cos t + \begin{bmatrix} -1 \\ 0 \end{bmatrix} \sin t = \begin{bmatrix} -\sin t \\ \cos t \end{bmatrix}$$

and

$$\mathbf{X}_2(t) = \begin{bmatrix} -1 \\ 0 \end{bmatrix} \cos t - \begin{bmatrix} 0 \\ 1 \end{bmatrix} \sin t = \begin{bmatrix} -\cos t \\ -\sin t \end{bmatrix}$$

are linearly independent solutions of (10–16). The general solution of this system therefore is

$$\mathbf{X}(t) = c_1 \mathbf{X}_1(t) + c_2 \mathbf{X}_2(t) = \begin{bmatrix} -c_1 \sin t - c_2 \cos t \\ c_1 \cos t - c_2 \sin t \end{bmatrix}.$$

**Repeated Eigenvalues**   Except for language and notation, the results we have obtained so far for first-order systems with constant coefficients are identical with those we obtained earlier for the corresponding cases involving constant-coefficient $n$th-order equations. However, if the characteristic equation for $\mathbf{A}$ has a repeated root, something new happens.

Suppose, for instance, that $\lambda_0$ is a (real) eigenvalue of multiplicity two for $\mathbf{A}$, by which we mean that $(\lambda - \lambda_0)^2$ is a factor of the characteristic equation for $\mathbf{A}$. Then, as we saw earlier, the system $\mathbf{X}' = \mathbf{A}\mathbf{X}$ has a nontrivial solution of the form

$$\mathbf{X}_1(t) = \mathbf{E}_{\lambda_0} e^{\lambda_0 t},$$

where $\mathbf{E}_{\lambda_0}$ is an eigenvector belonging to $\lambda_0$. Our experience with $n$th-order equations would suggest that the system also has a nontrivial solution of the form

$$\mathbf{X}_2(t) = \mathbf{B}t e^{\lambda_0 t}$$

which is linearly independent of $X_1(t)$. Unfortunately, this need not be true. What is true is that the system then has a nontrivial solution of the form

$$X_2(t) = Bte^{\lambda_0 t} + Ce^{\lambda_0 t}, \tag{10–17}$$

where **B** and **C** are vectors in $\mathscr{R}^n$ but **C** is not necessarily a scalar multiple of $E_{\lambda_0}$. When this happens the term $Ce^{\lambda_0 t}$ in (10–17) cannot be absorbed as part of $X_1(t)$, and therefore contributes a separate term to the general solution of the system.

More generally, we have the following result.

**Theorem 10–6**  *Let $\lambda_0$ be a real root of multiplicity $m$ of the characteristic equation for the $n \times n$ matrix $A$. Then the first-order system $X' = AX$ has $m$ linearly independent solutions of the form*

$$X_1(t) = B_{11}e^{\lambda_0 t}$$
$$X_2(t) = B_{21}te^{\lambda_0 t} + B_{22}e^{\lambda_0 t}$$

.
.
.

$$X_m(t) = B_{m1}t^{m-1}e^{\lambda_0 t} + B_{m2}t^{m-2}e^{\lambda_0 t} + \cdots + B_{mm}e^{\lambda_0 t},$$

*where the $B_{ij}$ are vectors in $\mathscr{R}^n$.*

An analogous result holds for repeated complex eigenvalues.

The proof of Theorem 10–6 uses results from the theory of linear transformations that are beyond the scope of this book, and is therefore omitted.

**Example 3**  Find the general solution of

$$x_1' = -2x_1 - 3x_2$$
$$x_2' = 3x_1 + 4x_2. \tag{10–18}$$

*Solution.*  The characteristic equation for the coefficient matrix of this system is

$$\begin{vmatrix} -2 - \lambda & -3 \\ 3 & 4 - \lambda \end{vmatrix} = (\lambda - 1)^2 = 0.$$

Thus $\lambda = 1$ is an eigenvalue of multiplicity two.

To find an eigenvector belonging to $\lambda$, we seek a nontrivial solution of

$$\begin{bmatrix} -2 & -3 \\ 3 & 4 \end{bmatrix} \begin{bmatrix} e_1 \\ e_2 \end{bmatrix} = \begin{bmatrix} e_1 \\ e_2 \end{bmatrix}.$$

Since this equation is satisfied if and only if $e_2 = -e_1$,

$$\begin{bmatrix} 1 \\ -1 \end{bmatrix}$$

is an eigenvector for the system, and

$$\mathbf{X}_1(t) = \begin{bmatrix} 1 \\ -1 \end{bmatrix} e^t$$

is a solution.

To continue, we seek a second solution of the form

$$\mathbf{X}_2(t) = \begin{bmatrix} B_1 \\ B_2 \end{bmatrix} te^t + \begin{bmatrix} C_1 \\ C_2 \end{bmatrix} e^t.$$

When this expression is substituted into (10–18) and like terms are collected, we find that

$$(3B_1 + 3B_2)te^t + (B_1 + 3C_1 + 3C_2)e^t = 0$$
$$(3B_1 + 3B_2)te^t + (-B_2 + 3C_1 + 3C_2)e^t = 0.$$

These equations imply that

$$B_2 = -B_1 \quad \text{and} \quad C_2 = -\tfrac{1}{3}B_1 - C_1,$$

where $B_1 \neq 0$ and $C_1$ are arbitrary. For instance, if we set $B_1 = 3$ and $C_1 = 0$, then

$$\mathbf{X}_2(t) = \begin{bmatrix} 3 \\ -3 \end{bmatrix} te^t + \begin{bmatrix} 0 \\ -1 \end{bmatrix} e^t.$$

Thus the general solution of (10–18) can be written as

$$\mathbf{X}(t) = c_1 \begin{bmatrix} 1 \\ -1 \end{bmatrix} e^t + c_2 \left( \begin{bmatrix} 3 \\ -3 \end{bmatrix} te^t + \begin{bmatrix} 0 \\ -1 \end{bmatrix} e^t \right).$$

### EXERCISES

Find the general solution of the systems in Exercises 1 through 10.

1. $x_1' = 3x_1 - 2x_2$
   $x_2' = 5x_1 - 3x_2$

2. $x_1' = 2x_1 + 4x_2$
   $x_2' = -5x_1 - 2x_2$

3. $x_1' = 2x_1 - x_2$
   $x_2' = 2x_1 + 4x_2$

4. $x_1' = -x_1 - x_2$
   $x_2' = 2x_1 - x_2$

5. $x_1' = x_1 - x_2$
   $x_2' = x_1 + 3x_2$

6. $x_1' = x_1 - 4x_2$
   $x_2' = 4x_1 - 7x_2$

7. $x_1' = 4x_1 + 2x_2$
   $x_2' = -8x_1 - 4x_2$

8. $x_1' = 6x_1 - 2x_2$
   $x_2' = 7x_1 - 2x_2$

9. $x_1' = x_1 + x_2 + x_3$
   $x_2' = 2x_1 + x_2 - x_3$
   $x_3' = -3x_1 + 2x_2 + 4x_3$

10. $x_1' = x_1 + x_2 + x_3$
    $x_2' = 2x_1 + x_2 - x_3$
    $x_3' = -x_2 + x_3$

11. A particle moves in the $xy$-plane in accordance with the equations

$$\frac{dx}{dt} = ay, \qquad \frac{dy}{dt} = -ax,$$

where $a$ is a positive constant. Find the trajectory of the particle.

*12. A particle moves in $\mathscr{R}^3$ in accordance with the equations

$$\frac{dx}{dt} = -y - 2z, \qquad \frac{dy}{dt} = x + z, \qquad \frac{dz}{dt} = 2x - y.$$

Show that the trajectory of the particle is a circle with its center on the line determined by the vector $-\mathbf{e}_1 - 2\mathbf{e}_2 + \mathbf{e}_3$.

## 10-5 AN APPLICATION TO BIOLOGY

In biology and medicine one is often interested in describing how a chemical compound, such as a drug, accumulates in cells as the compound diffuses across cell walls. One model that can be used to approximate this process consists of a sequence of $n$ compartments or boxes which correspond to the cells in which the compound is accumulating, and which are arranged linearly as shown in Fig. 10–1. We propose to derive equations that describe the diffusion process under the following assumptions.

1. The distribution of the compound within each cell is uniform at all times.
2. The rate of diffusion of the compound from Cell $i - 1$ to Cell $i$ (amount per unit time per unit of cell-wall area shared) is a constant, $\alpha_{i-1,i}$, times the concentration of the compound (amount per unit volume) in Cell $i - 1$. Similarly, the rate of diffusion from Cell $i$ back to Cell $i - 1$ is a constant, $\alpha_{i,i-1}$, times the concentration in Cell $i$.
3. The volume of each cell is equal to the area of each wall of the cell across which the compound is being diffused. (This would be the case, for example, if the cells were assumed to be unit cubes with shared faces.)

**Figure 10–1**

With these assumptions in force, we let $C_i = C_i(t)$ denote the concentration of the compound in the $i$th cell at time $t$. Then, as shown in Fig. 10–2, diffusion across the walls between Cells $i - 1$, $i$, and $i + 1$ causes $C_i$ to *increase* at the rate

$$\alpha_{i-1,i}C_{i-1} + \alpha_{i+1,i}C_{i+1}, \tag{10–19}$$

and *decrease* at the rate

$$(\alpha_{i,i-1} + \alpha_{i,i+1})C_i. \tag{10–20}$$

These expressions also apply to the cells at the ends of the sequence if we set

$$\alpha_{0,1} = \alpha_{1,0} = \alpha_{n,n+1} = \alpha_{n+1,n} = 0.$$

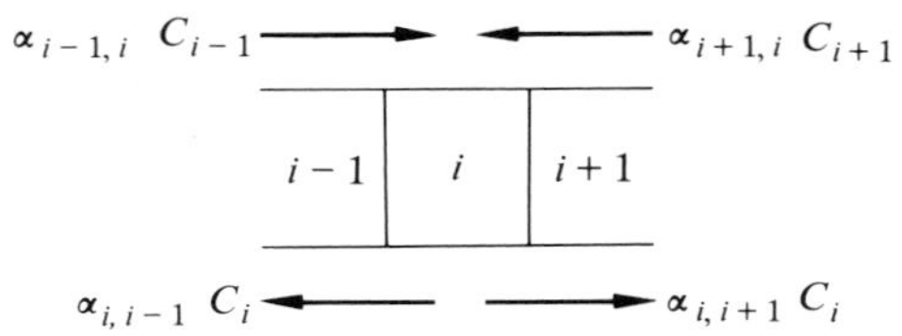

**Figure 10–2**

It follows from (10–19) and (10–20) that the rate of change of the concentration in the $i$th cell is

$$\frac{dC_i}{dt} = \alpha_{i-1,i}C_{i-1} - (\alpha_{i,i-1} + \alpha_{i,i+1})C_i + \alpha_{i+1,i}C_{i+1}. \qquad (10\text{--}21)$$

Thus the diffusion process is described mathematically by a system of $n$ first-order constant-coefficient linear differential equations in $n$ unknowns.

In the interest of further simplicity we now assume that we are confronted with only two types of cells, which alternate in sequence and which have respective diffusion rates $\alpha$ and $\beta$. Then if the sequence starts with a cell whose diffusion rate is $\alpha$, (10–21) becomes

$$\frac{dC_1}{dt} = -\alpha C_1 + \beta C_2$$

$$\frac{dC_2}{dt} = \alpha C_1 - 2\beta C_2 + \alpha C_3 \qquad (10\text{--}22)$$

$$\frac{dC_3}{dt} = \beta C_2 - 2\alpha C_3 + \beta C_4$$

$$\cdot$$
$$\cdot$$
$$\cdot$$

In actual practice the number of cells in the sequence is usually large, and (10–22) must be solved on a computer. Nevertheless, it is instructive to consider this system for small values of $n$, if only to determine whether the solutions conform to our expectations of what ought to happen.

**The Two-Cell Model**    Suppose that the sequence consists of only two cells and a unit amount of the compound is injected into the first cell at time $t = 0$, the concentration in the second cell being zero. In this case (10–22) reduces to the $2 \times 2$ system

$$\frac{dC_1}{dt} = -\alpha C_1 + \beta C_2$$

$$\frac{dC_2}{dt} = \alpha C_1 - \beta C_2, \qquad (10\text{--}23)$$

and is subject to the initial conditions

$$C_1(0) = 1, \qquad C_2(0) = 0. \qquad\qquad (10\text{--}24)$$

The characteristic equation of the coefficient matrix of (10–23) is

$$\lambda^2 + (\alpha + \beta)\lambda = 0,$$

and the eigenvalues for the system therefore are

$$\lambda = 0 \qquad \text{and} \qquad \lambda = -(\alpha + \beta).$$

A routine calculation (Exercise 1) reveals that

$$\mathbf{E}_0 = \begin{bmatrix} 1 \\ \alpha/\beta \end{bmatrix} \qquad \text{and} \qquad \mathbf{E}_{-(\alpha+\beta)} = \begin{bmatrix} 1 \\ -1 \end{bmatrix}$$

are eigenvectors belonging to these eigenvalues. It follows that the general solution of (10–23) is

$$C_1(t) = A + Be^{-(\alpha+\beta)t}$$

$$C_2(t) = \frac{\alpha}{\beta} A - Be^{-(\alpha+\beta)t}.$$

Finally, the given initial conditions imply that

$$A = \frac{\beta}{\alpha + \beta} \qquad \text{and} \qquad B = \frac{\alpha}{\alpha + \beta}$$

(Exercise 1) and hence that

$$C_1(t) = \frac{\beta}{\alpha + \beta} + \frac{\alpha}{\alpha + \beta} e^{-(\alpha+\beta)t}$$

$$C_2(t) = \frac{\alpha}{\alpha + \beta} (1 - e^{-(\alpha+\beta)t}).$$

Note that as $t \to \infty$, $C_1(t)$ and $C_2(t)$ approach the steady-state values

$$C_1 = \frac{\beta}{\alpha + \beta}, \qquad C_2 = \frac{\alpha}{\alpha + \beta},$$

and in the steady state the concentrations in Cells 1 and 2 are in the ratio $\beta/\alpha$. Both these results are just what one would expect.

## EXERCISES

1. a) Verify that the eigenvalues for the coefficient matrix of (10–23) are 0 and $-(\alpha + \beta)$.
   b) Show that

$$\mathbf{E}_0 = \begin{bmatrix} 1 \\ \alpha/\beta \end{bmatrix} \qquad \text{and} \qquad \mathbf{E}_{-(\alpha+\beta)} = \begin{bmatrix} 1 \\ -1 \end{bmatrix}$$

   are eigenvectors belonging to these eigenvalues.

c) Show that the initial conditions (10–24) are satisfied by letting

$$A = \frac{\beta}{\alpha + \beta} \quad \text{and} \quad B = \frac{\alpha}{\alpha + \beta}$$

in the general solution of (10–23).

2. Where in the derivation of (10–21) did we use the assumption that the volume of each cell is equal to the area of each cell wall across which the compound is being diffused?

3. a) Solve (10–22) for three cells with respective diffusion constants $\alpha$, $\beta$, $\alpha$, and initial conditions

$$C_1(0) = 1, \quad C_2(0) = C_3(0) = 0.$$

b) What are the steady-state solutions in this case?

## 10–6  AN APPLICATION TO ECONOMICS

During the 1940s the mathematical economist W. W. Leontief developed a description of the dynamic behavior of interindustry relations that is an especially fruitful example of a mathematical model involving systems of constant-coefficient first-order linear differential equations. In broad outline, the problem Leontief considered was this:

> Given n industries, each producing a single product and no two producing the same or equivalent products, find conditions under which the outputs and the prices charged for them will be in equilibrium.

When stated in these terms, the problem is uncomfortably vague. Who trades with whom and in what amounts? What conditions determine output, and what must be taken into consideration in setting prices? And most important of all, what constitutes "equilibrium" for such a system? In the discussion that follows we shall produce a mathematical model for this situation in stages, each stage taking account of facts that were not considered earlier. We begin with two models that ignore changes in time and are therefore called **static models**. Once they have been constructed we will turn to the dynamic case where changes in time are taken into account and derivatives appear.

**I.  The First Static Model: A Closed System of $n$ Industries**    The assumptions underlying the construction of this model are these:

1. Each industry uses a nonnegative amount of the output of the other industries in the system.

2. No industry consumes any of its own output.

3. The total output of each industry is consumed by the other industries in the system. (The system is **closed** with respect to production.)

4. The income received by each industry is equal to the amount it spends in buying the outputs from the other industries in the system. (The system is **closed** with respect to capital accumulation.)

5. Production, prices, purchases, and sales are constant over time.

The third and fourth conditions are the **equilibrium conditions** for the system: all output is consumed within the system, and all net profits are zero. Unrealistic, but an essential first step in the analysis.

To construct a mathematical model of this system, we introduce the variables $x_i$, $p_i$, and $z_{ij}$ where:

$x_i$ denotes the number of (pricing) units of output from industry $i$ per unit time, for example the number of tons of steel produced per day, with steel sold by the ton (the $x_i$ are assumed to be positive, of course);

$p_i$ denotes the price charged by industry $i$ for each unit of its output;

$z_{ij}$ denotes the number of units of output from industry $i$ used by industry $j$ per unit time. (At this point we are tacitly assuming that the output and pricing structures of the various industries in the system can be expressed in compatible units, a dubious assumption at best.)

Under the equilibrium conditions in effect, $z_{ij}$ is directly proportional to $x_j$, and each of the ratios

$$\alpha_{ij} = \frac{z_{ij}}{x_j}, \qquad i \neq j, \tag{10-25}$$

is therefore constant. These constants are called the **flow constants** of the system, and *we now assume that they are known.* The economic interpretation of the variables implies that all of the $\alpha_{ij}$ are nonnegative.

The equilibrium condition that requires all of the outputs of the system to be consumed within the system can now be expressed by the system of linear equations

$$x_1 = \widehat{z_{11}} + z_{12} + \cdots + z_{1i} + \cdots + z_{1n}$$
$$\vdots$$
$$x_i = z_{i1} + z_{i2} + \cdots + \widehat{z_{ii}} + \cdots + z_{in} \tag{10-26}$$
$$\vdots$$
$$x_n = z_{n1} + z_{n2} + \cdots + z_{ni} + \cdots + \widehat{z_{nn}},$$

where the circumflex, $\wedge$, over a term indicates that that term is missing from the equation. (The omission of the $z_{ii}$ corresponds to the assumption that the industry involved does not consume its own output.) Since $z_{ij} = \alpha_{ij}x_j$, (10–26) can be rewritten as

$$(-1)x_1 + \alpha_{12}x_2 + \cdots + \alpha_{1n}x_n = 0$$
$$\vdots$$
$$\alpha_{n1}x_1 + \alpha_{n2}x_2 + \cdots + (-1)x_n = 0,$$

or as

$$\mathbf{AX} = \mathbf{0}, \tag{10-27}$$

where

$$\mathbf{A} = \begin{bmatrix} -1 & \alpha_{12} \cdots & \alpha_{1n} \\ \alpha_{21} & -1 & \cdots & \alpha_{2n} \\ \cdot & \cdot & & \cdot \\ \cdot & \cdot & & \cdot \\ \cdot & \cdot & & \cdot \\ \alpha_{n1} & \alpha_{n2} & \cdots & -1 \end{bmatrix},$$

$$\mathbf{X} = \begin{bmatrix} x_1 \\ \cdot \\ \cdot \\ \cdot \\ x_n \end{bmatrix} \quad \text{and} \quad \mathbf{0} = \begin{bmatrix} 0 \\ \cdot \\ \cdot \\ \cdot \\ 0 \end{bmatrix}.$$

Similarly, the equilibrium condition for profits yields the system of equations

$$p_1 x_1 = \widehat{p_1 z_{11}} + p_2 z_{21} + \cdots + p_n z_{n1}$$

$$\begin{array}{c} \cdot \\ \cdot \\ \cdot \end{array}$$

$$p_n x_n = p_1 z_{1n} + p_2 z_{2n} + \cdots + \widehat{p_n z_{nn}}.$$

By using (10–25) we can rewrite this system, too, in terms of the $\alpha_{ij}$ as

$$p_1(-1) + p_2 \alpha_{21} + \cdots + p_n \alpha_{n1} = 0$$

$$\begin{array}{c} \cdot \\ \cdot \\ \cdot \end{array}$$

$$p_1 \alpha_{1n} + p_2 \alpha_{2n} + \cdots + p_n(-1) = 0,$$

or as

$$\mathbf{PA} = \mathbf{0}, \tag{10-28}$$

where $\mathbf{A}$ and $\mathbf{0}$ are as before and

$$\mathbf{P} = (p_1, \ldots, p_n).$$

Thus the two matrix equations

$$\mathbf{AX} = \mathbf{0}, \quad \mathbf{PA} = \mathbf{0}$$

describe the output and price structure of the industrial system under consideration. Note that this pair of equations will not have an economically meaningful solution unless $\mathbf{A}$ is singular, for only then do they admit nontrivial solutions. And even when nontrivial solutions do exist they yield only *relative* outputs and prices for the system, because whenever $\mathbf{X}_0$ and $\mathbf{P}_0$ satisfy (10–27) and (10–28), so do $\alpha \mathbf{X}_0$ and $\beta \mathbf{P}_0$ for any scalars $\alpha$ and $\beta$.

**II. The Second Static Model: An Open System of $n$ Industries** We continue our analysis of the $n$-industry system under the additional assumption that each industry also meets an external demand from consumers, who, in turn, supply labor to the industries at a fixed wage rate for each industry. We shall assume, in addition, that the amount of labor used by each industry is proportional to its total output. To describe this new situation, we introduce the variables $d_i$, $l_i$, and $w_i$ into our model, where

$d_i$ denotes the external demand filled by industry $i$,
$l_i$ denotes the amount of labor supplied to industry $i$, and
$w_i$ denotes the wage rate paid by industry $i$.

By assumption each ratio

$$\frac{l_i}{x_i} = c_i \tag{10–29}$$

is a constant; we also assume these constants to be known. Then the equations expressing the equilibrium conditions on the outputs and profits are

$$x_i = \alpha_{i1}x_1 + \cdots + \widehat{\alpha_{ii}x_i} + \cdots + \alpha_{in}x_n + d_i$$

and

$$p_i = p_1\alpha_{1i} + \cdots + \widehat{p_i\alpha_{ii}} + \cdots + p_n\alpha_{ni} + w_ic_i, \tag{10–30}$$

$i = 1, \ldots, n$. (See Exercise 1.) Thus if we set

$$\mathbf{W} = \begin{bmatrix} w_1 & 0 & \cdots & 0 \\ 0 & w_2 & \cdots & 0 \\ & \cdot & & \cdot \\ & \cdot & & \cdot \\ & \cdot & & \cdot \\ 0 & 0 & \cdots & w_n \end{bmatrix},$$

$$\mathbf{C} = \begin{bmatrix} c_1 \\ \cdot \\ \cdot \\ \cdot \\ c_n \end{bmatrix} \quad \text{and} \quad \mathbf{D} = \begin{bmatrix} d_1 \\ \cdot \\ \cdot \\ d_n \end{bmatrix},$$

then the behavior of the system is governed by the pair of nonhomogeneous matrix equations

$$\mathbf{AX} + \mathbf{D} = \mathbf{0} \tag{10–31}$$

$$\mathbf{PA} + \mathbf{WC} = \mathbf{0}.$$

These equations can be used to determine $\mathbf{X}$ and $\mathbf{P}$ when $\mathbf{A}, \mathbf{D}, \mathbf{W}$, and $\mathbf{C}$ are known. They can also be used, for example, to study the effect of different wage rates on production and prices by determining what effect changes in $\mathbf{W}$ have on $\mathbf{X}$ and $\mathbf{P}$. Note that in contrast to the closed system, (10–31) has a unique solution if and only if $\mathbf{A}$ is nonsingular (Theorem 9–10). This system is known as **Leontief's open system of $n$ industries**.

**III. Leontief's Dynamic Model for the Closed and Open Systems**   The dynamic model for the $n$-industry system is obtained by taking into account the process of capital formation over time, the various industries stockpiling some of their own output and the output of the other industries in the system. To describe this process, we let

$y_{ij}$ denote the amount of output from industry $i$ that is held as stock (as opposed to used) by industry $j$.

Once again we assume linearity by requiring that the ratios $y_{ij}/x_j$ be constant and set

$$\beta_{ij} = \frac{y_{ij}}{x_j}. \qquad (10\text{–}32)$$

The $\beta_{ij}$ are called the **capital coefficients** of the system, and we suppose that they too are known.

Now in this model the output of industry $i$ is not only distributed to the remaining $n - 1$ industries to meet their current production needs, but is also distributed to all $n$ industries for stockpiling. Thus the equilibrium condition governing the production for the $i$th industry in a *closed* system is

$$x_i = z_{i1} + \cdots + \widehat{z_{ii}} + \cdots + z_{in} + \frac{d}{dt}(y_{i1} + \cdots + y_{in}). \qquad (10\text{–}33)$$

Since (10–32) implies that

$$\frac{dy_{ij}}{dt} = \beta_{ij}\frac{dx_j}{dt},$$

(10–33) can be rewritten in terms of the flow constants and capital coefficients as

$$\alpha_{i1}x_1 + \cdots + (-1)x_i + \cdots + \alpha_{in}x_n + \frac{d}{dt}(\beta_{i1}x_1 + \cdots + \beta_{in}x_n) = 0.$$

It follows that the output vector $\mathbf{X}$ for the system must now satisfy the equation

$$\mathbf{AX} + \mathbf{BX}' = \mathbf{0}, \qquad (10\text{–}34)$$

where

$$\mathbf{B} = \begin{bmatrix} \beta_{11} \cdots \beta_{1n} \\ \cdot \qquad \cdot \\ \cdot \qquad \cdot \\ \cdot \qquad \cdot \\ \beta_{n1} \cdots \beta_{nn} \end{bmatrix}$$

and the prime denotes differentiation with respect to time. The equation for determining prices remains as it was in the static model.

Finally, in the case of an *open* system where there are external demands $d_1, \ldots, d_n$ for the outputs of the various industries, the output vector must satisfy the matrix equation

$$\mathbf{AX} + \mathbf{BX'} + \mathbf{D} = \mathbf{0}. \qquad (10\text{–}35)$$

Again the equation involving profits remains unchanged: $\mathbf{PA} + \mathbf{WC} = \mathbf{0}$.

**Example**  We conclude this discussion of Leontief's model by determining the output vector for the closed system in the dynamic case in which only two industries are involved. The equation to be solved, then, is

$$\mathbf{AX} + \mathbf{BX'} = \mathbf{0} \qquad (10\text{–}36)$$

or

$$\begin{bmatrix} -1 & \alpha_{12} \\ \alpha_{21} & -1 \end{bmatrix} \begin{bmatrix} x_1 \\ x_2 \end{bmatrix} + \begin{bmatrix} \beta_{11} & \beta_{12} \\ \beta_{21} & \beta_{22} \end{bmatrix} \begin{bmatrix} x_1' \\ x_2' \end{bmatrix} = \begin{bmatrix} 0 \\ 0 \end{bmatrix}. \qquad (10\text{–}37)$$

We solve it under the assumption that $\mathbf{B}$ has an inverse, $\mathbf{B}^{-1}$, for in this case (10–36) can be solved for $\mathbf{X'}$ by multiplying both sides of the equation by $\mathbf{B}^{-1}$ and using the fact that $\mathbf{B}^{-1}\mathbf{B} = \mathbf{I}$:

$$\begin{aligned}
\mathbf{AX} + \quad \mathbf{BX'} &= \quad \mathbf{0} \\
\mathbf{B}^{-1}\mathbf{AX} + \mathbf{B}^{-1}\mathbf{BX'} &= \mathbf{B}^{-1}\mathbf{0} \\
\mathbf{B}^{-1}\mathbf{AX} + \quad \mathbf{IX'} &= \quad \mathbf{0} \\
\mathbf{B}^{-1}\mathbf{AX} + \quad \mathbf{X'} &= \quad \mathbf{0} \\
\mathbf{X'} &= -\mathbf{B}^{-1}\mathbf{AX}. \qquad (10\text{–}38)
\end{aligned}$$

Thus our first task is to find the eigenvalues of the $2 \times 2$ matrix $-\mathbf{B}^{-1}\mathbf{A}$. Although this can be done by a straightforward calculation (Exercise 2), it turns out to be much easier to take advantage of the following two properties of determinants. If $\mathbf{M}$ and $\mathbf{N}$ are $n \times n$ matrices and $\alpha$ is a scalar, then

$$(\text{Det } \mathbf{M})(\text{Det } \mathbf{N}) = \text{Det } (\mathbf{MN}) \qquad (10\text{–}39)$$

and

$$\text{Det } \alpha\mathbf{M} = \alpha^n \text{ Det } \mathbf{M}. \qquad (10\text{–}40)$$

In particular, if $\mathbf{M}$ has an inverse, $\mathbf{M}^{-1}$, then (10–39) implies that

$$(\text{Det } \mathbf{M})(\text{Det } \mathbf{M}^{-1}) = \text{Det } \mathbf{I} = 1$$

or

$$\text{Det } \mathbf{M}^{-1} = \frac{1}{\text{Det } \mathbf{M}}. \qquad (10\text{–}41)$$

Returning now to (10–38) and the matrix equation

$$-\mathbf{B}^{-1}\mathbf{A} - \lambda\mathbf{I} = \mathbf{0}$$

for the eigenvalues of $-\mathbf{B}^{-1}\mathbf{A}$, we observe that

$$-\mathbf{B}^{-1}\mathbf{A} - \lambda\mathbf{I} = -\mathbf{B}^{-1}(\mathbf{A} + \lambda\mathbf{B}).$$

Hence by (10–41)

$$\text{Det}\,(-\mathbf{B}^{-1}\mathbf{A} - \lambda\mathbf{I}) = \text{Det}\,(\mathbf{A} + \lambda\mathbf{B})/\text{Det}\,\mathbf{B}$$

and the eigenvalues of $-\mathbf{B}^{-1}\mathbf{A}$ are the roots of the equation

$$\text{Det}\,(\mathbf{A} + \lambda\mathbf{B}) = 0.$$

In the case under consideration this equation is

$$\begin{vmatrix} -1 + \lambda\beta_{11} & \alpha_{12} + \lambda\beta_{12} \\ \alpha_{21} + \lambda\beta_{21} & -1 + \lambda\beta_{22} \end{vmatrix} = 0$$

or

$$(\beta_{11}\beta_{22} - \beta_{12}\beta_{21})\lambda^2 - (\beta_{11} + \beta_{22} + \alpha_{21}\beta_{12} + \alpha_{12}\beta_{21})\lambda$$
$$+ (1 - \alpha_{12}\alpha_{21}) = 0. \qquad (10\text{–}42)$$

A straightforward (though fussy) computation shows that the discriminant of (10–42) is

$$[(\beta_{11} + \alpha_{12}\beta_{21}) - (\beta_{22} + \alpha_{21}\beta_{12})]^2$$
$$+ 4(\beta_{12}\beta_{21} + \alpha_{12}\beta_{21}\beta_{22} + \alpha_{21}\beta_{11}\beta_{12} + \alpha_{12}\alpha_{21}\beta_{11}\beta_{12}), \qquad (10\text{–}43)$$

which is nonnegative because the $\alpha_{ij}$ and $\beta_{ij}$ are themselves nonnegative. Hence the eigenvalues $\lambda_1$ and $\lambda_2$ for this problem are real, and a routine calculation reveals that up to constant multiples the eigenvectors belonging to these eigenvalues are

$$\begin{bmatrix} 1 \\ k_1 \end{bmatrix} \quad \text{and} \quad \begin{bmatrix} 1 \\ k_2 \end{bmatrix}, \qquad (10\text{–}44)$$

where

$$k_i = \frac{1 - \lambda_i\beta_{11}}{\alpha_{12} + \lambda_i\beta_{12}}, \qquad i = 1, 2. \qquad (10\text{–}45)$$

Thus when $\lambda_1 \neq \lambda_2$ the production vector for the system is

$$\begin{bmatrix} x_1(t) \\ x_2(t) \end{bmatrix} = c_1 \begin{bmatrix} 1 \\ k_1 \end{bmatrix} e^{\lambda_1 t} + c_2 \begin{bmatrix} 1 \\ k_2 \end{bmatrix} e^{\lambda_2 t},$$

where $c_1$ and $c_2$ are constants whose values are determined by the initial conditions $x_1(0)$ and $x_2(0)$.

The economic interpretation of this result falls into three cases:

i) If either $\lambda_1$ or $\lambda_2$ is positive, production "explodes" and the model is unrealistic.

ii) If both eigenvalues are negative, production diminishes to zero, which is also unrealistic.

iii) If one solution of (10–42), say $\lambda_1$, is zero while the other is negative, then production eventually tends to

$$c_1 \begin{bmatrix} 1 \\ k_1 \end{bmatrix} = c_1 \begin{bmatrix} 1 \\ 1/\alpha_{12} \end{bmatrix}. \qquad (10\text{–}46)$$

Clearly (iii) is the only economically viable case. Moreover, it follows from (10–42) that this can happen if and only if

$$1 - \alpha_{12}\alpha_{21} = \text{Det } \mathbf{A} = 0,$$

a result which is eminently satisfactory in view of the fact that the corresponding static model then also has a nontrivial solution. When this condition is satisfied, the eigenvalue $\lambda_2$ is

$$\frac{\beta_{11} + \beta_{12} + \alpha_{21}\beta_{12} + \alpha_{12}\beta_{21}}{\beta_{11}\beta_{22} - \beta_{12}\beta_{21}}$$

and is negative if and only if

$$\text{Det } \mathbf{B} = \beta_{11}\beta_{22} - \beta_{12}\beta_{21} < 0.$$

Finally, the case $\lambda_1 = \lambda_2$ also turns out to be of no interest in economic terms. We leave the investigation of this situation, as well as other aspects of the dynamic model, as exercises.

### EXERCISES

1. Verify Eq. (10–30) for the open static system.

2. a) Given the invertible matrix

$$\mathbf{B} = \begin{bmatrix} \beta_{11} & \beta_{12} \\ \beta_{21} & \beta_{22} \end{bmatrix},$$

   show that

$$\mathbf{B}^{-1} = \frac{1}{\beta_{11}\beta_{22} - \beta_{12}\beta_{21}} \begin{bmatrix} \beta_{22} & -\beta_{21} \\ -\beta_{12} & \beta_{11} \end{bmatrix}.$$

   b) Use the result in (a) to show that the characteristic equation for $\mathbf{B}^{-1}\mathbf{A}$ is

$$(\beta_{11}\beta_{22} - \beta_{12}\beta_{21})\lambda^2 - (\beta_{11} + \beta_{22} + \alpha_{21}\beta_{12} + \alpha_{12}\beta_{21})\lambda + (1 - \alpha_{12}\alpha_{21}) = 0$$

   when

$$\mathbf{A} = \begin{bmatrix} -1 & \alpha_{12} \\ \alpha_{21} & -1 \end{bmatrix}.$$

3. Let

$$\mathbf{A} = \begin{bmatrix} \alpha_{11} & \alpha_{12} \\ \alpha_{21} & \alpha_{22} \end{bmatrix} \quad \text{and} \quad \mathbf{B} = \begin{bmatrix} \beta_{11} & \beta_{12} \\ \beta_{21} & \beta_{22} \end{bmatrix}.$$

   Prove that

   a) Det $(\mathbf{AB}) = (\text{Det } \mathbf{A})(\text{Det } \mathbf{B})$,
   b) Det $(\alpha\mathbf{A}) = \alpha^2 \text{ Det } \mathbf{A}$.

4. Show that the discriminant of (10–42) can be written as (10–43).

5. a) Show that the solutions of (10–42) cannot both be 0. [*Hint*: If the solutions of a quadratic equation $ax^2 + bx + c = 0$ are both 0, what can you say about $b$ and $c$?]

   b) Show that if (10–42) has solutions $\lambda_1 = \lambda_2 \neq 0$, then $\alpha_{12} = \alpha_{21} = 0$. [*Hint*: Remember that $\alpha_{ij}, \beta_{ij} \geqslant 0$ and Det **B** $\neq 0$.]

   c) What is the economic interpretation of $\alpha_{12} = \alpha_{21} = 0$?

6. Show that the asymptotic production vector given by (10–46) is the production vector obtained by solving the corresponding static model.

7. Show that in the dynamic case the solution of the open system of two industries with fixed demands $d_1$ and $d_2$ is

$$\begin{bmatrix} x_1(t) \\ x_2(t) \end{bmatrix} = \begin{bmatrix} d_1 \\ d_2 \end{bmatrix} + c_1 \begin{bmatrix} 1 \\ k_1 \end{bmatrix} e^{\lambda_1 t} + c_2 \begin{bmatrix} 1 \\ k_2 \end{bmatrix} e^{\lambda_2 t},$$

where $\lambda_1, \lambda_2, k_1$ and $k_2$ are as in the solution of the closed system. (Assume Det **A** $\neq 0$.)

8. To interpret the result of Exercise 7 in economic terms, assume that Det **B** $< 0$. (This assumption is reasonable because in practice an industry will keep only a small amount of its own output in stock so that $\beta_{11}\beta_{22}$ will be less than $\beta_{12}\beta_{21}$.)

   a) Show that in this case the following sign table will hold.

   |                       | Det A $> 0$ | Det A $< 0$ |
   |-----------------------|:-----------:|:-----------:|
   | $\lambda_1 + \lambda_2$ |      −      |      −      |
   | $\lambda_1\lambda_2$    |      −      |      +      |
   | $d_1, d_2$              |      +      |      −      |

   b) Use the table in (a) to conclude that Det **A** $< 0$ is economically unreasonable.

   c) Use the table in (a) to conclude that the output functions "explode" when Det **A** $> 0$, and hence that this case is also economically unreasonable.

   (*Remark*. As this exercise indicates, the model constructed in this section is inadequate in certain cases. But the obvious remedy, which is to allow the $\alpha_{ij}$ and $\beta_{ij}$ to vary with time, introduces great technical difficulties.)

9. Consider the constant-coefficient system

$$\mathbf{X}' = \mathbf{S}\mathbf{X} + \mathbf{D},$$

   where **D** is a constant vector and Det **S** $= 0$.

   a) Show that

$$\mathbf{S} = \begin{bmatrix} a & ka \\ b & kb \end{bmatrix}.$$

   b) Show that the system has a constant solution if and only if **D** is a constant multiple of a column of **S**.

c) Use the results of parts (a) and (b) to show that constant solutions of the open system

$$\mathbf{AX} + \mathbf{BX'} + \mathbf{D} = \mathbf{0}$$

exist when Det $\mathbf{A} = 0$.

10. Discuss the nature of the production vector for the closed dynamic system

$$\mathbf{AX} + \mathbf{BX'} = \mathbf{0}$$

for three industries when the eigenvalues for the system are 0, $a + bi$, and $a - bi$ with $a < 0$ and $b \neq 0$.

## 10–7  AN APPLICATION TO PHYSICS

Imagine two masses $m_1$ and $m_2$ coupled as shown in Fig. 10–3 and constrained to vibrate horizontally. We seek equations of motion for this system under the assumptions that it is immersed in a viscous medium which gives rise to retarding forces directly proportional to the velocities of the masses, and that each mass is subject to an external force that varies with time.

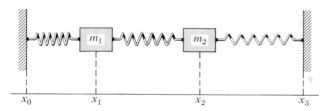

**Figure 10–3**

We begin our analysis by letting $l_1$, $l_2$, and $l_3$ denote the natural lengths of the springs as they appear from left to right in the figure, and $k_1$, $k_2$, and $k_3$ their respective spring constants. Then there are three kinds of forces acting on the first mass:

1. *restoring forces* $k_1(x_1 - x_0 - l_1)$ and $k_2(x_2 - x_1 - l_2)$ (see Fig. 10–4);

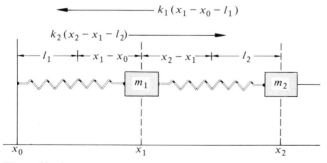

**Figure 10–4**

2. a *retarding force* $a_1 \dfrac{dx_1}{dt}$, $a_1$ a constant;

3. an *external force* $F_1 = F_1(t)$.

Similar forces act on the second mass, and therefore Newton's second law implies that the motion of the system is governed by the pair of second-order linear differential equations

$$m_1 \frac{d^2 x_1}{dt^2} = -k_1(x_1 - x_0 - l_1) + k_2(x_2 - x_1 - l_1) - a_1 \frac{dx_1}{dt} + F_1(t)$$

$$(10\text{--}47)$$

$$m_2 \frac{d^2 x_2}{dt^2} = -k_2(x_2 - x_1 - l_2) + k_3(x_3 - x_2 - l_3) - a_2 \frac{dx_2}{dt} + F_2(t).$$

In the interest of simplicity we now assume that

$$l_1 = l_2 = l_3 = l \quad \text{and} \quad k_1 = k_2 = k_3 = k.$$

Then (10–47) can be rewritten as

$$m_1 \frac{d^2 x_1}{dt^2} + a_1 \frac{dx_1}{dt} - k(x_0 - 2x_1 + x_2) = F_1(t)$$

$$(10\text{--}48)$$

$$m_2 \frac{d^2 x_2}{dt^2} + a_2 \frac{dx_2}{dt} - k(x_1 - 2x_2 + x_3) = F_2(t).$$

Up to this point we have allowed $x_0$ and $x_3$ to vary with time. We now assume, however, that they are fixed, with $x_0 = A_0$ and $x_3 = A_3$. Then when $F_1$ and $F_2$ are identically zero, (10–48) has the unique time-independent solution

$$x_1 = \tfrac{2}{3}A_0 + \tfrac{1}{3}A_3$$

$$(10\text{--}49)$$

$$x_2 = \tfrac{1}{3}A_0 + \tfrac{2}{3}A_3$$

found by setting the derivatives in each equation equal to zero and solving for $x_1$ and $x_2$ (Exercise 1). Thus, in the absence of external forces and moving boundaries the spring-mass system is in equilibrium with the masses equally spaced between the walls at $A_0$ and $A_3$. This suggests that we make a change of variables and measure displacements from these equilibrium positions, which we now denote by $A_1$ and $A_2$, respectively. Hence we set

$$y_1 = x_1 - A_1 \quad \text{and} \quad y_2 = x_2 - A_2.$$

In addition, we introduce the momentum variables

$$y_3 = m_1 \frac{dx_1}{dt} \quad \text{and} \quad y_4 = m_2 \frac{dx_2}{dt},$$

thereby converting (10–48) into the first-order system

$$\frac{dy_1}{dt} = \frac{1}{m_1} y_3$$

$$\frac{dy_2}{dt} = \frac{1}{m_2} y_4$$

$$\frac{dy_3}{dt} = -2ky_1 + ky_2 - \frac{a_1}{m_1} y_3 + F_1(t)$$

$$\frac{dy_4}{dt} = ky_1 - 2ky_2 - \frac{a_2}{m_2} y_4 + F_2(t)$$

(Exercise 3). This system can be written in matrix form as

$$Y' = AY + B, \tag{10–51}$$

where

$$A = \begin{bmatrix} 0 & 0 & 1/m_1 & 0 \\ 0 & 0 & 0 & 1/m_2 \\ -2k & k & -a_1/m_1 & 0 \\ k & -2k & 0 & -a_2/m_2 \end{bmatrix}, \qquad B = \begin{bmatrix} 0 \\ 0 \\ F_1(t) \\ F_2(t) \end{bmatrix}.$$

The characteristic polynomial of $A$ is (see Exercise 4)

$$\lambda^4 + \left( \frac{a_1}{m_1} + \frac{a_2}{m_2} \right) \lambda^3 + \left[ \frac{a_1 a_2}{m_1 m_2} + 2k \left( \frac{1}{m_1} + \frac{1}{m_2} \right) \right] \lambda^2$$
$$+ \frac{2k}{m_1 m_2} (a_1 + a_2)\lambda + \frac{3k^2}{m_1 m_2}. \tag{10–52}$$

We now simplify the problem further by assuming that the viscous forces acting on the masses can be ignored. Then $a_1 = a_2 = 0$, and the characteristic equation of $A$ becomes

$$\lambda^4 + 2k \left( \frac{1}{m_1} + \frac{1}{m_2} \right) \lambda^2 + \frac{3k^2}{m_1 m_2} = 0.$$

We find that

$$\lambda^2 = k[-(\mu_1 + \mu_2) \pm \sqrt{(\mu_1 + \mu_2)^2 - 3\mu_1 \mu_2}],$$

where

$$\mu_1 = \frac{1}{m_1}, \qquad \mu_2 = \frac{1}{m_2},$$

and

$$(\mu_1 + \mu_2)^2 - 3\mu_1 \mu_2 > 0.$$

Hence the eigenvalues for this problem have the form

$$\lambda_1 = \omega i, \qquad \lambda_2 = -\omega i,$$
$$\lambda_3 = v i, \qquad \lambda_4 = -v i, \tag{10-53}$$

with $\omega > 0$ and $v > 0$. To find eigenvectors for these eigenvalues, we must find a nontrivial solution of

$$\begin{bmatrix} 0 & 0 & \mu_1 & 0 \\ 0 & 0 & 0 & \mu_2 \\ -2k & k & 0 & 0 \\ k & -2k & 0 & 0 \end{bmatrix} \begin{bmatrix} y_1 \\ y_2 \\ y_3 \\ y_4 \end{bmatrix} = \lambda_i \begin{bmatrix} y_1 \\ y_2 \\ y_3 \\ y_4 \end{bmatrix}$$

as $\lambda_i$ assumes each of the values $\lambda_1, \lambda_2, \lambda_3, \lambda_4$. We omit the computational details and simply assert that the vectors

$$\mathbf{E}_{\lambda_i} = \begin{bmatrix} k \\ 2k + \dfrac{\lambda_i^2}{\mu_1} \\ \dfrac{k\lambda_i}{\mu_1} \\ \dfrac{\lambda_i}{\mu_2}\left(2k + \dfrac{\lambda_i^2}{\mu_1}\right) \end{bmatrix} \tag{10-54}$$

satisfy the system. Thus, when there are no viscous forces, the homogeneous equation

$$\mathbf{Y}' = \mathbf{A}\mathbf{Y}$$

has the four linearly independent solutions

$$\mathbf{Y}_i(t) = \mathbf{E}_{\lambda_i} e^{\lambda_i t},$$

and

$$y_1(t) = A_1 k \cos \omega t + A_2 k \sin \omega t + A_3 k \cos v t + A_4 k \sin v t$$

$$y_2(t) = A_1\left(2k - \frac{\omega^2}{\mu_1}\right)\cos \omega t + A_2\left(2k - \frac{\omega^2}{\mu_1}\right)\sin \omega t$$
$$+ A_3\left(2k - \frac{v^2}{\mu_1}\right)\cos v t + A_4\left(2k - \frac{v^2}{\mu_1}\right)\sin v t,$$

where $A_1, A_2, A_3$, and $A_4$ are arbitrary constants.

**EXERCISES**

1. Verify that (10–49) is the unique time-independent solution of (10–48) when $x_0 = A_0$, $x_3 = A_3$, and $F_1 = F_2 = 0$.

2. Show that the general spring–mass system described by (10–47) always has a unique equilibrium position when $x_0$ and $x_3$ are constants and $F_1 = F_2 = 0$.

3.  Verify that when $x_0 = A_0$ and $x_3 = A_3$ the change of variables given in the text transforms (10–48) into (10–50).

4.  Verify that (10–52) is the characteristic polynomial of **A**.

5.  a) Prove that

$$(\mu_1 + \mu_2)^2 - 3\mu_1\mu_2 > 0$$

    whenever $\mu_1 > 0$ and $\mu_2 > 0$.

    b) Use the result in (a) to show that when $k > 0$ the eigenvalues of the matrix

$$\begin{bmatrix} 0 & 0 & \mu_1 & 0 \\ 0 & 0 & 0 & \mu_2 \\ -2k & k & 0 & 0 \\ k & -2k & 0 & 0 \end{bmatrix}$$

    are of the form $\pm\omega i$ and $\pm v i$, where $\omega > 0$ and $v > 0$.

    c) What are the eigenvalues for the matrix in (b) when $\mu_1 = \mu_2 = k = 1$?

6.  Verify that the vectors (10–54) indeed satisfy the given system.

7.  Solve (10–48) for $m_1 = m_2 \to 0$, $F_1(t) = F_2(t) = 0$, and $a_1 > 0$, $a_2 > 0$, given that $x_0 = A_0$ and $x_3 = A_3$ are constants.

8.  Consider the masses and springs shown in Fig. 10–5, where $s_1$ and $s_2$ give the positions of the masses when the system is at rest.

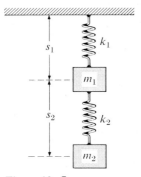

**Figure 10–5**

a) Find a system of linear differential equations that describes the motion of the system, using the displacements $x_1$ and $x_2$ from the equilibrium position as time-dependent variables.

b) Solve the system when

$$k_1 = k_2 = k, \qquad m_1 = m_2 = m.$$

c) Find the solution in (b) that satisfies the initial conditions

$$x_1(0) = L, \qquad x_1'(0) = x_2(0) = x_2'(0) = 0.$$

d) Modify the system in (a) to allow for the presence of forcing functions $F_1(t)$ and $F_2(t)$ and damping factors that are proportional to the velocity.

# Direction Fields and Numerical Methods

**11**

## 11–1 DIRECTION FIELDS

In Chapter 1 we developed a number of techniques for solving special forms of the first-order differential equation

$$y' = f(x, y), \tag{11–1}$$

where $f$ is continuous in a region $R$ of the $xy$-plane. We have made no attempt, however, either then or since, to discuss this equation systematically or describe the behavior of its solutions—if, indeed, it has solutions. (It does, as we shall soon see.) We now turn our attention to these questions, both because of their intrinsic interest and because their answers lead to new and powerful techniques for studying the solutions of differential equations. For the moment our approach will be geometric and informal; proofs of most of the assertions made in this discussion will be given in the next chapter.

We begin by observing that Eq. (11–1) assigns a slope to each point in $R$, and that a solution curve of the equation is a differentiable curve whose slope at a point in $R$ agrees with the slope already assigned to that point. Imagine, then, that through each point $(x_0, y_0)$ in $R$ a small line segment with slope $f(x_0, y_0)$ is drawn. The resulting collection of segments is called the **direction field** of (11–1), and the task of finding solutions of the equation can now be described as an attempt to "thread" these segments together into smooth (differentiable) curves. In Fig. 11–1 we have sketched the direction field and one of the solution curves of the equation

$$y' = 2x.$$

Note that it is clear from the direction field alone that the solution curves of this equation are a one-parameter family of parabolas opening upward about the $y$-axis.

The examples which follow illustrate the geometric information that can be obtained about the solutions of first-order differential equations from their direction fields.

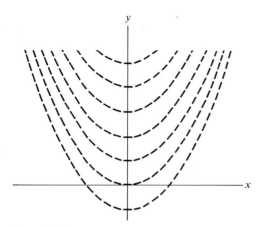

**Figure 11–1**

**Example 1**    Sketch the direction field and solution curves of

$$y' = \frac{x - y}{x}. \tag{11–2}$$

*Solution.*   This equation is linear and has

$$y = \frac{x}{2} + \frac{c}{x}, \qquad x \neq 0,$$

as its general solution. We shall proceed, however, as though this information were not available, and determine the behavior of the solution curves directly from (11–2) and its associated direction field.

First we observe that $y' = 0$ along the line $y = x$, and $y'$ is undefined along the line $x = 0$. These lines divide the plane into four regions, in two of which (the shaded regions in Fig. 11–2) $y'$ is negative and in two of which $y'$ is positive. A solution curve falls as it crosses a shaded region, rises as it crosses an unshaded region, and moves from one to the other with zero slope.

We now look for curves along which $y'$ is constant. Such curves are called **isoclines** of the direction field and are invaluable in obtaining accurate sketches. (The word "isocline" means "same slope.") In the present case $y' = k$ along the line

$$y = (1 - k)x$$

(set the right-hand side of (11–2) equal to $k$ and solve). It follows that $y' = 1$ along the x-axis, $y' = 2$ along the line $y = -x$, and so on.

Finally, since the isoclines in this example are lines and therefore have constant slopes, it is natural to ask whether any of them are also solutions of the differential equation. Since this will happen if and only if the isocline $y = (1 - k)x$ has slope $k$,

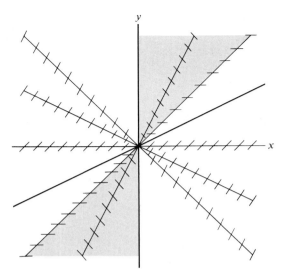

**Figure 11–2**

the condition for an isocline to be a solution is

$$1 - k = k.$$

Thus $k = \frac{1}{2}$ and the line $y = x/2$ is a solution of (11–2). We now have more than enough information to sketch the direction field and solution curves shown in Figs. 11–2 and 11–3.

**Example 2**    Sketch the solution curves of

$$y' = \frac{4x + 3y}{3x + y}. \tag{11–3}$$

*Solution.* Although this equation can be solved without difficulty (its right-hand side is homogeneous of degree 0), it is not easy to sketch its solution curves from the general integral $(y - 2x)^5 = c(y + 2x)$. Instead, we use the direction field.

As in Example 1, we begin by observing that $y' = 0$ along the line

$$y = -\tfrac{4}{3}x,$$

and that $y'$ is undefined along the line

$$y = -3x.$$

Therefore, an integral curve of (11–3) will have a horizontal tangent as it crosses the line $y = -4x/3$, and a vertical tangent as it crosses $y = -3x$ if in fact it does cross that line. This time $y'$ is negative in the shaded region shown in Fig. 11–4, and positive outside the shaded region.

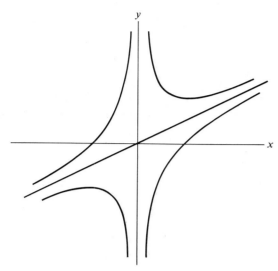

**Figure 11–3**

The isoclines for this equation, found by setting $y' = k$ in (11–3), are the lines

$$(3 - k)y = (3k - 4)x, \qquad (11-4)$$

so that in particular,

$$y' = 1 \quad \text{along the line} \quad y = -\tfrac{1}{2}x,$$
$$y' = \tfrac{4}{3} \quad \text{along the line} \quad y = 0,$$
$$y' = 3 \quad \text{along the line} \quad x = 0.$$

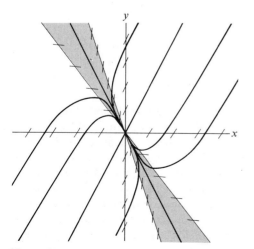

**Figure 11–4**

Finally, we look for isoclines that are also solutions of the equation. Since this can happen only when (11–4) has slope $k$, we have

$$\frac{3k - 4}{3 - k} = k,$$

or $k = \pm 2$. Thus the lines $y = \pm 2x$ are solution curves of (11–3), and we are now in a position to make an accurate sketch of the direction field and its associated solution curves. The results are shown in Fig. 11–4.

### EXERCISES

1. Prove that the equation

$$y' = \frac{ax + by}{cx + dy}$$

   has at least one solution of the form $y = kx$ if and only if $(b - c)^2 + 4ad \geqslant 0$.

2. Which lines through the origin are solutions of the following equations?

   a) $y' = \dfrac{2x - 3y}{3x + 8y}$    b) $y' = \dfrac{-2x + 3y}{x - 3y}$

   c) $y' = \dfrac{x + 3y}{2x - y}$    d) $y' = \dfrac{x + 3y}{x - y}$

3. Sketch the direction field and solution curves of each equation in Exercise 2.

4. Sketch the direction field and solution curves of each of the following equations.

   a) $y' = \dfrac{y - 2}{x - 1}$    b) $y' = xy$

   c) $y' = x - y$    d) $y' = \dfrac{x^2 - y^2}{xy}$

   e) $y' = x^2 + y^2$    f) $y' = (x^2 + y^2)^{1,000,000}$

## 11–2  AN EXISTENCE THEOREM

In the preceding section we saw that the equation

$$y' = f(x, y)$$

determines a direction field in its domain of definition $R$, and that its solution curves, if any, are found by "threading" the segments of this field together into smooth curves that have these segments as tangents. In these terms the *existence problem* for this equation appears as one of imposing conditions on $f$ which will guarantee that the elements of the direction field can be threaded together in this way. It is clear that, at the very least, $f$ must be *continuous* in $R$ if solution curves are to exist. But is continuity by itself enough? Geometric intuition suggests that it is, and this suggestion becomes a virtual conviction if we endow the direction field with physical significance and imagine, for instance, that it represents the vectors in a force field. For then a solution curve will be the *trajectory* or *path* of a particle moving through $R$ under the influence of that field. Since physical intuition demands the existence of such trajectories, the differential equation describing them *must* have solutions; even more important, there must be a solution curve through any preassigned point in $R$. (See Fig. 11–5.) This is the content of the following fundamental theorem, known as the **Peano existence theorem**.

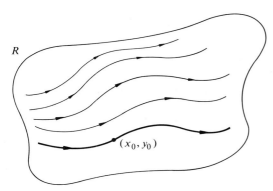

**Figure 11–5**

**Theorem 11–1**   *Let $y' = f(x, y)$ be continuous in an open region $R$ of the xy-plane, and let $(x_0, y_0)$ be any point in R. Then there exists at least one function $y = y(x)$ defined in an interval I about $x_0$ such that*

*i) $y'(x) = f(x, y(x))$ for all x in I, and*

*ii) $y(x_0) = y_0$.*

In short, every initial-value problem involving a normal first-order differential equation whose right-hand side is continuous has at least one solution.

The Peano existence theorem can be viewed, among other things, as a generalization of the fundamental theorem of calculus, which makes the same assertion for the first-order equation $y' = f(x)$. It also generalizes the results obtained earlier for the first-order linear equation $y' = -P(x)y + Q(x)$. Note, however, that Theorem 11–1 contains no assertion concerning the uniqueness of the solution $y = y(x)$. Indeed, it cannot, for without additional restrictions on $f$ initial-value problems involving the equation $y' = f(x, y)$ need not have unique solutions.

## 11–3 A UNIQUENESS THEOREM

Consider the initial-value problem

$$y' = 3y^{2/3}, \qquad y(-2) = -1. \tag{11–5}$$

Since $3y^{2/3}$ is continuous in the entire $xy$-plane, Theorem 11–1 guarantees that this problem has at least one solution. Indeed, a solution is easily found for if we separate variables and integrate, we find that the general integral of the equation is

$$y^{1/3} = x + c,$$

and its general solution is

$$y = (x + c)^3,$$

where $c$ is an arbitrary constant. Finally, since the solution

$$y = (x + 1)^3$$

satisfies the initial condition $y(-2) = -1$, we seem to have solved the problem. But have we? In particular, have we found *all* the solutions? Unfortunately, the answer is no; the problem has infinitely many solutions in addition to the one just found. All of them can be obtained by "piecing together" members of the one-parameter family $y = (x + c)^3$ and the solution $y = 0$ that was suppressed in the process of separating variables and that is not a member of this family. They are

$$y = \begin{cases} (x + 1)^3, & x \leqslant -1, \\ 0, & -1 < x, \end{cases}$$

and

$$y = \begin{cases} (x + 1)^3, & x \leqslant -1, \\ 0, & -1 < x < a, \\ (x - a)^3, & a \leqslant x, \end{cases}$$

where $a > -1$ is an arbitrary constant. Several of these solutions are sketched in Fig. 11–6. As we shall see shortly, we now have a complete catalog of solutions of (11–5).

The preceding example shows that continuity is not enough to ensure that initial-value problems for the first-order equation

$$y' = f(x, y)$$

have unique solutions. A stronger condition must be imposed on $f$ to prevent the equation from having "singular" solutions, like the solution $y = 0$ in the example,

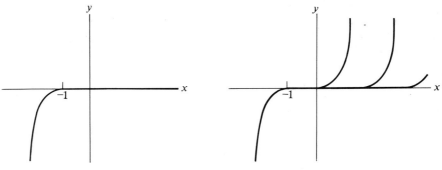

**Figure 11–6**

that are not included in the general integral of the equation.* As we shall see in Chapter 12, this can be accomplished by requiring $f$ to have a *continuous partial derivative with respect to* $y$ in a neighborhood of the point where the initial-value problem is being solved. Specifically, we have the following important result, known as **Picard's theorem**.

> **Theorem 11–2**   *Let $f(x, y)$ and $\partial f/\partial y$ be continuous in a rectangle $R$ of the xy-plane, and let $(x_0, y_0)$ be any point in the interior of $R$. Then there exists an interval $I$ containing the point $x_0$ in which the initial-value problem*
>
> $$y' = f(x, y), \qquad y(x_0) = y_0,$$
>
> *has a unique solution.*

If we analyze initial-value problems involving the equation

$$y' = 3y^{2/3}$$

in light of Picard's theorem, we can see precisely where uniqueness was lost when we solved (11–5). Since

$$\frac{\partial}{\partial y}(3y^{2/3}) = 2y^{-1/3},$$

the hypothesis of Picard's theorem concerning $\partial f/\partial y$ is not satisfied in any region $R$ containing points on the $x$-axis, because $2y^{-1/3}$ is not defined when $y = 0$. However, if $R$ does not intersect the $x$-axis, then Picard's theorem applies and initial-value problems involving $y' = 3y^{2/3}$ do have unique solutions. For (11–5), the solution is

$$y = (x + 1)^3, \qquad x \leqslant -1.$$

### EXERCISES

1. a) Describe the solution curves of $y' = \sqrt{y}$.
   b) Discuss the uniqueness problem for this equation.

2. Repeat Exercise 1 for the equation $y' = -\sqrt{y}$.

3. Use the results of Exercises 1 and 2 to show that no initial-value problem involving the first-order differential equation $(y')^2 = y$ can have a unique solution.

4. a) Show that $y = cx + c^2/4$, where $c$ is an arbitrary constant, is an equation for the family of lines tangent to the parabola $y = x^2$.
   b) Find a differential equation for the family of lines in (a).
   c) Show that in addition to the lines $y = cx + c^2/4$, the differential equation in (b) has $y = x^2$ as a solution. For obvious geometric reasons the parabola is called the envelope of the family of lines.
   d) Discuss the uniqueness problem for the differential equation in (b).

---

\* The solution $y = 0$ of $y' = 3y^{2/3}$ is the *envelope* of the general solution $y = (x + c)^3$. For a discussion of singular solutions and envelopes, consult *Ordinary Differential Equations* by W. Kaplan, Addison-Wesley, Reading, Mass., 1958.

5. The preceding exercise presented a special case of the differential equation

$$y = xy' + f(y'),$$

known as **Clairaut's equation**. Set $y' = p$ in Clairaut's equation, differentiate, and deduce that the lines of the one-parameter family

$$y = cx + f(c)$$

are among the solutions. (Assume that $f$ is differentiable.)

## 11–4 NUMERICAL METHODS (I): THE EULER METHOD

Most differential equations, even first-order equations in normal form, do not have solutions that can be expressed in closed form in terms of known functions. Even when they do, these solutions are often so complicated that they are nearly useless for computational purposes. In such situations, it is often necessary to produce numerical approximations to the solutions.

Students approaching numerical approximations for the first time sometimes view such approximations as inferior substitutes for truth. This attitude is unfortunate, because a good approximation is frequently more useful than an exact expression. If someone asks what $e$ is, for instance, there is only one exact answer: $e$. But this answer will get no thanks, because it begs the issue; the questioner is almost certainly seeking a rational approximation, such as 2.718. Not only is it more useful in computations than the exact expression, but the approximation also conveys information that the exact form conceals. It tells the approximate location of $e$ on the real line.

These remarks also apply to numerical approximations to solutions of initial-value problems or, as they are called, **numerical solutions**. But what constitutes a satisfactory numerical solution of a differential equation? In outline, a solution should consist of *a table of approximate values* of the (exact) solution, computed for specified values of $x$, *accompanied by an estimate of the error* incurred at each tabulated value. Moreover, there should be a guarantee that if the process used to obtain the approximations were refined, the errors in the tabulated values would diminish; that is, the successively refined approximations should *converge* to the actual solution. And finally, any technique developed to obtain such approximations should be efficient and computationally simple enough so that it can be programmed on a high-speed computer. In this section we shall use the concept of the direction field of a first-order differential equation to develop one such technique, known as the **Euler method**. Various refinements of this technique will be discussed in the sections which follow.

Consider, then, the initial-value problem

$$y' = f(x, y), \qquad y(x_0) = y_0, \tag{11-6}$$

which we assume has a unique solution $y = y(x)$ in an interval $I$ about $x_0$. We seek approximations to the values of this solution at the points

$$x_0, \quad x_1, \quad x_2, \quad \ldots, \quad x_n,$$

where $x_1 = x_0 + h$, $x_2 = x_1 + h$, ..., $x_n = x_{n-1} + h$, and all of the $x_k$ belong
to $I$. (To facilitate computer computation, the distance between successive points
at which the approximations are computed is made a constant $h$. This constant
is known as the *step size* of the numerical solution.) The Euler method proceeds
as follows.

The point $(x_0, y_0)$ lies on the solution curve $y = y(x)$ of (11-6), and at that
point the slope of the solution is

$$y'(x_0) = f(x_0, y_0).$$

Thus, *if $h$ is small*, the solution curve can be approximated with reasonable accuracy in the interval $[x_0, x_0 + h]$ by the line segment through $(x_0, y_0)$ with slope
$f(x_0, y_0)$. (See Fig. 11-7.) An equation for this line is

$$y - y_0 = f(x_0, y_0)(x - x_0).$$

Hence when $x = x_1 = x_0 + h$ we obtain the approximation

$$y_1 = y_0 + hf(x_0, y_0)$$

to the exact value $y(x_1)$ of the solution. Note that $y_1$ is the linear approximation
to $y(x_1)$ obtained by using the differential of $y = y(x)$. Now we repeat the process,
starting at the point $(x_1, y_1)$ and using the slope $f(x_1, y_1)$ assigned to that point
by the differential equation. This gives

$$y_2 = y_1 + hf(x_1, y_1)$$

as the approximation to the value of the solution at $x_2 = x_1 + h$. Continuing in
this way, we obtain the sequence of approximations

$$y_1 = y_0 + hf(x_0, y_0)$$
$$y_2 = y_1 + hf(x_1, y_1)$$
$$\cdot$$
$$\cdot$$
$$\cdot$$
$$y_n = y_{n-1} + hf(x_{n-1}, y_{n-1}).$$

Of course, since the graph of $y = y(x)$ is not in general a straight line in the
interval $[x_0, x_1]$, the point $(x_1, y_1)$ will not lie on the solution curve, and an error

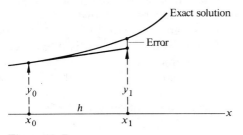

Figure 11-7

is incurred at the first step of the approximation. Additional errors arise at each successive step, so even when $h$ is small, we must not be surprised if the approximations $y_k$ diverge ever further from the exact value of the solution with increasing $k$, as suggested by Fig. 11–8.

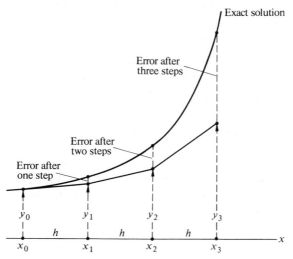

**Figure 11–8**

The simplicity of the Euler method is reflected by the simplicity of the flow-chart in Fig. 11–9, which shows how this method could be implemented on a computer. In the flowchart the READ statement specifies the input data that must be given to the computer; instructions such as $x_k \leftarrow x_k + h$ tell the computer to replace the current value of $x_k$ by $x_k + h$; and the question "Is $x_k > x_0 + N$?" serves to terminate the computation when $x_k$ exceeds the preassigned value $x_0 + N$.

It can be shown that if all the computations involved in using the Euler method could be carried out exactly, the error in each approximation would not exceed a constant multiple of $h$. Thus, if round-off error is neglected (something which cannot be done in practice), the Euler method converges. Unfortunately, to obtain acceptable results by this method in a preassigned interval $[x_0, x_n]$, the value of $h$ must be so small that the amount of computation required is often excessive for hand computation. The reason for this is that the constant appearing in the error estimate usually depends on $x_n - x_0 = nh$ in such a way that it increases with increasing $n$.

Table 11–1 shows the approximations to the solution $y = e^x$ of

$$y' = y, \qquad y(0) = 1$$

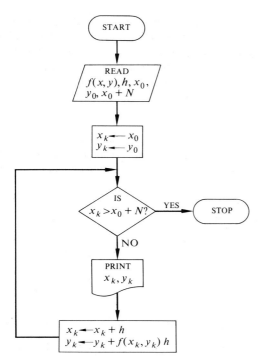

**Figure 11–9**

Table 11–1 Approximations to
$y = e^x$, the solution
of $y' = y$, $y(0) = 1$

| $x_k$ | $y_k$ (Euler) | $y_k$ (exact) |
|-------|---------------|---------------|
| 0.0 | 1.000 | 1.000 |
| 0.1 | 1.100 | 1.105 |
| 0.2 | 1.210 | 1.221 |
| 0.3 | 1.331 | 1.350 |
| 0.4 | 1.464 | 1.492 |
| 0.5 | 1.610 | 1.649 |
| 0.6 | 1.771 | 1.822 |
| 0.7 | 1.948 | 2.014 |
| 0.8 | 2.143 | 2.226 |
| 0.9 | 2.357 | 2.460 |
| 1.0 | 2.593 | 2.718 |

obtained by the Euler method in the interval $[0, 1]$, with $h = 0.1$ and all computations rounded to three decimal places. The table also gives the exact three-place value of the solution corresponding to each approximation. It is clear that these approximations are too crude to be of practical use in computation; the error in the approximate value of $e$, for instance, is nearly 4.6%. They will be used, however, in the improved techniques that will be developed in the next two sections.

## EXERCISES

Use the Euler method with the given value of $h$ to find a numerical solution for each of the following initial-value problems on the indicated interval. Round all computations to three decimal places.

1. $y' = 2x$, $y(0) = 0$, on $[0, 1]$ with $h = 0.1$

2. $y' = x + y$, $y(0) = 0$, on $[0, 1]$ with $h = 0.1$

3. $y' = \dfrac{x + y}{x}$, $y(1) = 1.5$, on $[1, 2]$ with $h = 0.2$

4. $y' = x^2 + y^2$, $y(0) = 1$, on $[0, 1]$ with $h = 0.2$

5. $y' = (x + y)^2$, $y(0) = -1$, on $[0, 1]$ with $h = 0.2$

## 11–5  NUMERICAL METHODS (II): THE MODIFIED EULER METHOD

The remaining methods that we shall develop for obtaining numerical solutions of the initial-value problem

$$y' = f(x, y), \qquad y(x_0) = y_0, \tag{11–7}$$

are based on the observation that if $f$ is continuous in a region $R$ of the $xy$-plane that contains the point $(x_0, y_0)$, then $y = y(x)$ is a solution of $(11-7)$ *if and only if* $y$ is also a solution of the integral equation

$$y(x) = y_0 + \int_{x_0}^{x} f(t, y(t))\, dt. \tag{11–8}$$

(See Exercise 1.) Thus, if $x_1 = x_0 + h$,

$$y(x_1) = y_0 + \int_{x_0}^{x_0 + h} f(t, y(t))\, dt,$$

and more generally, if $x_{k+1} = x_k + h$, then

$$
\begin{aligned}
y(x_{k+1}) &= y_0 + \int_{x_0}^{x_k + h} f(t, y(t))\, dt \\
&= y_0 + \int_{x_0}^{x_k} f(t, y(t))\, dt + \int_{x_k}^{x_k + h} f(t, y(t))\, dt \\
&= y(x_k) + \int_{x_k}^{x_k + h} f(t, y(t))\, dt.
\end{aligned}
\tag{11–9}
$$

It follows that the problem of approximating the values of the $y(x_k)$ is equivalent to that of approximating the definite integrals

$$\int_{x_k}^{x_k+h} f(t, y(t))\, dt.$$

In fact, if $y_k$ denotes the approximation to $y(x_k)$ obtained at the $k$th step of the approximation process, then (11–9) yields the formula

$$y_{k+1} = y_k + \int_{x_k}^{x_k+h} f(t, y(t))\, dt \qquad (11\text{–}10)$$

for the $(k + 1)$-st approximation.

**The Euler Method**    Among the many ways to approximate the integral in (11–10), the simplest is to replace the integrand $f(t, y(t))$ by its value $f(x_k, y(x_k))$ at the left-hand endpoint of the interval $[x_k, x_k + h]$, yielding the formula

$$y_{k+1} = y_k + \int_{x_k}^{x_k+h} f(x_k, y(x_k))\, dt$$
$$= y_k + hf(x_k, y(x_k)).$$

To convert this result into a workable formula, we need a replacement for $y(x_k)$. The choice is obvious: we use $y_k$, the approximation to $y(x_k)$ that has already been computed. We then obtain

$$y_{k+1} = y_k + hf(x_k, y_k),$$

which we recognize as the $(k + 1)$st approximation in the Euler method.

**The Modified Euler Method**    Perhaps the most obvious way to attempt to improve the approximation to the integral in (11–10) is to use the (approximate) value of $f(t, y(t))$ at the *midpoint* of the interval of integration rather than at the left-hand endpoint. (This is the so-called *midpoint method* of approximating an integral.) To retain the step size $h$ and the points

$$x_0, \quad x_1 = x_0 + h, \quad x_2 = x_1 + h, \quad \ldots$$

used in our earlier computations, we integrate over (overlapping) double intervals, as shown in Fig. 11–10. We then obtain the approximations

$$y_2 = y_0 + \int_{x_0}^{x_2} f(x_1, y_1)\, dt = y_0 + 2hf(x_1, y_1),$$
$$y_3 = y_1 + \int_{x_1}^{x_3} f(x_2, y_2)\, dt = y_1 + 2hf(x_2, y_2),$$

and in general

$$y_{k+2} = y_k + 2hf(x_{k+1}, y_{k+1}) \qquad (11\text{–}11)$$

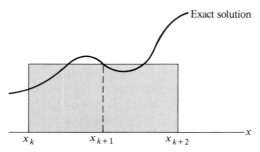

Figure 11-10

for $k = 0, 1, 2, \ldots, n - 2$. This method of approximation is known as the **modified Euler method**. It is an example of a **two-step method**, so called because it uses the approximations $y_k$ and $y_{k+1}$ from the two preceding steps in the computation to obtain $y_{k+2}$.

Note that to compute $y_2$, the first approximation generated by (11-11), we need to know both $y_0$ and $y_1$. The first, of course, causes no trouble; it is the known

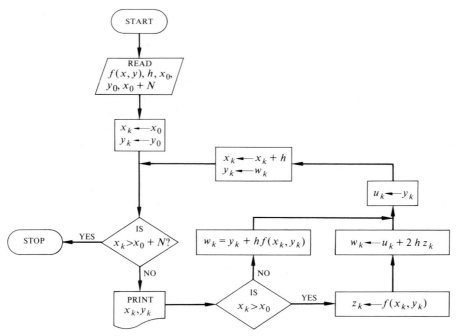

Figure 11-11

initial value of the solution. But how do we find $y_1$? One way is to use the original Euler method and let $y_1 = y_0 + hf(x_0, y_0)$. We then obtain the sequence of approximations

$$y_1 = y_0 + hf(x_0, y_0)$$
$$y_2 = y_0 + 2hf(x_1, y_1)$$
$$y_3 = y_1 + 2hf(x_2, y_2)$$
$$\cdot$$
$$\cdot$$
$$\cdot$$

The flowchart in Fig. 11.11 shows how the modified Euler method might be implemented on a computer.

If round-off error is neglected, the approximations obtained by this method differ from the exact values by less than a constant multiple of $h^2$. Thus, as Table 11–2 shows, when $h$ is small the modified Euler method yields significantly better approximations than the Euler method. For instance, the approximation of $e$ given by the modified Euler method has an error of less than 0.4%. For comparison, the results of both methods are given in the table. You may find it instructive to carry out these computations on a hand calculator.

**Table 11–2**    **Approximations to $y = e^x$, the solution of $y' = y$, $y(0) = 1$**

| $x_k$ | $y_k$ (Euler) | $y_k$ (modified Euler) | $y_k$ (exact) |
|-------|---------------|------------------------|---------------|
| 0.0 | 1.000 | 1.000 | 1.000 |
| 0.1 | 1.100 | 1.100 | 1.105 |
| 0.2 | 1.210 | 1.220 | 1.221 |
| 0.3 | 1.331 | 1.344 | 1.350 |
| 0.4 | 1.464 | 1.489 | 1.492 |
| 0.5 | 1.610 | 1.642 | 1.649 |
| 0.6 | 1.771 | 1.817 | 1.822 |
| 0.7 | 1.948 | 2.005 | 2.014 |
| 0.8 | 2.143 | 2.218 | 2.226 |
| 0.9 | 2.357 | 2.449 | 2.460 |
| 1.0 | 2.593 | 2.708 | 2.718 |

**EXERCISES**

1. Suppose that $f(x, y)$ is continuous in a region $R$ of the $xy$-plane that contains the point $(x_0, y_0)$. Prove that $y = y(x)$ is a solution of

$$y' = f(x, y), \qquad y(x_0) = y_0,$$

if and only if it is a solution of

$$y(x) = y_0 + \int_{x_0}^{x} f(t, y(t)) \, dt.$$

[*Hint*: Use the fundamental theorem of calculus.]

Use the modified Euler method to find a numerical solution of each of the following initial-value problems on the indicated interval. Round all computations to three decimal places.

2. $y' = 2x$, $y(0) = 0$, on $[0, 1]$ with $h = 0.1$

3. $y' = x + y$, $y(0) = 0$, on $[0, 1]$ with $h = 0.1$

4. $y' = \dfrac{x + y}{x}$, $y(1) = 1.5$, on $[1, 2]$ with $h = 0.2$

5. $y' = x^2 + y^2$, $y(0) = 1$, on $[0, 1]$ with $h = 0.2$

6. $y' = (x + y)^2$, $y(0) = -1$, on $[0, 1]$ with $h = 0.2$

## 11–6 NUMERICAL METHODS (III): RUNGE-KUTTA METHODS

In this section we shall discuss two techniques that yield much more accurate numerical solutions of initial-value problems than either the Euler or modified Euler method. The first is based on using the *average* of the values of $f(t, y(t))$ at the left and right endpoints of the interval $[x_k, x_{k+1}]$ to approximate the integrand in

$$y_{k+1} = y_k + \int_{x_k}^{x_{k+1}} f(t, y(t)) \, dt.$$

This average is

$$\tfrac{1}{2}[f(x_k, y(x_k)) + f(x_{k+1}, y(x_{k+1}))],$$

and it yields the approximation

$$y_{k+1} = y_k + \frac{h}{2}[f(x_k, y(x_k)) + f(x_{k+1}, y(x_{k+1}))].$$

(Once again we assume step size $h$ for each interval.)

To convert this expression into a workable formula, we need approximations to $y(x_k)$ and $y(x_{k+1})$. The first is easy; we use $y_k$, the approximation already computed for $y(x_k)$. To obtain the second, we apply the Euler method in the interval $[x_k, x_{k+h}]$ starting with the approximation $y_k$. Denoting the result by $z_{k+1}$ to avoid confusion with $y_{k+1}$, we have the two-step formula

$$y_{k+1} = y_k + \frac{h}{2}[f(x_k, y_k) + f(x_{k+1}, z_{k+1})],$$

$$z_{k+1} = y_k + hf(x_k, y_k).$$

This technique is known as the **two-step Runge-Kutta** or **improved Euler method**. It is an example of a **predictor-corrector** method of approximation, the name being derived from the fact that one method (in this case the Euler method) is used to *predict* the value of the $(k + 1)$st approximation, and the prediction is then *corrected* by a second computation. The corrected approximation $y_{k+1}$ is used in the next stage of the computation.

Table 11-3 shows the results of approximating by this method the solution of

$$y' = y, \quad y(0) = 1,$$

to three decimal places with a step size of 0.1. The predictor used at each step in the computation is given in the last column of the table. The results are now reasonably satisfactory, as a comparison between the approximated and exact values reveals. You are again invited to carry out these computations on a hand calculator.

**Table 11-3**   **Approximations to $y = e^x$, the solution of $y' = y, y(0) = 1$**

| $x_k$ | $y_k$ (Runge-Kutta) | $y(x_k)$ (exact) | $z_{k+1}$ |
|-------|---------------------|------------------|-----------|
| 0.0   | 1.000               | 1.000            | 1.100     |
| 0.1   | 1.105               | 1.105            | 1.216     |
| 0.2   | 1.221               | 1.221            | 1.343     |
| 0.3   | 1.349               | 1.350            | 1.484     |
| 0.4   | 1.491               | 1.492            | 1.640     |
| 0.5   | 1.648               | 1.649            | 1.813     |
| 0.6   | 1.821               | 1.822            | 2.003     |
| 0.7   | 2.012               | 2.014            | 2.213     |
| 0.8   | 2.223               | 2.226            | 2.445     |
| 0.9   | 2.456               | 2.460            | 2.702     |
| 1.0   | 2.714               | 2.718            | ———       |

The clear improvement over the Euler method so easily achieved by the Runge-Kutta method suggests that further modification might achieve even better approximations of the integral in the formula for $y_{k+1}$. The following four-step Runge-Kutta method will do just that, producing results of surprising accuracy.

As usual, we begin with the approximation $y_k$ to $y(x_k)$ obtained in the preceding step of the computation and use it to obtain the approximation

$$m_1 = f(x_k, y_k)$$

to the value of $f(x, y(x_k))$. (The letter $m$ is used here to serve as a reminder that $m_1$ is an approximation to the *slope* of the *solution curve* $y = y(x)$ when $x = x_k$.) We now use the Euler method to estimate the value of $f(x, y(x))$ at the *midpoint*

$x + h/2$ of the interval $[x_k, x_k + h]$, under the assumption that the slope of the solution curve is $m_1$ in that interval. This yields the approximation

$$y_k + \frac{hm_1}{2}$$

to the value of $y = y(x)$ at that point, and hence a second approximation to the value of $f$ in the interval:

$$m_2 = f(x_k + h/2, y_k + hm_1/2).$$

Now we take $m_2$ as a revised estimate of the slope of $y = y(x)$ in the interval and repeat the calculation with $m_2$ in place of $m_1$, obtaining a third approximation:

$$m_3 = f(x + h/2, y_k + hm_2/2).$$

Finally, we use the Euler method to make a fourth estimate of the value of $f$, this time at the right-hand endpoint of the interval, under the assumption that the slope of the solution curve is $m_3$ in the entire interval:

$$m_4 = f(x_k + h, y_k + hm_3).$$

We now have four estimates of the value of $f$ in the interval:

$$m_1 = f(x_k, y_k),$$
$$m_2 = f(x_k + h/2, y_k + hm_1/2),$$
$$m_3 = f(x_k + h/2, y_k + hm_2/2),$$
$$m_4 = f(x_k + h, y_k + hm_3).$$

The approximation $y_{k+1}$ is now obtained by adding to $y_k$ a weighted average of these approximations:

$$y_{k+1} = y_k + \frac{h}{6}(m_1 + 2m_2 + 2m_3 + m_4).*$$

This four-step method is remarkably accurate, as you can see from the following table of approximations to $y = e^x$. The error is now less than a constant multiple of $h^5$. This time the approximations have been computed to five decimal places, and the step size has even been *increased*, to 0.2. The intermediate values of various $m_i$ are shown in the middle column of the table, and the corresponding exact (five place) values of $e^x$ appear in the last column.

---

\* The rationale behind using this particular average to estimate the value of

$$\int_{x_k}^{x_k+h} f(x, y(x)) \, dx$$

is provided by Simpson's rule for approximating definite integrals.

**Table 11-4    Approximations to $y = e^x$, the
solution of $y' = y$, $y(0) = 1$**

| $x_k$ | $y_k$ (four-step Runge-Kutta) | $m_i$ | $y_k$ (exact) |
|-------|-------------------------------|-------|---------------|
| 0.0 | 1.00000 | $m_1 = 1.00000$<br>$m_2 = 1.10000$<br>$m_3 = 1.11000$<br>$m_4 = 1.22200$ | 1.00000 |
| 0.2 | 1.22140 | $m_1 = 1.22140$<br>$m_2 = 1.34354$<br>$m_3 = 1.35575$<br>$m_4 = 1.49255$ | 1.22140 |
| 0.4 | 1.49182 | $m_1 = 1.49182$<br>$m_2 = 1.64100$<br>$m_3 = 1.65592$<br>$m_4 = 1.82300$ | 1.49182 |
| 0.6 | 1.82211 | $m_1 = 1.82211$<br>$m_2 = 2.00432$<br>$m_3 = 2.02254$<br>$m_4 = 2.22662$ | 1.82212 |
| 0.8 | 2.22552 | $m_1 = 2.22552$<br>$m_2 = 2.44807$<br>$m_3 = 2.47033$<br>$m_4 = 2.71859$ | 2.22554 |
| 1.0 | 2.71822 | —————— | 2.71828 |

## EXERCISES

1. Draw a flowchart for the two-step Runge-Kutta method.

2. Use the two-step Runge-Kutta method to obtain a numerical solution for each of the following initial-value problems on the indicated interval. Round all computations to three decimal places.
   a) $y' = 2x$, $y(0) = 0$, on $[0, 1]$ with $h = 0.2$
   b) $y' = x + y$, $y(0) = 0$, on $[0, 1]$ with $h = 0.2$
   c) $y' = (x + y)^2$, $y(0) = -1$, on $[0, 1]$ with $h = 0.2$

3. Use the four-step Runge-Kutta method with a step size of 0.5 to obtain a numerical solution of:
   a) $y' = 2x$, $y(0) = 0$, on $[0, 1]$,
   b) $y' = x + y$, $y(0) = 0$, on $[0, 1]$.

   Round all computations to three decimal places.

* 4. Draw a flowchart for the four-step Runge-Kutta method.

### 11–7 SYSTEMS AND HIGHER-ORDER EQUATIONS

As might be expected, the numerical techniques developed in the preceding sections can be extended to apply to systems of first-order differential equations, and hence by implication to normal equations of order higher than one. For instance, let

$$\mathbf{X}' = F(t, \mathbf{X}), \qquad \mathbf{X}(t_0) = \mathbf{X}_0, \tag{11–12}$$

be the matrix version of the first-order system

$$x'_1 = f_1(t; x_1, \ldots, x_n)$$
$$\vdots$$
$$x'_n = f_n(t; x_1, \ldots, x_n),$$

with initial condition

$$x_1(t_0) = c_1, \quad \ldots, \quad x_n(t_0) = c_n.$$

Then if $x_{i,j}$ denotes the approximation to $x_i(t_j)$, and if

$$\mathbf{X}_j = \begin{bmatrix} x_{1,j} \\ \vdots \\ x_{n,j} \end{bmatrix},$$

a straightforward generalization of the Euler method applied to (11–12) yields the approximation formulas

$$\mathbf{X}_{k+1} = \mathbf{X}_k + h\mathbf{F}(t_k, \mathbf{X}_k), \tag{11–13}$$

where $h$ is the step size between successive points on the $t$-axis at which the approximations are to be computed. Similar generalizations of the other numerical methods introduced in this chapter can also be derived.

The following example illustrates how the Euler method can be used to obtain a numerical solution of a second-order differential equation.

**Example** Use the Euler method with $h = 0.2$ to find a numerical solution of

$$x'' - x = 0, \qquad x(0) = 2, \qquad x'(0) = 0, \tag{11–14}$$

on the interval $[0, 1]$.

*Solution.* We have again chosen a problem whose solution is known in order to be able to compare the approximations with the exact values.

The substitution $z = x'$ converts (11–14) into the system

$$x' = z, \qquad z' = x,$$

with initial condition

$$x(0) = 2, \qquad z(0) = 0.$$

By making the obvious changes of variables this problem can be rewritten in matrix form as

$$\begin{bmatrix} x'_1 \\ x'_2 \end{bmatrix} = \begin{bmatrix} x_2 \\ x_1 \end{bmatrix}, \qquad \begin{bmatrix} x_1(0) \\ x_2(0) \end{bmatrix} = \begin{bmatrix} 0 \\ 2 \end{bmatrix}. \tag{11-15}$$

Then (11–13) becomes

$$\begin{bmatrix} x_{1,k+1} \\ x_{2,k+1} \end{bmatrix} = \begin{bmatrix} x_{1,k} \\ x_{2,k} \end{bmatrix} + 0.2 \begin{bmatrix} x_{2,k} \\ x_{1,k} \end{bmatrix},$$

from which the approximations can be computed with ease. Table 11–5 shows the results to three decimal places and, for comparison, the three-place value of the exact solution

$$x = e^t + e^{-t}.$$

Note that in this table the $x_{2,k}$ are approximations to the solution, while the $x_{1,k}$ are approximations to $x' = e^t - e^{-t}$.

**Table 11–5    Approximations to the solution of**
$x'' + x = 0,\ x(0) = 2,\ x'(0) = 0,$
**or of**

$$\begin{bmatrix} x'_1 \\ x'_2 \end{bmatrix} = \begin{bmatrix} x_2 \\ x_1 \end{bmatrix}, \begin{bmatrix} x_1(0) \\ x_2(0) \end{bmatrix} = \begin{bmatrix} 0 \\ 2 \end{bmatrix}.$$

| $t_k$ | Approximate $\begin{bmatrix} x_{1,k} \\ x_{2,k} \end{bmatrix}$ | Exact $\begin{bmatrix} x_1(t_k) \\ x_2(t_k) \end{bmatrix}$ | Exact $x = e^t + e^{-t}$ |
|---|---|---|---|
| 0.0 | $\begin{bmatrix} 0.000 \\ 2.000 \end{bmatrix}$ | $\begin{bmatrix} 0.000 \\ 2.000 \end{bmatrix}$ | 2.000 |
| 0.2 | $\begin{bmatrix} 0.400 \\ 2.000 \end{bmatrix}$ | $\begin{bmatrix} 0.402 \\ 2.040 \end{bmatrix}$ | 2.040 |
| 0.4 | $\begin{bmatrix} 0.800 \\ 2.080 \end{bmatrix}$ | $\begin{bmatrix} 0.822 \\ 2.162 \end{bmatrix}$ | 2.162 |
| 0.6 | $\begin{bmatrix} 1.216 \\ 2.240 \end{bmatrix}$ | $\begin{bmatrix} 1.273 \\ 2.371 \end{bmatrix}$ | 2.371 |
| 0.8 | $\begin{bmatrix} 1.664 \\ 2.483 \end{bmatrix}$ | $\begin{bmatrix} 1.777 \\ 2.675 \end{bmatrix}$ | 2.675 |
| 1.0 | $\begin{bmatrix} 2.161 \\ 2.816 \end{bmatrix}$ | $\begin{bmatrix} 2.348 \\ 3.086 \end{bmatrix}$ | 3.086 |

## EXERCISES

1.  Use the modified Euler method to obtain a second numerical solution of the initial-value problem

    $$x'' - x = 0, \qquad x(0) = 2, \qquad x'(0) = 0,$$

    on $[0, 1]$. Let $h = 0.2$, and round all computations to three decimal places.

2.  Use the Euler method with $h = 0.2$ to find a numerical solution of

    $$x'' + x = 0, \qquad x(0) = x'(0) = 1,$$

    on $[0, 1]$. Round all computations to three decimal places.

3.  Use the Euler method with $h = 0.2$ to find a numerical solution of

    $$\begin{bmatrix} x' \\ y' \end{bmatrix} = \begin{bmatrix} 2y \\ x - y \end{bmatrix}, \qquad \begin{bmatrix} x(0) \\ y(0) \end{bmatrix} = \begin{bmatrix} 1 \\ 2 \end{bmatrix}$$

    on $[0, 1]$. Round all computations to three decimal places.

4.  Repeat Exercise 3 using the modified Euler method.

# Existence and Uniqueness Theorems        **12**

## 12–1 INTRODUCTION

In the preceding chapter we gave an informal argument based on physical considerations to support the assertion that the initial-value problem

$$y' = f(x, y), \qquad y(x_0) = y_0, \tag{12–1}$$

has at least one solution whenever $f$ is continuous in an open region $R$ of the
$xy$-plane containing the point $(x_0, y_0)$. We also saw, however, that the continuity
of $f$ alone is not enough to ensure that a solution of (12–1) be unique. A stronger
condition, such as the continuity in $R$ of $\partial f/\partial y$, must be imposed to obtain uniqueness. In this chapter we shall investigate the questions of existence and uniqueness
in greater detail and prove several of the existence and uniqueness theorems that
are of basic theoretical importance in the study of differential equations.

We have already observed that $y = y(x)$ is a solution of (12–1) if and only if it
is a solution of the integral equation

$$y(x) = y_0 + \int_{x_0}^{x} f(t, y(t))\, dt \tag{12–2}$$

(Exercise 1, Section 11–5). We will find it easier to work with (12–2) than with
(12–1), because the existence problem for (12–2) admits a particularly suggestive
formulation. If $z = z(x)$ is a function with the property that $f(x, z(x))$ is continuous
in an interval $I$ containing the point $x_0$, then the expression

$$y_0 + \int_{x_0}^{x} f(t, z(t))\, dt \tag{12–3}$$

defines a continuously differentiable function $Z = Z(x)$ in $I$, with $Z(x_0) = y_0$.
From this point of view (12–3) can be interpreted as the definition of a mapping or
operator $F$ which sends the function $z$ to the function $Z$ according to the formula

$$F: z \rightarrow y_0 + \int_{x_0}^{x} f(t, z(t))\, dt = Z(t).$$

In particular, if $y$ is a solution of (12–2), then

$$F(y) = y,$$

so that $y$ is a **fixed point** of $F$ in the sense that $F$ maps $y$ to itself. This fact suggests that we investigate fixed points for arbitrary mappings before considering the existence and uniqueness problem for (12–2). As so often happens in mathematics, this general problem is actually easier to handle than the specific problem from which it arose. In fact, once we have established a modest number of results about fixed points of mappings, the existence and uniqueness theorems we propose to prove will follow with relative ease.

### EXERCISES

1. Let $g$ be a continuous function with domain $[0, 1]$ and range a subset of $[0, 1]$. Use the graphs of $g$ and the function $y = x$ to give an intuitive explanation of why $g$ must have at least one fixed point.

2. Apply the intermediate-value theorem to the function

$$G(x) = g(x) - x,$$

where $g$ is as in Exercise 1, to deduce that there exists a point in $[0, 1]$ at which $G$ has the value 0, thereby proving that $g$ has at least one fixed point in $[0, 1]$.

3. a) Let $x_0$ be a fixed point for $F$, and let

$$F^{[2]}(x) = F(F(x)).$$

   Prove that $x_0$ is also a fixed point for $F^{[2]}$.
   b) Generalize the result in (a) to $F^{[m]}$ for any integer $m > 2$.

4. Discuss the existence and uniqueness of fixed points for the constant-coefficient operator

$$D^n + a_{n-1}D^{n-1} + \cdots + a_1 D + a_0$$

   given that

   a) $n = 1$,            b) $n = 2$,            c) $n$ is arbitrary.

5. Let $T: \mathscr{R}^2 \rightarrow \mathscr{R}^2$ be defined by

$$T\begin{bmatrix} x \\ y \end{bmatrix} = \begin{bmatrix} a & b \\ c & d \end{bmatrix}\begin{bmatrix} x \\ y \end{bmatrix}.$$

   Under what conditions does $T$ have a fixed point other than $\begin{bmatrix} 0 \\ 0 \end{bmatrix}$?

6. The methods of this section show that the initial-value problem

$$y' = x - y, \qquad y(0) = 1,$$

   leads to the operator

$$F: z(x) \rightarrow 1 + \int_0^x [t - z(t)]\, dt.$$

   a) Let $z_0(x) = 1$, and define

$$z_n(x) = F[z_{n-1}(x)], \qquad n = 1, 2, \ldots.$$

   Compute $z_1, \ldots, z_5$.

b) Repeat (a) with $z_0(x) = 2 - x$.

c) Compare the results in (a) and (b) with the exact solution of the differential equation.

7. Let $f$ be a continuous function with domain $[0, 1]$ and range a subset of $[0, 1]$, and suppose that $f[f(x)] = f(x)$.

a) Illustrate the meaning of the statement $f[f(x)] = f(x)$ graphically.

b) Show that if in addition $f$ is assumed to be differentiable, then either $f$ is a constant or $f(x) = x$.

## 12-2 METRIC SPACES, CONVERGENCE, AND CONTINUITY

In order to make the discussion of the preceding section precise, we must begin by determining a domain for the operator

$$F: z \rightarrow y_0 + \int_{x_0}^{x} f(t, z(t))\, dt.$$

Our first observation in this regard is the negative one that a vector space *per se* will not do, because $F$ is not necessarily linear. Rather, the appropriate setting is a space in which it is possible to measure distance.

> **Definition 12–1**    A **metric space** is a (*nonempty*) set M together with a **distance function** or **metric** which assigns a real number d(x, y) to each pair of points x and y in M in such a way that
>
> i) $d(x, y) \geq 0$,
> ii) $d(x, y) = 0$ if and only if $x = y$,
> iii) $d(x, y) = d(y, x)$,
> iv) $d(x, z) \leq d(x, y) + d(y, z)$.

For each pair $(x, y)$ the number $d(x, y)$ is called the **distance** between $x$ and $y$. Condition (iv) is known as the **triangle inequality** for the metric $d$.

   **Example 1**    The most familiar example of a metric space is $\mathscr{R}^n$ with its *standard* or *Euclidean distance*

$$d(\mathbf{x}, \mathbf{y}) = \left( \sum_{i=1}^{n} (x_i - y_i)^2 \right)^{1/2}, \tag{12–4}$$

where $\mathbf{x} = (x_1, \ldots, x_n)$ and $\mathbf{y} = (y_1, \ldots, y_n)$. See Exercise 7 at the end of this section.

   When $n = 2$ (12–4) reduces to the familiar distance formula of analytic geometry:

$$d(\mathbf{x}, \mathbf{y}) = \sqrt{(x_1 - y_1)^2 + (x_2 - y_2)^2},$$

and when $n = 1$ the standard distance is given by the usual absolute value formula

$$d(x, y) = |x - y|.$$

   **Example 2**    $\mathscr{R}^n$ also becomes a metric space when distance is defined by the formula

$$d(\mathbf{x}, \mathbf{y}) = \sum_{i=1}^{n} |x_i - y_i|, \tag{12–5}$$

where **x** and **y** are as in Example 1. (See Exercise 2.) This distance is sometimes called the "city-block" metric, because in $\mathscr{R}^2$ it measures the distance between points as though it were computed by traveling along a rectangular grid of streets (Fig. 12–1).

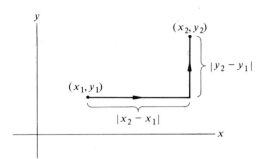

**Figure 12–1**

**Example 3** Show that $\mathscr{C}[a, b]$ becomes a metric space when distance is defined by

$$d(f, g) = \max_{x \in [a,b]} |f(x) = g(x)|, \tag{12–6}$$

where $\max_{x \in [a,b]} |f(x) - g(x)|$ denotes the maximum of the quantity $|f(x) - g(x)|$ for $x$ in $[a, b]$. (See Fig. 12–2.)

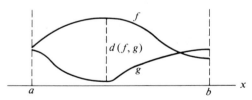

**Figure 12–2**

*Solution.* We first observe that since $f$ and $g$ are continuous on $[a, b]$, so is the absolute value of their difference. Hence by the extreme-value theorem, $|f(x) - g(x)|$ does assume a maximum in $[a, b]$, and (12–6) is a meaningful definition. This maximum is clearly nonnegative and is 0 if and only if $f(x) = g(x)$ for all $x$ in $[a, b]$. Thus the first two requirements of Definition 12–1 are satisfied. The third condition obviously holds, and the fourth can be established as follows. If $f$, $g$, and $h$ belong to $\mathscr{C}[a, b]$, then

$$\begin{aligned}
|f(x) - h(x)| &= |f(x) - g(x) + g(x) - h(x)| \\
&\leqslant |f(x) - g(x)| + |g(x) - h(x)|.
\end{aligned}$$

Hence

$$\max_{x\in[a,b]} |f(x) - h(x)| \leq \max_{x\in[a,b]} (|f(x) - g(x)| + |g(x) - h(x)|)$$
$$\leq \max_{x\in[a,b]} |f(x) - g(x)| + \max_{x\in[a,b]} |g(x) - h(x)|,$$

so that

$$d(f, h) \leq d(f, g) + d(g, h)$$

as required.

The space $\mathscr{C}[a, b]$, with this particular metric, will play an important role throughout this chapter.

Metric spaces are significant in mathematics because they provide a setting in which convergence and continuity can be defined. The definition of convergence is as follows.

**Definition 12–2**   *A sequence $\{x_n\}$ of points in a metric space $M$ is said to* ***converge*** *to a point $x$ in $M$ if for each $\varepsilon > 0$ there exists an integer $N$ such that $d(x_n, x) < \varepsilon$ whenever $n > N$. As usual, we write*

$$\lim_{n\to\infty} \{x_n\} = x \qquad \text{or} \qquad \{x_n\} \to x,$$

*when $\{x_n\}$ converges to $x$. A sequence that does not converge is said to* ***diverge***.

Before this definition can be accepted as a reasonable description of convergence, we must show that the limit of a convergent sequence is unique. This is the content of the following lemma

**Lemma 12–1**   *If $\lim_{n\to\infty} x_n = x$ and $\lim_{n\to\infty} x_n = y$, then $x = y$.*

*Proof.*   Let $\varepsilon > 0$ be given. Then there exists an integer $N$ such that

$$d(x_n, x) < \varepsilon/2 \qquad \text{and} \qquad d(x_n, y) < \varepsilon/2$$

whenever $n > N$. The triangle inequality implies that

$$d(x, y) \leq d(x, x_n) + d(x_n, y) < \frac{\varepsilon}{2} + \frac{\varepsilon}{2} = \varepsilon.$$

In other words, $d(x, y)$ is a nonnegative real number that is less than every positive $\varepsilon$. The only number with this property, of course, is 0. Hence $d(x, y) = 0$, and $x = y$ as asserted.   ■

**Example 4**   Let $\{x_n\}$ be a sequence of real numbers and let $d(x, y)$ be the standard distance $|x - y|$ in $\mathscr{R}^1$. Then the assertion that $\{x_n\}$ converges to $x$ with respect to the distance $d$ reduces to the familiar definition of sequential convergence given in calculus.

**Example 5**   Let $\{f_n\}$ be a sequence of functions in $\mathscr{C}[a, b]$ and suppose that $\{f_n\}$ converges to $f$ with respect to the distance defined in Example 3. Then for each $\varepsilon > 0$ there exists an integer $N$ such that whenever $n > N$,

$$|f_n(x) - f(x)| < \varepsilon$$

simultaneously for *all* x in [a, b]. This type of convergence plays a special role in the theory of continuous functions, as we shall see in the proof of Theorem 12–1, and is known as **uniform convergence**.

As a working tool, Definition 12–1 has a serious defect: it cannot be used to show that a sequence converges unless you know in advance what its limit is. Thus it is natural to seek a convergence criterion that can be applied directly to the terms of the sequence themselves, a criterion that does not require any fore-knowledge of whether the limit exists, or of what the limit might be if it does. This is the motive for introducing the following definition.

**Definition 12–3**   *A sequence* $\{x_n\}$ *in a metric space is said to be a **Cauchy** sequence if for each* $\varepsilon > 0$ *there exists an integer N such that* $d(x_m, x_n) < \varepsilon$ *whenever m and n are greater than N.*

Thus a Cauchy sequence is one in which the distance $d(x_m, x_n)$ between terms becomes *and remains* arbitrarily small for large m and n. The next lemma shows that this always happens in a convergent sequence.

**Lemma 12–2**   *Every convergent sequence in a metric space is a Cauchy sequence.*

*Proof.*   Let $\varepsilon > 0$ be given, and suppose that $\{x_n\}$ converges to x. Then there exists an integer N such that $d(x_n, x) < \varepsilon/2$ for all $n > N$. Thus when m and n are both greater than N, the triangle inequality implies that

$$d(x_m, x_n) \leqslant d(x_m, x) + d(x, x_n) < \frac{\varepsilon}{2} + \frac{\varepsilon}{2} = \varepsilon. \quad \blacksquare$$

This result implies that a sequence which fails to be a Cauchy sequence will not converge. But what of the converse? Is every Cauchy sequence necessarily convergent? Unfortunately the answer is no, and several examples illustrating this fact are given in the exercises at the end of the section. Thus it seems as though our search for a convergence test that can be applied to the terms of a sequence has failed. But we need not give up in despair, for there are two ways of circum-venting this unhappy turn of events. One consists of enlarging the space in question by adjoining new points to it until all Cauchy sequences in the enlarged space converge. This process is known as *completing* the space and is studied in advanced courses in analysis. The other way consists of adopting the ostrich-like behavior of ignoring those spaces that give trouble and considering only those that do not. Such spaces are said to be complete, a term which is important enough to merit a formal definition.

**Definition 12–4**   *A metric space M is said to be **complete** if every Cauchy sequence in M converges to a point in M.*

As we said, there exist metric spaces which are not complete. But all the spaces we shall encounter, in particular all those introduced earlier in this section, *are* complete. The completeness of $\mathcal{R}^n$ is a consequence of the completeness of the real number system $\mathcal{R}^1$ with respect to the usual distance. The fact that $\mathcal{R}^1$ is complete is usually proved in advanced calculus, and we shall assume it here. Actually, for our purposes, the completeness of $\mathcal{C}[a, b]$ with respect to the distance introduced in Example 3 is more important. It, too, is a consequence of the completeness of $\mathcal{R}^1$, as we shall now verify.

**Theorem 12–1**    *The metric space $\mathcal{C}[a, b]$ with distance defined by*

$$d(f, g) = \max_{x \in [a,b]} |f_m(x) - f_n(x)|$$

*is complete.*

*Proof.*    Let $\{f_n\}$ be a Cauchy sequence in $\mathcal{C}[a, b]$. Then $d(f_m, f_n) \to 0$ with increasing $m$ and $n$, and it follows that

$$\max_{x \in [a,b]} |f_m(x) - f_n(x)| \to 0$$

as $m, n \to \infty$. In particular, if $x_0$ is a fixed point in $[a, b]$, then $|f_m(x_0) - f_n(x_0)| \to 0$ as $m, n \to \infty$. Thus $\{f_n(x_0)\}$ is a Cauchy sequence in $\mathcal{R}^1$ and, by the completeness of $\mathcal{R}^1$, has a limit. Since $x_0$ is arbitrary in $[a, b]$, we can define a real-valued function $f$ on $[a, b]$ by the rule:

$$\text{for each } x \in [a, b], \quad f(x) = \lim_{n \to \infty} \{f_n(x)\}.$$

We now show that $f$ is the desired limit of the sequence $\{f_n\}$ in $\mathcal{C}[a, b]$.

The first step toward establishing this fact consists in showing that $\{f_n\}$ converges *uniformly* to $f$, by which we mean that given any $\varepsilon > 0$ there exists an integer $N$ such that $|f_n(x) - f(x)| < \varepsilon$ simultaneously for all $n > N$ and all $x$ in $[a, b]$. (The point here is that a single $N$ can be chosen which works simultaneously for the whole interval $[a, b]$.) To do so, we use the assumption that $\{f_n\}$ is a Cauchy sequence to find an integer $N$ such that $|f_m(x) - f_n(x)| < \varepsilon/2$ for all $m$, $n > N$ and all $x$ in $[a, b]$. We then rewrite this inequality as

$$f_n(x) - \frac{\varepsilon}{2} < f_m(x) < f_n(x) + \frac{\varepsilon}{2},$$

and let $m \to \infty$ to obtain

$$f_n(x) - \frac{\varepsilon}{2} \leqslant f(x) \leqslant f_n(x) + \frac{\varepsilon}{2}.$$

Thus

$$|f_n(x) - f(x)| \leqslant \frac{\varepsilon}{2} < \varepsilon$$

for all $n > N$ and all $x$ in $[a, b]$, which is precisely what we wished to show.

This proves that $\{f_n\}$ converges to $f$ in the metric of $\mathcal{C}[a, b]$, but we have yet to prove that $f$ actually belongs to $\mathcal{C}[a, b]$. That is, we have yet to prove that

$f$ is continuous. To do so, let $\varepsilon > 0$ be given, and let $x_0$ be any point in $[a, b]$. We will show that there exists a $\delta > 0$ such that $|f(x) - f(x_0)| < \varepsilon$ whenever $a \leqslant x \leqslant b$ and $|x - x_0| < \delta$. Now,

$$|f(x) - f(x_0)| = |f(x) - f_n(x) + f_n(x) - f_n(x_0) + f_n(x_0) - f(x_0)|$$
$$\leqslant |f(x) - f_n(x)| + |f_n(x) - f_n(x_0)| + |f_n(x_0) - f(x_0)|.$$

Hence by the uniform convergence argument given a moment ago, we can find an integer $n$ such that the first and last terms on the right of this inequality are each less than $\varepsilon/3$ for *every* $x$ in $[a, b]$. Moreover, since the function $f_n$ thus chosen is continuous, we can also find a $\delta > 0$ such that $|f_n(x) - f_n(x_0)| < \varepsilon/3$ whenever $a \leqslant x \leqslant b$ and $|x - x_0| < \delta$. With these choices made we have $|f(x) - f(x_0)| < \varepsilon$, as required. ∎

Finally, we introduce the notion of continuity for functions from one metric space to another. The definition is the same as the one given in calculus for real-valued functions, except that it is now phrased in terms of distance rather than absolute values.

**Definition 12–5**  *Let $M_1$ and $M_2$ be metric spaces with distance functions $d_1$ and $d_2$, respectively, and let $F: M_1 \to M_2$. Then $F$ is **continuous at a point** $x_1$ in $M_1$ if, given any $\varepsilon > 0$ there exists a $\delta > 0$ such that*

$$d_2[F(x), F(x_1)] < \varepsilon$$

*whenever $d_1(x, x_1) < \delta$.*
       *A function that is continuous at every point in $M_1$ is said to be **continuous on** $M_1$, or just **continuous** when $M_1$ is understood.*

Several examples of continuous functions are given in Exercises 16 and 17, and still others will be encountered in the sections that follow. We therefore omit examples here, and conclude with a lemma that will be needed soon.

**Lemma 12–3**  *Let $F: M_1 \to M_2$ be a continuous mapping between metric spaces, and suppose that $\{x_n\} \to x$ in $M_1$. Then $\{F(x_n)\} \to F(x)$ in $M_2$.*

In short, the image of a convergent sequence under a continuous mapping is convergent, or even more briefly, continuity preserves convergence.

*Proof.*  Let $\varepsilon > 0$ be given. Then since $F$ is continuous at $x$ there exists a $\delta > 0$ such that

$$d_2[F(x_n), F(x)] < \varepsilon$$

whenever $d_1(x_n, x) < \delta$. But $\{x_n\} \to x$. Hence there exists an integer $N$ such that $d_1(x_n, x) < \delta$ for all $n > N$. Together, these two facts imply that $\{F(x_n)\} \to F(x)$, as asserted. ∎

## EXERCISES

1. a) Prove that for all real numbers $a$ and $b$

$$|a + b| \leqslant |a| + |b|.$$

   [*Hint*: First deduce that $-(|a| + |b|) \leqslant a + b \leqslant |a| + |b|$.]
   b) Establish the triangle inequality in $\mathscr{R}^1$ by making the substitutions $x - y = a$ and $y - z = b$ in the inequality in (a).

2. a) Prove that $\mathscr{R}^n$ is a metric space under the definition of Example 2:

$$d(\mathbf{x}, \mathbf{y}) = \sum_{i=1}^{n} |x_i - y_i|.$$

   [*Hint*: Use the triangle inequality in $\mathscr{R}^1$ (Exercise 1) to prove the triangle inequality in $\mathscr{R}^n$ for $n > 1$.]
   b) Draw a graph of the set of all points in $\mathscr{R}^2$ that are one unit from the origin when distance is measured by the metric in (a).

3. a) Prove that

$$d(\mathbf{x}, \mathbf{y}) = \max_{1 \leqslant i \leqslant n} |x_i - y_i|$$

   where $\mathbf{x} = (x_1, \ldots, x_n)$ and $\mathbf{y} = (y_1, \ldots, y_n)$, is a metric for $\mathscr{R}^n$.
   b) Draw a graph of the set of all points in $\mathscr{R}^2$ that are one unit from the origin when distance is measured as in (a).

4. Show that the distance functions on $\mathscr{R}^n$ and $\mathscr{C}[a, b]$ defined in Examples 1, 2, and 3 all satisfy the equalities
   a) $d(x + z, y + z) = d(x, y)$,
   b) $d(\alpha x, \alpha y) = |\alpha| d(x, y)$, where $\alpha$ is a real number.
   These conditions are known as *translation invariance* and *homogeneity*, respectively.

5. Prove that in any metric space

$$|d(x, y) - d(x, z)| \leqslant d(y, z).$$

6. Use the discriminant of the quadratic function

$$f(t) = (|x_1|t - |y_1|)^2 + \cdots + (|x_n|t - |y_n|)^2$$

   (i.e., its discriminant as a quadratic in $t$) to deduce that for any real numbers $x_1, \ldots, x_n$, $y_1, \ldots, y_n$

$$\sum_{i=1}^{n} |x_i y_i| \leqslant \left( \sum_{i=1}^{n} x_i^2 \right)^{1/2} \left( \sum_{i=1}^{n} y_i^2 \right)^{1/2}.$$

   This result is known as **Cauchy's inequality**.

7. a) Expand the expression $(a_1 + b_1)^2 + (a_2 + b_2)^2$, and use Cauchy's inequality for $n = 2$ (Exercise 6), to deduce that

$$\sqrt{(a_1 + b_1)^2 + (a_2 + b_2)^2} \leqslant \sqrt{a_1^2 + a_2^2} \cdot \sqrt{b_1^2 + b_2^2}.$$

b) Set $x_1 - y_1 = a_1$, $x_2 - y_2 = a_2$, $y_1 - z_1 = b_1$, $y_2 - z_2 = b_2$ in the preceding inequality to establish the triangle inequality in $\mathscr{R}^2$.

c) Generalize the results of (a) and (b) to $\mathscr{R}^n$, $n > 2$.

8. a) Show that $\{(x_n, y_n)\} \to (x_0, y_0)$ with respect to the standard distance in $\mathscr{R}^2$ if and only if

$$\{x_n\} \to x_0 \quad \text{and} \quad \{y_n\} \to y_0$$

in the usual sense of convergence in $\mathscr{R}^1$.

b) Repeat (a) with

$$d(\mathbf{x}, \mathbf{y}) = |x_1 - y_1| + |x_2 - y_2|.$$

c) Repeat (a) with

$$d(\mathbf{x}, \mathbf{y}) = \max(|x_1 - y_1|, |x_2 - y_2|).$$

9. Use the fact that $\mathscr{R}^1$ is complete with respect to the standard distance to prove that $\mathscr{R}^n$ is also complete with respect to that distance when $n > 1$. [*Hint*: $\{\mathbf{x}_k\}$ is a Cauchy sequence in $\mathscr{R}^n$ if and only if the sequence formed from the components of the $\mathbf{x}_k = (x_{k1}, \ldots, x_{kn})$ are all Cauchy sequences.]

10. The open unit interval $(0, 1)$ with the standard distance is a metric space. Prove that the sequence $\{1/n\}$, $n = 2, 3, \ldots$, is a Cauchy sequence, and hence that a Cauchy sequence in a metric space need not converge.

11. Show that the set of rational numbers with the standard metric is not complete.

12. It is known that if $f$ is continuous on $[a, b]$, then there exists a sequence $\{p_n\}$ of polynomials that converges uniformly to $f$ on $[a, b]$. (This is one version of the *Weierstrass approximation theorem*.) Use this fact to deduce that the metric space $\mathscr{P}[a, b]$ with

$$d(p, q) = \max_{x \in [a,b]} |p(x) - q(x)|$$

is not complete.

13. a) Show that $\mathscr{R}^2$ becomes a metric space when

$$d(\mathbf{x}, \mathbf{y}) = |x_1 - y_1| + g(x_2, y_2),$$

where $\mathbf{x} = (x_1, x_2)$, $\mathbf{y} = (y_1, y_2)$, and

$$g(x_2, y_2) = \begin{cases} 1 & \text{if } x_2 \neq y_2, \\ 0 & \text{if } x_2 = y_2. \end{cases}$$

That is, show that $d$ is a metric for $\mathscr{R}^2$.

b) Draw a graph of the set of all points in $\mathscr{R}^2$ a unit distance from the origin when distance is defined as in (a).

c) Give a graphical description of convergence in $\mathscr{R}^2$ with respect to the distance in (a).

14. Prove that the formula

$$d_1(f, g) = \int_a^b |f(x) - g(x)| \, dx$$

defines a metric on $\mathscr{C}[a, b]$. [*Hint*: To prove that $d_1(f, g) = 0$ implies that $f = g$, prove that if $h$ is in $\mathscr{C}[a, b]$ and $h(x) \geqslant 0$ everywhere in $[a, b]$, then $\int_a^b h(x) \, dx = 0$ implies that $h(x) = 0$ for all $x$ in $[a, b]$.]

15. In this exercise we compare the two distance functions

$$d(f, g) = \max_{x \in [a,b]} |f(x) - g(x)|$$

and

$$d_1(f, g) = \int_a^b |f(x) - g(x)| \, dx$$

on $\mathscr{C}[a, b]$. (See Exercise 14.)

a) Use the inequality

$$\int_a^b |f(x) - g(x)| \, dx \leqslant (b - a) \max |f(x) - g(x)|$$

to prove that

$$d(f_n, f) \to 0 \qquad \text{implies} \qquad d_1(f_n, f) \to 0,$$

and hence that convergence with respect to $d$ implies convergence with respect to $d_1$.

b) Show that the converse of the statement in (a) is not true by considering the sequence $\{f_n\}$, where $f_n$ is the "broken-line" function whose graph is shown in Fig. 12–3.

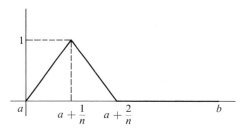

**Figure 12–3**

16. Show that each of the following functions $F: \mathscr{C}[a, b] \to \mathscr{R}^1$ is continuous with respect to the metrices

$$d(f, g) = \max_{x \in [a,b]} |f(x) - g(x)|$$

and

$$d(x, y) = |x - y|.$$

a) $F: f \to f(x_0)$, where $x_0$ is a fixed point in $[a, b]$

b) $F: f \to \int_a^b f(x) \, dx$

c) $F: f \to \left| \int_a^b f(x) \, dx \right|$

d) $F: f \to \int_a^b |f(x)| \, dx$

17. Show that each of the following functions $F: \mathscr{C}[a, b] \to \mathscr{C}[a, b]$ is continuous with respect to the metric

$$d(x, y) = \max_{x \in [a,b]} |f(x) - g(x)|,$$

given that $f_0$ is a fixed function in $\mathscr{C}[a, b]$.

a) $F: f \to f + f_0$                     b) $F: f \to f_0 \cdot f$

c) $F: f \to f^2$                            d) $F: f \to \int_a^x f(t) \, dt$

## 12–3  CONTRACTION MAPPINGS AND FIXED POINTS

Our reason for introducing metric spaces in the preceding section was to provide a setting in which to study fixed points of functions. We begin with a definition.

**Definition 12–6**  *A function F that maps a metric space M into itself is called a **contraction mapping** on M if there exists a real number $\alpha$ with $0 \leqslant \alpha < 1$ such that*

$$d[F(x_1), F(x_2)] \leqslant \alpha\, d(x_1, x_2) \tag{12–7}$$

*for all $x_1$ and $x_2$ in M.*

In other words, $F: M \to M$ is a contraction mapping if it decreases distances between points by at least the factor $\alpha$.

**Example 1**  Let $F: \mathscr{R}^1 \to \mathscr{R}^1$ be the linear function defined by

$$F(x) = ax + b.$$

(In this context a "linear" function is one whose graph is a line.) Then since

$$|F(x_2) - F(x_1)| = |a|\,|x_2 - x_1|,$$

$F$ is a contraction mapping if and only if $|a| < 1$. Thus the only linear functions on $\mathscr{R}^1$ that are contraction mappings are those whose graphs have slopes strictly between $-1$ and $1$.

**Example 2**  An arbitrary function $F: \mathscr{R}^1 \to \mathscr{R}^1$ is a contraction mapping if for all $x_1$ and $x_2$

$$|F(x_2) - F(x_1)| \leqslant \alpha|x_2 - x_1|, \tag{12–8}$$

where $0 \leqslant \alpha < 1$. This inequality is called a **Lipschitz condition** with **Lipschitz constant** $\alpha$ for the function $F$, and $F$ is said to be **Lipschitzian** on $\mathscr{R}^1$. Geometrically, (12–8) states that the slope

$$\frac{F(x_2) - F(x_1)}{x_2 - x_1}$$

of any chord drawn on the graph of $F$ is bounded in absolute value by $\alpha < 1$. In particular, if $F$ is differentiable with $|F'(x)| \leqslant \alpha$ for all $x$, then the mean-value theorem implies that $F$ satisfies (12–8). Hence *a differentiable function F on $\mathscr{R}^1$ is a contraction mapping whenever $|F'(x)| \leqslant \alpha < 1$ for all x.*

Our first result concerning contraction mappings is hardly more than a remark, but since it will be used to prove one of our basic fixed-point theorems, we state it formally.

**Lemma 12–4**  *If F is a contraction mapping on a metric space M, then F is continuous on M.*

*Proof.*  Let $\varepsilon > 0$ be given, and let $x_1$ be an arbitrary point in $M$. Then if $\alpha = 0$ in (12–7),

$$d[F(x_1), F(x_2)] = 0 < \varepsilon$$

for all $x_2$ in $M$, and $F$ is continuous at $x_1$. Otherwise, let $\delta = \varepsilon/\alpha$, and let $x_2$ be any point in $M$ within distance $\delta$ of $x$. Then

$$d[F(x_1), F(x_2)] \leqslant \alpha\, d(x_1, x_2) < \varepsilon,$$

and $F$ is again continuous at $x_1$.   ∎

With this result in hand we are in a position to prove the following fundamental theorem.

**Theorem 12–2**   *Every contraction mapping on a complete metric space has a unique fixed point.*

*Proof.*   We must show that if $F$ is a contraction mapping on a complete metric space $M$, then there exists *precisely one* point $\bar{x}$ in $M$ such that $F(\bar{x}) = \bar{x}$. To do so, let $x_0$ be an arbitrary point in $M$ and let $\{x_n\}$ be the sequence defined by

$$x_1 = F(x_0), \qquad x_2 = F(x_1), \quad \ldots.$$

Then for all $m > 0$,

$$d(x_{m+1}, x_m) = d[F(x_m), F(x_{m-1})] \leqslant \alpha\, d(x_m, x_{m-1}),$$

and it follows that

$$\begin{aligned}
d(x_{m+1}, x_m) &\leqslant \alpha\, d(x_m, x_{m-1}) \\
&\leqslant \alpha^2\, d(x_{m-1}, x_{m-2}) \\
&\phantom{\leqslant}\ \cdot \\
&\phantom{\leqslant}\ \cdot \\
&\phantom{\leqslant}\ \cdot \\
&\leqslant \alpha^{m-1}\, d[F(x_1), x_1].
\end{aligned}$$

But by the triangle inequality,

$$d(x_{m+p}, x_m) \leqslant d(x_{m+p}, x_{m+p-1}) + d(x_{m+p-1}, x_{m+p-2}) + \cdots + d(x_{m+1}, x_m).$$

Hence

$$\begin{aligned}
d(x_{m+p}, x_m) &\leqslant (\alpha^{m+p-2} + \alpha^{m+p-3} + \cdots + \alpha^{m-1})\, d[F(x_1), x_1] \\
&\leqslant \left( \sum_{k=m-1}^{\infty} \alpha^k \right) d[F(x_1), x_1],
\end{aligned}$$

and since $\sum_{k=1}^{\infty} \alpha^k$ converges when $0 \leqslant \alpha < 1$, we can make $d(x_{m+p}, x_m)$ arbitrarily small for all $p \geqslant 0$ by taking $m$ sufficiently large. This implies that $\{x_n\}$ is a Cauchy sequence and hence $\{x_n\}$ converges to some $\bar{x}$ in $M$, because $M$ is complete.

Now that we know that $\{x_n\}$ converges to $\bar{x}$, we also know that $\{F(x_n)\}$ converges to $F(\bar{x})$ because $F$, being a contraction mapping, is continuous. On the other hand, $F(x_n) = x_{n+1}$ for every $n$, so that $\{x_n\}$ converges to $F(\bar{x})$. Since a convergent sequence can have only one limit, $F(\bar{x}) = \bar{x}$ and $\bar{x}$ is a fixed point for $F$.

Finally, to show that $\bar{x}$ is the only fixed point for $F$, we suppose that $F(\bar{y}) = \bar{y}$ for some $\bar{y}$ in $M$. Then

$$d(\bar{x}, \bar{y}) = d[F(\bar{x}), F(\bar{y})] \leqslant \alpha\, d(\bar{x}, \bar{y}),$$

and since $\alpha < 1$, we have $d(\bar{x}, \bar{y}) = 0$ and $\bar{x} = \bar{y}$. ∎

**Example 3**     Let $F: \mathscr{R}^1 \to \mathscr{R}^1$ be defined by

$$F(x) = \frac{x}{2} + 1.$$

Then, as in Example 1, $F$ is a contraction mapping and therefore has a unique fixed point $\bar{x}$. Geometrically, the number $\bar{x}$ is both the $x$- and the $y$-coordinate of the point where the graph of $F$ intersects the line $y = x$, as shown in Fig. 12–4.

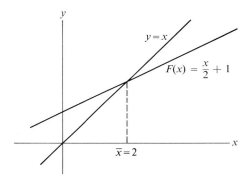

**Figure 12–4**

If we construct the sequence $\{x_n\}$ with $x_0 = 0$ and $x_n = F(x_{n-1})$. we obtain

$$x_0 = 0$$
$$x_1 = F(0) = 1$$
$$x_2 = F(1) = \tfrac{3}{2}$$
$$x_3 = F(\tfrac{3}{2}) = \tfrac{7}{4}$$
$$\vdots$$
$$x_n = \frac{2^n - 1}{2^{n-1}}.$$

This sequence converges to 2, the fixed point of $F$.

Theorem 12–2 and the argument used to prove it yield the following result, which is also of considerable importance.

***Theorem 12–3***    *Let $F$ be a contraction mapping on a complete metric space $M$, let $\bar{x}$ be the unique fixed point for $F$, and let $x_0$ be an arbitrary point in $M$.*

*Then the sequence* $\{x_n\}$ *defined by*

$$x_1 = F(x_0), \qquad x_2 = F(x_1), \quad \ldots$$

*converges to* $\bar{x}$.

Although Theorem 12–2 is sufficient to yield a number of existence and uniqueness theorems for differential equations, other theorems require a somewhat stronger result involving the "iterates" of a function mapping a metric space into itself. Specifically, if $F$ maps $M$ into itself, then so do the composites

$$F[F(x)], \quad F\{F[F(x)]\}, \quad \ldots.$$

We shall denote these composites by $F^{[2]}, F^{[3]}, \ldots$, calling $F^{[m]}$ the **mth iterate** of $F$. Then we have

**Theorem 12–4**   *Let $F$ map a complete metric space into itself and suppose that $F^{[m]}$ has a unique fixed point $\bar{x}$ for some $m > 0$. Then $\bar{x}$ is also the unique fixed point for $F$.*

*Proof.*   If $m = 1$, there is nothing to prove. Otherwise, since $F^{[m]}(\bar{x}) = \bar{x}$, we have

$$F(\bar{x}) = F[F^{[m]}(\bar{x})] = F^{[m]}[F(\bar{x})].$$

Hence $F(\bar{x})$ is a fixed point for $F^{[m]}$, and the uniqueness of $\bar{x}$ implies that $F(\bar{x}) = \bar{x}$. Finally, since every fixed point for $F$ is also a fixed point for $F^{[m]}$, $\bar{x}$ is the only fixed point for $F$.   ∎

When Theorems 12–2 and 12–4 are combined, they yield

**Corollary 12–1**   *Let $F$ map a complete metric space into itself and suppose that $F^{[m]}$ is a contraction mapping for some $m > 0$. Then $F$ has a unique fixed point.*

It is interesting to note that while $F^{[m]}$ is continuous because it is a contraction mapping, the function $F$ itself need not be continuous for the conclusion of Corollary 12–1 to hold. An example is given in Exercise 4.

## EXERCISES

1. Which of the following functions are contraction mappings on $\mathcal{R}^1$? Give reasons for your answers.

   a) $F(x) = \sin x$                     b) $F(x) = \dfrac{1}{1 + x^2}$

   c) $F(x) = (1 + x^2)^{-1/2}$            d) $F(x) = x^3$

2. Prove that a polynomial function of degree greater than one cannot be a contraction mapping on $\mathcal{R}^1$.

3. Let

$$F(x) = \begin{cases} 1, & x \leqslant 0, \\ \sqrt{1 + x^2}, & x > 0. \end{cases}$$

a) Show that $|F'(x)| < 1$ for all $x$.
b) Show that $F$ does not have a fixed point and hence is not a contraction mapping.
c) Compare the result in (b) with the conclusion of Example 2 in this section.

4. Let $F: \mathscr{R}^1 \to \mathscr{R}^1$ be defined as follows:

$$F(x) = \begin{cases} \frac{1}{2}, & \text{if } x \text{ is rational} \\ 0, & \text{if } x \text{ is irrational.} \end{cases}$$

a) Calculate $F^{[2]}$ and show that it is a contraction mapping.
b) Show that $F$ is not a contraction mapping.

5. Prove that if $F$ is everywhere differentiable, with $|F'(x)| < 1$ for all $x$, then $F$ has at most one fixed point.

6. The function

$$F(x) = \frac{1}{1 + x^2}$$

is a contraction mapping on $\mathscr{R}^1$. (See Exercise 1.)
a) Find the first four terms in the sequence $\{x_n\}$, $n \geqslant 1$, where $x_n = F(x_{n-1})$ and $x_0 = 0$.
b) Estimate the value of the fixed point for $F$.

7. Let $F$ be a continuous nondecreasing function mapping $[0, 1]$ into itself, and note that $[0, 1]$ is a complete metric space with respect to the standard distance in $\mathscr{R}^1$.
a) Give an example to show that $F$ need not be a contraction mapping.
b) Deduce that the sequence $\{x_n\}$, in which $x_0$ is an arbitrary point in $[0, 1]$ and $x_n = F(x_{n-1})$ for $n \geqslant 1$, converges [*Hint*: A bounded monotonic sequence of real numbers converges.]
c) Show that the limit point in (b) is a fixed point for $F$.
d) Give an example to show that the limit point in (b) need not be the only fixed point for $F$.

8. Consider $\mathscr{R}^2$ as a metric space with the standard distance, and let $F: \mathscr{R}^2 \to \mathscr{R}^2$ be defined by

$$F: \begin{bmatrix} x_1 \\ x_2 \end{bmatrix} \to \begin{bmatrix} a_{11} & a_{12} & b_1 \\ a_{21} & a_{22} & b_2 \end{bmatrix} \begin{bmatrix} x_1 \\ x_2 \\ 1 \end{bmatrix}.$$

a) Show that for all $\mathbf{x}$ and $\mathbf{y}$ in $\mathscr{R}^2$

$$d[F(\mathbf{x}), F(\mathbf{y})]^2 \leqslant (a_{11}^2 + a_{12}^2 + a_{21}^2 + a_{22}^2)d(\mathbf{x}, \mathbf{y})^2.$$

[*Hint*: Use Cauchy's inequality (Exercise 6, Section 12–2).]
b) Find a condition which will guarantee that $F$ is a contraction mapping.
c) Show that the mapping

$$\begin{bmatrix} x_1 \\ x_2 \end{bmatrix} \to \begin{bmatrix} \dfrac{x_1}{3} - \dfrac{x_2}{4} + 3 \\ \dfrac{x_1}{2} + \dfrac{x_2}{2} - 8 \end{bmatrix}$$

is a contraction mapping on $\mathscr{R}^2$, and find its fixed point.

9. a) Find a condition which will guarantee that the mapping $F$ of the preceding exercise is a contraction mapping on $\mathscr{R}^2$ with respect to the metric

$$d\left(\begin{bmatrix} x_1 \\ x_2 \end{bmatrix}, \begin{bmatrix} y_1 \\ y_2 \end{bmatrix}\right) = |x_1 - y_1| + |x_2 - y_2|.$$

b) Repeat (a) for the metric

$$d\left(\begin{bmatrix} x_1 \\ x_2 \end{bmatrix}, \begin{bmatrix} y_1 \\ y_2 \end{bmatrix}\right) = \max\left(|x_1 - y_1|, |x_2 - y_2|\right).$$

## 12–4  AN EXISTENCE AND UNIQUENESS THEOREM FOR $y' = f(x, y)$

At the beginning of this chapter we saw that

$$y' = f(x, y), \qquad y(x_0) = y_0, \tag{12–9}$$

can be solved when $f$ is continuous in a region $R$ of the $xy$-plane; we just look for the fixed points of the mapping

$$F: y \rightarrow y_0 + \int_{x_0}^{x} f(t, y(t))\, dt.$$

This fact, together with the results of the preceding section, implies that we can establish existence and uniqueness theorems for (12–9) by restricting the domain of $F$ to a complete metric space in which it or one of its iterates is a contraction mapping. To do so, we proceed as follows.

**I. Restrictions on $R$**  Throughout this discussion we shall require that $R$ be an *open* subset of the plane containing the point $(x_0, y_0)$. As suggested by Fig. 12–5, this means that each point of $R$ is the center of a circle whose interior lies entirely in $R$.

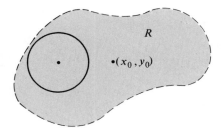

**Figure 12–5**

**II. Restrictions on $f$**  In addition to being continuous in $R$, we shall require the function $f$ appearing in (12–9) to satisfy a **Lipschitz condition** of the type

$$|f(x, y_1) - f(x, y_2)| \leqslant \beta |y_1 - y_2| \tag{12–10}$$

in every closed bounded subset of $R$. (Recall that a subset of the plane is *closed* if it contains all of its boundary points, and *bounded* if there exists a constant $B$ such that all points in the set are a distance less than $B$ from the origin.) A simpler version of the theorem we are about to prove can be obtained by replacing these restrictions on $f$ and $R$ by the requirement that *$f$ and $\partial f/\partial y$ be continuous on a closed rectangular region $R$ whose sides are parallel to the x- and y-axes and which contains* $(x_0, y_0)$ *in its interior.* For $f$ will then satisfy (12–10) in $R$, and the argument we are about to give will go through with only minor changes.

**III. Description of $M$**   The fact that $R$ is open implies that given any point $(x_0, y_0)$ in $R$, there exists a closed disc $D$ in $R$ centered at $(x_0, y_0)$. Then since $f$ is continuous in $R$, it is bounded on $D$, and we can find a positive constant $B$ such that

$$|f(x, y)| < B \qquad (12–11)$$

for all $(x, y)$ in $D$.

This done, we let $R_1$ be any closed region bounded by the lines through $(x_0, y_0)$ with slopes $\pm B$ and the vertical lines $x = a$ and $x = b$, where $a$ and $b$ are chosen so that

i)  $\beta \max (x_0 - a, b - x_0) < 1$, and

ii)  $R_1$ lies entirely in $D$.

(See Fig. 12–6.) With these choices in effect, we let $M$ be the metric space of all continuous functions on $[a, b]$ whose graphs lie in $R_1$ and pass through $(x_0, y_0)$, where the distance in $M$ is given by

$$d(y_1, y_2) = \max_{x \in [a,b]} |y_1(x) - y_2(x)|.$$

The following lemma shows that this choice of $M$ is just what we want.

**Lemma 12–5**   *$M$ is a complete metric space with respect to $d$, and*

$$F : y \rightarrow y_0 + \int_{x_0}^{x} f(t, y(t))\, dt$$

*is a contraction mapping on $M$.*

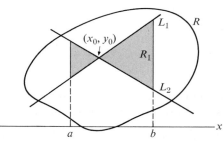

**Figure 12–6**

*Proof.* The fact that $M$ is complete is an easy consequence of Theorem 12–1, and the proof is left as an exercise.

To show that $F$ maps $M$ into itself, let $y$ be a function in $M$. Then since $[F(y)](x_0) = y_0$ and

$$\left| y_0 + \int_{x_0}^{x} f(t, y(t))\, dt \right| \leq |y_0| + \left| \int_{x_0}^{x} f(t, y(t))\, dt \right|$$
$$\leq |y_0| + B|x - x_0|,$$

$F(y)$ also belongs to $M$ (Exercise 1).

To show that $F$ is a contraction mapping, let $y_1$ and $y_2$ be functions in $M$ and observe that because of (12–10),

$$|f(x, y_1(x)) - f(x, y_2(x))| \leq \beta |y_1(x) - y_2(x)|$$
$$\leq \beta \max_{x \in [a,b]} |y_1(x) - y_2(x)|$$
$$= \beta\, d(y_1, y_2).$$

Thus

$$d[F(y_1), F(y_2)] = \max_{x \in [a,b]} \left| \int_{x_0}^{x} f(t, y_1(t))\, dt - \int_{x_0}^{x} f(t, y_2(t))\, dt \right|$$
$$= \max_{x \in [a,b]} \left| \int_{x_0}^{x} [f(t, y_1(t)) - f(t, y_2(t))]\, dt \right|$$
$$\leq \max_{x \in [a,b]} \left| \int_{x_0}^{x} |f(t, y_1(t)) - f(t, y_2(t))|\, dt \right|$$
$$\leq \max_{x \in [a,b]} \left| \int_{x_0}^{x} \beta\, d(y_1, y_2)\, dt \right|$$
$$= \beta \max_{x \in [a,b]} \left| \int_{x_0}^{x} dt \right| \cdot d(y_1, y_2)$$
$$= \beta \max (x_0 - a, b - x_0) \cdot d(y_1, y_2).$$

The assertion that $F$ is a contraction mapping on $M$ is now a consequence of the fact that $a$ and $b$ were chosen to make $\beta \max (x_0 - a, b - x_0) < 1$. ∎

With this we have proved our first existence and uniqueness theorem.

***Theorem 12–5*** *Let $R$ be an open region of the xy-plane, and suppose that $f$ is continuous on $R$ and satisfies a Lipschitz condition of the type*

$$|f(x, y_1) - f(x, y_2)| \leq \beta |y_1 - y_2|$$

*in every closed bounded subset of $R$. Then given any point $(x_0, y_0)$ in $R$, there exists a closed interval $[a, b]$ containing $x_0$ in its interior on which the initial-value problem*

$$y' = f(x, y), \qquad y(x_0) = y_0,$$

*has a unique solution.*

As we have already remarked, the hypotheses of this theorem are satisfied whenever $f$ and $\partial f/\partial y$ are continuous on a closed rectangular region $R$ of the $xy$-plane and $(x_0, y_0)$ is a point in the interior of $R$. Thus Theorem 12–5 includes Picard's existence and uniqueness theorem (Theorem 11–2) as a special case.

We can use Theorem 12–3 to obtain a sequence of approximations that converges to the solution of

$$y' = f(x, y), \qquad y(x_0) = y_0. \tag{12–12}$$

It is customary to begin the approximation process with the constant function

$$y_0(x) = y_0.$$

Then we obtain, in turn,

$$y_1(x) = y_0 + \int_{x_0}^{x} f(t, y_0)\, dt,$$

$$y_2(x) = y_0 + \int_{x_0}^{x} f(t, y_1(t))\, dt,$$

$$\cdot$$
$$\cdot$$
$$\cdot$$

$$y_n(x) = y_0 + \int_{x_0}^{x} f(t, y_{n-1}(t))\, dt.$$

This technique for approximating the solution of (12–12) is known as **Picard's method of successive approximations**.

When Picard's method is applied to the initial-value problem

$$y' = y, \qquad y(0) = 1,$$

it yields the sequence

$$y_0 = 1,$$

$$y_1 = 1 + \int_{0}^{x} dt = 1 + x,$$

$$y_2 = 1 + \int_{0}^{x} (1 + t)\, dt = 1 + x + \frac{x^2}{2},$$

$$\cdot$$
$$\cdot$$
$$\cdot$$

$$y_n = 1 + \int_{x_0}^{x} \left( 1 + t + \frac{t^2}{2} + \cdots + \frac{t^{n-1}}{2 \cdot 3 \cdots (n-1)} \right) dt$$

$$= 1 + x + \frac{x^2}{2!} + \cdots + \frac{x^n}{n!}.$$

In this case it is clear that the sequence $\{y_n\}$ converges to the solution $y = e^x$ of the problem.

## EXERCISES

1. This exercise refers to the proof of Lemma 12–5.
   a) Use Theorem 12–1 to show that the metric space $M$ is complete.
   b) Show that

   $$[F(y)](x_0) = y_0$$

   and

   $$\left| y_0 + \int_{x_0}^{x} f(t, y(t)) \, dt \right| \leq |y_0| + B|x - x_0|$$

   imply that $F(y)$ is in $M$ for all $y$ in $M$.

2. Prove that if $f$ and $\partial f / \partial y$ are continuous in an open region $R$ of the $xy$-plane, then $f$ satisfies a Lipschitz condition of the type

   $$|f(x, y_1) - f(x, y_2)| \leq \beta |y_1 - y_2|$$

   in every closed bounded subset of $R$.

3. Show that each of the following functions satisfies a Lipschitz condition in the given region of the $xy$-plane.
   a) $f(x, y) = ax^2 + by^2$, $|x| \leq 1, |y| \leq 1$
   b) $f(x, y) = x^3 e^{-xy^2}$, $0 \leq x \leq a$
   c) $f(x, y) = -P(x)y + Q(x)$, $|x| \leq a$

4. Let $f(x, y) = x^2 |y|$.
   a) Show that $f$ satisfies a Lipschitz condition on $|x| \leq 1, |y| \leq 1$.
   b) Show that $\partial f / \partial y$ does not exist at $(x, 0)$ if $x \neq 0$.

5. Use the method of successive approximations to obtain a series expansion of the solution of

   $$y' = xy, \qquad y(0) = 1.$$

   What is the closed-form solution of this problem?

6. Find the approximations $y_0, y_1, y_2$, and $y_3$ to the solution of

   $$y' = 1 + y^2, \qquad y(0) = 0.$$

   What is the closed-form solution of this problem?

7. Find the approximations $y_0, y_1, y_2$, and $y_3$ to the solutions of:
   a) $y' = y^2, \; y(0) = 0$,
   b) $y' = y^2, \; y(0) = 1$.

8. Repeat Exercise 7 for
   a) $y' = x^2 + y^2, \; y(0) = 0$,
   b) $y' = x^2 + y^2, \; y(0) = 1$.

9. Show that $y = 0$ and $y = x^3/27$ are solutions of

   $$y' = y^{2/3}, \qquad y(0) = 0.$$

   b) Show that $f(x, y) = y^{2/3}$ does not satisfy a Lipschitz condition in any region containing points on the $x$-axis.

## *12–5  AN IMPROVED THEOREM

Theorem 12–5 is an example of what is known as a "local" existence and unique-ness theorem because it guarantees its conclusion only in the immediate neigh-borhood of $x_0$. But all our experience with initial-value problems suggests that the problem $y' = f(x, y)$, $y(x_0) = y_0$ ought to have a solution whose graph extends all the way to the boundary (if any) of the region in which $f$ satisfies the hypotheses of Theorem 12–5. The following theorem supports this conjecture and gives a precise description of the "global" solution of the problem.

> **Theorem 12–6**   *Let R be an open region of the xy-plane, and suppose that f is continuous on R and satisfies a Lipschitz condition in every closed bounded subset of R. Then given any point $(x_0, y_0)$ in R, there exists an interval $\hat{I}$ con-taining $x_0$ and a function $\hat{y} = \hat{y}(x)$ defined in $\hat{I}$ with the following properties:*
>
> i) *$\hat{y}$ is the unique solution of*
>
> $$y' = f(x, y), \qquad y(x_0) = y_0, \qquad\qquad (12\text{–}13)$$
>
> *in $\hat{I}$;*
>
> ii) *$\hat{y}$ is a "maximal" solution of (12–13) in the sense that if $y = y(x)$ is a solution of (12–13) in an interval I about $x_0$, then $\hat{I}$ contains I and $\hat{y}(x) = y(x)$ for all x in I;*
>
> iii) *the graph of $\hat{y}$ is in R;*
>
> iv) *$\hat{I}$ is an open interval;*
>
> v) *the graph of $\hat{y}$ leaves every closed bounded subset of R that contains $(x_0, y_0)$.*

In other words, every solution $y = y(x)$ can be extended out to the boundary (if any) of $R$, as suggested in Fig. 12–7.

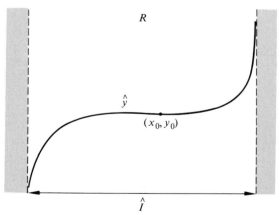

**Figure 12–7**

*Proof.* Let $\hat{y}$ be the set-theoretic union of all those solutions of (12–13) whose domains are intervals containing the point $x_0$ and whose graphs are in $R$. Then $\hat{y}$ is a function whose domain $\hat{I}$ is an interval containing $x_0$, and it is clear that $\hat{y}$ and $\hat{I}$ satisfy the first three conclusions of the theorem.

To establish (iv), we first observe that if $\hat{I} = (-\infty, \infty)$, then there is nothing to prove. Otherwise, let $a$ be an endpoint of $\hat{I}$ and suppose that $a$ belonged to $\hat{I}$. Then the point $(a, \hat{y}(a))$ would be in $R$, and the unique solution of $y' = f(x, y)$ through the point $(a, \hat{y}(a))$ could be used to extend $\hat{y}$ outside $\hat{I}$. Since this contradicts the maximality of $\hat{y}$, the endpoint $a$ cannot belong to $\hat{I}$, and $\hat{I}$ is therefore open.

To prove (v), let $S$ be a closed bounded subset of $R$ containing the point $(x_0, y_0)$. Since the graph of $\hat{y}$ must "escape" from $S$ whenever $\hat{I}$ is unbounded, we need only consider the case in which $\hat{I} = (a, b)$. In this situation let $\{(x_n, \hat{y}(x_n))\}$ be any sequence of points on the graph of $\hat{y}$ with the property that $x_n \to b$ as $n \to \infty$. (We could equally well have chosen the sequence so that $x_n \to a$ as $n \to \infty$.) Then if $\hat{y}(x_n) \to \infty$ or $-\infty$ with increasing $n$, the graph of $\hat{y}$ must again escape from $S$. Otherwise, the sequence $\{\hat{y}(x_n)\}$ contains a bounded and therefore convergent subsequence.* If $Y_0$ is the limit of the convergent subsequence, then we can find a sequence of points on the graph of $\hat{y}$ that converges to $(b, Y_0)$. We now propose to show that $(b, Y_0)$ is on the boundary of $R$, thereby proving that the graph of $\hat{y}$ eventually leaves $S$ in this case as well.

To do so, we observe that since the graph of $\hat{y}$ is in $R$, $(b, Y_0)$ is either on the boundary of $R$ or in the interior of $R$. Suppose the latter. Then by Theorem 12–5 there would exist an interval $I$ containing $b$ *in its interior* in which the initial-value problem

$$y' = f(x, y), \qquad y(b) = Y_0,$$

had a unique solution $Y = Y(x)$. The function $Y$ would then extend $\hat{y}$ to all of $I$. But this is impossible, because $I$ contains points that are not in $\hat{I}$. ∎

## EXERCISES

1. At the very end of the proof of Theorem 12–6 we said that $Y$ would extend $\hat{y}$ to $I$. Describe this extension precisely, and show that it does indeed yield a function which contradicts the choice of $\hat{y}$ and $\hat{I}$.

2. Let $R_1$ be the closed bounded region shown in Fig. 12–5 and used implicitly in the proof of Theorem 12–5. Prove that the maximal solution $\hat{y}$ of $y' = f(x, y), y(x_0) = y_0$, leaves $R_1$ through the vertical ends of the region. [*Hint:* Use the mean-value theorem.]

3. Discuss the behavior of the solutions of the general first-order linear equation

$$y' + P(x)y = Q(x)$$

when $P$ and $Q$ are continuous in an interval $I$.

---

\* This, too, is a theorem from advanced calculus.

### 12–6 FIRST-ORDER SYSTEMS

In this section we shall generalize the basic existence and uniqueness theorem for a single first-order equation to a theorem for a system of $n$ normal first-order equations in $n$ unknowns:

$$\frac{dy_1}{dx} = f_1(x; y_1, \ldots, y_n)$$

$$\cdot$$
$$\cdot \qquad\qquad\qquad (12\text{–}14)$$
$$\cdot$$

$$\frac{dy_n}{dx} = f_n(x; y_1, \ldots, y_n).$$

In physical problems in which the independent variable $x$ represents time, $(12\text{–}14)$ can be interpreted as a velocity field in $n$-space, the various derivatives $y_1', \ldots, y_n'$ being the components of the velocity vectors of the field. A solution $(y_1(x), \ldots, y_n(x))$ of $(12\text{–}14)$ will then describe the trajectory of a particle moving in $n$-space in such a way that its velocity vector at time $x$ is $(y_1'(x), \ldots, y_n'(x))$. Thus an initial-value problem for such a system consists of selecting an "initial time" $x = x_0$ and a point $(a_1, \ldots, a_n)$ in $\mathscr{R}^n$, and seeking a solution whose graph passes through the point $(a_1, \ldots, a_n)$ when $x = x_0$.

By setting

$$\mathbf{Y}(x) = (y_1(x), \ldots, y_n(x)),$$

$$f(x, \mathbf{Y}) = (f_1(x, \mathbf{Y}), \ldots, f_n(x, \mathbf{Y})),$$

and agreeing that differentiation and integration are to be performed component-wise according to the rules

$$\frac{d\mathbf{Y}}{dx} = (y_1'(x), \ldots, y_n'(x)),$$

$$\int_a^b \mathbf{Y}(x)\, dx = \left( \int_a^b y_1(x)\, dx, \ldots, \int_a^b y_n(x)\, dx \right),$$

we can rewrite $(12\text{–}14)$ in vector form as

$$\frac{d\mathbf{Y}}{dx} = f(x, \mathbf{Y}). \qquad\qquad (12\text{–}15)$$

In such an equation $\mathbf{Y}$ and $\mathbf{Y}'$ are functions defined on an interval of $\mathscr{R}^1$, with values in $\mathscr{R}^n$, while $f(x, \mathbf{Y})$ is a function from a region in $\mathscr{R}^{n+1}$ to $\mathscr{R}^n$.

As before we shall prove the basic existence and uniqueness theorem by appealing to the fixed-point theorems established earlier. This time, however, the relevant metric spaces will be $\mathscr{R}^n$ with the city block metric (Example 2, Section 12–2), and the space of continuous functions from $[a, b]$ to $\mathscr{R}^n$ with distance defined by

$$d(\mathbf{Y}, \mathbf{Z}) = \max_{x \in [a,b]} \left[ |y_1(x) - z_1(x)| + \cdots + |y_n(x) - z_n(x)| \right], \qquad (12\text{–}16)$$

where

$$\mathbf{Y}(x) = (y_1(x), \ldots, y_n(x)), \qquad \mathbf{Z}(x) = (z_1(x), \ldots, z_n(x)).$$

In particular, in each of these spaces we shall speak of the **norm** or **length** of a vector when we wish to refer to the distance of that vector from the origin, or zero vector. We introduce the notation $\|\mathbf{x}\|$ for the norm of a vector $\mathbf{x}$ in $\mathscr{R}^n$, and $|\mathbf{Y}|$ for the norm of a vector $\mathbf{Y}$ in the space of continuous functions from $[a, b]$ to $\mathscr{R}^n$. (Boldface and double bars are used here to avoid confusion of the norms in the two spaces with each other and with absolute values.) Thus if $\mathbf{x} = (x_1, \ldots, x_n)$ is a point in $\mathscr{R}^n$, then

$$\|\mathbf{x}\| = |x_1| + |x_2| + \cdots + |x_n|.$$

Similarly, if $\mathbf{Y}(x) = (y_1(x), \ldots, y_n(x))$ is a continuous function from $[a, b]$ to $\mathscr{R}^n$, then

$$|\mathbf{Y}| = \max_{x \in [a,b]} \left[ |y_1(x)| + \cdots + |y_n(x)| \right].$$

We begin by establishing a lemma that will be needed in the proof of the main existence and uniqueness theorem.

**Lemma 12–6**  Let $\mathbf{Y} = \mathbf{Y}(x)$ be a continuous function from $[a, b]$ to $\mathscr{R}^n$. Then

$$\left\| \int_a^b \mathbf{Y}(x)\, dx \right\| \leq \int_a^b \|\mathbf{Y}(x)\|\, dx.$$

*Proof.*  We simply apply the relevant definitions:

$$\left\| \int_a^b \mathbf{Y}(x)\, dx \right\| = \left\| \left( \int_a^b y_1(x)\, dx, \ldots, \int_a^b y_n(x)\, dx \right) \right\|$$

$$= \sum_{k=1}^n \left| \int_a^b y_k(x)\, dx \right|$$

$$\leq \sum_{k=1}^n \left( \int_a^b |y_k(x)|\, dx \right)$$

$$= \int_a^b \left( \sum_{k=1}^n |y_k(x)| \right) dx$$

$$= \int_a^b \|\mathbf{Y}(x)\|\, dx. \quad \blacksquare$$

*Remark.*  Note that when $n = 1$, $\mathbf{Y}$ is a real-valued continuous function on $[a, b]$ and the preceding inequality then becomes

$$\left| \int_a^b \mathbf{Y}(x)\, dx \right| \leq \int_a^b |\mathbf{Y}(x)|\, dx.$$

We now turn to the initial-value problem

$$\frac{d\mathbf{Y}}{dx} = f(x, \mathbf{Y}), \qquad \mathbf{Y}(x_0) = \mathbf{Y}_0, \tag{12-17}$$

which we shall discuss under the assumption that $f$ is continuous in an open region $R$ of $\mathscr{R}^{n+1}$ containing the point $(x_0, \mathbf{Y}_0)$ and satisfies a Lipschitz condition of the type

$$\|f(x, \mathbf{Y}_1) - f(x, \mathbf{Y}_2)\| \leqslant \beta\|\mathbf{Y}_1 - \mathbf{Y}_2\|$$

on every closed bounded subset of $R$. (Note that $f$ is continuous and satisfies a Lipschitz condition if and only if the same is true of all of its coordinate functions. See Exercise 1.) Then just as in the one-dimensional case, $\mathbf{Y} = \mathbf{Y}(x)$ is a solution of (12–17) if and only if it is a fixed point for the mapping

$$F: \mathbf{Y} \rightarrow \mathbf{Y}_0 + \int_{x_0}^{x} f(t, \mathbf{Y}(t))\, dt.$$

To obtain a complete metric space as a domain for $F$, let $R_1$ be the set of points $(x, \mathbf{Y})$ in $\mathscr{R}^{n+1}$ such that

$$\|\mathbf{Y} - \mathbf{Y}_0\| \leqslant B|x - x_0|, \qquad a \leqslant x \leqslant b,$$

where $[a, b]$ is chosen small enough to place $R_1$ in $R$, and $B$ is a bound for $f$ in $R_1$. This done, let $M$ be the set of all continuous functions from $[a, b]$ to $\mathscr{R}^n$ whose graphs lie in $R_1$. (See Fig. 12–8, drawn for $n = 2$.) Then $M$ is a complete metric space with respect to the distance given by (12–16), and $F$ is defined on $M$. With $M$ as the domain of $F$ we can now prove

> **Theorem 12–7**  Let $f = f(x, \mathbf{Y})$ be continuous in an open region $R$ of $\mathscr{R}^{n+1}$ and satisfy a Lipschitz condition with Lipschitz constant $\beta$ in every closed bounded subset of $R$. Then given any point $(x_0, \mathbf{Y}_0)$ in $R$, there exists a closed interval

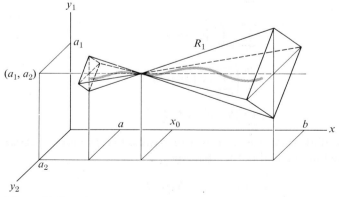

**Figure 12–8**

$[a, b]$ *containing* $x_0$ *in its interior on which the initial-value problem*

$$\frac{d\mathbf{Y}}{dx} = f(x, \mathbf{Y}), \qquad \mathbf{Y}(x_0) = \mathbf{Y}_0$$

*has a unique solution.*

*Proof.* According to Corollary 12–1, we will be done if we can show that $F^{[m]}$ is a contraction mapping on $M$ for some $m > 0$. To do so, we first observe that $F$ maps $M$ into itself since (12–17) is satisfied whenever $\mathbf{Y}$ is replaced by $F(\mathbf{Y})$ for any $\mathbf{Y}$ in $M$. Indeed, this is an immediate consequence of Lemma 12–6, since

$$\begin{aligned}
\|[F(\mathbf{Y})](x) - \mathbf{Y}_0\| &= \left\| \int_{x_0}^x f(t, \mathbf{Y}(t)) \, dt \right\| \\
&\leqslant \left| \int_{x_0}^x \|f(t, \mathbf{Y}(t))\| \, dt \right| \\
&\leqslant B|x - x_0|.
\end{aligned}$$

Next, suppose that $\mathbf{Y}$ and $\mathbf{Z}$ are any two functions in $M$. Then using Lemma 12–6 and the Lipschitz condition on $f$, we obtain

$$\begin{aligned}
\|[F(\mathbf{Y})](x) - [F(\mathbf{Z})](x)\| &= \left\| \int_{x_0}^x [f(t, \mathbf{Y}(t)) - f(t, \mathbf{Z}(t))] \, dt \right\| \\
&\leqslant \left| \int_{x_0}^x \|f(t, \mathbf{Y}(t)) - f(t, \mathbf{Z}(t))\| \, dt \right| \\
&\leqslant \beta \left| \int_{x_0}^x \|\mathbf{Y}(t) - \mathbf{Z}(t)\| \, dt \right| \\
&\leqslant \beta |x - x_0| \, |\mathbf{Y} - \mathbf{Z}| \\
&\leqslant \beta (b - a) |\mathbf{Y} - \mathbf{Z}|.
\end{aligned}$$

Since this inequality holds for every $x$ in $[a, b]$, we conclude that

$$|F(\mathbf{Y}) - F(\mathbf{Z})| \leqslant \beta(b - a)|\mathbf{Y} - \mathbf{Z}|.$$

Thus if $\beta(b - a) < 1$, then $F$ is a contraction mapping, and Theorem 12–2 applies. Otherwise,

$$\begin{aligned}
\|F^{[2]}(\mathbf{Y})(x) - F^{[2]}(\mathbf{Z})(x)\| &= \left\| \int_{x_0}^x \{f(t, [F(\mathbf{Y})](t)) - f(t, [F(\mathbf{Z})](t))\} \, dt \right\| \\
&\leqslant \beta \left| \int_{x_0}^x \|[F(\mathbf{Y})](t) - [F(\mathbf{Z})](t)\| \, dt \right| \\
&\leqslant \beta^2 |\mathbf{Y} - \mathbf{Z}| \left| \int_{x_0}^x |t - x_0| \, dt \right| \\
&= \frac{\beta^2 |x - x_0|^2}{2} |\mathbf{Y} - \mathbf{Z}|, \quad \text{etc.}
\end{aligned}$$

In general,

$$|F^{[m]}(\mathbf{Y}) - F^{[m]}(\mathbf{Z})| \leqslant \frac{\beta^m (b - a)^m}{m!} |\mathbf{Y} - \mathbf{Z}|,$$

and since $\beta^m(b - a)^m/m! < 1$ for sufficiently large $m$, some $F^{[m]}$ is a contraction mapping, and the theorem is proved.  ∎

As in the preceding section, there is a "global" analog of the theorem we have just proved. Its statement and proof are essentially the same as those for Theorem 12–6 and are therefore omitted. Finally, we recall that every initial-value problem involving a normal $n$th-order linear differential equation can be converted into an $n \times n$ system of normal first-order equations. Thus the theorem just mentioned includes the following basic result as a special case.

**Theorem 12–8**  *Every initial-value problem involving a normal linear differential equation*

$$y^{(n)} + a_{n-1}(x)y^{(n-1)} + \cdots + a_0(x)y = h(x),$$

*where $a_0, \ldots, a_{n-1}$, and $h$ are continuous on an open interval $I$ of the $x$-axis, has a unique solution on $I$.*

### EXERCISES

1. Let $f(x, \mathbf{Y}) = (f_1(x, \mathbf{Y}), \ldots, f_n(x, \mathbf{Y}))$ be a function from $\mathscr{R}^{n+1}$ to $\mathscr{R}^n$.
   a) Prove that $f$ is continuous at $(x_0, \mathbf{Y}_0)$ if and only if all of the $f_k$, $1 \leqslant k \leqslant n$, are continuous there.
   b) Prove that $f$ satisfies a Lipschitz condition in a region of $\mathscr{R}^{n+1}$ if and only if all of the $f_k$, $1 \leqslant k \leqslant n$, satisfy a Lipschitz condition there.

2. Find the function $f$ for the system

$$y_1' = xy_1y_2 + x^2$$
$$y_2' = y_1^2 + y_2^2 + x.$$

3. a) Find a Lipschitz constant for the system

$$y_1' = a_{11}y_1 + a_{12}y_2$$
$$y_2' = a_{21}y_1 + a_{22}y_2.$$

   b) Find a Lipschitz constant for the system

$$y_1' = y_2^2 + 1$$
$$y_2' = y_1^2 + x$$

   on $\|\mathbf{Y}\| \leqslant 1$.

### 12–7  CONTINUITY OF SOLUTIONS

We shall conclude this chapter by investigating the way in which the solutions of initial-value problems for a normal first-order system $\mathbf{Y}' = f(x, \mathbf{Y})$ vary with changes in $f$ and in the initial data. Specifically, we propose to show that when $f$ is continuous, bounded, and Lipschitzian in a region of $\mathscr{R}^{n+1}$, the unique solution of

$$\mathbf{Y}' = f(x, \mathbf{Y}), \qquad \mathbf{Y}(x_0) = \mathbf{Y}_0, \tag{12–18}$$

depends continuously on $f$, and on $x_0$ and $\mathbf{Y}_0$. In other words, small changes in $f$ or in $x_0$ and $\mathbf{Y}_0$ will produce only small changes in the solution of (12–18). This information is of critical importance in physical problems whose solutions are obtained from such a system, since experimental errors always affect the initial conditions and the values assigned to any physical constants appearing in the equation, and also since $f$ itself may vary because of the presence of variable parameters, such as density, temperature, pressure, etc. In fact, without the assurance that small changes in $(x_0, \mathbf{Y}_0)$ or in $f$ will not produce drastic changes in the solution of (12–18), such a solution would be physically useless.

To fix notation, we assume that $f$ is continuous in a region $R$ of $\mathscr{R}^{n+1}$, and that there exist constants $B$ and $\beta$ such that

$$\|f(x, \mathbf{Y})\| \leqslant B,$$
$$\|f(x, \mathbf{Y}_1) - f(x, \mathbf{Y}_2)\| \leqslant \beta\|\mathbf{Y}_1 - \mathbf{Y}_2\|,$$

whenever $(x, \mathbf{Y}), (x, \mathbf{Y}_1)$, and $(x, \mathbf{Y}_2)$ belong to $R$. As we saw earlier, these conditions are sufficient to guarantee that (12–8) has a unique solution $\mathbf{Z} = \mathbf{Z}(x)$ which belongs to the complete metric space $M$ introduced prior to the proof of Theorem 12–7.

Now let $\varepsilon > 0$ be given, and suppose that $g(x, \mathbf{Y})$ is continuous and bounded in $R$ with $\|g(x, \mathbf{Y})\| < \varepsilon$. Then it can be shown that the initial-value problem

$$\mathbf{Y}' = f(x, \mathbf{Y}) + g(x, \mathbf{Y}), \qquad \mathbf{Y}(x_0) = \mathbf{Y}_0,$$

has a solution $\mathbf{X}_0 = \mathbf{X}_0(x)$ in an interval about $x_0$, although under the given assumption this solution may not be unique. Furthermore, we can assume without loss of generality that both $\mathbf{X}_0$ and $\mathbf{Z}$ are defined on an interval $[a, b]$ about $x_0$, and that $\mathbf{X}_0$ belongs to $M$.* Our objective is to obtain an estimate for $\|\mathbf{Z}(x) - \mathbf{X}_0(x)\|$.

To begin, we note that $\mathbf{X}_0$ and $\mathbf{Z}$ satisfy the integral equations

$$\mathbf{Z}(x) = \mathbf{Y}_0 + \int_{x_0}^{x} f(t, \mathbf{Z}(t))\, dt,$$
$$\mathbf{X}_0(x) = \mathbf{Y}_0 + \int_{x_0}^{x} \left[ f(t, \mathbf{X}_0(t)) + g(t, \mathbf{X}_0(t)) \right] dt.$$

Now let $\{\mathbf{X}_n\}$ be the sequence in $M$ defined by

$$\mathbf{X}_1 = F(\mathbf{X}_0), \quad \mathbf{X}_2 = F(\mathbf{X}_1), \quad \ldots, \quad \mathbf{X}_n = F(\mathbf{X}_{n-1}), \quad \ldots,$$

where, as usual,

$$F \colon \mathbf{Y} \to \mathbf{Y}_0 + \int_{x_0}^{x} f(t, \mathbf{Y}(t))\, dt.$$

Then $F$ or one of its iterates is a contraction mapping on $M$, and $\{\mathbf{X}_n\}$ is a Cauchy sequence in $M$ which converges to the (unique) fixed point $\mathbf{Z}$ for $F$.† In addition,

---

\* In the definition of $M$ in the preceding section, $B$ must be replaced by $B + \varepsilon$.

† Strictly speaking we have only shown that if $F$ is a contraction, then $\{F^{(n)}(\mathbf{X}_0)\} \to \mathbf{Z}$; but it is sufficient to assume that $F^{[m]}$ is a contraction for some $m$. See Exercise 4.

we have

$$\|\mathbf{X}_1(x) - \mathbf{X}_0(x)\| = \left\|\int_{x_0}^{x} g(t, \mathbf{X}_0(t)) \, dt\right\| \leqslant \varepsilon |x - x_0|;$$

$$\|\mathbf{X}_2(x) - \mathbf{X}_0(x)\| = \left\|\int_{x_0}^{x} \left[ f(t, \mathbf{X}_1(t)) - f(t, \mathbf{X}_0(t)) - g(t, \mathbf{X}_0(t)) \right] dt\right\|$$

$$\leqslant \left\|\int_{x_0}^{x} \left[ f(t, \mathbf{X}_1(t)) - f(t, \mathbf{X}_0(t)) \right] dt\right\|$$

$$+ \left\|\int_{x_0}^{x} g(t, \mathbf{X}_0(t)) \, dt\right\|$$

$$\leqslant \varepsilon |x - x_0| + \beta \left|\int_{x_0}^{x} \|\mathbf{X}_1(t) - \mathbf{X}_0(t)\| \, dt\right|$$

$$\leqslant \varepsilon |x - x_0| + \beta \varepsilon \frac{|x - x_0|^2}{2}$$

$$= \frac{\varepsilon}{\beta} \left( \beta |x - x_0| + \frac{\beta^2 |x - x_0|^2}{2} \right).$$

Continuing, we find that

$$\|\mathbf{X}_n(x) - \mathbf{X}_0(x)\| \leqslant \frac{\varepsilon}{\beta} \sum_{k=1}^{n} \frac{\beta^k |x - x_0|^k}{k!}.$$

But since

$$e^{\beta|x - x_0|} = 1 + \beta|x - x_0| + \cdots + \frac{\beta^k |x - x_0|^k}{k!} + \cdots,$$

it follows that

$$\|\mathbf{X}_n(x) - \mathbf{X}_0(x)\| \leqslant \frac{\varepsilon}{\beta} \left[ e^{\beta|x - x_0|} - 1 \right]$$

for all $n$ and all $x$ in $[a, b]$. Hence since $\{\mathbf{X}_n\}$ converges to $\mathbf{Z}$, we have

$$\|\mathbf{Z}(x) - \mathbf{X}_0(x)\| \leqslant \frac{\varepsilon}{\beta} \left[ e^{\beta|x - x_0|} - 1 \right]$$

or

$$\|\mathbf{Z}(x) - \mathbf{X}_0(x)\| \leqslant \frac{\varepsilon}{\beta} \left[ e^{\beta(b - a)} - 1 \right] \tag{12–19}$$

for all $x$ in $[a, b]$. As it stands, (12–19) implies that the solution $\mathbf{Z}(x)$ of (12–18) varies continuously with $f$ in the sense that the quantity $\|\mathbf{Z}(x) - \mathbf{X}_0(x)\|$ is small for *all* $x$ in $[a, b]$ when $\varepsilon$ is small. In addition, (12–19) can be used to show that small changes in the initial conditions specified in (12–18) will produce only small changes in the solution of the problem.

Indeed, let $\bar{\mathbf{Z}}(x)$ be the solution of

$$\mathbf{Y}' = f(x, \mathbf{Y}), \qquad \mathbf{Y}(x_0) = \mathbf{Y}_0 + \mathbf{E}_0, \tag{12–20}$$

where the point $(x_0, \mathbf{Y}_0 + \mathbf{E}_0)$ still belongs to $R$. Then

$$\bar{\mathbf{Z}}(x) = \mathbf{Y}_0 + \mathbf{E}_0 + \int_{x_0}^{x} f(t, \bar{\mathbf{Z}}(t))\, dt$$

or, equivalently,

$$\bar{\mathbf{Z}}(x) - \mathbf{E}_0 = \mathbf{Y}_0 + \int_{x_0}^{x} \{f(t, \bar{\mathbf{Z}}(t) - \mathbf{E}_0) + [f(t, \bar{\mathbf{Z}}(t)) - f(t, \bar{\mathbf{Z}}(t) - \mathbf{E}_0)]\}\, dt.$$

$$(12\text{–}21)$$

But

$$\|f(x, \mathbf{Y}) - f(x, \mathbf{Y} - \mathbf{E}_0)\| \leqslant \beta \|\mathbf{E}_0\|$$

whenever $(x, \mathbf{Y})$ and $(x, \mathbf{Y}_0 - \mathbf{E}_0)$ belong to $R$. Thus if we set

$$g(x, \mathbf{Y}) = f(x, \mathbf{Y}) - f(x, \mathbf{Y} - \mathbf{E}_0),$$

and assume that $\|\mathbf{E}_0\| \leqslant \varepsilon/\beta$, then (12–21) implies that $\bar{\mathbf{Z}}(x) - \mathbf{E}_0$ is a solution of the initial-value problem

$$\mathbf{Y}' = f(x, \mathbf{Y}) + g(x, \mathbf{Y}), \qquad \mathbf{Y}(x_0) = \mathbf{Y}_0,$$

and it follows from (12–19) that

$$\|\mathbf{Z}(x) - [\bar{\mathbf{Z}}(x) - \mathbf{E}_0]\| \leqslant \frac{\varepsilon}{\beta} \left[ e^{\beta(b-a)} - 1 \right].$$

Hence

$$\begin{aligned}
\|\mathbf{Z}(x) - \bar{\mathbf{Z}}(x)\| &= \|\mathbf{Z}(x) - \bar{\mathbf{Z}}(x) + \mathbf{E}_0 - \mathbf{E}_0\| \\
&\leqslant \|\mathbf{Z}(x) - [\bar{\mathbf{Z}}(x) - \mathbf{E}_0]\| + \|\mathbf{E}_0\| \\
&\leqslant \frac{\varepsilon}{\beta} \left[ e^{\beta(b-a)} - 1 \right] + \frac{\varepsilon}{\beta} \\
&= \frac{\varepsilon}{\beta} e^{\beta(b-a)},
\end{aligned}$$

and we have established the following result: If $\mathbf{Z}$ and $\bar{\mathbf{Z}}$ are, respectively, the solutions in $[a, b]$ of

$$\mathbf{Z}(x) = \mathbf{Y}_0 + \int_{x_0}^{x} f(t, \mathbf{Z}(t))\, dt$$

and

$$\bar{\mathbf{Z}}(x) = \bar{\mathbf{Y}}_0 + \int_{x_0}^{x} f(t, \bar{\mathbf{Z}}(t))\, dt,$$

where $\|\mathbf{Y}_0 - \bar{\mathbf{Y}}_0\| \leqslant \varepsilon/\beta$, then

$$\|\mathbf{Z}(x) - \bar{\mathbf{Z}}(x)\| \leqslant \frac{\varepsilon}{\beta} e^{\beta(b-a)} \qquad\qquad (12\text{–}22)$$

for all $x$ in $[a, b]$. As before, this inequality can be viewed as asserting that $\mathbf{Z}(x)$ depends continuously on $\mathbf{Y}_0$.

Finally, suppose that $x_0$ is changed to, say $x_0 + \delta$, and let $\bar{\mathbf{Z}}(x)$ be the (unique) solution of

$$\bar{\mathbf{Z}}(x) = \mathbf{Y}_0 + \int_{x_0 + \delta}^{x} f(t, \bar{\mathbf{Z}}(t))\, dt.$$

Then

$$\bar{\mathbf{Z}}(x) = \left( \mathbf{Y}_0 - \int_{x_0}^{x_0 + \delta} f(t, \bar{\mathbf{Z}}(t))\, dt \right) + \int_{x_0}^{x} f(t, \bar{\mathbf{Z}}(t))\, dt,$$

and if we set

$$\mathbf{E}_0 = - \int_{x_0}^{x_0 + \delta} f(t, \bar{\mathbf{Z}}(t))\, dt,$$

we have $\|\mathbf{E}_0\| \leqslant B|\delta|$, and the results just obtained apply. In particular, if $\delta$ is chosen so that $x_0 + \delta$ belongs to $[a, b]$ and $|\delta| < \varepsilon/(B\beta)$, then (12–22) implies that

$$\|\mathbf{Z}(x) - \bar{\mathbf{Z}}(x)\| \leqslant \frac{\varepsilon}{\beta}\, e^{\beta(b-a)}$$

for all $x$ in $[a, b]$. In other words, $\mathbf{Z}(x)$ also depends continuously on $x_0$, and we have proved

**Theorem 12–9**   *The (unique) solution of any initial-value problem*

$$\mathbf{Y}' = f(x, \mathbf{Y}), \qquad \mathbf{Y}(x_0) = \mathbf{Y}_0,$$

*in which $f$ is continuous, bounded, and Lipschitzian, varies continuously with $f$ and with $(x_0, \mathbf{Y}_0)$.*


**EXERCISES**

1. Assume that

$$\mathbf{Y}' = f(x, \mathbf{Y}), \qquad \mathbf{Y}(x_0) = \mathbf{Y}_0,$$
$$\mathbf{Z}' = g(t, \mathbf{Z}), \qquad \mathbf{Z}(t_0) = \mathbf{Z}_0,$$

where $|x_0 - t_0| = \delta_0$, $\|\mathbf{Y}_0 - \mathbf{Z}_0\| = \delta_1$ and $\|f(x, \mathbf{Y}) - g(x, \mathbf{Y})\| \leqslant \varepsilon$ for all $(x, \mathbf{Y})$. Under the assumption that $f$ and $g$ are well behaved, estimate $\|\mathbf{Y}(x) - \mathbf{Z}(x)\|$.

2. Consider the systems

$$\text{(A)} \quad y_1' = y_1 + \varepsilon y_2, \qquad \text{(B)} \quad y_1' = y_1,$$
$$\qquad y_2' = \varepsilon y_1 + y_2; \qquad \qquad \quad y_2' = y_2.$$

   a) Find all solutions of (B).
   b) Find all solutions of (A).
   c) Using the results of parts (a) and (b), estimate $\|\mathbf{Y}_1(x) - \mathbf{Z}_1(x)\|$, where $\mathbf{Y}_1(x)$ and $\mathbf{Z}_1(x)$ are solutions of (A) and (B) and where $\mathbf{Y}_1(0) = \mathbf{Z}_1(0) = (1, -1)$.
   d) Estimate $\|\mathbf{Y}_1(x) - \mathbf{Z}_1(x)\|$ by using the theory of this section. [Again assume that $\mathbf{Y}_1(0) = \mathbf{Z}_1(0) = (1, -1)$.]

3. Consider $y^{(n)} + a_{n-1}(x)y^{(n-1)} + \cdots + a_1(x)y^{(1)} + a_0(x)y = 0$ subject to the conditions $y^{(k)}(0) = b_k,\ k = 0, 1, \ldots, n - 1$.

a) Suppose that the coefficient functions $a_k(x)$ are known only to within $\varepsilon$. What can you say about the accuracy of the solution?

b) Repeat (a) given that the initial constants $b_k$ are known only to within $\varepsilon$.

c) Combine (a) and (b) to allow for both possibilities.

4. Show that if $F$ is a function from a metric space $M$ into itself such that, for some $m$, $F^{[m]}$ is a contraction, then $\{F^{[n]}(\mathbf{X}_0)\} \to \mathbf{Z}$ for any $\mathbf{X}_0$, where $\mathbf{Z}$ is the fixed point of $F^{[m]}$. (Initially we only know that $F^{[m]}(\mathbf{X}_0), F^{[2m]}(\mathbf{X}_0), \ldots$ tend to $\mathbf{Z}$.)

# *Stability*   13

## 13–1  INTRODUCTION

At the end of the last chapter we saw that whenever the vector function $f(x, \mathbf{Y})$ is sufficiently well behaved in a region $\Omega$ of $\mathcal{R}^{n+1}$ the initial-value problem

$$\mathbf{Y}' = f(x, \mathbf{Y}), \qquad \mathbf{Y}(x_0) = \mathbf{Y}_0, \tag{13–1}$$

has a unique solution $\mathbf{Y} = \mathbf{Y}(x)$ in $\Omega$ which depends continuously on $f$ and on the initial data $x_0, \mathbf{Y}_0$. In particular, we proved that small changes, or *perturbations*, in $\mathbf{Y}_0$ will produce small changes in $\mathbf{Y}(x)$ in an interval about $x_0$. Important as this fact is, it is usually of even greater importance to have information concerning the long-term behavior of these perturbed solutions when the independent variable represents time, as it so often does in applications. Does *every* solution of (13–1) which is near $\mathbf{Y}_0$ at time $x_0$ remain near the solution $\mathbf{Y} = \mathbf{Y}(x)$ ever afterward, or are there solutions which eventually stray from $\mathbf{Y}$ no matter how close they are initially to $\mathbf{Y}_0$? Questions like this belong to the branch of mathematics known as **stability theory**—an area which has been intensively studied in recent years because of its importance in the design of automatic controls. In this chapter we shall answer the question posed above, and others as well, for an important class of vector differential equations: the so-called "autonomous" or time-independent systems, which arise in a wide variety of physical applications.

## 13–2  AUTONOMOUS SYSTEMS AND EQUILIBRIUM

To emphasize the physical significance of the ideas we are about to consider; we replace the independent variable in (13–1) by $t$ (to denote time) and rewrite the system as

$$\dot{\mathbf{X}} = F(t, \mathbf{X}), \qquad \mathbf{X}(t_0) = \mathbf{X}_0, \tag{13–2}$$

where $\dot{\mathbf{X}}$ denotes the time derivative of the vector function

$$\mathbf{X}(t) = \left( x_1(t), \ldots, x_n(t) \right).$$

Such a system defines a time-dependent vector field in a region of $n$-space, such as the velocity field associated with a fluid flow, and a solution of (13–2) describes the path or trajectory of a particle moving in $n$-space under the influence of this field given that the particle was at $\mathbf{X}_0$ when $t = t_0$.

We are particularly interested in those vector fields which are **stationary**, or time independent, where the systems assume the form

$$\dot{\mathbf{X}} = F(\mathbf{X}). \tag{13–3}$$

Such systems are said to be **autonomous**. The points in $n$-space at which $F$ vanishes are of special importance, since they are the points at which any physical process described by that system is in equilibrium. For instance, if (13–3) represents the velocity field of a fluid flow in 3-space and in $\mathbf{X}_0$ is a point in $\mathscr{R}^3$ for which $F(\mathbf{X}_0) = \mathbf{0}$, then the vector field vanishes at $\mathbf{X}_0$, and a particle at rest at $\mathbf{X}_0$ will remain at rest there. Thus the constant function $\mathbf{X}(t) = \mathbf{X}_0$ will be a solution of (13–3) and its associated "trajectory" will be a single point. These remarks motivate the following definition.

**Definition 13–1**  *A vector $\mathbf{X}_0$ in $\mathscr{R}^n$ is called an **equilibrium** or **critical point** for the autonomous system $\dot{\mathbf{X}} = F(\mathbf{X})$ if $F(\mathbf{X}_0) = \mathbf{0}$.*

In these terms, the first problem of stability theory is to determine the conditions under which those solutions of (13–3) originating near an equilibrium point for that system remain thereafter near that point, in which case we say that the equilibrium point is **stable**, and the conditions under which they do not, in which case we say that the equilibrium point is **unstable**.

An excellent example of a physical system which involves both types of equilibrium is provided by a simple pendulum of mass $m$ swinging without friction on a slender weightless rod of length $l$ (Fig. 13–1). By Newton's second law, the equation of motion of the pendulum is

$$-mg \sin \theta = ml \frac{d^2\theta}{dt^2},$$

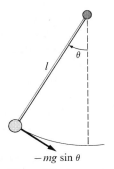

**Figure 13–1**

or

$$\ddot{\theta} = -g/l \sin \theta. \tag{13-4}$$

By setting $g/l = k^2$ and $\dot{\theta} = \omega$, we can rewrite (13–4) as

$$\begin{aligned} \dot{\theta} &= \omega, \\ \dot{\omega} &= -k^2 \sin \theta, \end{aligned} \tag{13-5}$$

in which form it is a $2 \times 2$ or plane autonomous system. It is clear from physical considerations alone that the pendulum is in equilibrium whenever it comes to rest at any of the positions $\theta = n\pi, n = 0, \pm 1, \pm 2, \ldots$. Moreover, these positions are stable or unstable according as $n$ is even or odd; that is, according as the pendulum is suspended vertically downward or balanced vertically upward from its point of rotation.

A mathematical analysis of (13–5) will sustain these conclusions. Indeed, each of the points $(n\pi, 0)$, where $n$ is an integer, in the $\theta\omega$-plane is an equilibrium point for this system in the sense of Definition 13–1. Furthermore, they are the only such points. To determine their stability or instability, we solve (13–5) as follows. Since

$$\frac{d\omega}{dt} = \frac{d\omega}{d\theta} \frac{d\theta}{dt} = \omega \frac{d\omega}{d\theta},$$

the second equation in (13–5) may be rewritten as

$$\omega \frac{d\omega}{d\theta} = -k^2 \sin \theta.$$

Separating the variables, we find that

$$\omega^2 = 2k^2 \cos \theta + c,$$

where $c \geqslant -2k^2$ is a constant. Several of these solution curves, or trajectories, are sketched in Fig. 13–2.

A study of this figure reveals that there is a region about each equilibrium point of the form $(2n\pi, 0)$ with the property that a trajectory which originates within that region can never escape from it. (These are the shaded portions of the figure.) The regions provide geometric evidence of the fact that these points are points of stable equilibrium for the pendulum. The corresponding values of $\theta$ are those about which the pendulum can oscillate in normal clocklike fashion. By contrast, no such region exists about any of the remaining equilibrium points. Indeed, when $n$ is odd, *every* region about $(n\pi, 0)$ contains trajectories which eventually depart arbitrarily far from that point. Physically these errant trajectories correspond to the circular motion of the pendulum, and their existence implies that the equilibrium points in question are unstable.

To carry the analysis of this problem one step further, let us suppose that damping forces due to friction and air resistance are taken into account. Although

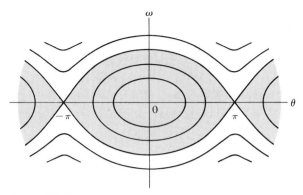

**Figure 13–2**

the mathematics of the problem is now more involved, it is clear that the pendulum has the same equilibrium positions, and that these positions are stable or unstable as before. This time, however, the behavior of trajectories about a point of stable equilibrium is rather different. The damping forces cause the oscillations to die out in time, and the corresponding trajectories in the $\theta\omega$-plane spiral inward toward the equilibrium point in the manner indicated in Fig. 13–3. For this reason we say that these points are *asymptotically stable*.

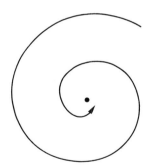

**Figure 13–3**

Guided by these observations, we now formulate working definitions of these concepts for an arbitrary autonomous system

$$\dot{\mathbf{X}} = F(\mathbf{X}), \tag{13–6}$$

which we assume to be sufficiently well behaved in a region $\Omega$ of $n$-space so that initial-value problems involving this system admit unique solutions in $\Omega$. For convenience we shall denote the solution of (13–6) which satisfies the initial condition $\mathbf{X}(0) = \mathbf{Y}_0$ by $\mathbf{X}(t, \mathbf{Y}_0)$. We then have

**Definition 13–2**   *Let* $\mathbf{X}_0$ *be an equilibrium point for* (13–6). *Then* $\mathbf{X}_0$ *is said to be*

  i) **stable** *if, given any* $\varepsilon > 0$, *there exists a* $\delta > 0$ *such that whenever* $\mathbf{Y}_0$ *is in* $\Omega$ *and* $\|\mathbf{Y}_0 - \mathbf{X}_0\| < \delta$, *then* $\|\mathbf{X}(t, \mathbf{Y}_0) - \mathbf{X}_0\| < \varepsilon$ *for all* $t \geqslant 0$;*

 ii) **asymptotically stable** *if, in addition to being stable, there exists a* $\delta > 0$ *such that whenever* $\mathbf{Y}_0$ *is in* $\Omega$ *with*

$$\|\mathbf{Y}_0 - \mathbf{X}_0\| < \delta,$$

*we have*

$$\lim_{t \to \infty} \|\mathbf{X}(t, \mathbf{Y}_0) - \mathbf{X}_0\| = 0;$$

*iii)* **unstable** *otherwise.*

Geometrically, $\mathbf{X}_0$ is a point of stable equilibrium for (13–6) if, given any sphere $S_\varepsilon$ of radius $\varepsilon$ about $\mathbf{X}_0$, there exists a sphere $S_\delta$ of radius $\delta$ about $\mathbf{X}_0$ with the property that *every* trajectory which is in $S_\delta$ at time $t = 0$ remains within $S_\varepsilon$ for all $t \geqslant 0$. If, in addition, all trajectories originating sufficiently close to $\mathbf{X}_0$ eventually approach $\mathbf{X}_0$ as a limiting position, then $\mathbf{X}_0$ is asymptotically stable. Finally, $\mathbf{X}_0$ is a point of unstable equilibrium if *every* spherical region lying in $\Omega$ with center $\mathbf{X}_0$ contains trajectories which are arbitrarily close to $\mathbf{X}_0$ at time $t = 0$, but which eventually escape from that region. Figure 13–4 illustrates these ideas for trajectories in the plane.

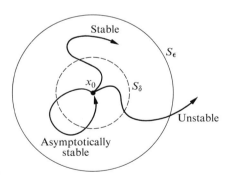

**Figure 13–4**

---

*  Throughout this chapter we shall use the standard norm or distance in $\mathscr{R}^n$, namely

$$\|\mathbf{X} - \mathbf{Y}\| = \left[ \sum_{i=1}^{n} (x_i - y_i)^2 \right]^{1/2}, \qquad \mathbf{X} = (x_1, \ldots, x_n), \qquad \mathbf{Y} = (y_1, \ldots, y_n).$$

## 13–3 STABILITY FOR CONSTANT-COEFFICIENT LINEAR SYSTEMS

We shall begin the study of stability by considering autonomous systems of constant-coefficient linear equations; that is, systems of the form

$$\dot{\mathbf{X}} = A\mathbf{X}, \tag{13–7}$$

where $A$ is an $n \times n$ matrix of real numbers. The origin of $\mathscr{R}^n$ is an equilibrium point for such a system, and when zero is not an eigenvalue for $A$ (i.e., when $A$ is *nonsingular*) it is the only one. For simplicity we shall assume that this is the case from now on.

In Chapter 10 we saw that the solutions of (13–7) can be described in terms of the eigenvalues of $A$ as follows.

When $\alpha \pm \beta i$, $\beta > 0$, are complex conjugate eigenvalues of multiplicity $m$, then (13–7) has $2m$ linearly independent "fundamental" solutions constructed from functions of the form

$$\mathbf{G}_\lambda e^{\alpha t} t^k \cos \beta t$$

and

$$\mathbf{H}_\lambda e^{\alpha t} t^k \sin \beta t,$$

where $k$ is an integer ranging from 0 to $m - 1$, and $\mathbf{G}_\lambda$ and $\mathbf{H}_\lambda$ are vectors in $\mathscr{R}^n$. Now suppose that $t \geqslant 0$. Then

$$\left\| \mathbf{G}_\lambda e^{\alpha t} t^k \cos \beta t \right\| = \left\| \mathbf{G}_\lambda \right\| e^{\alpha t} t^k \left| \cos \beta t \right| \leqslant \left\| \mathbf{G}_\lambda \right\| e^{\alpha t} t^k,$$

and since $\left\| \mathbf{G}_\lambda \right\|$ is a (positive) constant,

$$\lim_{t \to \infty} \left\| \mathbf{G}_\lambda e^{\alpha t} t^k \cos \beta t \right\| = 0 \qquad \text{if } \alpha < 0.$$

But

$$\left\| \mathbf{G}_\lambda e^{\alpha t} t^k \cos \beta t \right\|$$

is unbounded if $\alpha > 0$. Similarly,

$$\lim_{t \to \infty} \left\| \mathbf{H}_\lambda e^{\alpha t} t^k \sin \beta t \right\| = 0 \qquad \text{if } \alpha < 0,$$

while

$$\left\| \mathbf{H}_\lambda e^{\alpha t} t^k \sin \beta t \right\|$$

is unbounded if $\alpha > 0$.

Thus *every* trajectory of (13–7) arising from a pair of complex conjugate eigenvalues for $A$ will approach the origin of $\mathscr{R}^n$ as $t \to \infty$ if the real part of these eigenvalues is negative, and will depart arbitrarily far from the origin if the real part of these eigenvalues is positive.* In the first case the trajectories exhibit the properties required for asymptotic stability, and even asymptotic stability *in the*

---

\* Note that if the norms of $\mathbf{X}_1(t), \ldots, \mathbf{X}_r(t)$ are bounded, or approach 0 or $\infty$ as $t \to \infty$, the same will be true of $\left\| \mathbf{X}(t) \right\|$ whenever $\mathbf{X}(t)$ is a linear combination of $\mathbf{X}_1(t), \ldots, \mathbf{X}_r(t)$.

*large* in the sense that no matter where such a trajectory begins it will eventually approach the origin as a limiting position. In the second case, we have instability.

The reasoning just used applies equally well to trajectories arising from real eigenvalues for $A$. Thus to complete the analysis, it remains for us to consider the case where $A$ admits a pair of purely imaginary eigenvalues $\pm \beta i$. If these eigenvalues have multiplicity one, then the corresponding fundamental solutions of (13–7) are constructed from the functions

$$\mathbf{G}_\lambda \cos \beta t \qquad \text{and} \qquad \mathbf{H}_\lambda \sin \beta t,$$

and the fact that $\cos \beta t$ and $\sin \beta t$ are bounded as $t \to \infty$ but do not tend to zero implies that the resulting trajectories have the properties required for stability, but not asymptotic stability. On the other hand, if the multiplicity of $\pm \beta i$ is greater than one, then the solutions of (13–7) involve the functions

$$\mathbf{G}_\lambda t^k \cos \beta t \qquad \text{and} \qquad \mathbf{H}_\lambda t^k \sin \beta t$$

with $k \geq 1$. When this happens, the origin will be unstable because

$$|t^k \cos \beta t| \qquad \text{and} \qquad |t^k \sin \beta t|$$

are unbounded. Since this exhausts all possible combinations of eigenvalues for $A$, we have proved

**Theorem 13–1**   *If* $\dot{\mathbf{X}} = A\mathbf{X}$ *is an* $n \times n$ *linear autonomous system whose coefficient matrix is nonsingular, then the origin in* $\mathscr{R}^n$ *is*:

i) *asymptotically stable if the real parts of all of the eigenvalues of $A$ are negative;*

ii) *stable, but not asymptotically stable, if $A$ has at least one pair of purely imaginary eigenvalues of multiplicity one, no pure imaginary eigenvalues of multiplicity exceeding one, and no eigenvalues with positive real parts;*

iii) *unstable otherwise.*

Consider for instance the plane autonomous systems

$$\text{(i)} \ \dot{x} = y \qquad \text{(ii)} \ \dot{x} = y \qquad \text{(iii)} \ \dot{x} = -x$$
$$\dot{y} = x; \qquad\qquad \dot{y} = -x; \qquad\qquad \dot{y} = -y,$$

in which $x$ and $y$ represent the usual rectangular coordinates in $\mathscr{R}^2$.

The coefficient matrix for the first system is

$$\begin{bmatrix} 0 & 1 \\ 1 & 0 \end{bmatrix},$$

and its eigenvalues are the solutions of

$$\begin{bmatrix} -\lambda & 1 \\ 1 & -\lambda \end{bmatrix} = 0 \qquad \text{or} \qquad \lambda^2 - 1 = 0.$$

Thus $\lambda = \pm 1$, and the origin is unstable since one of the eigenvalues has a positive real part.

For the second system

$$A = \begin{bmatrix} 0 & 1 \\ -1 & 0 \end{bmatrix},$$

$\lambda = \pm i$, and by part (ii) of the theorem the origin is stable, but not asymptotically stable.

Finally, the origin is asymptotically stable for the third system since the matrix

$$\begin{bmatrix} -1 & 0 \\ 0 & -1 \end{bmatrix}$$

has $\lambda = -1$ as its only eigenvalue.

The trajectories of any *plane* autonomous system of constant-coefficient linear equations can be found by eliminating the parameter $t$ from the system and solving the resulting first-order equation. The most general system of this type is

$$\dot{x} = ax + by, \tag{13–8}$$
$$\dot{y} = cx + dy,$$

where $a$, $b$, $c$, and $d$ are constants. Since $\dot{y}/\dot{x} = dy/dx$, (13–8) is equivalent to the single first-order equation

$$\frac{dy}{dx} = \frac{cx + dy}{ax + by} \tag{13–9}$$

at all points in the $xy$-plane where $ax + by \neq 0$. Equation (13–9) can therefore be solved by the techniques introduced in Chapter 1.

The first-order equations corresponding to the systems in the example above are

(i) $y' = x/y$,     (ii) $y' = -x/y$,     (iii) $y' = y/x$.

In the first case the trajectories are the family of hyperbolas

$$x^2 - y^2 = c,$$

and the origin is an (unstable) **saddle point**. In the second case the trajectories are the family of circles

$$x^2 + y^2 = c^2,$$

and the origin is a (stable) **vortex point**. In the last case the trajectories are the family of straight lines

$$y = cx,$$

and the origin is an (asymptotically stable) **star point**. These trajectories, oriented by increasing $t$, appear as shown in Fig. 13–5.

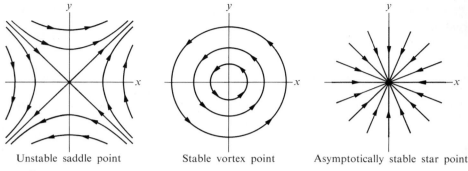

Unstable saddle point          Stable vortex point          Asymptotically stable star point

**Figure 13–5**

## 13–4 PLANE AUTONOMOUS LINEAR SYSTEMS

The results of the preceding section can be used to obtain a complete description
of the trajectories of a plane autonomous system of the form

$$\dot{x} = ax + by,$$
$$\dot{y} = cx + dy,$$

(13–10)

when the matrix

$$A = \begin{bmatrix} a & b \\ c & d \end{bmatrix}$$

is nonsingular. This description depends on the nature of the eigenvalues $\lambda_1, \lambda_2$
of $A$, which in this case are the roots of the quadratic equation

$$\lambda^2 - (a + d)\lambda + (ad - bc) = 0.$$

(13–11)

Since the discriminant of this equation is

$$\Delta = (a + d)^2 - 4(ad - bc) = (a - d)^2 + 4bc,$$

$\lambda_1$ and $\lambda_2$ are

 i) real and distinct if $\Delta > 0$,
 ii) real and equal if $\Delta = 0$,
iii) complex conjugates if $\Delta < 0$.

Moreover, when $\lambda_1$ and $\lambda_2$ are the roots of (13–11), we have $(\lambda - \lambda_1)(\lambda - \lambda_2) = 0$;
whence

$$\lambda^2 - (\lambda_1 + \lambda_2)\lambda + \lambda_1\lambda_2 = 0,$$

and

$$\lambda_1 + \lambda_2 = a + d,$$
$$\lambda_1\lambda_2 = ad - bc = \text{Det } A.$$

Thus when $\Delta \geqslant 0$, $\lambda_1$ and $\lambda_2$ will have the same sign if and only if $ad - bc > 0$, and will then be positive or negative according as $a + d$ is positive or negative. On the other hand, if $\Delta < 0$, then $\lambda_1 = \alpha + \beta i$, $\lambda_2 = \alpha - \beta i$, $\beta > 0$, and

$$a + d = 2\alpha.$$

In this case $\lambda_1$ and $\lambda_2$ are pure imaginary if and only if $a + d = 0$. Otherwise they have a nonzero real part which agrees in sign with $a + d$. Combining these observations with Theorem 13–1, we obtain the following description of the origin for a (nonsingular) plane autonomous linear system with constant coefficients:

asymptotically stable if $\begin{cases} \Delta \geqslant 0,\ ad - bc > 0,\ a + d < 0,\ \text{or} \\ \Delta < 0,\ a + d < 0; \end{cases}$

stable (but not asymptotically stable) if $\Delta < 0$, $a + d = 0$;

unstable if $\begin{cases} \Delta \geqslant 0,\ ad - bc < 0,\ \text{or} \\ \Delta < 0,\ a + d > 0. \end{cases}$

Notice that *all* of this information is contained in the coefficient matrix for the system.

We can now describe the trajectories which arise in each of these cases. To do so we rewrite (13–10) in matrix form as

$$\dot{\mathbf{X}} = A\mathbf{X}, \tag{13–12}$$

with

$$\mathbf{X} = \begin{bmatrix} x \\ y \end{bmatrix}, \qquad \dot{\mathbf{X}} = \begin{bmatrix} \dot{x} \\ \dot{y} \end{bmatrix},$$

and introduce the change of coordinates $\mathbf{Y} = B\mathbf{X}$, where $B$ is a $2 \times 2$ nonsingular matrix. Specifically, if $x^*$, $y^*$ denote the new coordinates in $\mathcal{R}^2$, then $B$ is the coefficient matrix of the equations

$$x^* = \beta_{11}x + \beta_{21}y$$
$$y^* = \beta_{12}x + \beta_{22}y,$$

which express the new coordinates in terms of the old.† Since these equations can be solved for $x$ and $y$ in terms of $x^*$, $y^*$, $B$ must be nonsingular. Hence $B^{-1}$ exists. We now multiply (13–12) on the left by $B$, and use the relations $B\dot{\mathbf{X}} = \dot{\mathbf{Y}}$ and $\mathbf{X} = B^{-1}\mathbf{Y}$ to obtain

$$\dot{\mathbf{Y}} = (BAB^{-1})\mathbf{Y}. \tag{13–13}$$

Thus (13–12) and (13–13) are the same autonomous system with respect to different coordinates in $\mathcal{R}^2$, and we can solve the original system by choosing $B$

---

† Some authors take $B$ to be the matrix expressing $x$, $y$ in terms of $x^*$, $y^*$. This has the effect of interchanging $B$ and $B^{-1}$ throughout the following computations. Otherwise, everything remains the same.

so that $BAB^{-1}$ is as simple as possible. The degree of simplification which can be obtained in any given case depends upon the nature of the eigenvalues for $A$, as follows.

CASE 1 ($\lambda_1$, $\lambda_2$ *real and distinct*). In this case $B$ can be chosen so that

$$BAB^{-1} = \begin{bmatrix} \lambda_1 & 0 \\ 0 & \lambda_2 \end{bmatrix},$$

and the simplified system is

$$\dot{x}^* = \lambda_1 x^*$$
$$\dot{y}^* = \lambda_2 y^*.$$

(Question: What is $B$?) Thus

$$x^* = c_1 e^{\lambda_1 t}$$
$$y^* = c_2 e^{\lambda_2 t},$$

and the parameter $t$ now can be eliminated to yield

$$y^* = c(x^*)^{\lambda_2/\lambda_1},$$

where $c$ is an arbitrary constant. (Note that the trajectory $x^* = 0$ is suppressed at this step.) It is now an easy matter to describe the trajectories in the $x^*y^*$-plane. When $\lambda_1$ and $\lambda_2$ are of the same sign, they are "parabolic" curves of the type shown in Fig. 13–6. Then the origin is a **nodal point**, unstable if $\lambda_1$ and $\lambda_2$ are positive, asymptotically stable if they are negative. (The two cases are related by the transformation $t \to -t$.) When $\lambda_1$ and $\lambda_2$ are of opposite signs, $\lambda_2/\lambda_1$ is negative, and

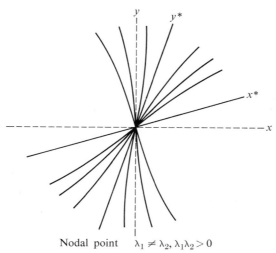

Nodal  point    $\lambda_1 \neq \lambda_2$, $\lambda_1 \lambda_2 > 0$

**Figure 13–6**

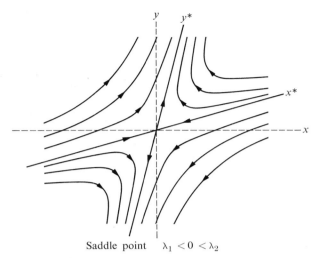

Saddle point     $\lambda_1 < 0 < \lambda_2$

**Figure 13–7**

the trajectories are "hyperbolic" curves in the $x^*y^*$-plane. (See Fig. 13–7, drawn for $\lambda_1 < 0 < \lambda_2$.) The origin is now a **saddle point**, and is unstable.

CASE 2 ($\lambda_1 = \alpha + \beta i$, $\lambda_2 = \alpha - Bi$, $\beta > 0$). Here the argument must be divided into subcases depending on the value of $\alpha$.

i) $\alpha = 0$. In this case $\lambda_1$ and $\lambda_2$ are pure imaginary, and $A$ is of the form

$$\begin{bmatrix} a & b \\ c & -a \end{bmatrix}$$

with the characteristic equation $\lambda^2 - (a^2 + bc) = 0$. Thus $\beta = (-a^2 - bc)^{1/2}$, and we have $a^2 + bc + \beta^2 = 0$. Set

$$B = \begin{bmatrix} 0 & \beta \\ -c & a \end{bmatrix}.$$

Then $BA = CB$, where

$$C = \begin{bmatrix} 0 & -\beta \\ \beta & 0 \end{bmatrix},$$

and since $B$ is nonsingular (Det $B \neq 0$), it has an inverse. Then $BAB^{-1} = C$, and the above choice of $B$ reduces the original system to

$$\dot{x}^* = -\beta y^*$$
$$\dot{y}^* = \beta x^*.$$

Thus the functions

$$\mathbf{X}_1^*(t) = \begin{bmatrix} \cos \beta t \\ \sin \beta t \end{bmatrix} \quad \text{and} \quad \mathbf{X}_2^*(t) = \begin{bmatrix} \sin \beta t \\ -\cos \beta t \end{bmatrix}$$

are a basis for the solution space of the system, and the general solution is

$$x^* = c_1 \cos \beta t + c_2 \sin \beta t$$
$$y^* = c_1 \sin \beta t - c_2 \cos \beta t,$$

where $c_1$ and $c_2$ are arbitrary. Since

$$(x^*)^2 + (y^*)^2 = (c_1^2 + c_2^2),$$

the trajectories are circles in the $x^*y^*$-plane centered at the origin, and the origin is a stable **vortex point**. (See Fig. 13–8.)

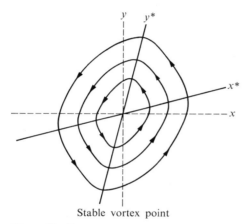

Stable vortex point

**Figure 13–8**

ii) $\alpha \neq 0$. In this case we form the matrix $A - \alpha I$, where $I$ is the $2 \times 2$ identity matrix. Since $A - \alpha I$ has $\pm \beta i$ as eigenvalues, we can use the argument just given to obtain

$$B(A - \alpha I)B^{-1} = C,$$

with $B$ and $C$ as before. Thus

$$BAB^{-1} = \begin{bmatrix} \alpha & -\beta \\ \beta & \alpha \end{bmatrix},$$

and the transformed system is

$$\dot{x}^* = \alpha x^* - \beta y^*$$
$$\dot{y}^* = \beta x^* + \alpha y^*.$$

Using the functions

$$\mathbf{X}_1^*(t) = \begin{bmatrix} e^{\alpha t} \cos \beta t \\ e^{\alpha t} \sin \beta t \end{bmatrix} \quad \text{and} \quad \mathbf{X}_2^*(t) = \begin{bmatrix} e^{\alpha t} \sin \beta t \\ -e^{\alpha t} \cos \beta t \end{bmatrix}$$

as a basis for the solution space, we have

$$x^* = e^{\alpha t}(c_1 \cos \beta t + c_2 \sin \beta t)$$
$$y^* = e^{\alpha t}(c_1 \sin \beta t - c_2 \cos \beta t).$$

These equations define a family of spirals in the $x^*y^*$-plane which diverge from the origin with increasing $t$ if $\alpha > 0$ and converge toward the origin if $\alpha < 0$. In this case the origin is a **focal point** and is unstable or asymptotically stable depending on the sign of $\alpha$. (See Fig. 13–9.)

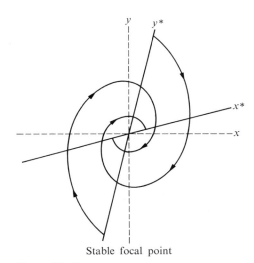

Stable focal point

**Figure 13–9**

CASE 3 ($\lambda_1$ *and* $\lambda_2$ *real and equal*). Here one of two things can happen: either $A = \lambda_1 I$ or $A \neq \lambda_1 I$. In the first case, the system is

$$\dot{x} = \lambda_1 x,$$
$$\dot{y} = \lambda_1 y,$$

and its coefficient matrix is already as simple as one could wish. The functions

$$\mathbf{X}_1(t) = \begin{bmatrix} e^{\lambda_1 t} \\ 0 \end{bmatrix}, \qquad \mathbf{X}_2(t) = \begin{bmatrix} 0 \\ e^{\lambda_1 t} \end{bmatrix}$$

are a basis for the solution space of the system, and the trajectories are rays about the origin as shown in Fig. 13–10. This is the case of a **star point** encountered earlier.

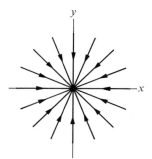

Asymptotically stable star point

**Figure 13–10**

When $A \neq \lambda_1 I$, the situation is considerably more complicated. Without going into the details, we assert that in this case $B$ can always be chosen so that

$$BAB^{-1} = \begin{bmatrix} \lambda_1 & 1 \\ 0 & \lambda_1 \end{bmatrix}.$$

When this is done, we obtain the system

$$\dot{x}^* = \lambda_1 x^* + y^*$$
$$\dot{y}^* = \lambda_1 y^*,$$

whose general solution is

$$x^* = (c_1 + c_2 t)e^{\lambda_1 t}$$
$$y^* = c_2 e^{\lambda_1 t}.$$

The trajectories are now of the type suggested in Fig. 13–11, and the origin is again a node, asymptotically stable when $\lambda_1 < 0$ and unstable when $\lambda_1 > 0$.

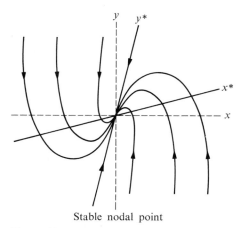

Stable nodal point

**Figure 13–11**

## EXERCISES

1. a) Discuss the stability of the equilibrium points for the plane autonomous system $\dot{\mathbf{X}} = A\mathbf{X}$ when $A$ is the $2 \times 2$ zero matrix.
   b) Discuss the more general $2 \times 2$ case where zero is the only eigenvalue.

2. a) Show that the matrix

$$A = \begin{bmatrix} a & b \\ c & d \end{bmatrix}$$

has eigenvalues $\lambda_1 = 0$, $\lambda_2 \neq 0$ if and only if $ad - bc = 0$ and $a + d \neq 0$.
   b) Discuss the stability of the equilibrium points and describe the solution curves of the plane autonomous system $\dot{\mathbf{X}} = A\mathbf{X}$ when $A$ has eigenvalues $\lambda_1 = 0$, $\lambda_2 \neq 0$.

3. Discuss the stability of the equilibrium points for each of the following systems. (When the coefficient matrix is singular, use the results of Exercises 1 and 2.)
   a) $\dot{x} = x + y$          b) $\dot{x} = x - 3y$          c) $\dot{x} = 4x - 6y$
      $\dot{y} = x - 3y$            $\dot{y} = 4x - 6y$            $\dot{y} = 6x - 9y$
   d) $\dot{x} = x + y$          e) $\dot{x} = -x + y + z$     f) $\dot{x} = x + y + z$
      $\dot{y} = x - y$            $\dot{y} = -2y$               $\dot{y} = 2y$
                                   $\dot{z} = -3z$               $\dot{z} = 3z$

4. a) Show that if $A$ is nonsingular, then any trajectory for the plane autonomous system $\dot{\mathbf{X}} = A\mathbf{X}$ different from $X(t) \equiv 0$ can be mapped onto any other trajectory by a linear transformation whose matrix with respect to the standard basis in $\mathscr{R}^2$ is of the form

$$\begin{bmatrix} k & 0 \\ 0 & k \end{bmatrix},$$

where $k$ is real. (When $k$ is positive a linear transformation with matrix of this type is said to be a *similarity transformation* of $\mathscr{R}^2$.) What is the geometric meaning of this result?
   b) Is the statement in (a) true when $A$ is singular? [*Hint*: See Exercise 2.]

5. Let $A$ be a $2 \times 2$ matrix with distinct real eigenvalues $\lambda_1$ and $\lambda_2$. Find a nonsingular matrix $B$ such that

$$BAB^{-1} = \begin{bmatrix} \lambda_1 & 0 \\ 0 & \lambda_2 \end{bmatrix}.$$

6. a) Show that the matrix

$$A = \begin{bmatrix} a & b \\ c & -a \end{bmatrix}$$

has eigenvalues $\pm\beta i$, with $a^2 + bc + \beta^2 = 0$, provided that $a^2 + bc < 0$.
   b) With $\beta$ as above, show that the matrix

$$B = \begin{bmatrix} 0 & \beta \\ -c & a \end{bmatrix}$$

is nonsingular, find its inverse, and compute $BAB^{-1}$.

7. Verify that the functions

$$\mathbf{X}_1(t) = \begin{bmatrix} e^{\alpha t} \cos \beta t \\ e^{\alpha t} \sin \beta t \end{bmatrix} \quad \text{and} \quad \mathbf{X}_2(t) = \begin{bmatrix} -e^{\alpha t} \sin \beta t \\ -e^{\alpha t} \cos \beta t \end{bmatrix}$$

are a basis for the solution space of

$$\dot{x} = \alpha x - \beta y$$
$$\dot{y} = \beta x + \alpha y.$$

8. Let $\dot{\mathbf{X}} = A\mathbf{X}$ be a plane autonomous system whose coefficient matrix

$$A = \begin{bmatrix} a & b \\ c & d \end{bmatrix}$$

has $\lambda_1 \neq 0$ as a real eigenvalue of multiplicity two, and let

$$\begin{bmatrix} A_1 \\ B_1 \end{bmatrix}$$

be an eigenvector belonging to $\lambda_1$. Apply the theory of Chapter 10 to show that the solution space of this system has a basis of the form

$$\mathbf{X}_1(t) = \begin{bmatrix} A_1 \\ B_1 \end{bmatrix} e^{\lambda_1 t}, \qquad \mathbf{X}_2(t) = \begin{bmatrix} A_2 t + A_3 \\ B_2 t + B_3 \end{bmatrix} e^{\lambda_1 t}.$$

9. Convert the constant-coefficient equation $y'' + ay' + by = 0$ into a plane autonomous system and discuss the trajectories and stability for this system.

## 13–5 STABILITY FOR NONLINEAR SYSTEMS; LIAPUNOV FUNCTIONS

In this section we return to the general case of an autonomous system

$$\dot{\mathbf{X}} = F(\mathbf{X}) \tag{13–14}$$

defined in region $\Omega$ of $\mathscr{R}^n$ with an equilibrium point $\mathbf{X}_0$ in $\Omega$. As usual, we demand that initial-value problems for this system have unique solutions everywhere in $\Omega$. This can be ensured, for instance, by requiring $F$ to have continuous first partial derivatives everywhere in $\Omega$ or, as we shall say, to be of *class* $\mathscr{C}^1$ in $\Omega$. Although the condition is stronger than it need be, we shall assume that it is in force from now on since it will be used later and, in any event, is not unduly restrictive.

To simplify notation somewhat, we shall also assume that the equilibrium point $\mathbf{X}_0$ for (13–14) is at the origin $\mathbf{0}$ of $\mathscr{R}^n$. By making the change of variable $\mathbf{X}^* = \mathbf{X} - \mathbf{X}_0$ (i.e., by translating the origin to $\mathbf{X}_0$) this can always be done without changing the form of (13–14). Thus, without loss of generality, we can limit our study to stability at the origin.

The technique for determining the behavior of the trajectories of (13–14) about the origin when $F(\mathbf{0}) = \mathbf{0}$ was devised by the Russian mathematician A. A. Liapunov, and is known after him as **Liapunov's direct** (or **second**) **method**. It is based on the well-known fact that a physical system loses potential energy in a neighborhood of a point of stable equilibrium. More precisely, a point of stable equilibrium for a physical system is a point at which the potential energy of the system has a local minimum—a fact which is known in physics as Lagrange's theorem. Accordingly, we seek a real-valued function (an energy function) $E(\mathbf{X}) = E(x_1, \ldots, x_n)$ of class $\mathscr{C}^1$ in $\Omega$, or some subregion of $\Omega$ containing the origin, such that $E(\mathbf{X}) \geqslant 0$, and $E(\mathbf{X}) = 0$ if and only if $\mathbf{X} = \mathbf{0}$. Then the surfaces $E(\mathbf{X}) = c$, $c$ a positive constant, will in general be closed surfaces surrounding the origin in $\mathscr{R}^n$. Moreover, if $\mathbf{X} = X(t, \mathbf{X}_0)$ is the solution of

$$\dot{\mathbf{X}} = F(\mathbf{X}), \qquad \mathbf{X}(0) = \mathbf{X}_0,$$

with $\mathbf{X}_0$ in $\Omega$, the time rate of change of energy, $\partial E/\partial t$, along the trajectory defined by this solution is*

$$\mathbf{V}E \cdot \dot{\mathbf{X}} = \mathbf{V}E \cdot \mathbf{F} = \sum_{i=1}^{n} \frac{\partial E}{\partial x_i} F_i.$$

Thus if this expression is *negative* in $\Omega$, a particle moving along any of the trajectories defined by the system $\dot{\mathbf{X}} = F(\mathbf{X})$ will be dissipating energy as it intersects the surface $E(\mathbf{X}) = c$. Consequently a particle which enters the region enclosed by such a surface can never gather the energy needed to escape, and its trajectory remains in that region. This clearly implies that the origin will be stable, and even asymptotically stable, for the given system.

Guided by these remarks, we now give

**Definition 13–3**   *Let $\Omega$ be a region of $\mathscr{R}^n$ containing the origin, and let $E = E(\mathbf{X})$ be a real-valued function of class $\mathscr{C}^1$ in $\Omega$. Then $E$ is said to be* **positive definite** *if*

   i)  $E(\mathbf{X}) \geqslant 0$ *for all $\mathbf{X}$ in $\Omega$, and*

   ii)  $E(\mathbf{X}) = 0$ *if and only if $\mathbf{X} = \mathbf{0}$.*

*If, in addition,*

   iii)  $\mathbf{V}E \cdot \mathbf{F} = \displaystyle\sum_{i=1}^{n} \frac{\partial E}{\partial x_i} F_i \leqslant 0$

*everywhere in $\Omega$, then $E$ is said to be a* **Liapunov function** *for the autonomous system $\dot{\mathbf{X}} = F(\mathbf{X})$.*

---

* The symbol $\mathbf{V}$ (read "del") denotes the vector operator

$$\mathbf{V} = \frac{\partial}{\partial x_1} \mathbf{e}_1 + \cdots + \frac{\partial}{\partial x_n} \mathbf{e}_n.$$

Here $\{\mathbf{e}_1, \ldots, \mathbf{e}_n\}$ is the standard basis in $\mathscr{R}^n$, while $\mathbf{V}E \cdot \mathbf{F}$ is the usual inner or dot product of the vectors $\mathbf{V}E$ and $\mathbf{F}$, where $\mathbf{F} = (F_1, \ldots, F_n) = F_1\mathbf{e}_1 + \cdots + F_n\mathbf{e}_n$.

In these terms the first major theorem on stability reads as follows.

**Theorem 13–2**   *The origin is a point of stable equilibrium for*

$$\dot{\mathbf{X}} = F(\mathbf{X}), \qquad F(\mathbf{X}) = 0$$

*if there exists a Liapunov function E for the system.*

*Proof.*   Let $S_r$ denote the sphere of radius $r$ about the origin in $\mathscr{R}^n$ (i.e., $S_r$ consists of all points $\mathbf{X}$ in $\mathscr{R}^n$ such that $\|\mathbf{X}\| = r$), and let $r$ be chosen so that $S_r$ lies in $\Omega$. Then since $E$ is continuous and positive definite in $\Omega$, it assumes a *positive* minimum $m$ on $S_r$.* Appealing to the continuity of $E$ again, we can find an $\varepsilon > 0$ such that $E(\mathbf{X}) < m$ whenever $\mathbf{X}$ is within or on the surface of the sphere $S_\varepsilon$ of radius $\varepsilon$ about $0$. (See Fig. 13–12, drawn for $n = 2$.)

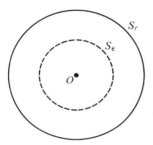

**Figure 13–12**

Now let $\mathbf{X}_0$ be any point within $S_\varepsilon$ and let $\mathbf{X}(t, \mathbf{X}_0)$ be the unique solution of $\dot{\mathbf{X}} = F(\mathbf{X})$ whose trajectory passes through $\mathbf{X}_0$ when $t = 0$. We assert that this trajectory remains within $S_r$ for all $t > 0$. For otherwise, $\|\mathbf{X}(t_1, \mathbf{X}_0)\| = r$ for some $t_1 > 0$; whence $E[\mathbf{X}(t_1, \mathbf{X}_0)] \geqslant m$. But this contradicts the assumption that $\partial E/\partial t = \nabla E \cdot \mathbf{F} \leqslant 0$ in $\Omega$, since this latter inequality implies that $E$ is a nonincreasing function of $t$ along the trajectory in question.   ∎

The argument used above suggests that if $E$ actually *decreased* along every trajectory of $\dot{\mathbf{X}} = F(\mathbf{X})$ in a neighborhood of the origin, these trajectories would have to spiral in toward the origin as required of asymptotic stability. Our next theorem shows that this is in fact the case.

**Theorem 13–3**   *If E is a Liapunov function for $\dot{\mathbf{X}} = F(\mathbf{X})$ with the property that $-\nabla E \cdot \mathbf{F}$ is positive definite in $\Omega$, then the origin is asymptotically stable.*

*Proof.*   We have seen that there exists an $\varepsilon > 0$ with the property that $E$ is a nonincreasing function along any trajectory $\mathbf{X}(t, \mathbf{X}_0)$ with $\|\mathbf{X}_0\| < \varepsilon$. Since $E$ is

---

* Any continuous real-valued function defined on a closed bounded set (such as $S_r$) in $\mathscr{R}^n$ assumes a maximum and minimum on that set.

positive definite everywhere in $\Omega$, $E$ must approach a nonnegative limit $E_0$ along this trajectory at $t \to \infty$. Thus we will be done if we can show that this limit is zero when $-\mathbf{V}E \cdot \mathbf{F}$ is positive definite in $\Omega$, for then the fact that $E$ vanishes only at the origin will imply that the trajectory in question must approach the origin with increasing $t$.

Suppose, to the contrary, that $E_0$ were positive. Then as we argued before, there would exist $\delta > 0$ such that $E(\mathbf{X}) < E_0$ for all $\mathbf{X}$ with $\|\mathbf{X}\| < \delta$. Hence the trajectory defined by $\mathbf{X}(t, \mathbf{X}_0)$ would never enter the sphere $S_\delta$ of radius $\delta$ about the origin. Now let $m$ denote the minimum value assumed by $-\mathbf{V}E \cdot \mathbf{F}$ in the "annular" region $\delta \leqslant \|\mathbf{X}\| \leqslant r$, where $r$ is as in the proof of the preceding theorem. Since $-\mathbf{V}E \cdot \mathbf{F}$ is positive definite, we have $m > 0$. Thus

$$\frac{\partial}{\partial t} E[\mathbf{X}(t, \mathbf{X}_0)] \leqslant -m$$

for all $t \geqslant 0$, and

$$E[\mathbf{X}(t, \mathbf{X}_0)] - E(\mathbf{X}_0) = \int_0^t \frac{\partial}{\partial t} E[\mathbf{X}(t, \mathbf{X}_0)] \, dt \leqslant -mt.$$

But then the right-hand member of the inequality $E[\mathbf{X}(t, \mathbf{X}_0)] \leqslant E(\mathbf{X}_0) - mt$ would become negative as $t \to \infty$. Hence the assumption $E_0 > 0$ is untenable, and we are done. ∎

The next theorem, whose proof we omit, asserts that these results are, in a sense, the best possible.

**Theorem 13–4**   *Let $E$ be a real-valued function of class $\mathscr{C}^1$ in $\Omega$, and suppose that*

   i) *$E(\mathbf{0}) = \mathbf{0}$,*

   ii) *$\mathbf{V}E \cdot \mathbf{F}$ is positive definite in $\Omega$, and*

   iii) *for each $\varepsilon > 0$ there exists an $\mathbf{X}_0$ in $\Omega$ with $\|\mathbf{X}_0\| < \varepsilon$ and $E(\mathbf{X}_0) > 0$ (that is, E assumes positive values in every neighborhood of the origin).*

   *Then the origin is unstable for the system $\dot{\mathbf{X}} = F(\mathbf{X})$, $F(\mathbf{0}) = \mathbf{0}$.*

The important feature of the theorems of this section is that they enable us to determine the stability or instability of equilibrium points for an autonomous system *without actually solving the system*. Thus in cases where the solutions are impossible to obtain in closed form or are difficult to analyze, we can still obtain valuable information on their stability provided, of course, that we can construct their Liapunov functions. Although there is no general method for constructing these functions, we shall see that it is fairly easy to do in many interesting cases.

For instance, the function $E(x, y) = x^2 + y^2$ is a Liapunov function for the system

$$\begin{aligned} \dot{x} &= -x + x^2 y \\ \dot{y} &= -y + xy^2. \end{aligned} \tag{13–15}$$

Indeed, $E$ is clearly positive definite and of class $\mathscr{C}^1$ in the entire $xy$-plane More-over, since

$$\mathbf{\nabla}E = \frac{\partial E}{\partial x}\,\mathbf{e}_1 + \frac{\partial E}{\partial y}\,\mathbf{e}_2 = 2x\mathbf{e}_1 + 2y\mathbf{e}_2$$

and

$$\mathbf{F}(x, y) = (-x + x^2 y)\mathbf{e}_1 + (-y + xy^2)\mathbf{e}_2,$$

we have

$$\begin{aligned}
\mathbf{\nabla}E \cdot \mathbf{F} &= -2(x^2 + y^2) + 2x^3 y + 2xy^3 \\
&= 2(x^2 + y^2)(xy - 1).
\end{aligned}$$

Hence $-\mathbf{\nabla}E \cdot \mathbf{F}$ is positive definite in the region $xy < 1$, and $E$ satisfies all of the requirements of Theorem 13–3 in this region. Thus the origin is an asymptotically stable equilibrium point for (13–15).

More generally, let $f(x, y)$ and $g(x, y)$ be functions of class $\mathscr{C}^1$ which vanish at the origin, and consider the system

$$\begin{aligned}
\dot{x} &= -x + f(x, y) \\
\dot{y} &= -y + g(x, y).
\end{aligned} \tag{13–16}$$

Then with $E$ as before,

$$\mathbf{\nabla}E \cdot \mathbf{F} = -2(x^2 + y^2) + 2xf(x, y) + 2yg(x, y).$$

Now suppose that the values of $f$ and $g$ are so small in a neighborhood $\Omega$ of the origin that the sign of $\mathbf{\nabla}E \cdot \mathbf{F}$ is determined by the first term in this expression. Then $-\mathbf{\nabla}E \cdot \mathbf{F}$ will be positive definite in $\Omega$, and the origin will again be asymptotically stable. This will always happen, for instance, whenever $f$ and $g$ have convergent power series expansions about $(0, 0)$ which only involve terms of degree two or higher.

## 13–6 LIAPUNOV FUNCTIONS

We shall begin this section by constructing a Liapunov function for the constant-coefficient linear autonomous system

$$\dot{\mathbf{X}} = A\mathbf{X}, \tag{13–17}$$

under the assumption that all of the eigenvalues of $A$ have negative real parts. The reason for undertaking this construction is not that it will tell us anything about such systems we do not already know but, rather, that it will allow us to extend some of our earlier results to nonlinear systems. The construction goes as follows.

Let $\mathbf{e}_1, \ldots, \mathbf{e}_n$ denote the standard basis vectors in $\mathscr{R}^n$ and, for each $i$, $1 \leqslant i \leqslant n$, let

$$\mathbf{X}_i(t) = \mathbf{X}_i(t, \mathbf{e}_i) = (x_{i1}(t), \ldots, x_{in}(t))$$

denote *the* solution of (13–17) which satisfies the initial condition $\mathbf{X}_i(0) = \mathbf{e}_i$.

Then if $\mathbf{Y} = y_1 \mathbf{e}_1 + \cdots + y_n \mathbf{e}_n$ is any vector in $\mathscr{R}^n$, the function

$$\mathbf{X}(t, \mathbf{Y}) = y_1 \mathbf{X}_1(t) + \cdots + y_n \mathbf{X}_n(t)$$

is *the* solution of (13–17) which satisfies the initial condition $\mathbf{X}(0) = \mathbf{Y}$. We now set

$$E(\mathbf{Y}) = \int_0^\infty \|\mathbf{X}(t, \mathbf{Y})\|^2 \, dt, \tag{13–18}$$

and note that *if* this integral converges, then

i) $E(\mathbf{Y}) \geqslant 0$, and

ii) $E(\mathbf{Y}) = 0$ if and only if $\mathbf{Y} = \mathbf{0}$. (Why?)

Next, the definition of $\mathbf{X}(t, \mathbf{Y})$ implies that

$$E(\mathbf{Y}) = \int_0^\infty \left\| \sum_{i=1}^n y_i X_i(t) \right\|^2 dt = \int_0^\infty \left\| \sum_{i=1}^n (y_i x_{i1}(t), \ldots, y_i x_{in}(t)) \right\|^2 dt$$

$$\int_0^\infty \left\| \left( \sum_{i=1}^n y_i x_{i1}(t), \ldots, \sum_{i=1}^n y_i x_{in}(t) \right) \right\|^2 dt.$$

But

$$\left\| \left( \sum_{i=1}^n y_i x_{i1}(t), \ldots, \sum_{i=1}^n y_i x_{in}(t) \right) \right\|^2 = \left( \sum_{i=1}^n y_i x_{i1}(t) \right)^2 + \cdots + \left( \sum_{i=1}^n y_i x_{in}(t) \right)^2$$

$$= \sum_{k=1}^n \left( \sum_{i=1}^n y_i x_{ik}(t) \right)^2$$

$$= \sum_{k=1}^n \left[ \left( \sum_{i=1}^n y_i x_{ik}(t) \right) \left( \sum_{j=1}^n y_j x_{jk}(t) \right) \right]$$

$$= \sum_{i,j=1}^n y_i y_j \left[ \sum_{k=1}^n x_{ik}(t) x_{jk}(t) \right].$$

Hence

$$E(\mathbf{Y}) = \sum_{i,j=1}^n y_i y_j \int_0^\infty \left[ \sum_{k=1}^n x_{ik}(t) x_{jk}(t) \right] dt,$$

and to prove that $E(\mathbf{Y})$ is defined for all $\mathbf{Y}$ in $\mathscr{R}^n$, we merely have to show that the integrals

$$\int_0^\infty x_{ik}(t) x_{jk}(t) \, dt$$

converge. This convergence, however, is an immediate consequence of the assumption that the eigenvalues of $A$ have negative real parts, since then each of the

$x_{pq}(t)$ has a factor $e^{\alpha t}$ with $\alpha < 0$ and no other factors save those of the form $t^k$, $\sin \beta t$, or $\cos \beta t$.

We now show that $E$ is in fact a Liapunov function for (13–17). To this end we compute the value of $E[\mathbf{X}(t, \mathbf{X}_0)]$, where $\mathbf{X}(t, \mathbf{X}_0)$ is the solution of $\dot{\mathbf{X}} = A\mathbf{X}$ that satisfies $\mathbf{X}(0, \mathbf{X}_0) = \mathbf{X}_0$. By (13–18), we have

$$E[\mathbf{X}(t, \mathbf{X}_0)] = \int_0^\infty \|\mathbf{X}[s, \mathbf{X}(t, \mathbf{X}_0)]\|^2 \, ds = \int_0^\infty \|\mathbf{X}(s + t, \mathbf{X}_0)\|^2 \, ds$$

$$= \int_t^\infty \|\mathbf{X}(u, \mathbf{X}_0)\|^2 \, du.$$

Hence

$$\frac{\partial}{\partial t} E[\mathbf{X}(t, \mathbf{X}_0)] = \frac{\partial}{\partial t} \int_t^\infty \|\mathbf{X}(u, \mathbf{X}_0)\|^2 \, du$$

$$= -\|\mathbf{X}(t, \mathbf{X}_0)\|^2,$$

as required. [Recall that in general $\mathbf{V}E \cdot \mathbf{F} = \partial E/\partial t$ along the trajectories of $\dot{\mathbf{X}} = F(\mathbf{X})$.] Finally, since the real-valued function $\|\mathbf{X}\|^2$ is positive definite on $\mathscr{R}^n$, $E$ satisfies all the hypotheses of Theorem 13–3, and we again have proved that the origin is a point of asymptotic stability for (13–17).

For instance, the system

$$\dot{x} = -x$$
$$\dot{y} = -2y$$

has coefficient matrix

$$\begin{bmatrix} -1 & 0 \\ 0 & -2 \end{bmatrix}$$

and eigenvalues $-1$, $-2$. In this case the solutions $\mathbf{X}_1(t)$, $\mathbf{X}_2(t)$ such that $\mathbf{X}_1(0) = \mathbf{e}_1$, $\mathbf{X}_2(0) = \mathbf{e}_2$ are

$$\mathbf{X}_1(t) = \begin{bmatrix} e^{-t} \\ 0 \end{bmatrix}, \qquad \mathbf{X}_2(t) = \begin{bmatrix} 0 \\ e^{-2t} \end{bmatrix}.$$

Thus, if $\mathbf{Y} = x\mathbf{e}_1 + y\mathbf{e}_2$, then

$$E(\mathbf{Y}) = \int_0^\infty \|xe^{-t}\mathbf{e}_1 + ye^{-2t}\mathbf{e}_2\|^2 \, dt$$

$$= \int_0^\infty (x^2 e^{-2t} + y^2 e^{-4t}) \, dt$$

$$= \frac{x^2}{2} + \frac{y^2}{4}.$$

Before this construction can be used to produce Liapunov functions for nonlinear systems, we must introduce the notion of a linear approximation to a vector function $F$. Specifically, we shall show that whenever $F$ is a function of class $\mathscr{C}^1$ in a region $\Omega$ of $\mathscr{R}^n$ containing the origin, and taking values in $\mathscr{R}^n$, we may replace $F$ by a function of the form

$$JX + G(X),$$

where $J$ is an $n \times n$-matrix and $G(X)$ is "small" in comparison with $JX$ when $\|X\|$ is small. To make these ideas precise, we need the following definition.

**Definition 13–4** *Let $F$ be as above and let $X_0$ be a point in $\Omega$. Then a function $L = L(X)$ is said to be a **linear approximation** to $F$ at $X_0$ if $L$ is linear on $\mathscr{R}^n$ and*

$$\lim_{\|X - X_0\| \to 0} \frac{[F(X) - F(X_0)] - [L(X) - L(X_o)]}{\|X - X_0\|} = 0. \qquad (13\text{–}19)$$

Geometrically, this definition asserts that by choosing a sphere of sufficiently small radius about $X_0$ we can make the difference between the norms of the values of $L$ and $F$ arbitrarily small everywhere within that sphere.

Note that since $L$ is linear, $(13\text{–}19)$ can also be written as

$$\lim_{\|X - X_0\| \to 0} \frac{F(X) - F(X_0) - L(X - X_0)}{\|X - X_0\|} = 0.$$

In particular, when $X_0 = 0$ and $F(0) = 0$, we have

$$\lim_{\|X\| \to 0} \frac{F(X) - L(X)}{\|X\|} = 0. \qquad (13\text{–}20)$$

Our next result shows that *if $F$ admits a linear approximation at $X_0$*, then that approximation is unique.

**Lemma 13–1** *If $L_1$ and $L_2$ are linear approximations to $F$ at $X_0$, then $L_1 = L_2$.*

*Proof.* By translating the origin of coordinates to $X_0$, if necessary, we may assume that $X_0 = 0$. In that case $(13\text{–}20)$ yields

$$\lim_{\|X\| \to 0} \frac{F(X) - L_1(X)}{\|X\|} = 0, \qquad \lim_{\|X\| \to 0} \frac{F(X) - L_2(X)}{\|X\|} = 0.$$

But by the triangle inequality,

$$\left\| L_1(X) - L_2(X) \right\| \leqslant \left\| L_1(X) - F(X) \right\| + \left\| F(X) - L_2(X) \right\|$$

or

$$\left\| (L_1 - L_2)(X) \right\| \leqslant \left\| L_1(X) - F(X) \right\| + \left\| F(X) - L_2(X) \right\|,$$

and it follows that

$$\lim_{\|X\| \to 0} \frac{(L_1 - L_2)(X)}{\|X\|} = 0.$$

In particular, this result holds when $X$ approaches the origin along *any* ray in $\mathscr{R}^n$ emanating from the origin. But in that case $X/\|X\|$ is a constant vector, and the linearity of $L_1 - L_2$ implies that $L_1 = L_2$ everywhere on that ray. ∎

This lemma entitles us to speak of *the* linear approximation to $F$ at $\mathbf{X}_0$ provided that one exists. Our next theorem guarantees the existence of this approximation whenever $F$ is of class $\mathscr{C}^1$.

***Theorem 13–5***    *Let $F(\mathbf{X}) = (\mathbf{F}_1(\mathbf{X}), \ldots, F_n(\mathbf{X}))$ be of class $\mathscr{C}^1$ in a region $\Omega$ of $\mathscr{R}^n$. Then $F$ has a linear approximation at every point $\mathbf{X}_0$ in $\Omega$, and the matrix of this approximation with respect to the standard basis in $\mathscr{R}^n$ is*

$$J = \begin{bmatrix} \dfrac{\partial F_1}{\partial x_1} \cdots \dfrac{\partial F_1}{\partial x_n} \\ \cdot \quad \cdot \\ \cdot \quad \cdot \\ \cdot \quad \cdot \\ \dfrac{\partial F_n}{\partial x_1} \cdots \dfrac{\partial F_n}{\partial x_n} \end{bmatrix}_{\mathbf{X}=\mathbf{X}_0}.$$

(The subscript $\mathbf{X} = \mathbf{X}_0$ means that each of the derivatives appearing in this matrix is evaluated at the point $\mathbf{X}_0 = (x_1^0, \ldots, x_n^0)$.)

*Remark.*    The matrix $J$ is known as the **Jacobian matrix** of $F$ at $\mathbf{X}_0$, and the linear approximation to $F$ at $\mathbf{X}_0$ is called the **differential** of $F$ at $\mathbf{X}_0$.

*Proof.*    Let $L: \mathscr{R}^n \to \mathscr{R}^n$ be the linear transformation defined by $J$. To prove the theorem, we must show that

$$\lim_{\|\mathbf{X}-\mathbf{X}_0\| \to 0} \frac{F(\mathbf{X}) - F(\mathbf{X}_0) - L(\mathbf{X} - \mathbf{X}_0)}{\|\mathbf{X} - \mathbf{X}_0\|} = \mathbf{0}.$$

To do so it suffices to show that the absolute value of each component of the vector in this expression tends to zero in the limit; that is,

$$\lim_{\|\mathbf{X}-\mathbf{X}_0\| \to 0} \frac{1}{\|\mathbf{X} - \mathbf{X}_0\|}$$
$$\times \left[ F_k(x_1, \ldots, x_n) - F_k(x_1^0, \ldots, x_n^0) - \sum_{i=1}^{n} \frac{\partial F_k(x_1^0, \ldots, x_n^0)}{\partial x_i}(x_i - x_i^0) \right] = 0$$

$$(13\text{–}21)$$

for $1 \leqslant k \leqslant n$. But

$$F_k(x_1, \ldots, x_n) - F_k(x_1^0, \ldots, x_n^0)$$
$$= [F_k(x_1, x_2, \ldots, x_n) - F_k(x_1^0, x_2, \ldots, x_n)]$$
$$+ [F_k(x_1^0, x_2, \ldots, x_n) - F_k(x_1^0, x_2^0, \ldots, x_n)] + \cdots$$
$$+ [F_k(x_1^0, \ldots, x_{n-1}^0, x_n) - F_k(x_1^0, \ldots, x_{n-1}^0, x_n^0)].$$

When we apply the mean-value theorem to each bracketed term in this expression we obtain

$$F_k(x_1, \ldots, x_n) - F_k(x_1^0, \ldots, x_n^0)$$

$$= \sum_{i=1}^{n} (x_i - x_i^0) \frac{\partial F_k[x_1^0, \ldots, x_i^0 + \theta_i(x_i - x_i^0), x_{i+1}, \ldots, x_n]}{\partial x_i},$$

where $0 < \theta_i < 1$ for all $i$. For simplicity we rewrite this equality as

$$F_k(x_1, \ldots, x_n) - F_k(x_1^0, \ldots, x_n^0) = \sum_{i=1}^{n} (x_i - x_i^0) \frac{\partial F_k[\ldots \theta_i \ldots]}{\partial x_i},$$

and substitute in (13–21) to get

$$\frac{1}{\|\mathbf{X} - \mathbf{X}_0\|} \sum_{i=1}^{n} (x_i - x_i^0) \left[ \frac{\partial F_k[\ldots \theta_i \ldots]}{\partial x_i} - \frac{\partial F_k(x_1^0, \ldots, x_n^0)}{\partial x_i} \right].$$

But the absolute value of this quantity does not exceed

$$\sum_{i=1}^{n} \left| \frac{\partial F_k[\cdots \theta_i \cdots]}{\partial x_i} - \frac{\partial F_k(x_1^0, \ldots, x_n^0)}{\partial x_i} \right|,$$

which approaches 0 as $\|\mathbf{X} - \mathbf{X}_0\| \to 0$ since $\partial F_k / \partial x_i$ is continuous.  ∎

To apply these results to the study of stability, let $F$ be a function of class $\mathscr{C}^1$ in a region $\Omega$ of $\mathscr{R}^n$ containing the origin, and suppose that

i)  $F(0) = 0$, and

ii) the real parts of all of the eigenvalues of the Jacobian matrix $J$ of $F$ are negative when evaluated at the origin.

We now show that under these conditions the Liapunov function $E = E(\mathbf{X})$ for the constant-coefficient linear system $\dot{\mathbf{X}} = J\mathbf{X}$ that was constructed earlier in this section is also a Liapunov function for the (nonlinear) system

$$\dot{\mathbf{X}} = F(\mathbf{X}).$$

Indeed, by our earlier results we know that $\nabla E \cdot (J\mathbf{X}) \leqslant -\|\mathbf{X}\|^2$. Hence

$$\nabla E(\mathbf{X}) \cdot F(\mathbf{X}) = \nabla E(\mathbf{X}) \cdot [J\mathbf{X} + G(\mathbf{X})]$$

$$= \nabla E(\mathbf{X}) \cdot J\mathbf{X} + \nabla E(\mathbf{X}) \cdot G(\mathbf{X})$$

$$\leqslant -\|\mathbf{X}\|^2 + \|\nabla E(\mathbf{X})\| \, \|G(\mathbf{X})\|.^*$$

---

\* The inequality $\nabla E(\mathbf{X}) \cdot G(\mathbf{X} \leqslant \|\nabla E(\mathbf{X})\| \, \|G(\mathbf{X})\|$ is a consequence of Cauchy's inequality, which states that for real numbers $p_1, \ldots, p_n$ and $q_1, \ldots, q_n$,

$$\left( \sum_{i=1}^{n} p_i q_i \right)^2 \leqslant \left( \sum_{i=1}^{n} p_i^2 \right) \left( \sum_{i=1}^{n} q_i^2 \right).$$

But by Exercise 7, $\nabla E(\mathbf{X}) = 2J\mathbf{X}$ and $\|J\mathbf{X}\| \leqslant k\|\mathbf{X}\|$ for some constant $k$ and and $\mathbf{X}$ in $\mathscr{R}^n$. Thus $\|\nabla E(\mathbf{X})\| \leqslant 2k\|\mathbf{X}\|$, and

$$\nabla E(\mathbf{X}) \cdot F(\mathbf{X}) \leqslant -\|\mathbf{X}\|^2 + 2k\|\mathbf{X}\|\,\|G(\mathbf{X})\|$$

We now use the fact that

$$\lim_{\|\mathbf{X}\| \to 0} \frac{F(\mathbf{X}) - J\mathbf{X}}{\|\mathbf{X}\|} = \lim_{\|\mathbf{X}\| \to 0} \frac{G(\mathbf{X})}{\|\mathbf{X}\|} = 0$$

to find a sphere $S_\delta$ of radius $\delta$ about the origin in $\mathscr{R}^n$ such that

$$\|G(\mathbf{X})\| \leqslant \frac{1}{4k}\|\mathbf{X}\|$$

for all $\mathbf{X}$ is $S_\delta$. Then within this sphere we have

$$\nabla E(\mathbf{X}) \cdot F(\mathbf{X}) \leqslant -\|\mathbf{X}\|^2 + (2k\|\mathbf{X}\|)\left(\frac{1}{4k}\|\mathbf{X}\|\right) = -\frac{1}{2}\|\mathbf{X}\|^2,$$

so that $E$ is indeed a Liapunov function for the system $\dot{\mathbf{X}} = F(\mathbf{X})$. Moreover, since this inequality also implies that $-\nabla E \cdot F$ is positive definite in $S_\delta$, the hypotheses of Theorem 13–3 are satisfied, and we have

> **Theorem 13–6**  Let $\mathbf{F} = F(\mathbf{X})$ be of class $\mathscr{C}^1$ in a region of $\mathscr{R}^n$ containing the origin, and suppose that $F(\mathbf{0}) = \mathbf{0}$. Then the origin is asymptotically stable for the system $\dot{\mathbf{X}} = F(\mathbf{X})$ whenever all of the eigenvalues of the Jacobian matrix of $F$ evaluated at the origin have negative real parts.

As before, there is a companion result when all of the eigenvalues of the Jacobian matrix of $\mathbf{F}$ have positive real parts. In that case the origin is a point of complete instability for $\dot{\mathbf{X}} = F(\mathbf{X})$ in the sense that there exists a spherical region about $\mathbf{X} = \mathbf{0}$ with the property that *all* trajectories originating in the region (except the trajectory $\mathbf{X}(t) \equiv \mathbf{0}$) eventually leave it never to return.

The results above show that the "linear part" of $\mathbf{F}$ is sometimes decisive for stability. The following example shows that unfortunately this is not always the case.

Consider the plane autonomous system

$$\dot{x} = y - xf(x, y)$$
$$\dot{y} = -x - yf(x, y),$$

where $f$ has a convergent power series expansion in a neighborhood of the origin in $\mathscr{R}^2$ and $f(0, 0) = \mathbf{0}$. Set $E(x, y) = x^2 + y^2$. Then

$$\nabla E \cdot \mathbf{F} = (2x, 2y) \cdot [(y - xf(x, y)), (-x - yf(x, y))]$$
$$= -2(x^2 + y^2)f(x, y).$$

Hence for this system the origin is

i) stable when $f(x, y) \geqslant 0$ in some neighborhood of $(0, 0)$;

ii) asymptotically stable when $f(x, y)$ is positive definite in a neighborhood of $(0, 0)$;

iii) unstable when $f(x, y) < 0$ in every neighborhood of $(0, 0)$.

In addition to showing that the linear part of **F** cannot always be used to settle questions of stability, this example illustrates some of the difficulties which arise when the Jacobian matrix of **F** admits pure imaginary eigenvalues. For results in this direction as well as an elementary treatment of some of the deeper problems of stability theory, such as estimating the extent of regions of asymptotic stability, etc., the reader should consult the excellent little book, *Stability by Liapunov's Direct Method with Applications*, by J. LaSalle and S. Lefschetz cited in the bibliography.

We now conclude this brief introduction to the subject of stability with two examples of a somewhat practical nature.

**Example 1**    In an earlier chapter we saw that the equation of motion for a mass $m$ bobbing on an elastic spring is $m\ddot{x} + kx = 0$ or, equivalently,

$$\dot{x} = y$$

$$\dot{y} = -\frac{k}{m}x,$$

where $k > 0$ is the spring constant and outside forces are neglected. This system describes what is known as a **linear spring**, so called because the restoring force is the linear function $G(x) = kx$.

We now consider the more general case of a restoring force $G$ which is odd $[G(-x) = -G(x)]$, and positive when $x$ is positive. The equations of motion then become

$$\dot{x} = y$$

$$\dot{y} = -\frac{1}{m}G(x), \tag{13–22}$$

and the work done against the restoring force to displace the spring $x$ units from equilibrium is

$$V(x) = \int_0^x G(t)\, dt.$$

This function gives the potential energy of the system, and since its kinetic energy is $(m/2)\dot{x}^2 = (m/2)y^2$, physical considerations suggest that

$$E(x, y) = \tfrac{1}{2}my^2 + V(x)$$

should be a Liapunov function for (13–22). This guess is indeed correct because $E$ is positive definite (recall that $G$ is odd and positive for $x > 0$), and

$$\mathbf{V}E \cdot \mathbf{F} = \frac{dV}{dx}y + my\left[-\frac{1}{m}G(x)\right] = yG(x) - yG(x) = 0.$$

Moreover, since $VE \cdot F$ is zero, the origin is stable but not asymptotically stable for this system. This, of course, agrees with the physics of the situation since all damping forces have been neglected.

*Remark.* The trajectories of (13–22) in the $xy$-plane are

$$y = 2[a^2 - V(x)]^{1/2}$$

(see Exercise 8). Notice the resemblance to the pendulum problem where $\theta$ is restricted to the interval $(-\pi/2, \pi/2)$.

**Example 2**  The differential equation

$$L\ddot{x} + R\dot{x} + \frac{1}{C}x + g(x, \dot{x}) = 0$$

governs the behavior of a simple *RLC* electric circuit when $x$ represents the charge on the capacitor, $\dot{x}$ the current in the loop, and $g(x, \dot{x})$ represents the nonlinearities in the loop, i.e.,

$$g(0, 0) = \frac{\partial g}{\partial x}(0, 0) = \frac{\partial g}{\partial \dot{x}}(0, 0) = 0.$$

The equivalent first-order system is

$$\dot{x} = y$$

$$\dot{y} = -\frac{1}{LC}x - \frac{R}{L}y - \frac{1}{L}g(x, y),$$

and the linear approximation is

$$\begin{bmatrix} \dot{x} \\ \dot{y} \end{bmatrix} = \begin{bmatrix} 0 & 1 \\ -\dfrac{1}{LC} & -\dfrac{R}{L} \end{bmatrix} \begin{bmatrix} x \\ y \end{bmatrix}.$$

The eigenvalues of the matrix of this system are

$$\lambda = \frac{1}{2}\left( -\frac{R}{L} \pm \sqrt{\frac{R^2}{L^2} - \frac{4}{LC}} \right),$$

and since they have negative real parts, Theorem 13–6 says that the origin is asymptotically stable.

**EXERCISES**

1. a) What is the approximating linear system for

$$\dot{x} = P_1(x, y)$$
$$\dot{y} = P_2(x, y),$$

when $P_1$ and $P_2$ are polynomials?

  b) What is the approximating linear system for

$$\dot{x} = F_1(x, y)$$
$$\dot{y} = F_2(x, y),$$

where $F_1$ and $F_2$ have convergent Taylor series expansions

$$F_1(x, y) = \sum a_{ij} x^i y^j,$$

$$F_2(x, y) = \sum b_{ij} x^i x^j$$

in a neighborhood of $(0, 0)$ which vanish at the origin?

2. Applying Exercise 1 to obtain approximating linear systems about $(0, 0)$ for each of the following.

   a) $\dot{x} = \sin x + e^y - 1$
      $\dot{y} = xy$

   b) $\dot{x} = \tan x + y^2$
      $\dot{y} = \ln (1 + x) + \cosh y - 1$

   c) $\dot{x} = -(\sinh x)(\cosh y)$
      $\dot{y} = -2ye^x(1 + y)$

3. Construct a Liapunov function for each of the following systems.

   a) $\dot{x} = -x$
      $\dot{y} = 4x - 3y$

   b) $\dot{x} = -2x - y$
      $\dot{y} = -2x - 5y$

   c) $\dot{x} = -x + y + z$
      $\dot{y} = -2y$
      $\dot{z} = -3z$

4. Convert the **Van der Pol equation**

$$y'' + \mu(y^2 - 1)y' + y = 0, \qquad \mu \text{ a constant,}$$

   into a plane autonomous system and show that the origin is asymptotically stable if $\mu < 0$.

5. Construct Liapunov functions for the systems of Exercise 2 if possible.

6. Investigate the stability of all equilibrium points for the system

$$\dot{x} = y$$
$$\dot{y} = -6x - y + y^2.$$

7. a) Let $E$ be the Liapunov function for $\dot{\mathbf{X}} = A\mathbf{X}$ constructed in this section. Prove that $\nabla E(\mathbf{X}) = 2A\mathbf{X}$.

   b) Let

$$A = \begin{bmatrix} a_{11} & a_{12} \\ a_{21} & a_{22} \end{bmatrix}, \qquad \mathbf{X} = \begin{bmatrix} x_1 \\ x_2 \end{bmatrix}.$$

   Use Cauchy's inequality to prove that

$$\|A\mathbf{X}\|^2 \leq (a_{11}^2 + a_{12}^2 + a_{21}^2 + a_{22}^2)(x_1^2 + x_2^2) \leq k^2 \|\mathbf{X}\|^2.$$

   c) Generalize the proof given for (b) to deduce that there exists a constant $k$ such that for all $\mathbf{X}$ in $\mathscr{R}^n$,

$$\|A\mathbf{X}\| \leq k \|\mathbf{X}\|,$$

   whenever $A$ is an $n \times n$ matrix.

8. Find the trajectories of the first-order system of Example 3.

# Recommendations for Further Reading

Bellman, R., *Modern Elementary Differential Equations*, Reading Mass.: Addison-Wesley, 1968.

Bellman, R., *Stability Theory of Differential Equations*, New York: McGraw-Hill, 1953.

Bentley, D., and K. Cooke, *Linear Algebra with Differential Equations*, New York: Holt, Rinehart, and Winston, 1973.

Birkhoff, G., and G. C. Rota, *Ordinary Differential Equations*, Boston: Ginn, 1962.

Boyce, W., and R. DiPrima, *Elementary Differential Equations and Boundary Value Problems*, New York: Wiley, 1965.

Brand, L., *Differential and Difference Equations*, New York: Wiley, 1966.

Brauer, F., and J. Nohel, *Ordinary Differential Equations: A First Course*, New York: W. A. Benjamin, 1967.

Coddington, E., *An Introduction to Ordinary Differential Equations*, Englewood Cliffs, N.J.: Prentice-Hall, 1961.

Coddington, E., and N. Levinson, *Theory of Ordinary Differential Equations*, New York: McGraw-Hill, 1955.

Crow, J., and M. Kimura, *Introduction to Population Genetics Theory*, New York: Harper and Row, 1970.

Finkbeiner, D., *Introduction to Matrices and Linear Transformation*, 2nd Ed., San Francisco: Freeman, 1966.

Ford, L., *Differential Equations*, 2nd Ed., New York: McGraw-Hill, 1955.

Gel'fand, I., *Lectures on Linear Algebra*, New York: Interscience, 1961.

Golomb, M., and M. Shanks, *Elements of Ordinary Differential Equations*, 2nd Ed., New York: McGraw-Hill, 1965.

Halmos, P., *Finite Dimensional Vector Spaces*, 2nd Ed., Princeton: Van Nostrand, 1958.

Hille, E., *Lectures on Ordinary Differential Equations*, Reading, Mass.: Addison-Wesley, 1969.

Hirsch, M., and S. Smale, *Differential Equations, Dynamical Systems, and Linear Algebra*, New York: Academic Press, 1974.

Hoffman, K., and R. Kunze, *Linear Algebra*, 2nd Ed., Englewood Cliffs, N.J., Prentice-Hall, 1971.

Hurewicz, W., *Lectures on Ordinary Differential Equations*, Cambridge, Mass.: M.I.T. Press, and New York: Wiley, 1958.

Ince, E., *Ordinary Differential Equations*, New York: Dover, 1956.

Kaplan, W., *Ordinary Differential Equations*, Reading, Mass.: Addison-Wesley, 1958.

Kreider, D., R. Kuller, D. Ostberg, and F. Perkins, *An Introduction to Linear Analysis*, Reading, Mass.: Addison-Wesley, 1966.

Lang, S., *Linear Algebra*, Reading, Mass.: Addison-Wesley, 1966.

La Salle, J., and S. Lefschetz, *Stability by Liapunov's Direct Method with Applications*, New York: Academic Press, 1961.

Leighton, W., *Ordinary Differential Equations*, 3rd Ed.; Belmont, Calif.: Wadsworth, 1970.

Lin, C.-C., and L. Segal, *Mathematics Applied to Deterministic Problems in the Natural Sciences*, New York: Macmillan, 1974.

Munkres, J., *Elementary Linear Algebra*, Reading, Mass.: Addison-Wesley, 1964.

Pontryagin, L., *Ordinary Differential Equations*, Reading, Mass.: Addison-Wesley, 1962.

Rabinstein, A., *Introduction to Ordinary Differential Equations*, 2nd Ed., New York: Academic Press, 1972.

Rabinstein, A., *Elementary Differential Equations with Linear Algebra*, 2nd Ed., New York: Academic Press, 1975.

Shilov, G., *Introduction to the Theory of Linear Spaces*, Englewood Cliffs, N.J.: Prentice-Hall, 1961.

Simmons, G., *Differential Equations with Applications and Historical Notes*, New York: McGraw-Hill, 1972.

Smith, J. Maynard, *Mathematical Ideas in Biology*, Cambridge: Cambridge University Press, 1971.

Tricomi, F., *Differential Equations*, New York: Hafner, 1961.

# Answers to Odd-Numbered Exercises

## Chapter 1

*Section 1–1*

1. $y = ce^{x^2/2}$

3. $y = e^{-x} + c$

5. $\dfrac{y^3}{3} + y = x - \dfrac{x^2}{2} + c$

7. $\dfrac{1}{y} = c - \dfrac{x^3}{3}$

9. a) $v = -32t + 40, \qquad y = -16t^2 + 40t + 144$

b) $\frac{5}{4}$ sec    c) 169 ft    d) $\frac{9}{2}$ sec

11. a) $y = y_0 e^{(k_1 - k_2)t}, \qquad y(0) = y_0$

b) $k_1 = \dfrac{\ln 2}{6}, \qquad k_2 = \dfrac{\ln 2}{8}$

13. a) $-\dfrac{\ln 2}{5700} \approx -0.0001216$

b) About 18,935 years

c) About 6658 years

15. About 2307 years

17. $y' = x^2 + y^2$

19. $y = \sqrt{1 + (y')^2}$

*Section 1–2*

1. $|x| < |c|$

3. $(-\infty, \infty)$

5. If $|c| > 1, \qquad (-\infty, \infty)$.

If $c = 1$, all intervals of the form $((2k - 1)\pi, (2k + 1)\pi)$, $k$ an integer.

If $c = -1$, all intervals of the form $(2k\pi, 2(k + 1)\pi)$, $k$ an integer.

If $|c| < 1$, all intervals of the form $(-a + 2k\pi, a + 2k\pi)$ and $(a + 2k\pi, -a + 2(k + 1)\pi)$, $k$ an integer and $a = \cos^{-1} c$.

7. a) $y = \dfrac{1}{2} \ln \left| \dfrac{x - 1}{x + 1} \right| + c, \qquad x \neq \pm 1$

b) $y = \dfrac{1}{2} \ln \dfrac{|x - 1|}{x + 1} + \dfrac{1}{2}, \qquad -1 < x < 1$

9. a) $y = \dfrac{x\sqrt{4 - x^2}}{2} + 2 \sin^{-1}\dfrac{x}{2} + c$,     $-2 \leqslant x \leqslant 2$

   b) $y = \dfrac{x\sqrt{4 - x^2}}{2} + 2 \sin^{-1}\dfrac{x}{2} + \pi$,     $-2 \leqslant x \leqslant 2$

11. a) $y = \pm\dfrac{1}{\sqrt{c - x^2}}$,     $c > 0$,     $|x| < \sqrt{c}$

   b) $y = \dfrac{1}{\sqrt{5 - x^2}}$,     $|x| < \sqrt{5}$

13. a) $y = \pm\tan^{-1}\sqrt{c^2 - x^2}$,     $x^2 \leqslant c^2$

   b) $y = \tan^{-1}\sqrt{1 - x^2}$,     $|x| \leqslant 1$

15. The general solution of $y' = 2x$ is

$$y = x^2 + c,$$

    $c$ an arbitrary constant, whereas

$$\int_a^x 2t \, dt = x^2 - a^2.$$

    Thus, for example, the solution with $c = 1$ cannot be written in integral form.

### Section 1–3

1. $y = c(x + 1) - 1$

3. $x^4 y^3 = ce^y$

5. $y = c\left(\dfrac{x + 2}{x - 2}\right)^{1/4} - 1$

7. $x(y + 1) = ce^{x+y}$

9. $x^3 - 2y^3 = cx$

11. $x(x^2 + 3y^2) = c$

13. $x \sin \dfrac{y}{x} = c$

15. $2x^2 y^2 = x^4 + c$

17. $y^2 - xy + x^2 - y - 4x = c$

19. $x^2 + 2x(y - 1) = c$

21. $y = 2 \cos x$

23. $y = 0$

25. If $m, n > 1$, $(1 - m)x^{m-1} = (1 - n)y^{n-1} + cx^{m-1}y^{n-1}$, and $y = 0$.
If $m = 1, n > 1$, $y^{n-1}[c + (1 - n) \ln |x|] = 1$, and $y = 0$.
If $m > 1, n = 1$, $y = ce^{1/(1-m)x^{1-m}}$.
If $m = n = 1$, $y = cx$.

27. b) If $a_2 = 0$ or $b_2 = 0$, no substitution is needed. If $a_2 \neq 0$ and $b_2 \neq 0$, let $u = a_2 x + b_2 y$.

29. $y = x + ce^x$

31. $y = \frac{1}{2} \ln |2x + y + \frac{1}{2}| + c$

### Section 1–4

1. $x^2 y + 2y^2 = c$

3. $\dfrac{x}{y} - \dfrac{y}{x} = c$

5. $\ln |y| - \dfrac{y}{x + y} = c$

7. $\ln |x| - \ln |y| + \ln (xy + \sqrt{1 + (xy)^2}) = c$

9. $y \ln x + x \ln y = c$

11. $\sin xy + \cos xy = c$

13. $x + \sec xy + x \tan xy = c$

15. $\tan^{-1} xy = c$

17. a) The general integral is

$$ax^2 + 2bxy - dy^2 = k,$$

$k$ an arbitrary constant. The solution curves form a family of conics, all of which are symmetric about the origin, have the same axes of symmetry, and the same eccentricity. If $ad - bc > 0$, the curves are hyperbolas; if $ad - bc < 0$ and $k > 0$ the curves are ellipses.

### Section 1–5

1. $x^3 - \dfrac{1}{xy} = c$, and $y = 0$

3. $\dfrac{y^2}{2} + \tan^{-1} \dfrac{x}{y} = c$

5. $\ln \left| \dfrac{x}{y} \right| - \dfrac{1}{xy} = c$, and $y = 0$

7. $\tan^{-1} \dfrac{y}{x} - \dfrac{1}{2(x^2 + y^2)} = c$

9. $\dfrac{x}{y^2} + xy = c$, and $y = 0$

11. $\dfrac{x^4}{4} + \dfrac{x^3}{3} + \dfrac{x^2 y^2}{2} = c$

13. $y \sec x + \tan x = c$

15. $\ln |y| + \dfrac{1}{2(xy)^2} = c$

17. $y = cx - x^2 - 1$

19. $x^2 y^2 (y^2 - x^2) = c$

21. $(x - y)y^2 = c(x + y)$

25. b) $\ln \dfrac{|x|}{y^2} - \dfrac{1}{x^2 y^2} = c$

### Section 1–6

1. $y = c_1 x^2 + c_2$

3. $y = -\dfrac{x^2}{2} + c_1 e^x + c_2 x + c_3$

5. $y = c_1 e^{kx} + c_2 e^{-kx}$

7. $c_1 y = \tan(c_1 x + c_2)$, and $y = c$

### Section 1–7

1. $y^2 = y' e^{xy'/y}$

3. $[(y')^2 + 1](y - xy') = 1$

5. $y - b = y'(x - a)$

7. $(y - xy')^2 = 1 + (y')^2$

9. $2xyy' + x^2 - y^2 = 0$

11. $y^2 = cx$

13. $y = x + 1 + ce^{-y}$

15. $x^2 + y^2 = c^2$

17. $\ln y^2 + \dfrac{4}{\sqrt{15}} \tan^{-1} \dfrac{4y - x}{x\sqrt{15}} = c$

19. $y = \cosh(x + c)$, and $y = 1$

21. The path of the bug at $(1, 0)$ is $r = e^{-\theta}$.

### Section 1–8

1. a) With the coordinate system chosen so that the center of the earth is at the origin and the hole coincides with the $y$-axis,

$$v^2 = -\dfrac{g}{R} y^2 + gR,$$

$$y = R \cos \sqrt{\dfrac{g}{R}}\, t$$

where $R$ is the radius of the earth.

b) $2\pi n \sqrt{\dfrac{R}{g}}, \qquad n = 1, 2, 3, \dots$

3. a) $v = \sqrt{\dfrac{mg}{k}} \cdot \dfrac{1 - e^{-2\sqrt{kg/m}\,t}}{1 + e^{-2\sqrt{kg/m}\,t}}$, or $v = \sqrt{\dfrac{mg}{k}} \tanh \sqrt{\dfrac{kg}{m}}\, t$

   b) $\sqrt{\dfrac{mg}{k}}$

   c) $y = \sqrt{\dfrac{mg}{k}}\, t + \dfrac{m}{k} \ln (1 + e^{-2\sqrt{kg/m}\,t}) + y_0 - \dfrac{m \ln 2}{k}$, or $y = \dfrac{m}{k} \ln \cosh \sqrt{\dfrac{kg}{m}}\, t$

5. $y = \dfrac{\rho}{2T_0} x^2 + y_0, \qquad y_0 = y(0)$

7. The surface generated by rotating the graph of $y = cx^4, c > 0$, about the $y$-axis.

## Chapter 2

*Section 2–2*

1. $(-1, 3), \qquad (-3, 9)$

3. $(-\frac{5}{2}, -\frac{2}{3}), \qquad (\frac{5}{2}, \frac{2}{3})$

5. $(-6, -3), \qquad (18, 9)$

7. $\tan^2 x + 1, \qquad -\tan^2 x - 1$

9. $\dfrac{10x + 5}{x^2 + x - 6}, \qquad \dfrac{2x + 1}{x^2 + x - 6}$

11. $(-3, -\frac{3}{2})$

13. $(-\frac{9}{5}, -\frac{106}{35})$

15. $\frac{1}{2}(1 + \sec^2 x)$

17. $\ln (x + 2)$

*Section 2–3*

7. No; $1\mathbf{x} \neq \mathbf{x}$

*Section 2–4*

1. a) No; not closed under addition.
   b) Yes.
   c) Yes.
   d) Yes.
   e) No; not closed under addition.

3. a) Yes.
   b) No; not closed under scalar multiplication.
   c) Yes.
   d) No; not closed under addition.
   e) Yes.

5. a, b, f

7. a, f

9. a) $x(\alpha x + \beta), \qquad \alpha, \beta$ arbitrary
   b) $(x + 1)(\alpha x + \beta), \qquad \alpha, \beta$ arbitrary
   c) $\alpha x^2 + \beta x + \gamma, \qquad \alpha, \beta, \gamma$ arbitrary

*Section 2-5*

3. a) Dependent    b) Independent    c) Dependent

5. a) $(6, -4, 7) = 3(4, -2, 1) + 2(-3, 1, 2)$

   b) $2(10, -4, -3) = 2(4, -2, 1) - 4(-3, 1, 2)$

*Section 2-7*

3. b) $(4, -3, -3) = 2(2, 1, -2) - 2(-1, 3, 1) + (-2, 1, 3)$

5. $(2, -2, 1, 3) = -1(1, 0, 0, 0) - 5(0, 1, 0, 0) - 2(0, 0, 1, 0) + 3(1, 1, 1, 1)$

   $= 0(1, 1, 0, 0) + 3(0, 0, 1, 1) - 2(-1, 0, 1, 1) + 2(0, -1, 0, 1)$

   $= 2(2, -1, 0, 1) + 0(1, 3, 2, 0) + 1(0, -1, -1, 0) + 1(-2, 1, 2, 1)$

   $= -1(1, -1, 2, 0) + 0(1, 2, 2, 0) + 3(3, 0, 0, 1) - 3(2, 1, -1, 0)$

7. $x^2 = \frac{1}{3} \cdot 1 + 0 \cdot x + \frac{2}{3}(\frac{3}{2}x^2 - \frac{1}{2})$

   $x^3 = 0 \cdot 1 + \frac{3}{5} \cdot x + 0(\frac{3}{2}x^2 - \frac{1}{2}) + \frac{2}{5}(\frac{5}{2}x^3 - \frac{3}{2}x)$

*Section 2-8*

1. a) $-1, 1, 0$    b) $-3, 0, 1$    c) $0, -1, 1$

   d) $-\frac{3}{2}, -\frac{1}{2}, 1$    e) $6, -4, 2$    f) $-2, 1, 2$

3. a) $(2, 1, 0)$    b) $(4, 3, 3)$    c) $(-2, -1, -5)$

   d) $(7, \frac{9}{2}, 3)$    e) $(6, 2, 0)$    f) $(10, 6, 3)$

5. One such basis is $(-3, 0, 0, 0), (0, 1, 0, 0), (0, 0, 2, 0), (0, 0, 0, -1)$.

7. One such basis is $(1, -\frac{1}{2}), (0, \frac{1}{2})$.

*Section 2-9*

1. a) 2    b) 3

3. a) One basis is $1, x$.

   b) One basis is $1, x, x^2, \ldots, x^{n-1}$.

5. a) 3    b) 3                                7. 2

9. One such pair is $\mathbf{x}_3 = (1, 0, 0, 0), \mathbf{x}_4 = (0, 1, 0, 0)$.

*Section 2-10*

1. a) $-5$    b) 0    c) $-4$    d) 0    e) $-35$    f) 40

3. a, c, d

**Chapter 3**

*Section 3-1*

1. Central reflection in the origin

3. Compression toward the $x_1$-axis by the factor $\frac{1}{3}$, followed by a magnification by the factor 2

5. Counterclockwise rotation through $45°$, followed by magnification by the factor 2

7. Reflection in the line $x_2 = -x_1$

9. The whole plane is mapped on (0,0).

11. Not linear                13. Linear                15. Linear

21. $A(x) = cx$, $c$ an arbitrary real number.

*Section 3–2*

5. b) If $\mathcal{X} = \{0\}$, then $\mathcal{A}(\mathcal{X})$ is the space of all linear transformations from $\mathcal{V}_1$ to $\mathcal{V}_2$. If $\mathcal{X} = \mathcal{V}_1$, then $\mathcal{A}(\mathcal{X})$ contains only the zero transformation.

*Section 3–3*

3. a) $AB(x_1, x_2) = (0, 0)$,    $BA(x_1, x_2) = (0, 0)$,    $AB = BA$

   b) $AB(x_1, x_2) = 2(x_1, -x_2)$,    $BA(x_1, x_2) = 2(x_2, x_1)$,    $AB \neq BA$

   c) $AB(x_1, x_2, x_3) = (x_1, x_2, -x_3)$,    $BA(x_1, x_2, x_3) = (x_1, x_2, -2x_1 - 2x_2 - x_3)$, $AB \neq BA$

   d) $AB(x_1, x_2, x_3) = (0, 0, 0)$,    $BA(x_1, x_2, x_3) = 2(x_1 + x_2 + x_3, 0, -x_1 - x_2 - x_3)$, $AB \neq BA$

5. $LDy = y(x) - y(a)$,    $DLy = y$

*Section 3–4*

1. a) $(D - 4)(D + 3)$            b) $D(D + 3)^2$

   c) $(D + 1)^2(D - 1)^2$          d) $(D + 2)(D + 3)(D - 1)$

   e) $[D + (1 + \sqrt{6})][D + (1 - \sqrt{6})]$

3. a) $x^2D^2 + 3xD + 1$           b) $2xD^2 + (1 - 2x)D - 1$

   c) $2xD^2 + (3 - 2x)D - 1$        d) $x^2D^2 + 2x(1 - x^2)D - 6x^2$

   e) $x^2D^2 + 2x(2 - x^2)D + 2 - 4x^2$

*Section 3–5*

1. a) $\mathcal{N}(A) = \mathcal{O}$; $A^{-1}(y_1, y_2) = \frac{1}{2}(y_1, -y_2)$

   b) $\mathcal{N}(A)$ is the line $x_2 = 0$; there is no inverse.

   c) $\mathcal{N}(A)$ is the line $x_1 + x_2 = 0$; there is no inverse.

   d) $\mathcal{N}(A) = \mathcal{O}$; $A^{-1}(y_1, y_2) = \frac{1}{2}(y_1 + y_2, y_1 - y_2)$

3. a) $\mathcal{N}(A)$ consists of all constant polynomials; there is no inverse.

   b) $\mathcal{N}(A) = \mathcal{O}$; there is no inverse because $A$ is not onto.

   c) $\mathcal{N}(A)$ consists of all constant polynomials; there is no inverse.

   d) $\mathcal{N}(A) = \mathcal{O}$ unless $q(x) = 0$, in which case $\mathcal{N}(A) = P$; there is no inverse because $A$ is not onto.

5. $\alpha_1\beta_2 - \alpha_2\beta_1 \neq 0$

*Section 3–7*

3. a) $\alpha(1, -2, 4, \ldots, (-2)^{n-1}, \ldots) + (0, -1, 3, -7, \ldots, (-1)^{n-1}(2^{n-1} - 1), \ldots)$, $\alpha$ arbitrary

   b) $\alpha(1, 2, 4, \ldots, 2^{n-1}, \ldots) + 4(0, 1, 3, 7, \ldots, (2^{n-1} - 1), \ldots)$, $\alpha$ arbitrary

c) $\alpha(1, -1, 1, \ldots, (-1)^{n-1}, \ldots) + \left(0, 1, 1, 2, 2, 3, \ldots, \left[\dfrac{n}{2}\right], \ldots\right)$, $\alpha$ arbitrary

d) $\alpha(1, 1, 1, \ldots) + \left(0, 0, 2, 8, \ldots, \dfrac{n(n-1)(n-2)}{3}, \ldots\right)$, $\alpha$ arbitrary

# Chapter 4

## Section 4–1

1. b) $y = c_1 e^{ax} \cos bx + c_2 e^{ax} \sin bx$

 c) $y = b e^{ax} \cos bx - \dfrac{a(b+1)}{b} e^{ax} \sin bx$

3. a) The solutions are not linearly independent.
 b) $y = c_1 \sin^3 x + c_2$

5. a) $y = c_1 \sinh x + c_2 \cosh x$
 b) $y = c_1 x^3 + c_2 x^3 \ln x$
 c) $y = c_1 \sin 2x + c_2 \cos 2x$
 d) $y = c_1 x + c_2 x \ln \dfrac{1+x}{1-x}$

7. a) $y = \dfrac{h}{a_0}$   b) $y = \dfrac{m}{a_0} x + \dfrac{a_0 b - a_1 m}{a_0^2}$

 c) $y = \dfrac{m}{2a_1} x^2 - \dfrac{a_2 m}{a_1^2} x$

## Section 4–2

1. $y = \dfrac{c}{x^2}$

3. $y = e^{-x} + c e^{-(3/2)x}$

5. $y = \dfrac{c e^x - (2x+1)e^{-x}}{4(x^2+1)}$

7. $y = \dfrac{e^x(x-1) + c}{x \sin x}$

9. $y = \left(\dfrac{1 + \sqrt{1-x^2}}{x}\right)(e^x + c)$

11. $y = -\dfrac{\sin x + \ln |1 - \sin x| + c}{(1 + \sin x)^2}$

13. $y = e^x[\frac{1}{2} + c(x^2+1)^{-3/2}]^{2/3}$

15. On $(-\infty, -1)$,    $y = \{(1-x)[c + \frac{1}{2} \ln(-x + \sqrt{x^2-1})] - \sqrt{x^2 - 1}\}^2$.
 On $(-1, 1)$,    $y = -\{(1-x)[c + \frac{1}{2} \cos^{-1} x] + \sqrt{1 - x^2}\}^2$.
 On $(1, \infty)$,    $y = \{(x-1)[c + \frac{1}{2} \ln(x + \sqrt{x^2-1})] - \sqrt{x^2 - 1}\}^2$.
 Also, $y = 0$ is a solution on $-\infty < x < \infty$.

19. $y = c e^{-x^2} + 1$

## Section 4–3

3. a) $\dfrac{dv}{dt} = k_1(1 - k_2)v$,    $\dfrac{dp}{dt} = k_2 v$

b) $v = v_0 e^{k_1(1-k_2)t}$    c) $p = \dfrac{k_2 v_0}{k_1(1-k_2)}\left[e^{k_1(1-k_2)t} - 1\right]$

5. If $k_1 \neq k_2$,    $B = \dfrac{k_1 A_0}{k_2 - k_1}(e^{-k_1 t} - e^{-k_2 t})$.

   If $k_1 = k_2$,    $B = k_1 A_0 t e^{-k_1 t}$.

7. $I = \left(I_0 + \dfrac{bE_0 L}{Y^2}\right)e^{-(R/L)t} + \dfrac{E_0}{Y}e^{-at}\sin(bt - \beta)$,

   where $Y = \sqrt{(R - aL)^2 + b^2 L^2}$,    $R - aL = Y\cos\beta$,    $bL = Y\sin\beta$,    $0 < \beta < \pi$.

11. $v' + v^2 - 1 = 0; v = 1 + \dfrac{2}{ce^{2x} - 1}$ and $v = 1; y = c_1 e^x + c_2 e^{-x}$

### Section 4–4

1. a) $y = -\dfrac{1}{x^2}$ on $(0, \infty)$      b) $y = \dfrac{\sqrt{2}}{\sin x}$ on $(0, \pi)$

   c) $y = e^{-x} - (e^{-3/2} + 3e^{-9/2})e^{-(3/2)x}$ on $(-\infty, \infty)$

   d) $y = \dfrac{1}{1 - x^2}$ on $(-1, 1)$

   e) $y = \dfrac{1}{\sin x}\left[1 + (1 + \sin x)\ln\dfrac{1 + \sin x}{2}\right]$ on $(0, \pi)$

   f) $y = (\ln x + \ln 2)\left(\dfrac{1 + \sqrt{1 - x^2}}{x}\right)$ on $(0, 1)$

   g) $y = \dfrac{1}{x\sin x}\displaystyle\int_{-1/2}^{x}\dfrac{e^t}{t}\,dt$ on $(-\pi, 0)$

5. $xy' - y = 0$ on any interval containing the origin.

7. $y = \dfrac{k}{a} + \left(c - \dfrac{k}{a}\right)e^{-ax}$,    $k = \dfrac{ac(1 - e^{-ab})}{a(a - b) - e^{-ab} + 1}$

### Section 4–5

1. a) $y_1 = e^x - xe^x$,     $y_2 = xe^x$

   b) $y_1 = 1$,     $y_2 = \frac{1}{2} - \frac{2}{3}e^{-x} + \frac{1}{6}e^{2x}$,     $y_3 = -\frac{1}{2} + \frac{1}{3}e^{-x} + \frac{1}{6}e^{2x}$

   c) $y_1 = 1$,     $y_2 = \frac{1}{2}e^x - \frac{1}{2}e^{-x}$,     $y_3 = -1 + \frac{1}{2}e^x + \frac{1}{2}e^{-x}$

   d) $y_1 = \dfrac{x^2}{3} + \dfrac{2}{3x}$,     $y_2 = \dfrac{x^2}{3} - \dfrac{1}{3x}$

   e) $y_1 = \dfrac{3x}{4} + \dfrac{1}{4x} - \dfrac{x\ln x}{2}$,     $y_2 = \dfrac{x}{4} - \dfrac{1}{4x} + \dfrac{x\ln x}{2}$,     $y_3 = -\dfrac{x}{4} + \dfrac{1}{4x} + \dfrac{x\ln x}{2}$

### Section 4–6

1. $W[1, e^{-x}, 2e^{2x}] = -12e^x$

3. $W[1, x, x^2, \ldots, x^n] = 1! \cdot 2! \cdot 3! \cdots n!$

5. $W[x^{1/2}, x^{1/3}] = -\frac{1}{6}x^{-1/6}$

7. $W[e^x, e^x \sin x] = e^{2x}\cos x$

9. $W[1, \sin^2 x, 1 - \cos x] = 2 \sin^3 x$

11. $W[\sqrt{1 - x^2}, x] = 1/\sqrt{1 - x^2}$

23. $a_1(x) = -\dfrac{u_1(x)u_2''(x) - u_2(x)u_1''(x)}{u_1(x)u_2'(x) - u_2(x)u_1'(x)},$

$a_2(x) = \dfrac{u_1'(x)u_2''(x) - u_2'(x)u_1''(x)}{u_1(x)u_2'(x) - u_2(x)u_1'(x)}$

25. a) $x^2y'' - x(x + 2)y' + (x + 2)y = 0$
    b) $x^2y'' - 2xy' + 2y = 0$
    c) $y'' + y = 0$
    d) $(x \cos x - \sin x)y'' + (x \sin x)y' - (\sin x)y = 0$
    e) $x^2(1 - \ln x)y'' + xy' - y = 0$

*Section 4–7*

1. a) $-1/x$     b) $-4/(1 - x^2)$     c) $2x^3$
   d) $-1$     e) $e^{1 - \cos x}$     f) $e^{(2/3)[\sqrt{1+x^3} - \sqrt{2}]}$

3. $xe^{ax}$          5. $\dfrac{-1}{5 \sin^2 x}$          7. $-1 + \dfrac{x}{2} \ln \left| \dfrac{1 + x}{1 - x} \right|$

# Chapter 5

*Section 5–1*

3. $x^2 - 2ax + (a^2 + b^2)$

*Section 5–2*

1. $y = c_1 e^{-2x} + c_2 e^x$

3. $y = c_1 e^{3x/4} + c_2 e^{-5x/2}$

5. $y = c_1 \cos 2x + c_2 \sin 2x$

7. $y = c_1 e^{-2x} \cos 2x + c_2 e^{-2x} \sin 2x$

9. $y = c_1 e^x \cos x + c_2 e^x \sin x$

11. $y = c_1 e^{-x} \cos \sqrt{3}x + c_2 e^{-x} \sin \sqrt{3}x$

13. $y = c_1 e^{2\sqrt{3}x} + c_2 e^{\sqrt{3}x/2}$

15. $y = c_1 e^{(3/8)x} \cos \dfrac{\sqrt{2}}{4} x + c_2 e^{(3/8)x} \sin \dfrac{\sqrt{2}}{4} x$

17. $y = (\frac{1}{2} - \frac{7}{3}x)e^{4x}$

19. $y = 2 \cos \sqrt{2}x + 2 \sin \sqrt{2}\, x$

21. $y = -\frac{2}{3}e^{-2x} \sin 3x$

23. $y = 3xe^{\sqrt{5}x}$

25. $y = \sqrt{2}e^{(\sqrt{2}/2)x} \left( \cos \dfrac{\sqrt{2}}{2} x - \sin \dfrac{\sqrt{2}}{2} x \right)$

29. a) $y'' + 6y' + 9y = 0$     b) $y'' - 2y' + 5y = 0$
    c) $y'' + 4y' + 4y = 4$     d) $y'' + 4y' + 3y = 3x + 16$
    e) $y'' + 9y = 3x$

31. b) $y = c_1 \cos(2x + c_2)$

33. a) $y = e^x \cos 5x + \frac{1}{5}e^x \sin 5x$     b) $y = (\frac{1}{2} - \frac{1}{10}i)e^{(1+5i)x} + (\frac{1}{2} + \frac{1}{10}i)e^{(1-5i)x}$

## Section 5–3

1. $y = c_1 e^x + c_2 e^{-x} + c_3 e^{-3x}$

3. $y = c_1 + (c_2 + c_3 x)e^{-(3/2)x}$

5. $y = c_1 e^{-2x} + c_2 e^{2x} + c_3 e^{-(1/2)x}$

7. $y = c_1 + c_2 x + c_3 e^{-x} + c_4 e^x$

9. $y = (c_1 + c_2 x) \cos 3x + (c_3 + c_4 x) \sin 3x$

11. $y = c_1 + c_2 x + c_3 e^{-(1/2)x} \cos \dfrac{\sqrt{3}}{2} x + c_4 e^{-(1/2)x} \sin \dfrac{\sqrt{3}}{2} x$

13. $y = (c_1 + c_2 x + c_3 x^2 + c_4 x^3)e^x$

15. $y = e^{(\sqrt{2}/2)x}\left(c_1 \cos \dfrac{\sqrt{2}}{2} x + c_2 \sin \dfrac{\sqrt{2}}{2} x\right) + e^{-(\sqrt{2}/2)x}\left(c_3 \cos \dfrac{\sqrt{2}}{2} x + c_4 \sin \dfrac{\sqrt{2}}{2} x\right)$

17. $(D^2 + 1)^3$     19. $(D - 1)^3$     21. $(D - 1)^2(D^2 - 2D + 5)^2$

23. $D(D - 1)^2(D - 2)^3(D - 3)^4$

## Section 5–4

(*Note.* Sometimes a simplified particular solution may be obtained by deleting from the solution given by variation of parameters, or the use of Green's functions, any terms which satisfy the homogeneous equation.)

1. $y = (c_1 + \ln|\cos x|) \cos x + (c_2 + x) \sin x$

3. $y = \dfrac{2x - 1}{32} e^{2x} + (c_1 + c_2 x)e^{-2x}$

5. $y = \frac{1}{4}e^{-(1/2)x}(-x \sin x - 2 \cos x + c_1 + c_2 x)$

7. $y = \frac{1}{84}(7e^{2x} - 14 - 3e^{-2x}) + c_1 e^{(-5+\sqrt{37})x} + c_2 e^{(-5-\sqrt{37})x}$

9. $y = \frac{1}{5}e^{2x}(\sin x - 2 \cos x) + c_1 e^x \cos x + c_2 e^x \sin x$

11. $y_p = x^3(\frac{1}{2} \ln x - \frac{3}{4})$

13. $y_p = \dfrac{1}{16}\left[\ln|\tan x| + (\cos 2x) \ln \dfrac{\cos^2 2x}{(1 + \cos 2x)|\sin 2x|}\right]$

15. $y_p = -x$     17. $K(x, t) = \frac{1}{3}[e^{2(x-t)} - e^{-(x-t)}]$

19. $K(x, t) = 2e^{x-t} \sin \frac{1}{2}(x - t)$     21. $K(x, t) = \dfrac{x(x - t)}{t}$

23. $K(x, t) = \dfrac{1 - t^2}{2} \ln \left|\dfrac{(1 + x)(1 - t)}{(1 - x)(1 + t)}\right|$

25. If $a \neq 0$ or if $\omega^2 \neq b^2 - a^2$, then

$$y = \frac{1}{2c}\left[\frac{\sin(\omega t + \varphi_1)}{D_1} - \frac{\sin(\omega t + \varphi_2)}{D_2}\right] + \frac{1}{2c}\left[e^{-at}\frac{\sin(ct - \varphi_1)}{D_1} + \frac{\sin(ct + \varphi_2)}{D_2}\right],$$

where

$$c = (b^2 - a^2)^{1/2}, \qquad D_1 = [a^2 + (\omega + c)^2]^{1/2}, \qquad D_2 = [a^2 + (\omega - c)^2]^{1/2},$$

and $\varphi_1$, $\varphi_2$ are determined by the relations

$$\cos \varphi_1 = (\omega + c)/D_1, \qquad \sin \varphi_1 = a/D_1, \qquad \cos \varphi_2 = (\omega - c)/D_2,$$
$$\sin \varphi_2 = a/D_2, \quad 0 \le \varphi_1, \varphi_2 < 2\pi.$$

If $a = 0$ and $\omega^2 = b^2$, then

$$y = \frac{1}{2\omega^2}(\sin \omega t - \omega t \cos \omega t).$$

## Section 5–5

1. $y = (\frac{1}{3}x^3 - \frac{1}{2}x^2 + \frac{1}{2}x - \frac{1}{4})e^x + c_1 e^{-x} + (c_2 + c_3 x)e^x$

3. $y = -(\frac{1}{3}x^3 + \frac{3}{2}x^2 + \frac{3}{2}x + \frac{3}{4}) + c_1 + c_2 x + c_3 e^{2x}$

5. $y = \frac{1}{25}(4 \cos x + 3 \sin x) + c_1 e^x + c_2 e^{2x} + c_3 e^{-3x}$

7. $y = \frac{1}{2}\cos x + c_1 + c_2 e^{-x} + c_3 e^x$

9. $y = \left(\frac{x^2}{2} - \frac{5}{2}x + c_1\right)e^x + c_2 + c_3 x + c_4 e^{-x}$

11. $K(x, t) = e^{x-t} - (x - t) - 1$

13. $K(x, t) = \frac{1}{2}[e^{x-t} - 2e^{2(x-t)} + e^{3(x-t)}]$

15. $K(x, t) = \sinh(x - t) - (x - t)$

## Section 5–6

1. $y = 3x + k_1 \int e^{-x^2/2}\, dx + k_2$

3. $y = k_1/x + k_2 x^2$

5. $y = (k_1 x + k_2)e^{x^2}$

7. $y = x^{3/2}(k_1 \ln x + k_2), \qquad x > 0$

9. $y = k_1(x - 1) + k_2 e^{-x}, \qquad x > 0$

13. b) One such solution is $y = x^{-1/2} \cos x$.

15. $y = \left(k_1 \int \frac{e^{-\int a_1(x)\, dx}}{y_1(x)^2}\, dx + k_2\right)y_1(x)$

17. $y = k_1 x + k_2 x^3 + k_3 x^{-1}$

## Section 5–7

1. $y_p = \frac{3}{2}e^x + x^2 - 2x$

3. $y_p = \frac{1}{2}\cos x - \frac{1}{2}\sin x$

5. $y_p = -\frac{1}{2}xe^{-2x}$

7. $y_p = (x^2 - 3x + \frac{30}{7})e^x$

9. $y_p = \frac{1}{625}(899 + 1305x + 675x^2 + 125x^3)$

11. $y_p = \frac{1}{6250}(2374 - 2940x + 3200x^2 - 1500x^3 + 625x^4)$

13. $y_p = \frac{1}{3}(x - 1)e^x$

15. $y_p = 2 + 2x + xe^{-x}$

17. $y_p = -(\sin x + \cos x)$

19. $y_p = \frac{1}{12}(39x + 9x^2 + 2x^3 + 12xe^x)$

21. $y_p = (c_1 + c_2 x)x^2 e^{2x} + (c_3 + c_4 x + c_5 x^2)\cos x + (c_6 + c_7 x + c_8 x^2)\sin x$

23. $y_p = c_1 e^{-x} + x^3[(c_2 + c_3 x + c_4 x^2)\cos x + (c_5 + c_6 x + c_7 x^2)\sin x]$

25. $y_p = x^2[(c_1 + c_2 x)e^x + (c_3 + c_4 x)e^{-x} + (c_5 + c_6 x + c_7 x^2 + c_8 x^3)\cos x$
$+ (c_9 + c_{10}x + c_{11}x^2 + c_{12}x^3)\sin x]$

27. $y_p = x^3(c_1 + c_2 x + c_3 x^2)e^x + (c_4 + c_5 x)\cos x + (c_6 + c_7 x)\sin x$

29. $y_p = c_1 + c_2 x + x^2 e^{x/2}[(c_3 + c_4 x)\cos x + (c_5 + c_6 x)\sin x]$
$+ (c_7 + c_8 x + c_9 x^2)\cos x + (c_{10} + c_{11}x + c_{12}x^2)\sin x$

### Section 5–8

1. $y = c_1|x| + c_2|x|^{-2}$
3. $y = c_1 \sin(3 \ln |x|) + c_2 \cos(3 \ln |x|)$
5. $y = c_1|x|^p + c_2|x|^{-p}$, if $p \neq 0$; $\quad y = c_1 + c_2 \ln |x|$, if $p = 0$
7. $y = |x|^{-1}(c_1 + c_2 \ln |x|) + c_3|x|^7$
9. $y = c_1|x|^{-1} + c_2|x|^{\sqrt{2}} + c_3|x|^{-\sqrt{2}}$
11. $y = c_1|x| + c_2|x|^{-1} + c_3 \cos(\ln |x|) + c_4 \sin(\ln |x|)$
13. $y = -\frac{1}{6}(\ln x)\cos(\ln x^3)$ $\qquad\qquad$ 15. $y = \frac{1}{2}x$

### Section 5–9

1. a) $Q = E_0 C[1 - e^{-(1/RC)t}]$,

$I = \dfrac{E_0}{R} e^{-(1/RC)t}$

b) $Q = Q_0 e^{-(1/RC)t}$, $\quad I = -\dfrac{Q_0}{RC} e^{-(1/RC)t}$

c) $Q = \dfrac{E_0 C}{1 + (RC\omega)^2}(\sin \omega t - RC\omega \cos \omega t)$,

$I = \dfrac{E_0 C\omega}{1 + (RC\omega)^2}(\cos \omega t + RC\omega \sin \omega t)$

d) As $\omega$ increases, $Z = \sqrt{R^2 + (1/\omega C)^2}$ approaches $R$ so that the amplitude of $I_s$ approaches $E_0/Z$. For small values of $\omega$, $Z$ is large, and the amplitude of $I_s$ is correspondingly small. In contrast to the $RLC$ circuit, this $RC$ circuit has no resonating frequency.

3. $Q = e^{\alpha t}(A_1 e^{\beta t} + A_2 e^{-\beta t})$,
$I = e^{\alpha t}[(\alpha + \beta)A_1 e^{\beta t} + (\alpha - \beta)A_2 e^{-\beta t}]$,
where $\quad \alpha = -R/2L, \quad \beta = \sqrt{(R/2L)^2 - (1/LC)}$,

$A_1 = \dfrac{(\beta - \alpha)Q_0}{2\beta}, \qquad A_2 = \dfrac{(\alpha + \beta)Q_0}{2\beta}$

## Chapter 6

### Section 6–1

1. No $\qquad\qquad$ 3. Yes $\qquad\qquad$ 5. No

7. No $\qquad\qquad$ 9. No $\qquad\qquad$ 11. $\dfrac{4}{\pi}$

13. 1

15. $\dfrac{1}{s^2}$

17. $\dfrac{a}{s^2 + a^2}$

19. $\dfrac{1}{s^2 - 1}$

21. $\dfrac{b - a}{(s - b)(s - a)}$

23. $\dfrac{(s + 1)e^{-s}}{s^2}$

## Section 6–2

5. b) The domain of $\mathscr{L}$ includes functions that are not piecewise continuous on $[0, \infty)$.

## Section 6–3

1. $\dfrac{2}{s^3 + 4s}$

3. $\dfrac{a}{s^2 - a^2}$

5. $\dfrac{a - b}{(s - a)(s - b)}$

7. $n!\left[\dfrac{a^n}{n!s} + \dfrac{a^{n-1}}{(n-1)!s^2} + \cdots + \dfrac{a}{1!s^n} + \dfrac{1}{s^{n+1}}\right]$

9. $\dfrac{s^2 - a^2}{(s^2 + a^2)^2}$

11. $\dfrac{2s(s^2 - 3a^2)}{(s^2 + a^2)^3}$

13. $\dfrac{1}{(s - 1)^2 + 1}$

15. $\dfrac{e^{-as}}{s}$

17. $y = t + 2\sin t - \cos t$

19. $y = -2 + 2\cos \tfrac{1}{2}t + \sin \tfrac{1}{2}t$

21. $y = -\tfrac{1}{2} + \tfrac{1}{5}e^{-t} + \tfrac{3}{10}\cos 2t + \tfrac{3}{5}\sin 2t$

23. $y = \sin 3t$

25. If $|f(t)| \leqslant Ce^{\alpha t}$, then $\lim_{t \to \infty} e^{-st}f(t) = 0$ when $s > \alpha$.

## Section 6–4

1. $\dfrac{3}{(s - 2)^2 + 9}$

3. $\dfrac{72s(s^2 - 9)}{(s^2 + 9)^4}$

5. $\dfrac{(s + 3)\cos 4 - 2\sin 4}{(s + 3)^2 + 4}$

7. If $\mathscr{L}[f] = \varphi(s)$, then $\mathscr{L}[te^{2t}f'(t)] = -\left[(s - 2)\dfrac{d}{ds}\varphi(s - 2) + \varphi(s - 2)\right]$.

9. $\dfrac{(3s + 2)e^{-s/2}}{2s^2}$

11. $\dfrac{1 - e^{-2\pi s}}{s^2 + 1}$

13. $\dfrac{1}{s^2}\left[1 - 4e^{-2s} + 4e^{-3s} - e^{-4s}\right]$

15. $\dfrac{2}{(s - 2)^2 + 4}$

17. $\dfrac{1}{s^2 + 1}\left(\dfrac{1 + e^{-\pi s}}{1 - e^{-\pi s}}\right)$

19. $\dfrac{2(s - 2)}{s[(s - 2)^2 + 1]^2}$

21. $\dfrac{8(5s^2 - 1)}{(s^2 + 1)^4} + \dfrac{2(\cos 1 - \sin 1)}{s^3}$

23. $\dfrac{3s^4 - 16s^3 + 96s - 108}{(s - 1)^2[(s - 3)^2 + 1]^3} + \dfrac{1}{5(s - 1)^2} + \dfrac{e^{2a}(2\sin a - 5a\sin a - \cos a)}{5(s - 1)^2}$

$= \dfrac{s^4 - 16s^3 + 120s^2 - 400s + 460}{5[(s - 3)^2 + 1]^3} + \dfrac{e^{2a}(2\sin a - 5a\sin a - \cos a)}{5(s - 1)^2}$

25. $1 - e^{-t}$

27. $\tfrac{1}{4} - \tfrac{1}{4}e^{-2t} - \tfrac{1}{2}te^{-2t}$

29. $\dfrac{1}{(n-1)!}t^{n-1}e^{at}$

31. $\tfrac{1}{5}e^{-2t}\sin 5t$

33. $t\sin t$

35. $\tfrac{3}{2}\sin t + \tfrac{3}{2}t\cos t$

37. $\sqrt{3}\, u_2(t) \sin \dfrac{1}{\sqrt{3}}(t-2) = \begin{cases} 0 & \text{if } t \leqslant 2, \\ \sqrt{3} \sin \dfrac{1}{\sqrt{3}}(t-2) & \text{if } t > 2 \end{cases}$

39. $\dfrac{1}{\sqrt{2}}\left[\cosh \dfrac{t}{\sqrt{2}} \sin \dfrac{t}{\sqrt{2}} - \sinh \dfrac{t}{\sqrt{2}} \cos \dfrac{t}{\sqrt{2}}\right]$

41. $1 + u_1(t) = \begin{cases} 1 & \text{if } t \leqslant 1, \\ 2 & \text{if } t > 1 \end{cases}$

43. $\dfrac{1}{3a^2} e^{-at} - \dfrac{1}{3a^2} e^{at/2} \cos \dfrac{\sqrt{3}}{2} at + \dfrac{4\sqrt{3}}{9a^2} e^{at/2} \sin \dfrac{\sqrt{3}}{2} at$

45. $\dfrac{1}{t}(e^{-2t} - e^{-3t})$

49. b) $\mathscr{L}^{-1}\left[\dfrac{s}{(s^2 + a^2)^{n+1}}\right] = \dfrac{1}{2^n an!}\, t \underbrace{\int_0^t t \int_0^t t \cdots \int_0^t}_{n-1 \text{ times}} t \sin at\, dt\, dt \cdots dt, \qquad n \geqslant 2$

## Section 6–5

3. $y = \frac{1}{4}e^t - \frac{1}{4}e^{-t} - \frac{1}{2}te^{-t}$

5. $y = \frac{2}{3}e^{-t} \cos \sqrt{2}t + \dfrac{\sqrt{2}}{3} e^{-t} \sin \sqrt{2}t + t - \frac{2}{3}$

7. $y = \dfrac{1}{k}[1 - e^{-kt}] + \dfrac{1}{k} \displaystyle\sum_{j=0}^{\infty} (-1)^j u_j(t)[1 - e^{-k(t-j)}]$

9. $y = -\frac{27}{50} \cos 2t + \frac{57}{25} \sin 2t - \frac{3}{5}te^t + \frac{1}{25}e^t + \frac{1}{2}e^{2t}$

11. $y = te^{-2t} + e^{-2t} + u_2(t)[e^{-(t-2)} - (t-1)e^{-2(t-2)}]$

$\quad = \begin{cases} te^{-2t} + e^{-2t} & \text{if } t \leqslant 2 \\ te^{-2t} + e^{-2t} + e^{-(t-2)} - (t-1)e^{-2(t-2)} & \text{if } t > 2 \end{cases}$

13. $y = 2 \sin 2t + \frac{3}{2}e^{-(t-\pi)} - \frac{5}{2}e^{t-\pi}$

15. $y = \begin{cases} 1 - e^{-t}, & 0 \leqslant t \leqslant 1 \\ te^{1-t} - e^{-t}, & t > 1 \end{cases}$

17. $y = \cos t - \cos(t-\pi) + u_\pi(t)$

21. b) $J_0(t) = \displaystyle\sum_{k=0}^{\infty} \dfrac{(-1)^k}{2^{2k}(k!)^2} t^{2k}$

## Section 6–6

1. $\displaystyle\int_0^t f(t-\xi) \sin \xi\, d\xi$

3. $e^{-t} + t - 1$

5. $\frac{3}{2}[t \cos t + \sin t]$

7. $\dfrac{e^{bt} - e^{at}}{b-a}$ if $a \neq b$;  $te^{at}$ if $a = b$

9. $\dfrac{a}{a^2 - b^2}(\cos bt - \cos at)$ if $a^2 \neq b^2$;  $\frac{1}{2}t \sin at$ if $b = \pm a$

11. $e \displaystyle\int_1^{t+1} f(t-\xi)e^{-\xi}g(\xi)\, d\xi$

17. $\dfrac{t^{n-1}}{(n-1)!}$

21. $\dfrac{\pi}{2} - \tan^{-1}\dfrac{s}{a}$

23. $\dfrac{3\pi}{2} - \dfrac{1}{2}s\ln\left(1 + \dfrac{9}{s^2}\right) - 3\tan^{-1}\dfrac{s}{3}$

25. a) $\phi(t) = -\frac{2}{3}\sin t + \frac{4}{3}\sin 2t$

   b) $\phi(t) = \cos t$

   c) $\phi(t) = \frac{3}{5}\cos t - \frac{1}{5}\sin t + \frac{2}{5}e^{-2t}$

   d) $\phi(t) = \frac{1}{4}\cos 2t + \frac{3}{5}\sin 2t - \frac{1}{4}e^{-2t}$

   e) $\phi(t) = \cos t$

## Section 6–7

3. $K(t, \xi) = \frac{1}{2}e^{-3(t-\xi)}\sin 2(t - \xi)$

5. $K(t, \xi) = (t - \xi)e^{(t-\xi)/6}$

7. $K(t, \xi) = \dfrac{1}{3}\left[e^{-(t-\xi)} - e^{(t-\xi)/2}\cos\dfrac{\sqrt{3}}{2}(t - \xi) + \sqrt{3}e^{(t-\xi)/2}\sin\dfrac{\sqrt{3}}{2}(t - \xi)\right]$

9. $K(t, \xi) = \dfrac{\sqrt{2}}{4}\left\{e^{-\sqrt{2}(t-\xi)/2}\left[\sin\dfrac{\sqrt{2}}{2}(t - \xi) + \cos\dfrac{\sqrt{2}}{2}(t - \xi)\right]\right.$

$$\left. + e^{\sqrt{2}(t-\xi)/2}\left[\sin\dfrac{\sqrt{2}}{2}(t - \xi) - \cos\dfrac{\sqrt{2}}{2}(t - \xi)\right]\right\}$$

11. $y = u_1(t)\{1 - t + \frac{1}{2}[e^{t-1} - e^{-(t-1)}]\}$

$$= \begin{cases} 0 & \text{if } 0 \leqslant t \leqslant 1, \\ 1 - t + \frac{1}{2}[e^{t-1} - e^{-(t-1)}] & \text{if } t > 1 \end{cases}$$

17. $y = \frac{1}{6}e^{-(t-\pi)} - \frac{4}{15}e^{(t-\pi)/2} - \frac{3}{10}\sin t - \frac{1}{10}\cos t$

19. $y = \frac{3}{5}e^{4(t-2)} + \frac{12}{5}e^{-(t-2)} + \frac{9}{25}e^{-t} + \frac{1}{25}e^{4t-10} - \frac{1}{5}te^{-t}$

21. $y = e^{t/2}[\frac{1}{24}(t - a)\sin 3t - \frac{1}{72}\sin 3a\sin 3(t - a)]$

$$+ e^{(t-a)/2}[7\cos 3(t - a) - \frac{11}{6}\sin 3(t - a)]$$

23. $y = t - a - \frac{3}{2}\sin(t - a) + \frac{1}{2}(t - a)\cos(t - a)$

## Section 6–8

3. $y = \frac{1}{2}e^t + \frac{1}{2}e^{-t} + \dfrac{2\sqrt{3}}{3}e^{t/2}\sin\dfrac{\sqrt{3}}{2}t$

$$+ u_1(t)\left[\frac{1}{3}e^{-(t-1)} - \frac{1}{3}e^{(t-1)/2}\cos\dfrac{\sqrt{3}}{2}(t - 1) + \dfrac{\sqrt{3}}{3}e^{(t-1)/2}\sin\dfrac{\sqrt{3}}{2}(t - 1)\right]$$

$$= \begin{cases} \frac{1}{2}e^t + \frac{1}{2}e^{-t} + \dfrac{2\sqrt{3}}{3}e^{t/2}\sin\dfrac{\sqrt{3}}{2}t & \text{if } 0 \leqslant t \leqslant 1, \\[2mm] \frac{1}{2}e^t + \frac{1}{2}e^{-t} + \dfrac{2\sqrt{3}}{3}e^{t/2}\sin\dfrac{\sqrt{3}}{2}t + \frac{1}{3}e^{-(t-1)} - \frac{1}{3}e^{(t-1)/2}\cos\dfrac{\sqrt{3}}{2}(t - 1) \\[2mm] \qquad\qquad\qquad\qquad + \dfrac{\sqrt{3}}{3}e^{(t-1)/2}\sin\dfrac{\sqrt{3}}{2}(t - 1) & \text{if } t > 1 \end{cases}$$

5. $y = \frac{1}{2}\sin 2t + \frac{1}{6}u_{\pi/2}(t)[2\cos t - \sin 2t]$

$$= \begin{cases} \frac{1}{2}\sin 2t & \text{if } 0 \leqslant t \leqslant \pi/2, \\ \frac{1}{2}\sin 2t + \frac{1}{3}\cos t - \frac{1}{6}\sin 2t & \text{if } t > \pi/2 \end{cases}$$

7. $y = 2t - \dfrac{\sqrt{3}}{3} \sin \sqrt{3}t$

*Section 6–9*

3. b) Between $F(t - v_k)R'(v_k)\,\Delta_k v$ and $F(t - u_{k+1})R'(u_k)\,\Delta_k u$. For $\Delta_k u$ small, the number will be approximately $F(t - v_k)R'(v_k)\,\Delta_k v$.

   d) $\mathscr{L}[R] = \dfrac{\mathscr{L}[N]}{s\mathscr{L}[F]} - \dfrac{N(0)}{s}$

   e) $R(t) = aN_0 t$

   f) $F(t) = 1 - u_L(t)$, so that

$$R = \mathscr{L}^{-1}\left[\frac{\mathscr{L}[N]}{1 - e^{-Ls}} - \frac{N(0)}{s}\right].$$

   g) $R(t) = N_0(f(t) - 1)$, where $f$ is the staircase function $f(t) = \displaystyle\sum_{n=0}^{\infty} u_{nL}(t - nL)$.

**Chapter 7**

*Section 7–1*

5. (a) and (d)

*Section 7–2*

1. $a_0 \displaystyle\sum_{n=0}^{\infty} \dfrac{x^n}{n!}$

3. a) $1 + \displaystyle\sum_{n=1}^{\infty} \dfrac{p(p-1)\cdots(p-n+1)}{n!}\, x^n$

   b) (i) infinite;    (ii) 1

   c) The solution is $y = (1 + x)^p$. Hence, when $p$ is an integer,

$$(1 + x)^p = 1 + px + \frac{p(p-1)}{2!}\, x^2 + \cdots + x^p,$$

   and when $p$ is not an integer,

$$(1 + x)^p = 1 + \sum_{n=1}^{\infty} \frac{p(p-1)\cdots(p-n+1)}{n!}\, x^n.$$

5. $y = a_0(1 - 3x^2) + a_1\left(x - \dfrac{x^3}{3}\right)$

7. $y = a_0 y_0 + a_1 y_1$, where

$$y_0 = \sum_{n=0}^{\infty} \frac{(-1)^n}{n!\,3^n}\, x^{3n}$$

$$y_1 = \sum_{n=0}^{\infty} \frac{(-1)^n}{1(1 + 3\cdot 1)(1 + 3\cdot 2)\cdots(1 + 3n)}\, x^{3n+1}$$

9. $y = a_0 y_0 + a_1 y_1$, where

$$y_0 = 1 + \sum_{n=1}^{\infty} \frac{1}{n![2 \cdot 5 \cdot 8 \cdots (3n-1)]} x^{3n}$$

$$y_1 = \sum_{n=0}^{\infty} \frac{1}{n![1 \cdot 4 \cdot 7 \cdots (3n+1)]} x^{3n+1}$$

11. $y \approx \dfrac{x^2}{2 \cdot 3} + \dfrac{x^3}{2 \cdot 3^3} - \dfrac{x^4}{2^2 \cdot 3^4} - \dfrac{4x^5}{5 \cdot 3^5}$

13. $y \approx \dfrac{1}{8} x^2 + \dfrac{x^3}{3(3!)} + \dfrac{x^4}{3(4!)} + \dfrac{x^5}{2(5!)}$

## Section 7–3

1. b)  Since $P_1(x) = x$ is a solution that is linearly independent of

$$\frac{x}{2} \ln\left[\frac{1+x}{1-x}\right] - 1,$$

the general solution is

$$y = c_1 x + c_2 \left[\frac{x}{2} \ln\left(\frac{1+x}{1-x}\right) - 1\right].$$

3. b)  $y = a_0 y_0 + a_1 y_1$, where

$$y_0 = 1 - \frac{2p}{2!} x^2 + \frac{2^2(p)(p-2)}{4!} x^4 - \frac{2^3(p)(p-2)(p-4)}{6!} x^6 + \cdots$$

$$y_1 = x - \frac{2(p-1)}{3!} x^3 + \frac{2^2(p-1)(p-3)}{5!} x^5 - \frac{2^3(p-1)(p-3)(p-5)}{7!} x^7 + \cdots$$

11. a)  $Q_i(x) = a_i \cdot \dfrac{\displaystyle\prod_{\substack{j=1 \\ j \neq i}}^{m} (x - x_j)}{\displaystyle\prod_{\substack{j=1 \\ j \neq i}}^{m} (x_i - x_j)}$

## Section 7–4

1.  Regular singular points $x = \pm 1$; irregular singular point $x = 0$

3.  Regular singular point $x = 1$; irregular singular point $x = 0$

5.  Regular singular points at $x = \pm 1$

11.  $y_1 = |x|^{1/3}(1 + \frac{1}{5}x)$

$$y_2 = |x|^{-1/3} \sum_{n=0}^{\infty} \frac{(-1)^n}{n!3^n(3n-5)(3n-2)} x^n$$

13.  $y_1 = |x|^{1/2}$

$$y_2 = |x|^{1/4} \sum_{n=0}^{\infty} \frac{1}{n!2^{2n}(4n-1)} x^n$$

*Section 7–5*

1. a) $v^2 - 3v - 1 = 0$
   b) $v^2 - 3v + 1 = 0$
   c) $v^2 = 0$
   d) $v^2 - v + \frac{1}{4} = 0$
   e) $v^2 = 0$

3. $y_1 = |x|^{1/2} \sum\limits_{k=0}^{\infty} a_k x^k$, where

$$a_0 = 1, \qquad a_k = \frac{2}{k(2k+1)} \sum_{j=0}^{k-1} \frac{2j - 15}{4^{k-j-1}} a_j, \qquad k \geqslant 1; \qquad 0 < |x| < 4;$$

$y_2 = \sum\limits_{k=0}^{\infty} a_k x^k$, where

$$a_0 = 1, \qquad a_k = \frac{1}{k(2k-1)} \sum_{j=0}^{k-1} \frac{j - 8}{4^{k-j}} a_j, \qquad k \geqslant 1; \qquad |x| < 4$$

5. $y_1 = |x|^{5/4} \sum\limits_{k=0}^{\infty} a_k x^k$, where

$$a_0 = 1, \qquad a_k = \frac{(-1)^k}{4^{k+1}k(4k+5)} \sum_{j=0}^{k-1} (-1)^j 4^j (4j + 7) a_j, \qquad k \geqslant 1; \qquad 0 < |x| < 4;$$

$y_2 = \sum\limits_{k=0}^{\infty} a_k x^k$, where

$$a_0 = 1, \qquad a_k = \frac{(-1)^k}{2^{2k+1}k(4k-5)} \sum_{j=0}^{k-1} (-1)^j 4^j (2^j + 1) a_j, \qquad k \geqslant 1; \qquad |x| < 4$$

7. $y_1 = x$

$y_2 = x^{-2/3} \sum\limits_{k=0}^{\infty} a_k x^k$, where

$$a_0 = 1, \qquad a_k = \frac{3k - 8}{3k} a_{k-1}, \qquad k \geqslant 1$$

9. $y_1 = x \sum\limits_{k=0}^{\infty} a_k x^k$, where

$$a_0 = 1, \qquad a_k = \frac{(-1)^k}{k(4k+3)} \sum_{j=0}^{k-1} (-1)^j (4j + 3) a_j, \qquad k \geqslant 1; \qquad |x| < 1;$$

$y_2 = |x|^{1/4} \sum\limits_{k=0}^{\infty} a_k x^k$, where

$$a_0 = 1, \qquad a_k = \frac{(-1)^k 4}{k(4k-3)} \sum_{j=0}^{k-1} (-1)^j j a_j, \qquad k \geqslant 1; \qquad 0 < |x| < 1$$

11.  $y_1 = x\left[1 + \sum_{k=1}^{\infty} \frac{1}{k![7 \cdot 10 \cdot 13 \cdots (3k+4)]} x^k\right], \qquad |x| < \infty$

$y_2 = |x|^{-1/3}\left[1 + \sum_{k=1}^{\infty} \frac{1}{k![(-1) \cdot 2 \cdot 5 \cdots (3k-4)]} x^k\right], \qquad 0 < |x| < \infty$

13.  $v_1 = i,\ a_0 = 1$ give

$$a_1 = -\tfrac{2}{5} - \tfrac{1}{5}i,$$
$$a_2 = \tfrac{1}{10} + \tfrac{1}{20}i,$$
$$a_3 = -\tfrac{17}{780} - \tfrac{1}{130}i;$$

$v_2 = -i,\ a_0 = 1$ give

$$a_1 = -\tfrac{2}{5} + \tfrac{1}{5}i,$$
$$a_2 = \tfrac{1}{10} - \tfrac{1}{20}i,$$
$$a_3 = -\tfrac{17}{780} + \tfrac{1}{130}i.$$

This yields $y = c_1 y_1 + c_2 y_2$, where

$y_1 = (1 - \tfrac{2}{5}x + \tfrac{1}{10}x^2 - \tfrac{17}{780}x^3 + \cdots) \cos(\ln|x|)$
$\qquad\qquad\qquad + (\tfrac{1}{5}x - \tfrac{1}{20}x^2 + \tfrac{1}{130}x^3 + \cdots) \sin(\ln|x|),$

$y_2 = (-\tfrac{1}{5}x + \tfrac{1}{20}x^2 - \tfrac{1}{130}x^3 + \cdots) \cos(\ln|x|)$
$\qquad\qquad\qquad + (1 - \tfrac{2}{5}x + \tfrac{1}{10}x^2 - \tfrac{17}{780}x^3 + \cdots) \sin(\ln|x|).$

This solution is valid for $0 < |x| < \infty$.

15.  $v_1 = i,\ a_0 = 1$ give

$$a_1 = \tfrac{1}{5}(-2 - i), \qquad a_2 = \tfrac{1}{80}(3 - i), \qquad a_3 = \tfrac{1}{9360}(67 + 81i),$$

$v_2 = -1,\ a_0 = 1$ give

$$a_1 = \tfrac{1}{5}(-2 + i), \qquad a_2 = \tfrac{1}{80}(3 + i), \qquad a_3 = \tfrac{1}{9360}(67 - 81i).$$

This yields

$$y = c_1 y_1 + c_2 y_2$$

where

$y_1 = (1 - \tfrac{2}{5}x + \tfrac{3}{80}x^2 + \tfrac{67}{9360}x^3 + \cdots) \cos(\ln|x|)$
$\qquad\qquad\qquad + (\tfrac{1}{5}x + \tfrac{1}{80}x^2 - \tfrac{9}{1040}x^3 + \cdots) \sin(\ln|x|),$

$y_2 = (-\tfrac{1}{5}x - \tfrac{1}{80}x^2 + \tfrac{9}{1040}x^3 + \cdots) \cos(\ln|x|)$
$\qquad\qquad\qquad + (1 - \tfrac{2}{5}x + \tfrac{3}{80}x^2 + \tfrac{67}{9360}x^3 + \cdots) \sin(\ln|x|).$

This solution is valid for $0 < |x| < \infty$.

17.  a)  $v = 0$ gives $y = \sum_{k=0}^{\infty} a_k(x-1)^k,$

where

$$a_k = \frac{(-1)^{k+1}}{2^k k^2} \sum_{j=0}^{k-1} (-1)^j 2^j [\lambda(\lambda+1) - j] a_j, \qquad k \geqslant 1.$$

(*Note.*  It can be shown that, with $a_0 = 1$, the formulas above imply that

$$y = 1 + \sum_{k=1}^{\infty} \frac{[\lambda(\lambda + 1) - (k - 1)k][\lambda(\lambda + 1) - (k - 2)(k - 1)] \cdots [\lambda(\lambda + 1) - 0 \cdot 1]}{2^k (k!)^2}$$

$$\times (x - 1)^k$$

so that if $\lambda$ is a nonnegative integer $n$, then $y$ is a polynomial of degree $n$.)

b)  $-1 < x < 3$. (If $\lambda$ is a nonnegative integer, then the solution is valid for all $x$.)

*Section 7–6*

1.  $y_1 = x$,  $\quad y_2 = \sum_{k=1}^{\infty} \frac{(-1)^k}{k! \cdot k} x^{k+1} + x \ln x$

3.  $y_1 = x^{-1} \left[ 1 + \sum_{k=1}^{\infty} (-1)^k \frac{1}{(k!)^2} x^k \right]$,

$y_2 = 2x^{-1} \sum_{k=1}^{\infty} (-1)^{k+1} \frac{\sum_{j=1}^{k} 1/j}{(k!)^2} x^k + x^{-1} \left[ 1 + \sum_{k=1}^{\infty} (-1)^k \frac{1}{(k!)^2} x^k \ \ln x \right]$

5.  $y_1 = 4! \sum_{k=0}^{\infty} \frac{k + 1}{(k + 4)!} x^{k+4}$,

$y_2 = 1 + \frac{2}{3}x + \frac{1}{6}x^2$

7.  $y_1 = x^2 + \sum_{k=1}^{\infty} (-1)^k \frac{1}{5 \cdot 8 \cdots (3k + 2)} x^{3k+2}$,

$y_2 = \sum_{k=0}^{\infty} (-1)^k \frac{1}{3^k k!} x^{3k}$

9.  $y_1 = 1 - 3x + \frac{3}{2}x^2 - \frac{1}{6}x^3$,

$y_2 = 7x - \frac{23}{4}x^2 + \frac{11}{12}x^3 - 6 \sum_{k=4}^{\infty} \frac{(k - 4)!}{(k!)^2} x^k + (1 - 3x + \frac{3}{2}x^2 - \frac{1}{6}x^3) \ln x$

11.  If $2p$ is not an integer, Case I; if $p = 0$, Case II; if $2p$ is an integer other than zero, Case III.

*Section 7–8*

3.  $J_3 = \left[ \frac{8}{x^2} - 1 \right] J_1 - \frac{4}{x} J_0$

$J_4 = \left[ \frac{48}{x^3} - \frac{8}{x} \right] J_1 - \left[ \frac{24}{x^2} + 1 \right] J_0$

**Chapter 8**

*Section 8–1*

3.  a)  $v = e^{-(1/2)\int P(x)\, dx}$

b)  $q(x) = -\frac{1}{2}P'(x) - \frac{1}{4}P^2(x) + Q(x)$

*Section 8–2*

1. a) $u'' + \dfrac{7 - 4x^2}{(1 - x^2)^2} u = 0$

   b) $\dfrac{d}{dx}\left[(1 - x^2)\dfrac{dy}{dx}\right] + 6y = 0$

3. a) $u'' + \dfrac{(-3x^4 - 12x^2 - 8x + 24)}{4(x^3 - 2)^2} u = 0$

   b) $\dfrac{d}{dx}\left[(x^3 - 2)^{-(1/3)}\dfrac{dy}{dx}\right] - 3(x^3 - 2)^{-(4/3)}y = 0$

5. a) $u'' + \dfrac{(8x^2 + 8x - 3)}{4(x + 1)^2} u = 0$

   b) $\dfrac{d}{dx}\left[\dfrac{1}{x + 1}\dfrac{dy}{dx}\right] + \dfrac{2x}{(x + 1)^2} y = 0$

*Section 8–3*

3. b) No. If $f(x) \equiv 0$ on $(0, \infty)$ except possibly on a finite closed interval, then the equation cannot have infinitely many solutions on $(0, \infty)$.

*Section 8–5*

5. a) Every solution has infinitely many zeros on every interval of the form $(-\infty, -a)$ or $(a, \infty)$, where $a > 1$, and does not oscillate on $(-1, 0)$ or $(0, 1)$.
   b) Every solution has infinitely many zeros on every interval of the form $(-\infty, a)$ and does not oscillate on $(1, \infty)$.
   c) Every solution has infinitely many zeros on every interval of the form $(a, \infty)$, where $a \geqslant 1$, but does not oscillate on $(-\infty, 0)$ or $(0, 1)$.

7. a) Every solution has infinitely many zeros on every interval of the form $(a, \infty)$, where $a \geqslant 0$, but does not oscillate on $(-\infty, 0)$.

## Chapter 9

*Section 9–1*

1. a) $A(2, -1) = (3, -1)$
   b) $A(x_1, x_2) = (2x_1 + x_2)\mathbf{e}_1 + x_2\mathbf{e}_2$
3. a) Rank $n - 1$; nullity 1
   b) Rank 1; nullity $n - 1$
5. a) Rank 2; nullity 1
   b) $\{\mathbf{e}_1 + \mathbf{e}_3\} = \{(1, 0, 1)\}$
9. a) The dimension of $\mathscr{V}$ is even.
   b) Define $A: \mathscr{R}^2 \to \mathscr{R}^2$ by $A(\mathbf{e}_1) = 0$ and $A(\mathbf{e}_2) = \mathbf{e}_1$.

*Section 9–2*

1. If $\mathscr{B}_1 = \{\mathbf{e}_1, \ldots, \mathbf{e}_n\}$ and $\mathscr{B}_2 = \{\mathbf{f}_1, \ldots, \mathbf{f}_m\}$, then

$$A(x) = \sum_{i=1}^{m}\left(\sum_{j=1}^{n}\alpha_{ij}x_j\right)\mathbf{f}_i.$$

3. a) $(1, \frac{1}{2}, \frac{1}{3})$     b) $(1, -\frac{1}{2}, -\frac{1}{6})$     c) $(\frac{1}{2}, -\frac{1}{4}, -\frac{1}{12})$

5. a) $\begin{bmatrix} 0 & -1 & 0 \\ 0 & 0 & -2 \\ 0 & 1 & 0 \\ 0 & 0 & 2 \end{bmatrix}$     b) $\begin{bmatrix} 0 & 0 & 0 \\ 0 & 2 & 4 \\ 0 & 1 & 6 \\ 0 & 0 & 2 \end{bmatrix}$     c) $\begin{bmatrix} 0 & 3 & 6 \\ 0 & 4 & 14 \\ 0 & 1 & 8 \\ 0 & 0 & 2 \end{bmatrix}$

7. The order of the columns in $[A: \mathcal{B}_1, \mathcal{B}_2]$ changes when the order of the vectors in $\mathcal{B}_1$ changed; the order of the rows in $[A: \mathcal{B}_1, \mathcal{B}_2]$ changes when the order of the vectors in $\mathcal{B}_2$ is changed.

*Section 9–3*

1. a) $\begin{bmatrix} -1 & 1 \\ 1 & -1 \end{bmatrix}$     b) $\begin{bmatrix} 0 & 0 \\ 0 & -2 \end{bmatrix}$

3. $\begin{bmatrix} 1 & 0 & 0 & 0 \\ 0 & 1 & 0 & 0 \\ 0 & 0 & 0 & 0 \\ 0 & 0 & 1 & 0 \\ 0 & 0 & 0 & 1 \\ 0 & 0 & 0 & 0 \end{bmatrix}$

5. a) 24     b) $mn$     c) $m^2 n^2$

*Section 9–4*

1. $\begin{bmatrix} 18 & -16 \\ 19 & 11 \\ 31 & -25 \end{bmatrix}$

3. $[3]$

5. They all have the form $\begin{bmatrix} a & b \\ 0 & a-b \end{bmatrix}$.

7. b) $\begin{bmatrix} 2 & 0 & 1 & 0 \\ 0 & 2 & 0 & 1 \\ 0 & 0 & -1 & 0 \\ 0 & 0 & 0 & -1 \end{bmatrix}$

c) $\begin{bmatrix} 0 & 4 \\ 2 & -2 \end{bmatrix}$

9. $\dfrac{1}{ad-bc} \begin{bmatrix} d & -b \\ -c & a \end{bmatrix}$

11. The matrix is invertible if and only if $\alpha \neq 0$, in which case the inverse is

$$\begin{bmatrix} 0 & \dfrac{\beta}{\alpha} & \dfrac{1}{\alpha} \\ 0 & -1 & 0 \\ \dfrac{1}{\alpha} & -\dfrac{\beta}{\alpha^2} & -\dfrac{1}{\alpha^2} \end{bmatrix}.$$

*Section 9–5*

1. $\begin{bmatrix} 1 & 0 & 0 \\ 0 & 1 & 0 \\ 0 & 0 & 1 \end{bmatrix}$

3. $\begin{bmatrix} 1 & 0 & 0 & 5 \\ 0 & 1 & 0 & -10 \\ 0 & 0 & 0 & 1 \\ 0 & 0 & 0 & 0 \end{bmatrix}$

5. $\begin{bmatrix} 0 & 0 & 0 \\ 0 & 0 & 0 \end{bmatrix}, \begin{bmatrix} 1 & \alpha & \beta \\ 0 & 0 & 0 \end{bmatrix}, \begin{bmatrix} 0 & 1 & \alpha \\ 0 & 0 & 0 \end{bmatrix}, \begin{bmatrix} 0 & 0 & 1 \\ 0 & 0 & 0 \end{bmatrix}, \begin{bmatrix} 1 & 0 & \alpha \\ 0 & 1 & \beta \end{bmatrix}, \begin{bmatrix} 0 & 1 & 0 \\ 0 & 0 & 1 \end{bmatrix}, \begin{bmatrix} 1 & \alpha & 0 \\ 0 & 0 & 1 \end{bmatrix},$
   for arbitrary $\alpha$ and $\beta$.

7. a) Rank 3; nullity 0    b) Rank 2; nullity 1

9. (a) and (d) are linearly independent; (b) and (c) are not.

*Section 9–6*

1. $x_1 = \frac{43}{23}, \qquad x_2 = \frac{5}{23}, \qquad x_3 = -\frac{27}{23}$

3. Inconsistent

5. Inconsistent

7. (a), (b), and (c): the dimension of the solution space may be 0, 1, 2 or 3.

*Section 9–7*

1. The null space, and the space $\mathscr{V}$ itself.

*Section 9–8*

1. Using standard basis $\mathbf{e}_1, \mathbf{e}_2$:
   a) $\lambda = 1$, with eigenvectors of the form $x_1\mathbf{e}_1$;
      $\lambda = 2$, with eigenvectors of the form $x_1(\mathbf{e}_1 + \mathbf{e}_2)$.
   b) $\lambda = 0$, with eigenvectors of the form $x_2\mathbf{e}_2$;
      $\lambda = 1$, with eigenvectors of the form $x_1\mathbf{e}_1$.
   c) $\lambda = 0$, with eigenvectors of the form $x_1(\mathbf{e}_1 - \mathbf{e}_2)$;
      $\lambda = 2$, with eigenvectors of the form $x_1(\mathbf{e}_1 + \mathbf{e}_2)$.
   d) $\lambda = 3$, with eigenvectors of the form $x_2(2\mathbf{e}_1 + \mathbf{e}_2)$;
      $\lambda = -1$, with eigenvectors of the form $x_2(-2\mathbf{e}_1 + \mathbf{e}_2)$.

3. Using standard basis $\mathbf{e}_1, \mathbf{e}_2, \mathbf{e}_3$:
   a) $\lambda = 0$, with eigenvectors of the form $x_1(\mathbf{e}_1 - \mathbf{e}_3)$;
      $\lambda = 1$, with eigenvectors of the form $x_2\mathbf{e}_2$;
      $\lambda = 2$, with eigenvectors of the form $x_1(\mathbf{e}_1 + \mathbf{e}_3)$.
   b) $\lambda = 2$, with eigenvectors of the form $x_2\mathbf{e}_2$;
      $\lambda = \sqrt{2}$, with eigenvectors of the form $x_1(\mathbf{e}_1 + [\sqrt{2} - 1]\mathbf{e}_3)$;
      $\lambda = -\sqrt{2}$, with eigenvectors of the form $x_1(\mathbf{e}_1 - [1 + \sqrt{2}]\mathbf{e}_3)$.
   c) $\lambda = 1$, with eigenvectors of the form $x_2(-3\mathbf{e}_1 + \mathbf{e}_2 - 3\mathbf{e}_3)$;
      $\lambda = 2$, with eigenvectors of the form $x_2(2\mathbf{e}_1 + \mathbf{e}_2) + x_3(2\mathbf{e}_1 + \mathbf{e}_3)$.

d) $\lambda = 0$, with eigenvectors of the form $x_3(3\mathbf{e}_1 - 2\mathbf{e}_2 + \mathbf{e}_3)$;

$\lambda = \dfrac{5 + \sqrt{5}}{2}$, with eigenvectors of the form $x_2\left(\mathbf{e}_1 + \mathbf{e}_2 + \dfrac{\sqrt{5} - 1}{2}\mathbf{e}_3\right)$;

$\lambda = \dfrac{5 - \sqrt{5}}{2}$, with eigenvectors of the form $x_2\left(\mathbf{e}_1 + \mathbf{e}_2 - \dfrac{1 + \sqrt{5}}{2}\mathbf{e}_3\right)$.

5. a) $\mathbf{x} = -\frac{2}{3}\mathbf{e}_1 + \mathbf{e}_2 + \frac{4}{3}\mathbf{e}_3$
   b) $\mathbf{x} = \frac{1}{3}\mathbf{e}_1 + \frac{1}{3}\mathbf{e}_3$
   c) $\mathbf{x} = -\frac{8}{3}\mathbf{e}_1 - 2\mathbf{e}_2 + \frac{10}{3}\mathbf{e}_3$.

9. b) $\lambda = 0$ is an eigenvalue of multiplicity $n + 1$. $\mathcal{S}_0$ is the subspace of constant functions; dim $\mathcal{S}_0 = 1$.

# Chapter 10

*Section 10–2*

1. a) $x'''(t) + 3x''(t) - x'(t) - 2x(t) = e^t$
   b) $x''(t) + p(t)x'(t) + q(t)x(t) = h(t)$

3. a)
$$
\begin{bmatrix} x_1' \\ x_2' \\ \vdots \\ x_{n-1}' \\ x_n' \end{bmatrix}
=
\begin{bmatrix}
0 & 1 & 0 & \cdots & 0 \\
0 & 0 & 1 & \cdots & 0 \\
\vdots & & & & \vdots \\
0 & 0 & 0 & \cdots & 1 \\
-a_0(t) & -a_1(t) & -a_2(t) & \cdots & -a_{n-1}(t)
\end{bmatrix}
\begin{bmatrix} x_1 \\ x_2 \\ \vdots \\ x_{n-1} \\ x_n \end{bmatrix}
+
\begin{bmatrix} 0 \\ 0 \\ \vdots \\ 0 \\ h(t) \end{bmatrix}
$$

5. a)
$$
\begin{bmatrix} x_1 \\ x_2 \end{bmatrix}
=
\begin{bmatrix} -c_1 \cos at + c_2 \sin at \\ c_1 \sin at + c_2 \cos at \end{bmatrix}
$$

b)
$$
\begin{bmatrix} x_1 \\ x_2 \end{bmatrix}
=
\begin{bmatrix}
-c_1 \cos at + c_2 \sin at + \dfrac{2a - 1}{a^2 - 1} \sin t \\[2mm]
c_1 \sin at + c_2 \cos at + \dfrac{2 - a}{a^2 - 1} \cos t
\end{bmatrix}
$$

d) $\mathbf{X}(\mathbf{X}_0, t) = \begin{bmatrix} \alpha \cos at + \beta \sin at \\ -\alpha \sin at + \beta \cos at \end{bmatrix}$

*Section 10–3*

1. $\mathbf{X}(t) = \begin{bmatrix} c_1 e^t + c_2 e^{-t} \\ c_1 e^t + 3c_2 e^{-t} \end{bmatrix}$

3. $\mathbf{X}(t) = \begin{bmatrix} 7c_1 e^{5t} + c_2 e^{-t} \\ c_1 e^{5t} + c_2 e^{-t} \end{bmatrix}$

5. $\mathbf{X}(t) = \begin{bmatrix} -c_1 e^t + c_2 e^{-2t} + c_3 e^{3t} \\ 4c_1 e^t - c_2 e^{-2t} + 2c_3 e^{3t} \\ c_1 e^t - c_2 e^{-2t} + c_3 e^{3t} \end{bmatrix}$

7. $\mathbf{X}(t) = \begin{bmatrix} c_1 e^t + c_2 e^{-t} - 1 \\ c_1 e^t + 3c_2 e^{-t} + t - 2 \end{bmatrix}$

9. $\mathbf{X}(t) = \begin{bmatrix} c_1 e^{(a+b)t} + c_2 e^{(a-b)t} \\ c_1 e^{(a+b)t} - c_2 e^{(a-b)t} \end{bmatrix}$

13. $\mathbf{X}(t) = \begin{bmatrix} c_1 t^2 + c_2 t^4 \\ 3c_1 t^2 + c_2 t^4 \end{bmatrix}$

*Section 10–4*

1. $\mathbf{X} = c_1 \mathbf{X}_1 + c_2 \mathbf{X}_2$, where

$$\mathbf{X}_1 = \begin{bmatrix} 3 \\ 5 \end{bmatrix} \cos t - \begin{bmatrix} 1 \\ 0 \end{bmatrix} \sin t$$

$$\mathbf{X}_2 = \begin{bmatrix} 1 \\ 0 \end{bmatrix} \cos t + \begin{bmatrix} 3 \\ 5 \end{bmatrix} \sin t$$

3. $\mathbf{X} = c_1 \mathbf{X}_1 + c_2 \mathbf{X}_2$, where

$$\mathbf{X}_1 = e^{3t} \left( \begin{bmatrix} -1 \\ 1 \end{bmatrix} \cos t + \begin{bmatrix} 0 \\ -1 \end{bmatrix} \sin t \right)$$

$$\mathbf{X}_2 = e^{3t} \left( \begin{bmatrix} 0 \\ -1 \end{bmatrix} \cos t - \begin{bmatrix} -1 \\ 1 \end{bmatrix} \sin t \right)$$

5. $\mathbf{X} = c_1 \mathbf{X}_1 + c_2 \mathbf{X}_2$, where

$$\mathbf{X}_1 = \begin{bmatrix} 1 \\ -1 \end{bmatrix} e^{2t}$$

$$\mathbf{X}_2 = \begin{bmatrix} t \\ -1-t \end{bmatrix} e^{2t} = \begin{bmatrix} 1 \\ -1 \end{bmatrix} te^{2t} + \begin{bmatrix} 0 \\ -1 \end{bmatrix} e^{2t}$$

7. $\mathbf{X} = c_1 \mathbf{X}_1 + c_2 \mathbf{X}_2$, where

$$\mathbf{X}_1 = \begin{bmatrix} 2 \\ -4 \end{bmatrix}$$

$$\mathbf{X}_2 = \begin{bmatrix} 2t \\ -4t+1 \end{bmatrix} = \begin{bmatrix} 2 \\ -4 \end{bmatrix} t + \begin{bmatrix} 0 \\ 1 \end{bmatrix}$$

9. $\mathbf{X} = c_1 \mathbf{X}_1 + c_2 \mathbf{X}_2 + c_3 \mathbf{X}_3$

$$\mathbf{X}_1 = \begin{bmatrix} 0 \\ 1 \\ -1 \end{bmatrix} e^{2t}$$

$$\mathbf{X}_2 = \begin{bmatrix} 0 \\ 1 \\ -1 \end{bmatrix} te^{2t} + \begin{bmatrix} 1 \\ 0 \\ 1 \end{bmatrix} e^{2t}$$

$$\mathbf{X}_3 = \begin{bmatrix} 0 \\ 1 \\ -1 \end{bmatrix} t^2 e^{2t} + \begin{bmatrix} 2 \\ 1 \\ 1 \end{bmatrix} te^{2t} + \begin{bmatrix} 3 \\ 0 \\ 5 \end{bmatrix} e^{2t}$$

11. $\mathbf{X} = \begin{bmatrix} c_1 \sin at + c_2 \cos at \\ c_1 \cos at - c_2 \sin at \end{bmatrix}$

*Section 10-5*

3. a) The general solution is

$$C_1(t) = \quad A + Be^{-\alpha t} + \quad De^{-(\alpha + 2\beta)t}$$

$$C_2(t) = \frac{\alpha}{\beta} A \qquad - 2De^{-(\alpha + 2\beta)t}$$

$$C_3(t) = \quad A - Be^{-\alpha t} + \quad De^{-(\alpha + 2\beta)t}$$

b) The initial conditions yield the particular solution with

$$A = \frac{\beta}{\alpha + 2\beta}, \quad B = \frac{\alpha + 2\beta}{2\beta}, \quad D = \frac{\alpha}{2(\alpha + 2\beta)}.$$

The steady-state solution in this case is

$$C_1(t) = \frac{\beta}{\alpha + 2\beta}, \quad C_2(t) = \frac{\alpha}{\alpha + 2\beta}, \quad C_3(t) = \frac{\beta}{\alpha + 2\beta}.$$

## Chapter 11

*Section 11-4*

1.

| $x_k$ | $y_k$ |
|-------|-------|
| 0.0 | 0.000 |
| 0.1 | 0.000 |
| 0.2 | 0.020 |
| 0.3 | 0.060 |
| 0.4 | 0.120 |
| 0.5 | 0.200 |
| 0.6 | 0.300 |
| 0.7 | 0.420 |
| 0.8 | 0.560 |
| 0.9 | 0.720 |
| 1.0 | 0.900 |

3.

| $x_k$ | $y_k$ |
|-------|-------|
| 1.0 | 1.500 |
| 1.2 | 2.000 |
| 1.4 | 2.533 |
| 1.6 | 3.095 |
| 1.8 | 3.682 |
| 2.0 | 4.291 |

5.

| $x_k$ | $y_k$ |
|-------|-------|
| 0.0 | $-1.000$ |
| 0.2 | $-0.800$ |
| 0.4 | $-0.728$ |
| 0.6 | $-0.706$ |
| 0.8 | $-0.704$ |
| 1.0 | $-0.702$ |

*Section 11-5*

3.

| $x_k$ | $y_k$ |
|-------|-------|
| 0.0 | 0.000 |
| 0.1 | 0.000 |
| 0.2 | 0.020 |
| 0.3 | 0.044 |
| 0.4 | 0.089 |
| 0.5 | 0.142 |
| 0.6 | 0.217 |
| 0.7 | 0.305 |
| 0.8 | 0.418 |
| 0.9 | 0.549 |
| 1.0 | 0.708 |

5.

| $x_k$ | $y_k$ |
|-------|-------|
| 0.0 | 1.000 |
| 0.2 | 1.200 |
| 0.4 | 1.592 |
| 0.6 | 2.278 |
| 0.8 | 3.812 |
| 1.0 | 8.347 |

*Section 11–6*

3. a)

| $x_k$ | $y_k$ | |
|-------|-------|---|
| 0.0 | 0.000 | $m_1 = 0.0$ <br> $m_2 = 0.5$ <br> $m_3 = 0.5$ <br> $m_4 = 1.0$ |
| 0.5 | 0.250 | $m_1 = 1.0$ <br> $m_2 = 1.5$ <br> $m_3 = 1.5$ <br> $m_4 = 2$ |
| 1.0 | 1.000 | — |

b)

| $x_k$ | $y_k$ | |
|-------|-------|---|
| 0.0 | 0.000 | $m_1 = 0$ <br> $m_2 = 0.25$ <br> $m_3 = 0.313$ <br> $m_4 = 0.656$ |
| 0.5 | 0.148 | $m_1 = 0.648$ <br> $m_2 = 1.060$ <br> $m_3 = 1.163$ <br> $m_4 = 1.729$ |
| 1.0 | 0.717 | — |

*Section 11–7*

1.

| $t_k$ | $\mathbf{X}_k$ |
|-------|------|
| 0.0 | $\begin{bmatrix} 0.000 \\ 2.000 \end{bmatrix}$ |
| 0.2 | $\begin{bmatrix} 0.400 \\ 2.000 \end{bmatrix}$ |
| 0.4 | $\begin{bmatrix} 0.800 \\ 2.160 \end{bmatrix}$ |
| 0.6 | $\begin{bmatrix} 1.264 \\ 2.320 \end{bmatrix}$ |
| 0.8 | $\begin{bmatrix} 1.728 \\ 2.666 \end{bmatrix}$ |
| 1.0 | $\begin{bmatrix} 2.330 \\ 3.011 \end{bmatrix}$ |

3.

| $t_k$ | $\mathbf{X}_k$ |
|-------|------|
| 0.0 | $\begin{bmatrix} 1.000 \\ 2.000 \end{bmatrix}$ |
| 0.2 | $\begin{bmatrix} 1.800 \\ 1.800 \end{bmatrix}$ |
| 0.4 | $\begin{bmatrix} 2.520 \\ 1.800 \end{bmatrix}$ |
| 0.6 | $\begin{bmatrix} 3.240 \\ 1.944 \end{bmatrix}$ |
| 0.8 | $\begin{bmatrix} 4.107 \\ 2.203 \end{bmatrix}$ |
| 1.0 | $\begin{bmatrix} 4.898 \\ 2.566 \end{bmatrix}$ |

# Chapter 12

*Section 12–1*

1. Note that any point $x_0$ where the graphs of $g$ and $y = x$ cross satisfies $g(x_0) = x_0$.

3. a) $F^{[2]}(x_0) = F[F(x_0)] = F(x_0) = x_0$

   b) $F^{[m]}(x_0) = F[F^{[m-1]}(x)] = F^{[m-1]}[F(x)]$ so that $F^{[m]}$ is defined inductively; $F^{[m]}(x_0) = F^{[m-1]}[F(x_0)] = F^{[m-1]}(x_0)$ for $m = 2, 3, \ldots$ . Thus $F^{[m]}(x_0) = F^{[m-1]}(x_0) = F^{[2]}(x_0) = F(x_0) = x_0$ and $x_0$ is a fixed point under $F^{[m]}$.

5. The fixed points of $T$ are the solutions of the system

$$ \begin{aligned} ax + by &= x \\ cx + dy &= y \end{aligned} \quad \text{or} \quad \begin{aligned} (a - 1)x + by &= 0 \\ cx + (d - 1)y &= 0 \end{aligned} $$

The latter has nontrivial solutions if and only if $(a - 1)(d - 1) - cb = 0$.

7. a) We mark the range of $f(x)$ on the $y$-axis, and on the $x$-axis. If a number $c$ is in the range of $f$, then $c = f(x_0)$ for some $x_0$; hence

$$f(c) = f[f(x_0)] = f(x_0) = c$$

and thus $c$ is a fixed point under $f$. Thus $f(c) = c$ whenever $a \leqslant c \leqslant b$, and the straight line from $(a, a)$ to $(b, b)$ must be part of the graph of $f$. The remainder of the graph must lie in the shaded regions because, for all $x$, $a \leqslant f(x) \leqslant b$.

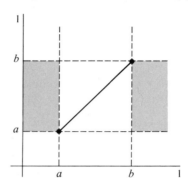

b) If $c$ is differentiable its graph can have no "corners." To avoid "corners" at $(a, a)$ and $(b, b)$ we must take either $a = b = f(x)$, $0 \leqslant x \leqslant 1$, or else $a = 0$, $b = 1$ and $f(x) = x$, $0 \leqslant x \leqslant 1$.

*Section 12–2*

3. b)

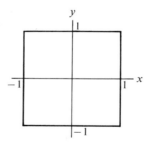

11. Construct a sequence of rational numbers that converges to an irrational number.

13. c) A sequence of points $(x_1^{(n)}, x_2^{(n)})$ in the plane approaches $(a, b)$ if and only if $x_1^{(n)} \to a$ and only finitely many $x_2^{(n)}$'s are different from $b$.

*Section 12–3*

1. a) Not a contraction mapping on $\mathscr{R}^1$ because

$$\lim_{x \to 0} \left| \frac{\sin x - \sin 0}{x - 0} \right| = \lim_{x \to 0} \left| \frac{\sin x}{x} \right| = 1,$$

so that there is no $\alpha < 1$ for which $|\sin x - \sin 0| \leqslant \alpha |x - 0|$.

b)  A contraction mapping because

$$|F'(x)| \leqslant \frac{3\sqrt{3}}{8} < 1.$$

c)  A contraction mapping because

$$|F'(x)| \leqslant \frac{2\sqrt{3}}{9} < 1.$$

d)  Not a contraction mapping because

$$\frac{F(x_2) - F(x_1)}{x_2 - x_1} = \frac{x_2^3 - x^3}{x_2 - x_1} = x_2^2 + x_2 x_1 + x_1^2$$

is unbounded.

3.  b)  $F$ has no fixed points because $F(x) > x$ for all $x$.

7.  a)  No polynomial of degree greater than 1 is a contraction mapping.

b)  Either $x_2 \geqslant x_1$ or $x_2 < x_1$. From $x_2 \geqslant x_1$ we obtain $f(x_2) \geqslant f(x_1)$ since $f$ is non-decreasing; thus $x_3 \geqslant x_2$. In like manner $x_{n+1} \geqslant x_n$ and hence $\{x_n\}$ is nondecreasing. Such a sequence is convergent ($x_n \leqslant 1$ for all $n$, and any bounded monotone sequence converges). If $x_2 < x_1$ it can be shown in the same way that $\{x_n\}$ is a monotone nonincreasing, bounded sequence. Thus it too would converge.

c)  If $x_n \to x_0$ then the continuity of $f$ implies that $f(x_n) \to f(x_0)$; but $f(x_n) = x_{n+1} \to x_0$ and so $f(x_0) = x_0$, by the uniqueness of the limit.

d)  The functions $x^n$, $n = 0, 1, 2, \ldots$ have 0 and 1 as fixed points.

9.  a)  Since $F$ is linear and $d[(x_1, x_2), (y_1, y_2)] = d[(x_1 - y_1, x_2 - y_2), (0, 0)]$ it is sufficient to find a condition which insures that

$$d[(a_{11}x_1 + a_{12}x_2, a_{21}x_1 + a_{22}x_2), (0, 0)] \leqslant \alpha d[(x_1, x_2), (0, 0)].$$

In other words, we must have

$$|a_{11}x_1 + a_{12}x_2| + |a_{21}x_1 + a_{22}x_2| \leqslant \alpha[|x_1| + |x_2|].$$

Since

$$|a_{11}x_1 + a_{12}x_2| + |a_{21}x_1 + a_{22}x_2| \leqslant [|a_{11}| + |a_{21}|]|x_1| + [|a_{12}| + |a_{22}|]|x_2|,$$

$F$ is a contraction if

$$\max [|a_{11}| + |a_{21}|, |a_{12}| + |a_{22}|] < 1.$$

b)  Reasoning as in (a), we can see that $F$ is a contraction if

$$\max [|a_{11}x_1 + a_{12}x_2|, |a_{21}x_1 + a_{22}x_2|] \leqslant \alpha \max [|x_1|, |x_2|].$$

Now,

$$|a_{11}x_1 + a_{12}x_2| \leqslant |a_{11}| |x_1| + |a_{12}| |x_2| \leqslant (|a_{11}| + |a_{12}|) \max (|x_1|, |x_2|)$$

and

$$|a_{21}x_1 + a_{22}x_2| \leqslant |a_{21}| |x_1| + |a_{22}| |x_2| \leqslant (|a_{21}| + |a_{22}|) \max (|x_1|, |x_2|)$$

so that

$$\max\left[|a_{11}x_1 + a_{22}x_2|, |a_{21}x_1 + a_{22}x_2|\right] \leqslant |a_{11}x_1 + a_{12}x_2| + |a_{21}x_1 + a_{22}x_2|$$
$$\leqslant (|a_{11}| + |a_{12}| + |a_{21}| + |a_{22}|)\max(|x_1|, |x_2|).$$

Thus $F$ is a contraction if

$$(|a_{11}| + |a_{12}| + |a_{21}| + |a_{22}|) < 1.$$

### Section 12–4

3.  a) $|f(x, y_2) - f(x, y_1)| = |b|\,|y_1 + y_2|\,|y_1 - y_2| \leqslant 2|b|\,|y_1 - y_2|$

  b) $\left|\dfrac{\partial}{\partial y}(x^3 e^{-xy^2})\right| = |-2x^4 y e^{-xy^2}| \leqslant \sqrt{\dfrac{2}{e}}\,a^{7/2}$  for $0 \leqslant x \leqslant a$, $|y| < \infty$.

  Take as Lipschitz constant $\sqrt{\dfrac{2}{e}}\,a^{7/2}$

  c) $|f(x, y_2) - f(x, y_1)| = p(x)|y_2 - y_1|$
  Take $\beta = \max|p(x)|, |x| \leqslant a$.

5.  $y_n(x) = 1 + \dfrac{x^2}{2} + \dfrac{1}{2!}\left(\dfrac{x^2}{2}\right)^2 + \cdots + \dfrac{1}{n!}\left(\dfrac{x^2}{2}\right)^n$

  $y = \lim_{n \to \infty} y_n(x) = e^{x^2/2}$

7.  a) $y_0(x) = y_1(x) = y_2(x) = y_3(x) = 0$

  b) $y_0(x) = 1$

  $y_1(x) = 1 + x$

  $y_2(x) = 1 + x + x^2 + \dfrac{x^3}{3}$

  $y_3(x) = 1 + x + x^2 + x^3 + \tfrac{2}{3}x^4 + \tfrac{1}{3}x^5 + \tfrac{1}{9}x^6 + \tfrac{1}{63}x^7$

9.  b) Where $y_2 = 0$,

$$\frac{|y_1^{2/3} - y_2^{2/3}|}{|y_1 - y_2|} = \frac{1}{|y_1^{1/3}|},$$

which is not bounded in any region that contains a point of the $x$-axis.

### Section 12–6

3.  a) Here we have

$$\|f(x, \mathbf{Y}_1) - f(x, \mathbf{Y}_2)\| \leqslant \max\left[|a_{11}| + |a_{21}|, |a_{12} + a_{22}|\right]\|\mathbf{Y}_1 - \mathbf{Y}_2\|$$

so that we can take the Lipschitz constant

$$\beta = \max\left[|a_{11}| + |a_{21}|, |a_{12}| + |a_{22}|\right]$$

  b) $\|f(x, \mathbf{Y}) - f(x, \mathbf{Z})\| = \|(y_2^2 - z_2^2, y_1^2 - z_1^2)\|$
  $= |y_2^2 - z_2^2| + |y_1^2 - z_1^2|$
  $= |y_2 + z_2|\,|y_2 - z_2| + |y_1 + z_1|\,|y_1 - z_1|$
  $\leqslant (|y_2| + |z_2|)\,|y_2 - z_2| + (|y_1| + |z_1|)\,|y_1 - z_1|$
  $\leqslant \{\max\left[|y_2| + |x_2|, |y_1| + |z_1|\right]\}\{|y_2 - z_2| + |y_1 - z_1|\}$
  $\leqslant 2\|\mathbf{Y} - \mathbf{Z}\|$

**Chapter 13**

*Section 13–4*

1. a) Every point is an equilibrium point.
   b) There is a line $L$ of equilibrium points through $(0, 0)$ and the trajectories are parallel to it, as shown below. The equilibrium points are unstable.

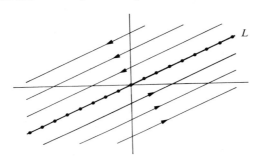

3. a) $\lambda_1, \lambda_2 = -1 \pm \sqrt{5}$
      $a + d = -2 = \lambda_1 + \lambda_2$
      $\Delta = 20$     Det $A = -4 = \lambda_1 \lambda_2$

   b) $\lambda_1, \lambda_2 = -3, -2$
      $a + d = -5 = \lambda_1 \times \lambda_2$
      $\Delta = 1$     Det $A = 6 = \lambda_1 \lambda_2$

   c) $\lambda_1, \lambda_2 = 0, -5$
      $a + d = -5 = \lambda_1 + \lambda_2$
      $\Delta = 25$     Det $A = 0 = \lambda_1 \lambda_2$

   d) $\lambda_1, \lambda_2 = \pm\sqrt{2}$
      $a + d = 0 = \lambda_1 + \lambda_2$
      $\Delta = 8$     Det $A = -2 = \lambda_1 \lambda_2$

   e) $\lambda_1, \lambda_2, \lambda_3 = -1, -2, -3$
      The origin is the only equilibrium point, and it is asymptotically stable.

   f) $\lambda_1, \lambda_2, \lambda_3 = 1, 2, 3$
      The origin is the only equilibrium point, and it is unstable.

5. The columns of $B^{-1}$ are the eigenvectors of $A$.

9. Let $z_1 = y$, $z_2 = y'$; then $z_1' = z_2$ and $z_2' = -bz_1 - az_2$, so that the equivalent $2 \times 2$ system is

$$\begin{bmatrix} z_1' \\ z_2' \end{bmatrix} = \begin{bmatrix} 0 & 1 \\ -b & -a \end{bmatrix} \begin{bmatrix} z_1 \\ z_2 \end{bmatrix}.$$

The quadratic equation for the eigenvalues is

$$\lambda^2 + a\lambda + b = 0$$

so that

$$\lambda_1 + \lambda_2 = -a, \qquad \lambda_1 \lambda_2 = b = \text{Det } A, \qquad \text{and} \qquad \Delta = a^2 - 4b.$$

*Section 13–6*

1.  a)  Assume $P_1(0, 0) = P_2(0, 0) = 0$; otherwise translate coordinates. Then

$$\dot{x} = ax + by + \cdots$$
$$\dot{y} = cx + dy + \cdots$$

and the approximating linear system is

$$\begin{bmatrix} \dot{x} \\ \dot{y} \end{bmatrix} = \begin{bmatrix} a & b \\ c & d \end{bmatrix} \begin{bmatrix} x \\ y \end{bmatrix}$$

b)  $\begin{bmatrix} \dot{x} \\ \dot{y} \end{bmatrix} = \begin{bmatrix} a_{10} & a_{01} \\ b_{10} & b_{01} \end{bmatrix} \begin{bmatrix} x \\ y \end{bmatrix}$

3.  a)  $\lambda_1, \lambda_2 = -1, -3$

$$\mathbf{X}_1(t) = \begin{bmatrix} e^{-t} \\ 0 \end{bmatrix}, \qquad \mathbf{X}_2(t) = \begin{bmatrix} 0 \\ e^{-3t} \end{bmatrix}$$
$$E(x, y) = \tfrac{1}{2}x^2 + \tfrac{1}{6}y^2$$

b)  $\lambda_1, \lambda_2 = -3, -4$

$$\mathbf{X}_1(t) = \begin{bmatrix} e^{-3t} \\ 0 \end{bmatrix}, \qquad \mathbf{X}_2(t) = \begin{bmatrix} 0 \\ e^{-4t} \end{bmatrix}$$
$$E(x, y) = \tfrac{1}{6}x^2 + \tfrac{1}{8}y^2$$

c)  $\lambda_1, \lambda_2, \lambda_3 = -1, -2, -3$

$$\mathbf{X}_1(t) = \begin{bmatrix} e^{-t} \\ 0 \\ 0 \end{bmatrix}, \qquad \mathbf{X}_2(t) = \begin{bmatrix} 0 \\ e^{-2t} \\ 0 \end{bmatrix}, \qquad \mathbf{X}_3(t) = \begin{bmatrix} 0 \\ 0 \\ e^{-3t} \end{bmatrix}$$
$$E(x, y, x) = \tfrac{1}{2}x^2 + \tfrac{1}{4}y^2 + \tfrac{1}{6}z^2$$

5.  a)  Impossible
    b)  Impossible
    c)  $\tfrac{1}{2}x^2 + \tfrac{1}{4}y^2$

# *Index*